Lecture Notes in Computer Science 9682

Commenced Publication in 1973
Founding and Former Series Editors:
Gerhard Goos, Juris Hartmanis, and Jan van Leeuwen

More information about this series at http://www.springer.com/series/7407

Quentin Louveaux · Martin Skutella (Eds.)

Integer Programming and Combinatorial Optimization

18th International Conference, IPCO 2016
Liège, Belgium, June 1–3, 2016
Proceedings

 Springer

Editors
Quentin Louveaux
Université de Liège
Liège
Belgium

Martin Skutella
Technische Universität Berlin
Berlin
Germany

ISSN 0302-9743   ISSN 1611-3349 (electronic)
Lecture Notes in Computer Science
ISBN 978-3-319-33460-8   ISBN 978-3-319-33461-5 (eBook)
DOI 10.1007/978-3-319-33461-5

Library of Congress Control Number: 2016937374

LNCS Sublibrary: SL1 – Theoretical Computer Science and General Issues

Printed on acid-free paper

This Springer imprint is published by Springer Nature
The registered company is Springer International Publishing AG Switzerland

Preface

This volume contains the 33 extended abstracts presented at IPCO 2016, the 18th Conference on Integer Programming and Combinatorial Optimization, held June 1–3, 2016, in Liège, Belgium.

The IPCO conference is run under the auspices of the Mathematical Optimization Society. It is held every year, except for those in which the International Symposium on Mathematical Programming takes place. The conference is a forum for researchers and practitioners working on various aspects of integer programming and combinatorial optimization. The aim is to present recent developments in theory, computation, and applications in these areas. The first IPCO conference took place at the University of Waterloo in May 1990. More information on IPCO and its history can be found at www.mathopt.org/?nav=ipco.

This year, there were 125 submissions, of which one was withdrawn before the review process started. Each submission was reviewed by at least three Program Committee members, often with the help of external reviewers. After an initial electronic discussions using the EasyChair conference management system, the Program Committee met in Aussois in January 2016. After two long nights of thorough discussions, conscious and aware that there is no "right" or "wrong" when it comes to the ultimate decisions, the Program Committee selected 33 papers to be presented at IPCO 2016 and included in this volume.

We would like to thank all authors who submitted extended abstracts to IPCO 2016, the members of the Program Committee, who graciously gave their time and energy, the external reviewers, the members of the local Organizing Committee, the speakers of the summer school preceding IPCO (Michel Goemans, Nicolas Stier-Moses, and Juan-Pablo Vielma), the Mathematical Optimization Society and in particular the members of its IPCO Steering Committee, and — last but not least — the Aussois barkeeper for providing after-hours refreshment.

March 2016

Quentin Louveaux
Martin Skutella

Organization

Program Committee

Karen Aardal	TU Delft/CWI, The Netherlands
Daniel Bienstock	Columbia University, USA
José Correa	Universidad de Chile, Chile
Oktay Günlük	IBM Research, USA
Satoru Iwata	University of Tokyo, Japan
Volker Kaibel	Otto-von-Guericke Universität Magdeburg, Germany
Jochen Könemann	University of Waterloo, Canada
Andrea Lodi	University of Bologna/Polytechnique Montréal, Italy/Canada
Quentin Louveaux	Université de Liège, Belgium
Gianpaolo Oriolo	Università degli Studi di Roma Tor Vergata, Italy
András Sebő	CNRS, Laboratoire G-SCOP, Grenoble, France
Bruce Shepherd	McGill University, Canada
Martin Skutella (Chair)	TU Berlin, Germany
Leen Stougie	VU Amsterdam/CWI, The Netherlands
Gerhard Woeginger	TU Eindhoven, The Netherlands

Organizing Committee

Quentin Louveaux (Chair)	Université de Liège, Belgium
Yves Crama	Université de Liège, Belgium
Mathieu Van Vyve	Université catholique de Louvain, Belgium
Laurence Wolsey	Université catholique de Louvain, Belgium
Michèle Delville	AIM, Liège, Belgium
Céline Dizier	AIM, Liège, Belgium

Additional Reviewers

Abdi, Ahmad	Baiou, Mourad	Bonami, Pierre
Agnetis, Alessandro	Bampis, Evripidis	Bonifaci, Vincenzo
Ahmed, Shabbir	Bansal, Nikhil	Bonomo, Flavia
Alvelos, Filipe	Barahona, Francisco	Buchheim, Christoph
An, Hyung-Chan	Bazzi, Abbas	Byrka, Jaroslaw
Angulo, Gustavo	Benchetrit, Yohann	Böhm, Martin
Atamtürk, Alper	Bienkowski, Marcin	Cacchiani, Valentina
Averkov, Gennadiy	Boland, Natashia	Calinescu, Gruia

Celaya, Marcel
Chalermsook, Parinya
Chandrasekaran, Karthekeyan
Cseh, Ágnes
D'Ambrosio, Claudia
Dash, Sanjeeb
De Loera, Jesús A.
Dey, Santanu
Ee, Martijn van
Eisenbrand, Friedrich
Eppstein, David
Epstein, Leah
Faenza, Yuri
Fiorini, Samuel
Friggstad, Zachary
Fujishige, Satoru
Fukunaga, Takuro
Furini, Fabio
Gao, Zhihan
Geelen, Jim
Gentile, Claudio
Georgiou, Konstantinos
Goycoolea, Marcos
Grandoni, Fabrizio
Granot, Daniel
Grigoriev, Alexander
Groß, Martin
Gupta, Anupam
Hartman, Irith
Hartvigsen, David
Henk, Martin
Hildebrand, Robert
Hirai, Hiroshi
Hoeksma, Ruben
Hooker, John
Huang, Chien-Chung
Iersel, Leo van
Iori, Manuel
Jansen, Klaus
Jansson, Jesper
Jeż, Łukasz
Kakimura, Naonori
Kamiyama, Naoyuki
Kang, Nano
Kapralov, Michael

Kesselheim, Thomas
Khan, Muhammad
Khanna, Sanjeev
Kleer, Pieter
Kling, Peter
Knust, Sigrid
Kreutzer, Stephan
Kumar, Amit
Lasserre, Jean B.
Laurent, Monique
Lee, James
Levin, Asaf
Li, Shi
Linderoth, Jeff
Loebl, Martin
Luedtke, James
Maffray, Frédéric
Malaguti, Enrico
Manlove, David
Mannino, Carlo
Marcos Alvarez, Alejandro
Mastrolilli, Monaldo
Matuschke, Jannik
McCormick, Tom
Megow, Nicole
Mehta, Aranyak
Michaels, Dennis
Miyashiro, Ryuhei
Miyazaki, Shuichi
Mnich, Matthias
Morris, Walter
Moseley, Benjamin
Mömke, Tobias
Nagarajan, Viswanath
Nannicini, Giacomo
Naor, Seffi
Naves, Guyslain
Newman, Alantha
Nikolov, Aleksandar
Norine, Serguei
Okamoto, Yoshio
Olver, Neil
Onak, Krzysztof
Ono, Hirotaka
Oosterwijk, Tim

Ostrowski, James
Otachi, Yota
Pacifici, Andrea
Pap, Gyula
Parotsidis, Nikos
Paulusma, Daniel
Peis, Britta
Penev, Irena
Pettie, Seth
Pfetsch, Marc
Pilipczuk, Michał
Poirrier, Laurent
Pokutta, Sebastian
Post, Ian
Proiettu, Guido
Radoszewski, Jakub
Rendl, Franz
Romeijnders, Ward
Rothvoß, Thomas
Rybicki, Bartosz
Saberi, Amin
Sadykov, Ruslan
Saitoh, Toshiki
Sanità, Laura
Santiago Torres, Richard
Schaudt, Oliver
Schmidt, Daniel R.
Schulz, Andreas S.
Schürmann, Achill
Shim, Sangho
Shmoys, David
Sitters, René
Skopalik, Alexander
Smith, Cole
Soma, Tasuku
Soto, José A.
Spoerhase, Joachim
Stee, Rob van
Ster, Suzanne van der
Stiller, Sebastian
Svensson, Ola
Swamy, Chaitanya
Syrgkanis, Vasilis
Szigeti, Zoltán
Takazawa, Kenjiro
Tanigawa, Shin-Ichi

Tramontani, Andrea
Tunçel, Levent
Urrutia, Sebastián
Ventura, Paolo
Verschae, José
Vielma, Juan Pablo
Vondrák, Jan

Vredeveld, Tjark
Végh, László
Ward, Justin
Weismantel, Robert
Weltge, Stefan
Wiese, Andreas
Wiese, Sven

Williamson, David
Wolsey, Laurence
Zambelli, Giacomo
Zenklusen, Rico
Zhang, Lisa
Zou, Chenglong
Zuylen, Anke van

Contents

On Approximation Algorithms for Concave Mixed-Integer Quadratic Programming

Alberto Del Pia$^{(\boxtimes)}$

Department of Industrial and Systems Engineering & Wisconsin
Institute for Discovery, University of Wisconsin-Madison,
Madison, WI, USA
delpia@wisc.edu

Abstract. We describe an algorithm that finds an ϵ-approximate solution to a concave mixed-integer quadratic programming problem. The running time of the proposed algorithm is polynomial in the size of the problem and in $1/\epsilon$, provided that the number of integer variables and the number of negative eigenvalues of the objective function are fixed. The running time of the proposed algorithm is expected unless $\mathcal{P} = \mathcal{NP}$.

1 Introduction

Mixed-Integer Quadratic Programming (MIQP) problems are optimization problems in which the objective function is a quadratic polynomial, the constraints are linear inequalities, and some of the variables are required to be integers:

$$
\begin{aligned}
\text{minimize} \quad & x^\top H x + h^\top x \\
\text{subject to} \quad & W x \geq w \\
& x \in \mathbb{Z}^p \times \mathbb{R}^{n-p}.
\end{aligned}
\tag{1}
$$

In this formulation, H is symmetric, and all the data is rational. *Concave MIQP* is the special case of MIQP when the objective is concave, which occurs when H is negative semidefinite. Concave quadratic cost functions are frequently encountered in real-world problems concerning economies of scale, which corresponds to the economic phenomenon of "decreasing marginal cost" (see [7, 21]). Concave MIQP is $\mathcal{NP}$-complete [6, 22]. This is even true in very restricted settings such as the problem to minimize $\sum_{i=1}^{n} (w_i^\top x)^2$ over $x \in \{0, 1\}^n$ [18].

If we assume that the dimension n is fixed, then Concave MIQP is polynomially solvable. Cook, Hartmann, Kannan, and McDiarmid [3] showed that in fixed dimension we can enumerate the vertices of the integer hull $\mathrm{conv}\{x \in \mathbb{Z}^n : x \in P\}$ of a polyhedron P in polynomial time, and this result can be extended to the mixed-integer hull $P_I = \mathrm{conv}\{x \in \mathbb{Z}^p \times \mathbb{R}^{n-p} : x \in P\}$ by discretization [4, 12]. Since the minimum of a Concave MIQP is always achieved at one of the vertices of P_I, Concave MIQP can now be solved in fixed dimension by evaluating all the vertices of P_I and by selecting one with lowest objective value. In this work, we will not assume that the dimension n of the problem is fixed.

© Springer International Publishing Switzerland 2016
Q. Louveaux and M. Skutella (Eds.): IPCO 2016, LNCS 9682, pp. 1–13, 2016.
DOI: 10.1007/978-3-319-33461-5_1

In order to state our result, we first give the definition of ϵ-approximation. Consider an instance of MIQP, and let $f(x)$ denote the objective function. Let x^* be an optimum point of the problem, and let $f_{\max}$ be the maximal value of $f(x)$ on the feasible region. For $\epsilon \in [0, 1]$, we say that x° is an ϵ-*approximate solution* if

$$|f(x^\circ) - f(x^*)| \leq \epsilon |f_{\max} - f(x^*)|.$$

If the problem is unbounded or infeasible, the definition fails, and we expect our algorithm to return an indicator that the problem is unbounded or infeasible. The definition of ϵ-approximation has some useful invariance properties. For instance, it is invariant under dilation and translation of the objective function, as well as under affine transformation of the feasible region. Our definition of approximation has been used in earlier works (see, e.g., [1, 5, 17, 25]).

We now state the main result of this paper.

Theorem 1. *There is an algorithm to find an ϵ-approximate solution to Concave MIQP* (1). *The running time of the proposed algorithm is polynomial in the size of the problem and in $1/\epsilon$, provided that the number p of integer variables, and the number k of negative eigenvalues of H are fixed.*

Concave Quadratic Programming (Concave QP) is the continuous version of Concave MIQP, and can be obtained by setting $p = 0$ in (1). Concave QP is also $\mathcal{NP}$-complete [22], even when the concave quadratic objective has only one concave direction [20]. In [25], Vavasis gives an algorithm to find an ϵ-approximate solution to Concave QP whose running time is polynomial in the size of the problem and in $1/\epsilon$, provided that the number of negative eigenvalues of H is fixed. Our algorithm can be seen as a direct extension of Vavasis' result to the mixed-integer case. This shows that the computational effort needed to find an ϵ-approximate solution to a Concave QP or to a Concave MIQP are essentially the same, as long as the number of integer variables is small.

This is not the first time that the same type of problem can be solved with the same complexity in the continuous case, and in the mixed-integer case with a fixed number of integer variables. In fact, celebrated results by Khachiyan [13] and by Lenstra [14] show that this is true also for linear problems: both Linear Programming (LP), and Mixed-Integer Linear Programming (MILP) with a fixed number of integer variables can be solved in polynomial time.

The proposed algorithm has running time polynomial in the size of the problem and in $1/\epsilon$, and exponential in both k and p. We now explain why this running time is expected unless $\mathcal{P} = \mathcal{NP}$. First, we consider the dependence on ϵ. Suppose we had a similar approximation algorithm, but with running time polynomial in $|\log \epsilon|$. Vavasis [25] showed that, for $k = 1$ and $p = 0$, such an algorithm could solve Concave QP with one concave direction in polynomial time, and the latter problem is $\mathcal{NP}$-complete [20]. Suppose now that we had an approximation algorithm with running time polynomial in $1/\epsilon$ and in k. Vavasis [25] showed that, for $p = 0$, such algorithm could solve 3SAT in polynomial time, again implying $\mathcal{P} = \mathcal{NP}$. Finally, the existence of an approximation algorithm with running time polynomial in p, even for $k = 0$ and ϵ fixed, would allow us to solve MAX3SAT in polynomial time, again implying $\mathcal{P} = \mathcal{NP}$.

The key idea of our algorithm consists in iteratively subdividing the feasible region into two parts: one inner region where the mixed-integer points are "dense", and an outer region where the mixed-integer points are "sparse". The geometry of the mixed-integer points then allows us to employ tools used in the continuous QP setting, like mesh partition and linear underestimators, in order to obtain an ϵ-approximate solution for the inner region. This part of the proof adapts classic algorithms introduced by Pardalos and Rosen [19] and analyzed by Vavasis [25]. In the outer region, we use the concept of flatness to subdivide the problem into a fixed number of lower-dimensional MIQPs. In this second part of our proof we will use lattice algorithms introduced by Lenstra [14].

2   Proof of Theorem 1

Diagonalization. Our starting point is a problem of the form (1) in which H is negative semidefinite with rank k. The first task of our algorithm is to perform a linear change of variables that transform the objective function of (1) in separable form, where the negative-definite part of the problem is confined to k variables.

In order to do so, we diagonalize H via a LDL^T decomposition. This is a decomposition of the form $\Pi H \Pi^\top = LDL^T$, where Π is a permutation matrix, the matrix D is diagonal, whose first k diagonal entries are negative and the rest are zero, and where L is lower triangular with normalized diagonal entries $L_{ii} = 1$. The importance of the LDL^T decomposition is that, in contrast to other factorizations like the Eigenvalue decomposition or the Cholesky decomposition, it is a rational decomposition; i.e., if the matrix H is rational then all numbers that appear in the decomposition are rational and polynomially sized (see, e.g., [9]).

We can now perform the change of basis $(y, z) = L^\top \Pi x$, and we end up with an equivalent problem of the form:

$$
\begin{aligned}
\text{minimize} \quad & y^\top D y + c^\top y + f^\top z \\
\text{subject to} \quad & Ay + Bz \geq b \\
& (y, z) \in \mathcal{L},
\end{aligned}
\tag{2}
$$

where y is a k-dimensional vector, z is a $(n-k)$-dimensional vector, D is diagonal and negative definite, and $\mathcal{L} = \{(y, z) \in \mathbb{R}^n : (y, z) = L^\top \Pi x, \text{ for } x \in \mathbb{Z}^p \times \mathbb{R}^{n-p}\}$. Note that, if we denote by $l^1, \ldots, l^n$ the columns of $L^\top \Pi$, the set $\mathcal{L}$ is the *mixed-integer lattice*

$$
\mathcal{L} = \{(y, z) = \sum_{i=1}^n \lambda_i l^i : \lambda_1, \ldots, \lambda_p \in \mathbb{Z}, \lambda_{p+1}, \ldots, \lambda_n \in \mathbb{R}\}.
$$

Boundedness of MIQP. The next task of our algorithm is to detect if our concave MIQP problem (2) is unbounded or not. We remark that detecting if an *indefinite* MIQP problem is unbounded is $\mathcal{NP}$-hard even in the pure continuous case (see [6, 16]).

We first need to define two functions. The first one is the part of the objective function that depends on y:

$$q(y) = y^\top D y + c^\top y.$$

The second one is a function that associates to each $\bar{y} \in \mathbb{R}^k$, the optimal value of the MILP problem obtained as the restriction of (2) to the set of points $(\bar{y}, z)$. Formally,

$$\phi(y) = \min\{f^\top z : Ay + Bz \geq b, (y, z) \in \mathcal{L}\}, \tag{3}$$

for all y for which the minimum exists. If, for a fixed y, the MILP on the right-hand is infeasible, we write $\phi(y) = \infty$. Similarly, if, for a fixed y, the MILP on the right-hand is unbounded, we write $\phi(y) = -\infty$. Our MIQP problem (2) is now equivalent to the unconstrained problem

$$\text{minimize} \quad \{q(y) + \phi(y) : y \in \mathbb{R}^k\}.$$

Given a set $S \subseteq \mathbb{R}^n$, we denote by $\pi(S) = \{y \in \mathbb{R}^k : \exists z \in \mathbb{R}^{n-k} \text{ with } (y, z) \in S\}$ the projection of S onto the space of the y variables. We also denote by P the rational polyhedron $P = \{(y, z) \in \mathbb{R}^n : Ay + Bz \geq b\}$, and we define $\mathcal{F} = P \cap \mathcal{L}$. The next proposition characterizes when our MIQP is unbounded.

Proposition 1. *Problem* (2) *is unbounded if and only if*

(i) For every $y \in \pi(\mathcal{F})$, $\phi(y) = -\infty$, or
(ii) Region $\pi(\mathcal{F})$ is unbounded.

Proof. Condition (i) trivially implies that (2) is unbounded. In fact, the existence of a single y such that $\phi(y) = -\infty$ means that the corresponding restriction (3) of (2) is unbounded.

To prove the sufficiency of condition (ii), we now assume that $\pi(\mathcal{F})$ is unbounded. As $\pi(\mathcal{F}) \subseteq \pi(\text{conv}\,\mathcal{F})$, we have that $\pi(\text{conv}\,\mathcal{F})$ is unbounded as well. By Meyer's theorem [15] we have that conv $\mathcal{F}$ is a rational polyhedron, and Fourier's method implies that $\pi(\text{conv}\,\mathcal{F})$ is a rational polyhedron as well. Let y^r be a nonzero rational vector in the recession cone of $\pi(\text{conv}\,\mathcal{F})$. It follows that there exists a rational vector z^r such that (y^r, z^r) is in the recession cone of conv $\mathcal{F}$. Let $(\bar{y}, \bar{z}) \in \mathcal{F}$, and consider the ray $\{(\bar{y}, \bar{z}) + t(y^r, z^r) : t \geq 0\}$. The objective function $y^\top D y + c^\top y + f^\top z$ evaluated on the ray is $t^2(y^{r\top} D y^r) + O(t)$. Since the leading term is negative, the objective function tends to $-\infty$ along the ray. It follows that (2) is unbounded because, for every $\bar{t} \in \mathbb{R}$, the ray $\{(\bar{y}, \bar{z}) + t(y^r, z^r) : t \geq \bar{t}\}$ contains (infinitely many) points in $\mathcal{F}$.

To prove necessity of the conditions, we now assume that (2) is unbounded, and show that at least one of the two conditions holds. Consider the relaxation of (2) obtained by dropping the constraint $(y, z) \in \mathcal{L}$. Also the latter continuous problem is unbounded, and in this case Vavasis [24] proved that there exists a rational ray $\{(\tilde{y}, \tilde{z}) + t(y^r, z^r) : t \geq 0\} \subseteq P$ along which the objective function of (2) tends to $-\infty$. At least one among y^r and z^r must be nonzero.

Suppose y^r is nonzero, and let $(\bar{y}, \bar{z}) \in \mathcal{F}$. The ray $\{(\bar{y}, \bar{z}) + t(y^r, z^r) : t \geq 0\}$ is contained in P. Moreover, for every $\bar{t} \in \mathbb{R}$, the ray $\{(\bar{y}, \bar{z}) + t(y^r, z^r) : t \geq \bar{t}\}$ contains points in $\mathcal{F}$. This implies that the ray $\{\bar{y} + ty^r : t \geq \bar{t}\}$ contains points in $\pi(\mathcal{F})$ for every $\bar{t} \in \mathbb{R}$, and hence $\pi(\mathcal{F})$ is unbounded.

The other case is that $y^r = 0$, in which case z^r is nonzero. Our ray $\{(\tilde{y}, \tilde{z}) + t(y^r, z^r) : t \geq 0\}$ can then be written as $\{(\tilde{y}, \tilde{z} + tz^r) : t \geq 0\}$. Also, $f^\top z^r < 0$ because the objective function decreases along the ray by assumption. Let $\bar{y} \in \pi(\mathcal{F})$, let $\bar{z}$ be a vector such that $(\bar{y}, \bar{z}) \in \mathcal{F}$, and consider the ray $\{(\bar{y}, \bar{z} + tz^r) : t \geq 0\}$. This new ray is contained in the polyhedron $\{(y, z) \in P : y = \bar{y}\}$. Moreover, since (y^r, z^r) is rational, the ray $\{(\bar{y}, \bar{z} + tz^r) : t \geq \bar{t}\}$ contains points in $\mathcal{F}$ for every $\bar{t} \in \mathbb{R}$. Finally, $f^\top(\bar{z} + tz^r) = t(f^\top z^r) + O(1)$ tends to $-\infty$ along the ray, implying $\phi(\bar{y}) = -\infty$. □

The characterization given in Proposition 1 allows us to determine whether problem (2) is unbounded or not. For every $j = 1, \ldots, k$, solve

$$\min\{y_j : (y, z) \in \mathcal{F}\}, \qquad \max\{y_j : (y, z) \in \mathcal{F}\}.$$

Each problem is a MILP with a fixed number p of integer variables, which can be solved in polynomial time [14]. If any of these MILPs is unbounded, then MIQP (2) is unbounded by Proposition 1. Otherwise, let $(\bar{y}, \bar{z})$ be a point in $\mathcal{F}$, which can be, for example, an optimal solution of one of the $2k$ MILPs just solved. We can now compute $\phi(\bar{y})$ in polynomial time, since it amounts to solving another MILP with a fixed number of integer variables. As the point $\bar{y}$ is in $\pi(\mathcal{F})$, Proposition 1 implies that $\phi(\bar{y}) = -\infty$ if and only if problem (2) is unbounded. From now on, we assume that Problem (2) is bounded.

Boundedness of the Feasible Region. We now show that we can assume without loss of generality that P is bounded.

In order to do so, we first show that there exists an optimal solution to (2) whose encoding size is polynomially bounded by the size of (2). Assume that there exists an optimal solution (y^*, z^*) to (2) that lies in the positive orthant $y \geq 0, z \geq 0$ (the other cases follow symmetrically). We define P^+ to be the polyhedral convex hull of $P \cap \{(y, z) : y \geq 0, z \geq 0\} \cap \mathcal{L}$. Consider the following continuous problem:

$$\text{minimize} \quad y^\top Dy + c^\top y + f^\top z \tag{4}$$
$$\text{subject to} \quad (y, z) \in P^+.$$

As (4) is a Concave QP, it is well-known that there exists an optimal solution of (4) at a vertex of the feasible set P^+ (see, e.g., [7]). Since vertices of P^+ are in $\mathcal{L}$, this implies that such optimal solution of (4) is also an optimal solution of (2). The vertices of P^+ have size polynomially bounded by the size of P, which is in turn polynomially bounded by the size of (2) (see, e.g., [2]).

Now let B be such a polynomial bound. (We will not give this bound explicitly, but it can be easily calculated following [2,23].) Our algorithm can now replace our polyhedron P with the polytope of polynomial size $P \cap \{(y, z) : \|(y, z)\|_\infty \leq 2^B\}$. In this way we are guaranteed that at least an optimal solution of the original problem (2) will be feasible for the new problem.

Full-Dimensionality of the Feasible Region. We now show that we can assume without loss of generality that P is full-dimensional.

To do so, we test whether the constraint set defining P is full dimensional. This can be done by solving a single LP [8]. If not, we can find a linear change of basis that does not change the format of (2), and that lowers the dimension of the problem, ensuring without loss of generality that the feasible set is full dimensional. Note that this change of basis does not increase the number of negative eigenvalues of the objective function, or the number of integer variables.

2.1   Approximation in the Inner Region

In this section we define a polyhedron $\bar{P} \subseteq P$, and we find an ϵ-approximate solution of MIQP problem (2) restricted to feasible region $\bar{P} \cap \mathcal{L}$. Later, in Sect. 2.2, we give an algorithm that decomposes the feasible points in $P \setminus \bar{P}$ into a fixed number of lower-dimensional polytopes P_i. Our algorithm will then be applied recursively to each polytope P_i.

In order to define $\bar{P}$ we introduce the "even" mixed-integer lattice

$$\mathcal{L}^2 = \{(y, z) = \sum_{i=1}^{n} \lambda_i l^i : \lambda_1, \ldots, \lambda_p \in 2\mathbb{Z}, \lambda_{p+1}, \ldots, \lambda_n \in \mathbb{R}\} \subseteq \mathcal{L}.$$

By solving $2k$ MILPs with a fixed number of integer variables we compute, for every $j = 1, \ldots, k$, the values

$$l_j = \min\{y_j : (y, z) \in P \cap \mathcal{L}^2\}, \qquad u_j = \max\{y_j : (y, z) \in P \cap \mathcal{L}^2\}.$$

Each of these MILPs is bounded, since P is bounded. Moreover, we assume that $l_j < u_j$ for all j, because otherwise y_j is uniquely determined and can be dropped from the problem. The polytope $\bar{P}$ is then defined as

$$\bar{P} = \{(y, z) \in P : l_j \leq y_j \leq u_j, j = 1, \ldots, k\}.$$

The algorithm that we propose in this section is similar to algorithms that have appeared in the continuous optimization literature, like those described in Pardalos and Rosen [19] and in Vavasis [25]. In particular, the analysis in Claims 1, 2, and 4 below follows the one given by Vavasis. These algorithms are not directly applicable to the mixed-integer case as they rely on the convexity of the feasible region. Our contribution is to adapt such algorithmic techniques to the mixed-integer case: the special properties of $\bar{P}$ will in fact allow us to relax the convexity requirement, as we will see in Claim 3.

In order to simplify the notation in the remainder of the section, we further assume that the coordinates of the vector y are translated and rescaled so that

$$[l_1, u_1] \times \cdots \times [l_k, u_k] = [0, 1]^k.$$

Note that this affine transformation can be found and applied in polynomial time, and it does not change the format of (2), except for the set $\mathcal{L}$. In fact, because of the translation, both sets $\mathcal{L}$ and $\mathcal{L}^2$ will no longer necessarily contain the origin, thus from now on they will be translated lattices.

We now place an $(m + 1) \times \cdots \times (m + 1)$ grid of points in the cube $[0, 1]^k$. The value of m is the ceiling of $\sqrt{k/\epsilon}$, and the reason behind this choice will be explained later. The coordinates of the points of the grids have the form $(i_1/m, i_2/m, \ldots, i_k/m)$, where $i_1, \ldots, i_k \in \{0, 1, \ldots, m\}$. The grid partitions $[0, 1]^k$ into m^k subcubes. Next, for each subcube C, we construct an affine underestimator of the restriction of $q(y)$ to C. In what follows, we denote by γ the absolute value of the most negative diagonal entry of D.

Claim 1. *For each subcube C, we can construct an affine function $\mu(y)$ such that for every $y \in C$ we have*

$$\mu(y) \leq q(y) \leq \mu(y) + \frac{\gamma k}{4m^2}.$$

Proof of Claim. Let C be a particular subcube, say $C = [r_1, s_1] \times \cdots \times [r_k, s_k]$, where $s_j - r_j = 1/m$ for every j. For each $j = 1, \ldots, k$, the affine univariate function

$$\lambda_j(y_j) = d_{jj}(r_j + s_j)y_j + c_j y_j - d_{jj} r_j s_j \qquad (5)$$

satisfies $\lambda_j(r_j) = d_{jj} r_j^2 + c_j r_j$, and $\lambda_j(s_j) = d_{jj} s_j^2 + c_j s_j$. We define the affine function from $\mathbb{R}^k$ to $\mathbb{R}$ given by

$$\mu(y) = \sum_{j=1}^{k} \lambda_j(y_j).$$

The separability of q implies that $\mu(y)$ and $q(y)$ attain the same values at all vertices of C. As q is concave, this in particular implies that $\mu(y) \leq q(y)$.

We now show that $q(y) - \mu(y)$ is bounded from above by $\gamma k/(4m^2)$. From the separability of q, we obtain

$$q(y) - \mu(y) = \sum_{j=1}^{k} (d_{jj} y_j^2 + c_j y_j - \lambda_j(y_j)).$$

Using the explicit formula for λ_j given in (5), it can be derived that:

$$d_{jj} y_j^2 + c_j y_j - \lambda_j(y_j) = -d_{jj}(y_j - r_j)(s_j - y_j).$$

The univariate function on the right-hand is concave, and its maximum is achieved at $y_j = (r_j + s_j)/2$. This maximum value is $-d_{jj}/(4m^2)$. Therefore, as $-d_{jj} \leq \gamma$, for $j = 1, \ldots, k$, we establish that $q(y) - \mu(y) \leq \gamma k/(4m^2)$.

For each subcube C_ℓ, $\ell = 1, \ldots, m^k$, our algorithm now constructs the corresponding affine function $\mu_\ell(y)$ described in Claim 1. Then, we minimize $\mu_\ell(y) + \phi(y)$ on each subcube C_ℓ. This can be done by solving the following MILP with a fixed number of integer variables:

$$\begin{aligned} \text{minimize} \quad & \mu_\ell(y) + f^\top z \\ \text{subject to} \quad & Ay + Bz \geq b \\ & y \in C_\ell \\ & (y, z) \in \mathcal{L}. \end{aligned}$$

Finally, our algorithm returns the best solution (y°, z°) among all the m^k optimum solutions just obtained.

We now show that (y°, z°) is an ϵ-approximate solution to the MIQP problem (2) restricted to $\bar{P}$. To do so, we will obtain two bounds. The first bound is an upper bound on the gap between the objective value at (y°, z°) and the objective value at an optimum solution (y^*, z^*) of (2), where P is replaced by $\bar{P}$.

Claim 2. *The objective value of the point (y°, z°) is at most $\gamma k/(4m^2)$ above the objective value of (y^*, z^*).*

Proof of Claim. Note that the objective value at (y°, z°) is $q(y^\circ) + \phi(y^\circ)$, while the objective value at (y^*, z^*) is $q(y^*) + \phi(y^*)$. Let $\tilde{\mu}$ be the piecewise linear function on $[0, 1]^k$ that coincides with μ_ℓ on each subcube C_ℓ. We have

$$q(y^\circ) + \phi(y^\circ) - \frac{\gamma k}{4m^2} \leq \tilde{\mu}(y^\circ) + \phi(y^\circ) \leq \tilde{\mu}(y^*) + \phi(y^*) \leq q(y^*) + \phi(y^*).$$

The first and third inequalities follow because, by Claim 1, we have $\tilde{\mu}(y) \leq q(y) \leq \tilde{\mu}(y) + \gamma k/(4m^2)$ for every $y \in [0, 1]^k$. The second inequality holds because y° is a minimum for function $\tilde{\mu} + \phi$ over $\bar{P} \cap \mathcal{L}$.

The second bound is a lower bound on the gap between the maximum and the minimum objective function values of the points in $\bar{P} \cap \mathcal{L}$. Without loss of generality, we now assume that the most negative entry of D, the one with absolute value γ, is d_{11}. By construction of $\bar{P}$, there exists a point $(y^0, z^0) \in \bar{P} \cap \mathcal{L}^2$ such that $y_1^0 = 0$. Similarly, there is a point $(y^1, z^1) \in \bar{P} \cap \mathcal{L}^2$ such that $y_1^1 = 1$. We define

$$(y^\bullet, z^\bullet) = \frac{1}{2}(y^0, z^0) + \frac{1}{2}(y^1, z^1).$$

Claim 3. *The point $(y^\bullet, z^\bullet)$ is in $\bar{P} \cap \mathcal{L}$.*

Proof of Claim. The vector $(y^\bullet, z^\bullet)$ is clearly in $\bar{P}$, since it is a convex combination of two points in $\bar{P}$. We now show that it is also in the translated mixed-integer lattice $\mathcal{L}$. Let w be a vector in $\mathcal{L}^2$. Then, we can write (y^β, z^β), for $\beta = 0, 1$, as:

$$(y^\beta, z^\beta) = w + \sum_{i=1}^{n} \lambda_i^\beta l^i \qquad \text{for } \lambda_1^\beta, \ldots, \lambda_p^\beta \in 2\mathbb{Z}, \ \lambda_{p+1}^\beta, \ldots, \lambda_n^\beta \in \mathbb{R}.$$

We obtain

$$(y^\bullet, z^\bullet) = \frac{1}{2}(y^0, z^0) + \frac{1}{2}(y^1, z^1) = w + \sum_{i=1}^{n} \left(\frac{\lambda_i^0}{2} + \frac{\lambda_i^1}{2}\right) l^i,$$

and the last vector is in $\mathcal{L}$ since $\frac{\lambda_i^0}{2} + \frac{\lambda_i^1}{2} \in \mathbb{Z}$ for every $i = 1, \ldots, p$.

We are now ready to derive our lower bound.

Claim 4. *The objective value of the point $(y^\bullet, z^\bullet)$ is at least $\gamma/4$ above the objective value of (y^*, z^*).*

Proof of Claim. Since $q(y^0) + \phi(y^0) \geq q(y^*) + \phi(y^*)$, and $q(y^1) + \phi(y^1) \geq q(y^*) + \phi(y^*)$, in order to prove the claim we just need to show the following bound:

$$y^{\bullet\top} D y^\bullet + c^\top y^\bullet + f^\top z^\bullet \geq \frac{1}{2}\Big(q(y^0) + \phi(y^0) + q(y^1) + \phi(y^1)\Big) + \gamma/4.$$

To do so, note that $y^{\bullet\top} D y^\bullet + c^\top y^\bullet + f^\top z^\bullet$ can be rewritten as

$$\frac{1}{2}\big(q(y^0) + f^\top z^0\big) + \frac{1}{2}\big(q(y^1) + f^\top z^1\big) - \frac{1}{4}(y^0 - y^1)^\top D(y^0 - y^1).$$

By definition of ϕ, the latter is at least

$$\frac{1}{2}\big(q(y^0) + \phi(y^0)\big) + \frac{1}{2}\big(q(y^1) + \phi(y^1)\big) - \frac{1}{4}(y^0 - y^1)^\top D(y^0 - y^1).$$

Therefore, we just need to show that $-\frac{1}{4}(y^0 - y^1)^\top D(y^0 - y^1) \geq \gamma/4$. We can write:

$$-\frac{1}{4}(y^0 - y^1)^\top D(y^0 - y^1) = \frac{1}{4}\sum_{j=1}^{k}(y_j^0 - y_j^1)^2(-d_{jj}).$$

All the terms of the summation are nonnegative, thus a lower bound is given by the first term $(y_1^0 - y_1^1)^2(-d_{11})$. By choice of y^0 and y^1, we obtain $(y_1^0 - y_1^1)^2 = 1$, and since $-d_{11} = \gamma$, we get the desired lower bound of $\gamma/4$.

By Claim 2, we can find a point $(y^\circ, z^\circ) \in \bar{P} \cap \mathcal{L}$ that is at most $\gamma k/(4m^2)$ above optimal. By Claims 3 and 4, there is a point in $\bar{P} \cap \mathcal{L}$ that is at least $\gamma/4$ above optimal. Therefore, (y°, z°) is an ϵ-approximate solution to the MIQP problem (2) restricted to $\bar{P}$ provided that

$$\frac{\gamma k}{4m^2} \leq \epsilon \cdot \frac{\gamma}{4},$$

and the latter condition holds if m is the ceiling of $\sqrt{k/\epsilon}$.

In this section, to find (y°, z°), we solved a total of

$$2k + m^k = 2k + \left\lceil \sqrt{\frac{k}{\epsilon}} \right\rceil^k$$

MILPs with p integer variables. For fixed k, the number of MILPs is polynomial in $1/\epsilon$. Moreover, for fixed p, each MILP can be solved in polynomial time. Therefore, for fixed k and p, the running time of the given approximation algorithm is polynomial in the size of the problem, and in $1/\epsilon$.

2.2   Decomposition of the Outer Region

In this section we explain how to decompose the feasible points in $P \backslash \bar{P}$ into a number of lower-dimensional polytopes P_i. For each of these polytopes P_i, we will then apply recursively the presented algorithm. The total number of times that we need to run our algorithm will be polynomial when the number of integer variables p is fixed.

At the end of the execution of the recursive algorithm we will have stored a polynomial number of ϵ-approximate solutions, each one corresponding to a particular polyhedral subset of the original polyhedron. (For example, the solution (y°, z°) obtained above is the one corresponding to $\bar{P}$.) By construction, each feasible solution will be contained in at least one of these polyhedral subsets.

Finally, we return as the approximate solution to the original MIQP the vector $(y^\triangle, z^\triangle)$ that achieves the minimum objective function value among all the ϵ-approximate solutions obtained. The objective value of $(y^\triangle, z^\triangle)$ is at most that of the ϵ-approximate solution, say (y°, z°), corresponding to the polyhedral subset which contains a global optimum solution of the problem. The vector (y°, z°) is an ϵ-approximate solution to the original MIQP, and so is the point $(y^\triangle, z^\triangle)$ returned by the algorithm.

In the remainder of this section, we will explain how to use Lenstra's [14] seminal algorithm in order to obtain a number of lower-dimensional polytopes P_i which contain all the feasible points that are in P but not in $\bar{P}$.

The points in $P \backslash \bar{P}$ are contained in the full-dimensional polytopes among the following:

$$\begin{aligned}
\{(y, z) \in P : y_j \leq l_j\} \qquad & j = 1, \dots, k \qquad\qquad (6)\\
\{(y, z) \in P : y_j \geq u_j\} \qquad & j = 1, \dots, k.
\end{aligned}$$

This is because, if $(\bar{y}, \bar{z}) \in P \backslash \bar{P}$, then there is at least one inequality among $y_j \leq l_j$ and $y_j \geq u_j$ that is satisfied strictly by $(\bar{y}, \bar{z})$, and the corresponding polytope among (6) is full-dimensional since so is P. We denote by $\mathcal{Q}$ the family of the full-dimensional polytopes among the $2k$ polytopes in (6). Each point in $\mathcal{L} \cap P \backslash \bar{P}$ is now contained in at least a polytope in $\mathcal{Q}$.

Let Q be one polytope in $\mathcal{Q}$. We now describe how to decompose Q into a fixed number of lower dimensional polytopes. In order to do so, we will essentially apply Lenstra's algorithm [14] to Q. To keep the notation simple, we go back to the space of the x-variables, which can be done via the change of variables $x = \Pi^\top L^{-\top}(y, z)$. Let $\bar{Q}$ be the image of Q in the space of the x-variables. Since Q is bounded and full-dimensional, so is $\bar{Q}$. Moreover, since Q contains points in $\mathcal{L}^2$ only on its boundary, we have that $\bar{Q}$ contains points in $2\mathbb{Z}^p \times \mathbb{R}^{n-p}$ only on its boundary. Let $\tilde{Q}$ be the projection of $\bar{Q}$ on the integer space:

$$\tilde{Q} = \{(x_1, \ldots, x_p) \in \mathbb{R}^p : \exists (x_{p+1}, \ldots, x_n) \in \mathbb{R}^{n-p} \text{ such that } (x_1, \ldots, x_n) \in \bar{Q}\}.$$

Clearly $\tilde{Q}$ has dimension p, and contains points in $2\mathbb{Z}^p$ only on its boundary. In order to apply Lenstra's algorithm, we do not need to explicitly construct the polytope $\tilde{Q}$, which generally has an exponential number of facets.

Since the number of integer variables p is fixed, following Sect. 5 in Lenstra [14] (see also Sect. 18 in [23] and Sect. 9 in [2]), with $2\mathbb{Z}$ instead of $\mathbb{Z}$, we obtain in polynomial time a nonzero vector $d \in \mathbb{Z}^p$ such that

$$\max\{d^\top x : x \in \tilde{Q}\} - \min\{d^\top x : x \in \tilde{Q}\} \leq 4p(p+1)2^{p(p-1)/4}.$$

Clearly, the value on the right-hand is a fixed number for p fixed. (Note that better bounds and running times can be obtained by using modern Lenstra-type algorithms, like the ones described in [10,11].)

In general, Lenstra's algorithm either finds a point in $\bar{Q} \cap (2\mathbb{Z}^p \times \mathbb{R}^{n-p})$, or a flat direction d as above. However, since $\bar{Q}$ contains points in $2\mathbb{Z}^p \times \mathbb{R}^{n-p}$ only on its boundary, the algorithm in any case finds a flat direction d.

We now construct a fixed number of subproblems with one less integer variable. Determine $\mu = \min\{dx : x \in \tilde{Q}\}$, and consider the polytopes

$$\bar{Q}_t = \{x \in \bar{Q} : (d, 0)^\top x = t\} \qquad \text{for } t = \lceil \mu \rceil, \ldots, \lceil \mu + 4p(p+1)2^{p(p-1)/4} \rceil.$$

Then each point in $\bar{Q} \cap (\mathbb{Z}^p \times \mathbb{R}^{n-p})$ is in one of these $\bar{Q}_t$. Now, via a linear change of basis we can lower by one the dimension of the space where each $\bar{Q}_t$ lives. In this new space, the lattice to be considered will be $\mathbb{Z}^{p-1} \times \mathbb{R}^{n-p}$. Since we have to consider each polytope $Q \in \mathcal{Q}$, in total we construct at most $2k \cdot 4p(p+1)2^{p(p-1)/4}$ polytopes with one less integer variable.

Our algorithm will now be applied recursively to each polytope $\bar{Q}_t$, until each obtained polytope will be considered in a purely continuous space. The algorithm will then be applied a total number of times upper bounded by

$$\sum_{j=0}^{p}(2k \cdot 4p(p+1)2^{p(p-1)/4})^j \leq (2k \cdot 4p(p+1)2^{p(p-1)/4})^{p+1},$$

which is a polynomial number for fixed p.   □

References

1. Bellare, M., Rogaway, P.: The complexity of aproximating a nonlinear program. In: Pardalos, P.M. (ed.), Complexity in Numerical Optimization. World Scientific (1993)
2. Conforti, M., Cornuéjols, G., Zambelli, G.: Integer Programming. Springer, Heidelberg (2014)
3. Cook, W., Hartman, M., Kannan, R., McDiarmid, C.: On integer points in polyhedra. Combinatorica **12**(1), 27–37 (1992)
4. Cook, W.J., Kannan, R., Schrijver, A.: Chvátal closures for mixed integer programming problems. Math. Program. **47**(1–3), 155–174 (1990)
5. de Klerk, E., Laurent, M., Parrilo, P.A.: A PTAS for the minimization of polynomials of fixed degree over the simplex. Theoret. Comput. Sci. **361**, 210–225 (2006)
6. Del Pia, A., Dey, S.S., Molinaro, M.: Mixed-integer quadratic programming is in NP, Manuscript (2014)
7. Floudas, C.A., Visweswaran, V.: Quadratic optimization. In: Horst, R., Pardalos, P.M. (eds.) Handbook of Global Optimization. Nonconvex Optimization and its Applications, vol. 2, pp. 217–269. Springer, New York (1995)
8. Freund, R.M., Roundy, R., Todd, M.J.: Identifying the set of always-active constraints in a system of linear inequalities by a single linear program. Working Paper, pp. 1674–85, Sloan School of Management, MIT, Cambridge, MA (1985)
9. Golub, G.H., Van Loan, C.F.: Matrix Computations, 3rd edn. Johns Hopkins University Press, Baltimore, MD, USA (1996)
10. Heinz, S.: Complexity of integer quasiconvex polynomial optimization. J. Complex. **21**, 543–556 (2005)
11. Hildebrand, R., Köppe, M.: A new lenstra-type algorithm for quasiconvex polynomial integer minimization with complexity $2^{O(n \log n)}$. Discrete Optim. **10**, 69–84 (2013)
12. Hildebrand, R., Oertel, T., Weismantel, R.: Note on the complexity of the mixed-integer hull of a polyhedron. Oper. Res. Lett. **43**, 279–282 (2015)
13. Khachiyan, L.G.: A polynomial algorithm in linear programming (in Russian). Doklady Akademii Nauk SSSR, 244, pp. 1093–1096 (1979). (English translation: Soviet Mathematics Doklady, 20, pp. 191–194, 1979)
14. Lenstra Jr., H.W.: Integer programming with a fixed number of variables. Math. Oper. Res. **8**(4), 538–548 (1983)
15. Meyer, R.R.: On the existence of optimal solutions to integer and mixed-integer programming problems. Math. Program. **7**(1), 223–235 (1974)
16. Murty, K.G., Kabadi, S.N.: Some NP-complete problems in quadratic and linear programming. Math. Program. **39**, 117–129 (1987)
17. Nemirovsky, A.S., Yudin, D.B.: Problem Complexity and Method Efficiency in Optimization. Wiley, Chichester (1983). Translated by E.R. Dawson from Slozhnost' Zadach i Effektivnost' Metodov Optimizatsii (1979)
18. Onn, S.: Convex discrete optimization. In: Chvátal, V. (ed.) Combinatorial Optimization: Observation of Strains. Infect Dis Ther. Methods Appl. **3**(1), 35–43, pp. 183–228. IOS Press (2011)
19. Pardalos, P.M., Rosen, J.B. (eds.): Constrained Global Optimization: Algorithms and Applications. LNCS, vol. 268. Springer, Heidelberg (1987)
20. Pardalos, P.M., Vavasis, S.A.: Quadratic programming with one negative eigenvalue is NP-hard. J. Global Optim. **1**(1), 15–22 (1991)

21. Rosen, J.B., Pardalos, P.M.: Global minimization of large-scale constrained concave quadratic problems by separable programming. Math. Program. **34**, 163–174 (1986)
22. Sahni, S.: Computationally related problems. SIAM J. Comput. **3**, 262–279 (1974)
23. Schrijver, A.: Theory of Linear and Integer Programming. Wiley, Chichester (1986)
24. Vavasis, S.A.: Quadratic programming is in NP. Inform. Proces. Lett. **36**, 73–77 (1990)
25. Vavasis, S.A.: On approximation algorithms for concave quadratic programming. In: Floudas, C.A., Pardalos, P.M. (eds.) Recent Advances in Global Optimization, pp. 3–18. Princeton University Press, Princeton (1992)

Centerpoints: A Link Between Optimization and Convex Geometry

Amitabh Basu[1] and Timm Oertel[2(✉)]

[1] Johns Hopkins University, Baltimore, USA
[2] Cardiff University, Cardiff, UK
oertelt@cardiff.ac.uk

Abstract. We introduce a concept that generalizes several different notions of a "centerpoint" in the literature. We develop an oracle-based algorithm for convex mixed-integer optimization based on centerpoints. Further, we show that algorithms based on centerpoints are "best possible" in a certain sense. Motivated by this, we establish several structural results about this concept and provide efficient algorithms for computing these points.

1 Introduction

Let μ be a Borel-measure[1] on $\mathbb{R}^n$ such that $0 < \mu(\mathbb{R}^n) < \infty$. Without any loss of generality, we normalize the measure to be a probability measure, i.e., $\mu(\mathbb{R}^n) = 1$. For $S \subset \mathbb{R}^n$ closed, we define the set of *centerpoints* $\mathcal{C}(S, \mu) \subset S$ as the set that attains the following maximum

$$\mathcal{F}(S, \mu) := \max_{x \in S} \inf_{u \in \mathcal{S}^{n-1}} \mu(H^+(u, x)), \tag{1}$$

where $\mathcal{S}^{n-1}$ denotes the $(n - 1)$-dimensional sphere and $H^+(u, x)$ denotes the half-space $\{y \in \mathbb{R}^n \mid u \cdot (y - x) \geq 0\}$. This definition unifies several definitions from convex geometry and statistics. Two notable examples are:

1. *Winternitz measure of symmetry.* Let μ be the Lebesgue measure restricted to a convex body K (i.e., K is compact and has a non-empty interior), or equivalently, the uniform probability measure on K, and let $S = \mathbb{R}^n$. $\mathcal{F}(S, \mu)$ in this setting is known in the literature as the *Winternitz measure of symmetry* of K, and the centerpoints $\mathcal{C}(S, \mu)$ are the "points of symmetry" of K. This notion was studied by Grünbaum in [9] and surveyed by the same author (along with other measures of symmetry for convex bodies) in [10].

---

[1] The reader who is unfamiliar with measure theory, may simply consider μ to be the volume measure or the mixed-integer measure on the mixed-integer lattice, i.e., $\mu(S)$ returns the volume of S or the "mixed-integer volume" of the mixed-integer lattice points inside S, where S will usually be a convex body. The "mixed-integer volume" reduces to the number of integer points when the lattice is the set of integer points. See (3) for a precise definition which generalizes both the standard volume and the "counting measure" for the integer lattice.

© Springer International Publishing Switzerland 2016
Q. Louveaux and M. Skutella (Eds.): IPCO 2016, LNCS 9682, pp. 14–25, 2016.
DOI: 10.1007/978-3-319-33461-5_2

2. *Tukey depth and median.* In statistics and computational geometry, the function $f_\mu : \mathbb{R}^n \to \mathbb{R}$ defined as

$$f_\mu(x) := \inf_{u \in \mathcal{S}^{n-1}} \mu(H^+(u, x)) \tag{2}$$

is known as the *halfspace depth function* or the *Tukey depth function* for the measure μ, first introduced by Tukey [20]. Taking $S = \mathbb{R}^n$, the centerpoints $\mathcal{C}(\mathbb{R}^n, \mu)$ are known as the *Tukey medians* of the probability measure μ, and $\mathcal{F}(\mathbb{R}^n, \mu)$ is known as the maximum *Tukey depth* of μ. Tukey introduced the concept when μ is a finite sum of Dirac measures (i.e., a finite set of points with the counting measure), but the concept has been generalized to other probability measures and analyzed from structural, as well as computational perspectives. See [15] for a survey of structural aspects and other notions of "depth" used in statistics, and [7] and the references therein for a survey and recent approaches to computational aspects of the Tukey depth.

Our Results. To the best of our knowledge, all related notions of centerpoints in the literature always insist on the set S being $\mathbb{R}^n$, i.e., the centerpoint can be any point from the Euclidean space. We consider more general S, as this captures certain operations performed by oracle based mixed-integer convex optimization algorithms. In Sect. 2, we elaborate on this connection between centerpoints and algorithms for mixed-integer optimization. We first give an algorithm for solving convex mixed-integer optimization given access to first-order (separation) oracles, based on centerpoints. Second, we show that oracle-based algorithms for convex mixed-integer optimization that use centerpoint information are "best possible" in a certain sense. This comprises our main motivation in studying centerpoints.

In Sect. 4, we show that when $S = \mathbb{R}^n$ and μ is the uniform measure on polytopes, centerpoints are unique, a question which was surprisingly not proved earlier. We also present a new technique to lower bound $\mathcal{F}(\mathbb{Z}^n \times \mathbb{R}^d, \nu)$ where ν is the "mixed-integer" uniform measure on $K \cap (\mathbb{Z}^n \times \mathbb{R}^d)$ and K is some convex body. Such bounds immediately imply bounds on the complexity of oracle-based convex mixed-integer optimization algorithms.

In Sect. 5, we present a number of exact and approximation algorithms for computing centerpoints. To the best of our knowledge, the computational study of centerpoints has only been done for measures μ that are a finite sum of Dirac measures, i.e., for finite point sets, or when μ is the uniform measure on two dimensional polygons (e.g. see [5] and the references therein). We initiate a study for other measures; in particular, the uniform measure on a convex body, the counting measure on the integer points in a convex body, and the mixed-integer volume of the mixed-integer points in a convex body. All our algorithms are exponential in the dimension n but polynomial in the remaining input data, so these are polynomial time algorithms if n is assumed to be a constant. Algorithms that are polynomial in n are likely to not exist because of the reduction to the so-called *closed hemisphere problem* – see Chap. 7 in Bremner, Fukuda and Rosta [14].

2   An Application to Mixed-Integer Optimization

We will be interested in the setting when the measure μ is based on the mixed-integer volume of the mixed-integer points in a convex body K, and S is the set of mixed-integer points in K. More precisely, let $K \subseteq \mathbb{R}^n \times \mathbb{R}^d$ be a convex set. Let vol_d be the d-dimensional volume (Lebesgue measure). We define the *mixed-integer volume with respect to K* as

$$\mu_{mixed,K}(C) := \frac{\sum_{z \in \mathbb{Z}^n} \text{vol}_d(C \cap K \cap (\{z\} \times \mathbb{R}^d))}{\sum_{z \in \mathbb{Z}^n} \text{vol}_d(K \cap (\{z\} \times \mathbb{R}^d))} \tag{3}$$

for any Lebesgue measurable subset $C \subseteq \mathbb{R}^n \times \mathbb{R}^d$. For later use we want to introduce the notation $\bar{\mu}_{mixed}(C) = \sum_{z \in \mathbb{Z}^n} \text{vol}_d(C \cap (\{z\} \times \mathbb{R}^d))$. The dimensions n and d will be clear from the context.

Our main motivation to study centerpoints comes from its natural connection to convex mixed-integer optimization. Consider the following unconstrained optimization problem

$$\min_{(x,y) \in \mathbb{Z}^n \times \mathbb{R}^d} g(x,y). \tag{4}$$

where $g : \mathbb{R}^n \times \mathbb{R}^d \mapsto \mathbb{R}$ is a convex function given by a first-order evaluation oracle. Queried at a point the oracle return the corresponding function value and an element from the subdifferential. We assume that the problem is bounded. Further, if $d \neq 0$, we assume that for every fixed $x \in \mathbb{Z}^n$, $g(x,y)$ is Lipschitz continuous in the y variables with Lipschitz constant L. We present a general cutting plane method based on centerpoints, i.e. the *centerpoint-method*. This can be interpreted as an extension of the well-known Method of Centers of Gravity or other cutting plane methods such as the Ellipsoid method or Kelly's cutting plane method (see [16, Sect. 3.2.6.]) for convex optimization. This type of idea was also explored by Bertsimas and Vempala in [4] for continuous convex optimization. Our approach bears similarities to Lenstra-type algorithms [8,13] for convex integer optimization problems. Most variations of Lenstra-type algorithms rely on a combination of the ellipsoid method and enumeration on lower dimensional subproblems. The key difference is that our algorithm avoids enumerating low dimensional subproblems.

We assume that we have access to (approximate) centerpoints of polytopes through an oracle. As in statistics, we introduce the notation

$$D_\mu(\alpha) := \{x \in \mathbb{R}^n : f_\mu(x) \geq \alpha\}. \tag{5}$$

We define the oracle for the case that we only have access to approximate centerpoints as follows.

Definition 1 (α-central-point-oracle). *For a polytope P and a given $\alpha > 0$, the oracle returns a point $z \in D_\mu(\alpha)$, where $\mu := \mu_{mixed,P}$.*

This way we hide the complexity of computing centerpoints in the oracle and keep the following discussion as general as possible. However, for several special cases the oracle can be realized as we discuss in subsequent sections.

The general algorithmic framework is as follows. We start with a bounding box, say $P^0 := [0, B]^{n+d}$ with $B \in \mathbb{Z}_+$, that is guaranteed to contain an optimum and initialize $x^* = 0, y^* = 0$. Then, we construct iteratively a sequence of polytopes $P^1, P^2, \ldots$ by intersecting P^k with the half-space defined by its (approximate) centerpoint and the corresponding subgradient arising from g. That is, let $(x_k, y_k) \in D_{\mu_{mixed}, P_k}(\alpha)$ and let $h_k \in \partial g(x_k, y_k)$ be an element of the subdifferential. Then we define $(x^*, y^*) := \text{argmin}\{g(x^*, y^*), g(x_k, y_k)\}$ and $P^{k+1} := \{(x, y) \in P^k \mid g(x^*, y^*) - g(x_i, y_i) \geq h_i \cdot (x - x_i, y - y_i), i = 1, \ldots, k\}$. It follows that

$$P^k \supset P^{k+1} \supset \underset{x \in \mathbb{Z}^n \times \mathbb{R}^d}{\text{argmin}} \, g(x)$$

for all $k \in \mathbb{N}$. Further, by the choice of (x_k, y_k), the measure of P^k decreases in each iteration by a fraction of at least $1 - \alpha$. With (x^*, y^*) we keep track of the (approximate) centerpoint x_k that has the smallest objective value $g^* := g(x^*, y^*)$ among all points we encounter.

Let $(\hat{x}, \hat{y}) \in \mathbb{Z}^n \times \mathbb{R}^d$ attain the optimal value $\hat{g}$ of Problem (4). We have $\bar{\mu}_{mixed}(\{0, \ldots, B\}^n \times [0, B]^d) \approx B^{n+d}$. By standard arguments, we can bound $\bar{\mu}_{mixed}(P_k)$ from below as follows

$$\bar{\mu}_{mixed}(P_k) \geq \bar{\mu}_{mixed}(\{(x, y) \in \mathbb{Z}^n \times \mathbb{R}^d \mid g((x, y)) - \hat{g} \leq g^* - \hat{g}\})$$

$$\geq \bar{\mu}_{mixed}\left(\left\{(\hat{x}, y) \in \{\hat{x}\} \times \mathbb{R}^d \, \middle| \, \|(\hat{x}, \hat{y}) - (\hat{x}, y)\|_2 \leq \frac{g^* - \hat{g}}{L}\right\}\right)$$

$$= \left(\frac{\hat{g} - g^*}{L}\right)^d \kappa_d,$$

where κ_d denotes the volume of the d-dimensional unit-ball. Then, it follows that after at most

$$k \leq \frac{d \ln\left(\frac{LB}{\epsilon}\right) + n \ln(B)}{\ln(1 - \alpha)}$$

iterations we have that $g(x^*, y^*) - g(\hat{x}, \hat{y}) \leq \epsilon$. Note that in the pure integer case when $d = 0$ we can actually solve the problem exactly.

It is not difficult to generalize this to the constrained optimization case:

$$\min_{\substack{x \in \mathbb{Z}^n \times \mathbb{R}^d, \\ h(x) \leq 0}} \, g(x).$$

where $g, h : \mathbb{R}^n \times \mathbb{R}^d \mapsto \mathbb{R}$ are convex functions given by first-order oracles. Further, the algorithm can be extended to handle quasi-convex functions, if one has access to separation oracles for their sublevel sets.

The main feature of this approach is that, from the point view of the number of function oracle calls, this algorithm is best possible. Assume that $d = 0$ and that we can compute exact centerpoints. Then one can prove the following theorem.

Theorem 1. *No algorithm can exist for solving general bounded convex integer optimization problems, that needs fewer function oracle calls than the exact centerpoint-method in the worst case.*

We omit the proof from this extended abstract.

3   General Properties

In this section we first establish some analytic properties of f_μ. This will justify the use of "maximum" in (1), instead of a supremum. The main result of this section is a bound on the quality of the centerpoints based on Helly numbers. We will denote the complement of a set X by X^c. We begin with a technical lemma whose proof is omitted from this extended abstract.

Lemma 1. *For any probability measure μ, $f_\mu(x)$ defined in (2) is quasi-concave on $\mathbb{R}^n$ and upper semicontinuous. Moreover, given $\bar{x} \in \mathbb{R}^n$ and $\delta > 0$, let $\bar{u} \in S^{n-1}$ be such that $\mu(H^+(\bar{u}, \bar{x})) \leq \inf_{u \in S^{n-1}} \mu(H^+(u, \bar{x})) + \frac{\delta}{2}$. Then $\bar{u}$ strongly separates the set $\{x \in \mathbb{R}^n \mid f(x) \geq f(\bar{x}) + \delta\}$ and $\bar{x}$, i.e., $\bar{u} \cdot x < \bar{u} \cdot \bar{x}$ for all x such that $f(x) \geq f(\bar{x}) + \delta$.*

Remark 1. Lemma 1 shows that $\sup_{x \in S} f_\mu(x)$ is always attained. See Proposition 7 in [18] where this is discussed for $S = \mathbb{R}^n$. The generalization to any closed subset S is easy; see also Proposition 5 in [18] which states the for every $\alpha \geq 0$, the set $D_\mu(\alpha)$ given by (5) is compact.

Next we generalize a theorem well-known in the literature on half-space (Tukey) depth [18, Proposition 9]; this was earlier stated by Grünbaum [9, Theorem 1] for uniform probability measures on convex bodies. In all of these works, the authors consider $S = \mathbb{R}^n$, as discussed in the introduction. We consider more general sets S. For this we define the Helly number of a set $S \subseteq \mathbb{R}^n$. Let $\mathcal{K} := \{S \cap K \mid K \subset \mathbb{R}^n \text{ convex}\}$. Then the Helly-number $h = h(S) \in \mathbb{N}$ of S is defined as the smallest number such that the following property is satisfied for all finite subsets $\{C_1, \ldots, C_m\} \subset \mathcal{K}$: If

$$C_{i_1} \cap \cdots \cap C_{i_h} \neq \emptyset \text{ for all } \{i_1, \ldots, i_h\} \subset \{1, \ldots, m\}$$

then

$$C_1 \cap \cdots \cap C_m \neq \emptyset.$$

If no such number exists, then $h(S) = \infty$. This extension of Helly's number was first considered by Hoffman [11], and has recently been studied in [1,2,6].

Theorem 2. *Let $S \subseteq \mathbb{R}^n$ be a closed subset and let μ be such that $\mu(\mathbb{R}^n \backslash S) = 0$. If $h(S) < \infty$, then $\mathcal{F}(S, \mu) \geq h(S)^{-1}$.*

The proof follows along similar lines as [18, Proposition 9]. By applying the well known bound for the mixed-integer Helly-number [2,6,11] we get the following Corollary.

Corollary 1. *$\mathcal{F}(\mathbb{Z}^n \times \mathbb{R}^d, \mu) \geq \frac{1}{2^n(d+1)}$ for any finite measure μ on $\mathbb{R}^{n+d}$ such that $\mu(\mathbb{R}^{n+d} \backslash (\mathbb{Z}^n \times \mathbb{R}^d)) = 0$. In particular, this holds for $\mu_{mixed, K}$ for any convex body $K \subseteq \mathbb{R}^n \times \mathbb{R}^d$.*

Remark 2. Let $K \subset \mathbb{R}^{n+d}$ be a convex body and let $\mu_{mixed, K}$ denote the mixed-integer volume with respect to K, as defined in (3). One can show that in this case the infimum in (1) and (2) is actually achieved.

4 Specialized Properties

For a general measure, the centerpoint may not be unique. One can show however that when $S = \mathbb{R}^n$ and μ is the uniform measure on a polytope, the centerpoint is unique. Surprisingly, this question had not been investigated before, and as far as we know the question of uniqueness for the centerpoint for general convex bodies is open. We omit the proof of the following proposition.

Proposition 1. *Let μ be the uniform measure on a full-dimensional polytope $P \subset \mathbb{R}^n$. Then $C(\mathbb{R}^n, \mu)$ is a singleton, i.e., the centerpoint is unique.*

Remark 3. With similar arguments one can show the proposition also holds for strictly convex sets, although the question remains open for general convex bodies. Another interesting open question is the following: Is the centerpoint of a rational polytope rational? If so, is the size of the centerpoint polynomially bounded in the size of an irredundant description of the rational polytope?

In the remaining section we want to improve the bound on $\mathcal{F}(\mathbb{Z}^n \times \mathbb{R}^d, \nu)$ coming from Helly numbers (Theorem 2 and Corollary 1) when ν is a mixed-integer measure. Better bounds have been obtained by Grünbaum by exploiting properties of the *centroid* of a convex body K, which is defined as $c_K := \int_K x \, d\mu(x)$, where the integral is taken with respect to the uniform measure μ on K. Grünbaum proved in [9] that $\mu(H^+(u, c_K)) \geq \left(\frac{n}{n+1}\right)^n \geq e^{-1}$ for any $u \in \mathcal{S}^{n-1}$, which immediately implies that $\mathcal{F}(\mathbb{R}^n, \mu) \geq e^{-1}$. This, of course, drastically improves the Helly bound of $\frac{1}{n+1}$. In the following we want to extend these improved bounds to the mixed-integer setting. Ideally, we would want to prove the following conjecture.

Conjecture 1. Let $S = \mathbb{Z}^n \times \mathbb{R}^d$ and let $\nu = \mu_{mixed,K}$ for some convex body $K \in \mathbb{R}^{n+d}$. Then $\mathcal{F}(\mathbb{Z}^n \times \mathbb{R}^d, \nu) \geq \frac{1}{2^n} \left(\frac{d}{d+1}\right)^d$.

In a first step we consider convex sets K that have a large *lattice-width*, where the lattice-width is defined as $\omega(K) := \min_{z \in \mathbb{Z}^n \setminus \{0\}} [\max_{x \in K} u \cdot x - \min_{x \in K} u \cdot x]$. As an auxiliary lemma, we show that for convex sets with large lattice width, the d-dimensional Lebesgue measure $\bar{\nu} := \bar{\mu}_{mixed}$ of $K \cap (\mathbb{Z}^n \times \mathbb{R}^d)$ can be bounded by the $(d+n)$-dimensional Lebesgue measure $\bar{\mu}$ of K and vice versa. Note that in this case we do not normalize the measures. In the pure integer setting, i.e., $d = 0$, this connection is well known. However, to the best of our knowledge, this kind of result has never been proven for the mixed-integer setting nor explicitly with respect to the lattice width. Again, we omit the proof in this extended abstract. We denote the projection of a set $X \subset \mathbb{R}^{n+d}$ onto the first n coordinates by $X|_{\mathbb{R}^n}$.

Lemma 2. *Let $K \subset \mathbb{R}^{n+d}$ be a closed convex set with non-empty interior. Let ω denote the lattice-width of $K|_{\mathbb{R}^n}$. If $\omega > cn(n + d)^{7/2} \alpha n^n \sqrt{n}$ for a universal constant α and a $c \in \mathbb{N}$, then*

$$e^{-\frac{1}{c}} \leq \frac{\bar{\nu}(K \cap (\mathbb{Z}^n \times \mathbb{R}^d))}{\bar{\mu}(K)} \leq e^{\frac{1}{c}}.$$

For the following theorem we introduce the following technical rounding procedure. Let K be a convex body with a sufficiently large lattice width, i.e., $\omega(K) > cn(n+d)^{7/2}\alpha n^n\sqrt{n}$ for some positive integer c, where α is the constant from Lemma 2. Let μ be the uniform measure on K and let $x^\star \in \mathcal{C}(\mathbb{R}^{n+d}, \mu)$. One can show that there exist $b_i \in (-x^\star + K) \cap (\mathbb{Z}^n \times \mathbb{R}^d)$ for $i = 1,\ldots,n$ such that $b_1|_{\mathbb{R}^n},\ldots,b_n|_{\mathbb{R}^n}$ is a Korkine-Zolotarev basis [12] of $\mathbb{Z}^n$ with respect to the maximum inscribed ellipsoid of $K|_{\mathbb{R}^n}$. However, in this extended abstract we omit the details. In addition we define for $i = n+1,\ldots,n+d$ $b_i \in \mathbb{R}^{n+d}$ as the i-th unit vector. Hence, $b_1,\ldots,b_{n+d}$ define a basis of $\mathbb{R}^{n+d}$.

Given $x = \sum_{i=1}^{n+d}\lambda_i b_i \in \mathbb{R}^{n+d}$ with $\lambda_i \in \mathbb{R}$ for all i, we define $[x]_K \in \mathbb{Z}^n \times \mathbb{R}^d$ as $\sum_{i=1}^n \lfloor \lambda_i \rceil b_i + \sum_{i=n+1}^{n+d}\lambda_i b_i$, i.e., we round x to a close mixed-integer point with respect to the norm induced by K.

Theorem 3. *Let $\nu := \mu_{mixed,K}$, where $K \subset \mathbb{R}^{n+d}$ is a convex body and $\nu(\mathbb{R}^{n+d}) \neq 0$, and let $x^\star$ be the centerpoint with respect to μ, the uniform measure on K. If $\omega > 2c(n+d)^{9/2}\alpha n^n\sqrt{n}$ for a universal constant α, then*

$$f_\nu([x^\star]_K) \geq e^{-1/c}(\mathcal{F}(\mathbb{R}^{d+n}, \mu) - e^{2/c} + 1).$$

Grünbaum's Theorem implies then, that $\mathcal{F}(\mathbb{Z}^n \times \mathbb{R}^d, \nu) \geq e^{-1/c-1} - e^{1/c} + e^{-1/c}$.

Proof. As before, let $\bar{\mu}$ denote the $(d+n)$-dimensional Lebesgue measure with respect to K and let $\bar{\nu}$ denote the d-dimensional Lebesgue measure with respect to $K \cap (\mathbb{Z}^n \times \mathbb{R}^d)$, i.e. they are not normalized.

In a first step we prove the following claim: $|\mu(H^+) - \nu(H^+)| \leq e^{2/c} - 1$ for any half-space H^+. This implies that $|\mathcal{F}(\mathbb{R}^{d+n}, \mu) - \mathcal{F}(\mathbb{R}^{d+n}, \nu)| \leq e^{2/c} - 1$, and, in particular, $|\mathcal{F}(\mathbb{R}^{d+n}, \mu) - f_\nu(x^\star)| \leq e^{2/c} - 1$.

Let H^+ be any half-space and let H^- denote its closed complement. The lattice-width of either $K \cap H^+$ or $K \cap H^-$ is larger or equal than $\omega/2$. Since both cases are similar, we only consider the case $\omega(K \cap H^-) \geq cn(n+d)\alpha n^n\sqrt{n}$. Then, by Lemma 2,

$$\begin{aligned}
\nu(H^+) = \frac{\bar{\nu}(K \cap H^+)}{\bar{\nu}(K)} &\leq \frac{e^{1/c}\bar{\mu}(K) - e^{-1/c}\bar{\mu}(K \cap H^-)}{e^{-1/c}\bar{\mu}(K)} \\
&= \frac{\bar{\mu}(K \cap H^+)}{\bar{\mu}(K)} + \frac{(e^{1/c} - e^{-1/c})\bar{\mu}(K)}{e^{-1/c}\bar{\mu}(K)} \\
&= \mu(H^+) + (e^{2/c} - 1).
\end{aligned}$$

Similarly we can derive a lower bound. This proves the claim.

In the second step we bound the error made by rounding the $x^\star$ to $[x^\star]_K$. By the choice of $x^\star$ and our rounding procedure one can prove that $[x^\star]_K$ is contained in $\frac{1}{c(n+d)}(K - x^\star) + x^\star$. The details will appear in the full version of the paper. Hence, $[x^\star]_K + \frac{c(n+d)-1}{c(n+d)}(K - x^\star) \subset K \subset [x^\star]_K + \frac{c(n+d)+1}{c(n+d)}(K - x^\star)$. We have for any $u \in \mathcal{S}^{n+d-1}$ that $e^{-1/c}\nu(H^+(u,x^\star)) \leq \nu(H^+(u,[x^\star]_K)) \leq e^{1/c}\nu(H^+(u,x^\star))$. Together with the previous claim it follows that

$$f_\nu([x^\star]_K) \geq e^{-1/c}(\mathcal{F}(\mathbb{R}^{d+n}, \mu) - e^{2/c} + 1). \qquad \square$$

5   Computational Aspects

All our algorithms are under the standard Turing machine model of computation. We say that $x \in S$ is an ϵ-centerpoint for S, μ, if $f_\mu(x) \geq \mathcal{F}(S, \mu) - \epsilon$ where $\mathcal{F}(S, \mu)$ is defined in (1) and f_μ is defined in (2).

5.1   Exact Algorithms

Uniform Measure on Polytopes. Since the rationality of the centerpoint for uniform measures on rational polytopes is an open question (see Remark 3), we consider an "exact" algorithm as one which returns an ϵ-centerpoint and runs in time polynomial in $\log(\frac{1}{\epsilon})$ and the size of the description of the rational polytope.

Theorem 4. *Let n be a fixed natural number. There is an algorithm which takes as input a rational polytope $P \subseteq \mathbb{R}^n$ and $\epsilon > 0$, and returns an ϵ-centerpoint for $\mathbb{R}^n, \mu$. The algorithms runs in time polynomial in the size of an irredundant description of P and $\log(\frac{1}{\epsilon})$.*

Proof. Since f_μ defined in (2) is quasi-concave by Lemma 1, a x^* satisfying $f_\mu(x) \geq \mathcal{F}(S, \mu) - \epsilon$ can be found, if one has an approximate evaluation oracle for f_μ, and an approximate separation oracle for the level sets $D_\mu(\alpha)$ [8]. Moreover, the number of oracle calls made is bounded by a polynomial in the size of an irredundant description of P and $\log(\frac{1}{\epsilon})$.

By Lemma 1, the problem boils down to the following:

Given $\bar{x} \in \mathbb{R}^n$ and $\delta > 0$, find $\bar{u} \in \mathcal{S}^{n-1}$ such that

$$\mu(H^+(\bar{u}, \bar{x})) \leq \inf_{u \in \mathcal{S}^{n-1}} \mu(H^+(u, \bar{x})) + \delta. \tag{6}$$

Given $\bar{x}$, let $\mathcal{P}$ be the set of all partitions of the vertices of P into two sets that can be achieved by a hyperplane through $\bar{x}$. This induces a covering of the sphere $\mathcal{S}^{n-1}$: For each $X \in \mathcal{P}$ define U_X to be the set of $u \in \mathcal{S}^{n-1}$ such that the hyperplane $u \cdot x = u \cdot \bar{x}$ induces the partition X on the vertices of P. The number of such partitions is closely related to the VC-dimension of hyperplanes, and in particular, is easily seen to be $O(M^n)$ where M is the number of vertices of P. Moreover, one can enumerate these partitions in the same amount of time, by picking $n-1$ vertices $\{v_1, \ldots, v_{n-1}\}$ of P such that $\{\bar{x}, v_1, \ldots, v_{n-1}\}$ are affinely independent.

To solve problem (6), we will proceed along these steps.

1. For each $X \in \mathcal{P}$, find $\bar{u}_X \in \mathcal{S}^{n-1}$ be such that

$$\mu(H^+(\bar{u}_X, \bar{x})) \leq \inf_{u \in U_X} \mu(H^+(u, \bar{x})) + \delta.$$

2. Pick X^* such that $\mu(H^+(\bar{u}_{X^*}, \bar{x})) \leq \mu(H^+(\bar{u}_X, \bar{x}))$ for all $X \in \mathcal{P}$ and report $\bar{u}_{X^*}$ as the solution to (6).

To complete the proof, we need to implement Step 1, above in polynomial time. This is done in Lemma 3.

Lemma 3. *For a fixed $X \in \mathcal{P}$, one can compute $\bar{u}_X \in \mathcal{S}^{n-1}$ such that*

$$\mu(H^+(\bar{u}_X, \bar{x})) \leq \inf_{u \in U_X} \mu(H^+(u, \bar{x})) + \delta,$$

using an algorithm whose running time is bounded by a polynomial in $\log(\frac{1}{\delta})$ and the size of an irredundant description of P.

This lemma can be proved using methods from real algebraic geometry for quantifier elimination in systems of polynomials inequalities. However, we omit the detailed proof.

Counting Measure on the Integer Points in Two Dimensional Polytopes. If we use the counting measure on the integer points in a polytope, the algorithm requires no accuracy parameter ϵ.

Theorem 5. *Let $P = \{x \in \mathbb{R}^2 \mid Ax \leq b\}$ be a rational polytope, where $A \in \mathbb{Z}^{m \times 2}$ and $b \in \mathbb{Z}^m$, such that $P \cap \mathbb{Z}^2 \neq \emptyset$. Let μ denote the uniform measure on $P \cap \mathbb{Z}^2$. Then in polynomial time in the input-size of A and b, one can compute a point*

$$z \in \mathcal{C}(\mathbb{Z}^2, \mu).$$

Proof. As already stressed in the previous section, it suffices to show that for a given $\bar{x} \in \mathbb{Z}$ one can compute in polynomial time

$$\bar{u} := \underset{u \in \mathcal{S}^1}{\operatorname{argmin}}\, \mu(H^+(u, \bar{x})).$$

Let $g : [0, 2\pi) \mapsto [0, 1]$ be defined as $g(\alpha) := \mu(H^+((\sin(\alpha), \cos(\alpha))^\mathsf{T}, \bar{x}))$. The key observations are that g is piecewise constant and that the domain $[0, 2\pi)$ can be partitioned into a polynomial number of intervals S_i such that g is monotone on each of them. This implies, that in order to compute $\bar{u}$, one only needs to evaluate g at the beginning and the end of each interval S_i.

Let $l^+(\alpha)$ denote the line segment $P \cap \{\bar{x} + \lambda(\sin(\alpha + \pi/2), \cos(\alpha + \pi/2))^\mathsf{T} \mid \lambda \geq 0\}$ and let $l^-(\alpha)$ denote $P \cap \{\bar{x} + \lambda(\sin(\alpha - \pi/2), \cos(\alpha - \pi/2))^\mathsf{T} \mid \lambda \geq 0\}$. Observe that $g(\alpha)$ is monotone increasing if the line segment $l^+(\alpha)$ is longer than the line segment $l^-(\alpha)$ and $g(\alpha)$ is monotone decreasing if the line segment $l^+(\alpha)$ is shorter than the line segment $l^-(\alpha)$. Hence, the monotonicity can only change when the two lengths are equal. All those critical α can be computed by comparing each pair of facets. $\qquad\square$

5.2   Approximation Algorithms

A Lenstra-Type Algorithm to Compute Approximate Centerpoints. As we already pointed out in Sect. 2, centerpoints can be used to design "optimal" oracle-based algorithms for convex mixed-integer optimization problems.

In turn, it is possible to employ linear mixed-integer optimization techniques to compute approximate centerpoints. However, this comes with a significant loss in the approximation guarantee.

Theorem 6. *Let $n, d \in \mathbb{N}$ be fixed and let P be a rational polytope. Then in polynomial time in the input-size of P, one can find a point*

$$z \in D_{\mu_{mixed,P}} \left(\frac{1}{2^{n^2} (d+1)^{(n+1)}} \right) \cap (\mathbb{Z}^n \times \mathbb{R}^d).$$

Recall the definition of $\mu_{mixed,P}$ from (3); the proof idea of Theorem 6 is that either one employs Theorem 3, or the lattice-width is small and one enumerates lower dimensional subproblems.

Computing Approximate Centerpoints with a Monte-Carlo Algorithm. In this section, we compute ϵ-centerpoints, but for any family of measures from which one can sample uniformly. However, now the algorithm's runtime depends polynomially on $\frac{1}{\epsilon}$, as opposed to $\log(\frac{1}{\epsilon})$ as for the uniform measure on rational polytopes from Sect. 5.1.

Suppose we have access to two black-box algorithms:

1. OPT is an algorithm which works for some family $\mathcal{S}$ of closed subsets of $\mathbb{R}^n$. OPT takes as input a closed set $S \in \mathcal{S}$ and computes $\text{argmax}_{x \in S} g(x)$ for any quasi-concave function g, given an evaluation oracle for g and a separation oracle for the sets $\{x \mid g(x) \geq \alpha\}_{\alpha \in \mathbb{R}}$. Let $T_1(S)$ be the number of calls that OPT makes to the evaluation and separation oracles, and $T_2(S)$ be the number of elementary arithmetic operations OPT makes during its execution.
2. SAMPLE is an algorithm which works for some family of probability measures Γ. SAMPLE takes as input a measure $\mu \in \Gamma$ and produces a sample point $x \in \mathbb{R}^n$ from the measure μ. Let $T(\mu)$ be the running time for SAMPLE.

We now show that with access to the above two algorithms, one can compute an ϵ-centerpoint for $(S, \mu) \in \mathcal{S} \times \Gamma$.

Theorem 7. *Let $\mathcal{S}$ be a family of closed subsets of $\mathbb{R}^n$ equipped with an algorithm OPT as described above, and let Γ be a family of measures on $\mathbb{R}^n$ equipped with an algorithm SAMPLE as described above.*

There exists a Monte Carlo algorithm which takes as input $(S, \mu) \in \mathcal{S} \times \Gamma$, real numbers $\epsilon, \delta > 0$ and computes an ϵ-approximate centerpoint for S, μ with probability at least $1 - \delta$. The running time of this algorithm is $T_1(S) \cdot N^d + T_2(S) + T(\mu) \cdot N$, where $N = O(\frac{1}{\epsilon^2}((n+1) + \log \frac{1}{\delta}))$.

To prove this theorem, we will need a deep result from probability theory that has resulted after a long line of research sparked by the seminal ideas of Vapnik and Chervonenkis [21], and culminated in a result of Talagrand [19]. The following theorem is a rewording of Talagrand's result [19], specialized for function classes with bounded VC-dimension.

Theorem 8. *Let (X, μ) be a probability space. Let $\mathcal{F}$ be a family of functions mapping X to $\{0, 1\}$ and let ν be the VC-dimension of the family $\mathcal{F}$. There exists a universal constant C, such that for any $\epsilon, \delta > 0$, if M is a sample of size $C \cdot \frac{1}{\epsilon^2}(\nu + \log \frac{1}{\delta})$ drawn independently from X according to μ, then with probability at least $1 - \delta$, for every function $f \in \mathcal{F}$, $\left| \frac{|\{x \in M \,|\, f(x) = 1\}|}{|M|} - \mu(\{x \in X \mid f(x) = 1\}) \right| \leq \epsilon$.*

Proof (Theorem 7). We call SAMPLE to create a sample M of size $C \cdot \frac{1}{\epsilon^2}((n+1) + \log \frac{1}{\delta})$ by drawing independently and uniformly at random from S (note that M may contain multiple copies of the same point from S). Since the VC-dimension of the family of half spaces in $\mathbb{R}^n$ is $n + 1$, we know from Theorem 8 that with probability at least $1 - \delta$, for every half space H^+, $\left| \frac{|H^+ \cap M|}{|M|} - \mu(H^+) \right| \leq \epsilon$. Let μ' be the counting measure on M. Then we obtain that $|f_{\mu'}(x) - f_\mu(x)| \leq \epsilon$ for all $x \in \mathbb{R}^n$. Therefore, any $x^* \in \arg\max_{x \in S} f_{\mu'}(x)$ is an ϵ-centerpoint for S. This can be achieved by calling OPT to compute $x^* \in \arg\max_{x \in S} f_{\mu'}(x)$. For this, we need to exhibit evaluation and separation oracles for $f_{\mu'}$. But notice that, by Lemma 1, this can be accomplished by simply implementing the following procedure: given $x \in R^d$, find the best hyperplane H through x such that $\frac{|H^+ \cap M|}{|M|}$ is minimized. This can be done in time $O(|M|^n)$ by simply enumerating all hyperplanes that contain $n - 1$ affinely independent points from M. $\qquad \square$

The following result is a consequence.

Theorem 9. *Let $n \geq 1$ and $d \geq 0$ be fixed integers. There exists a Monte Carlo algorithm which takes as input an integer $m \geq 1$, a matrix $A \in \mathbb{R}^{m \times (n+d)}$, a vector $b \in \mathbb{R}^m$, real numbers $\epsilon, \delta > 0$ and returns an ϵ-approximate centerpoint when $S = \mathbb{Z}^n \times \mathbb{R}^d$ and μ is the uniform measure on $\{x \in \mathbb{Z}^n \times \mathbb{R}^d \mid Ax \leq b\}$, with probability $1 - \delta$. The running time of the algorithm is a polynomial in $m, \log(\max\{|A_{i,j}|, |b_k|\}), \frac{1}{\epsilon}, \log \frac{1}{\delta}$.*

Proof. By using classical results on maximizing quasi-concave functions over the integer points in a polyhedron [8], OPT can be implemented for the family $\mathcal{S}$ which is the collection of all sets S that can be represented as the set of mixed-integer points in a rational polytope. SAMPLE can be implemented for the family Γ which is the uniform measure on the sets S from $\mathcal{S}$ by adapting a result of Igor Pak [17] on $d = 0$ to $d \geq 1$, using results on computing mixed-integer volumes in polynomial time for fixed dimensions [3]. $\qquad \square$

References

1. Averkov, G.: On maximal S-free sets and the helly number for the family of s-convex sets. SIAM J. Discrete Math. **27**(3), 1610–1624 (2013)
2. Averkov, G., Weismantel, R.: Transversal numbers over subsets of linear spaces. Adv. Geom. **12**(1), 19–28 (2012)
3. Baldoni, V., Berline, N., Koeppe, M., Vergne, M.: Intermediate sums on polyhedra: computation and real Ehrhart theory. Mathematika **59**(01), 1–22 (2013)

4. Bertsimas, D., Vempala, S.: Solving convex programs by random walks. J. ACM **51**(4), 540–556 (2004)
5. Braß, P., Heinrich-Litan, L., Morin, P.: Computing the center of area of a convex polygon. Int. J. Comput. Geom. Appl. **13**(05), 439–445 (2003)
6. De Loera, J.A., La Haye, R.N., Oliveros, D., Roldán-Pensado, E.: Helly numbers of algebraic subsets of $\mathbb{R}^d$. arXiv preprint arXiv:1508.02380 (2015)
7. Dyckerhoff, R., Mozharovskyi, P.: Exact computation of halfspace depth. arXiv preprint arXiv:1411.6927v2 (2015)
8. Grötschel, M., Lovász, L., Schrijver, A.: Geometric Algorithms and Combinatorial Optimization. Algorithms and Combinatorics: Study and Research Texts, vol. 2. Springer, Berlin (1988)
9. Grünbaum, B.: Partitions of mass-distributions and of convex bodies by hyperplanes. Pac. J. Math. **10**, 1257–1261 (1960)
10. Grünbaum, B.: Measures of symmetry for convex sets. In: Convexity: Proceedings of the Seventh Symposium in Pure Mathematics of the American Mathematical Society, vol. 7, p. 233. American Mathematical Society (1963)
11. Hoffman, A.J.: Binding constraints and helly numbers. Ann. N. Y. Acad. Sci. **319**, 284–288 (1979)
12. Korkine, A.N., Zolotareff, Y.I.: Sur les formes quadratiques. Math. Ann. **6**(3), 366–389 (1873)
13. Lenstra Jr., H.W.: Integer programming with a fixed number of variables. Math. Oper. Res. **8**(4), 538–548 (1983)
14. Liu, R.Y., Serfling, R.J., Souvaine, D.L.: Data Depth: Robust Multivariate Analysis, Computational Geometry, and Applications, vol. 72. American Mathematical Society, Providence (2006)
15. Mosler, K.: Depth statistics. In: Becker, C., Fried, R., Kuhnt, S. (eds.) Robustness and Complex Data Structures, pp. 17–34. Springer, Heidelberg (2013)
16. Nesterov, Y.: Introductory Lectures on Convex Optimization. Applied Optimization, vol. 87. Kluwer Academic Publishers, Boston (2004)
17. Pak, I.: On sampling integer points in polyhedra. In: Foundations of computational mathematics (Hong Kong, 2000), pp. 319–324. World Sci. Publ., River Edge (2002)
18. Rousseeuw, P.J., Ruts, I.: The depth function of a population distribution. Metrika **49**(3), 213–244 (1999)
19. Talagrand, M.: Sharper bounds for gaussian and empirical processes. Ann. Probab. **22**, 28–76 (1994)
20. Tukey, J.W.: Mathematics and the picturing of data. In: Proceedings of the International Congress of Mathematicians, vol. 2, pp. 523–531 (1975)
21. Vapnik, V.N., Chervonenkis, A.Ya.: On the uniform convergence of relative frequencies of events to their probabilities. Theory Probab. Appl. **16**(2), 264–280 (1971)

Rescaled Coordinate Descent Methods for Linear Programming

Daniel Dadush[1], László A. Végh[2]([✉]), and Giacomo Zambelli[2]

[1] Centrum Wiskunde & Informatica, Amsterdam, The Netherlands
dadush@cwi.nl
[2] London School of Economics, London, UK
{l.vegh,g.zambelli}@lse.ac.uk

Abstract. We propose two simple polynomial-time algorithms to find a positive solution to $Ax = 0$. Both algorithms iterate between coordinate descent steps similar to von Neumann's algorithm, and rescaling steps. In both cases, either the updating step leads to a substantial decrease in the norm, or we can infer that the condition measure is small and rescale in order to improve the geometry. We also show how the algorithms can be extended to find a solution of maximum support for the system $Ax = 0$, $x \geq 0$. This is an extended abstract. The missing proofs will be provided in the full version.

1 Introduction

Let $A = [a_1, \dots, a_n]$ be an integral $m \times n$ matrix with rank m, and let L denote the encoding size of A. We propose two simple polynomial algorithms for the linear feasibility problem, that is, to find a solution to systems of the form

$$Ax = 0$$
$$x > 0. \tag{1}$$

Our main contributions are: *(i)* new simple iterative methods for (1) with guaranteed finite convergence, *(ii)* a new geometric potential for these systems together with a rescaling method for improving it.

The algorithms we propose fit into a line of research developed over the past 10 years [2–6,8,15,16,20], where simple iterative updates, such as variants of perceptron [17] or of the relaxation method [1,11], are combined with some form of rescaling in order to get polynomial time algorithms for linear programming.

While these methods are slower than current interior point methods, they nevertheless yield important insights into the structure of linear programs. In particular, rescaling methods provide geometric potentials associated with a linear system which quantify how "well-conditioned" the system is, together with rescaling procedures for improving these potentials. Importantly, these potentials often provide more fine grained measures of the complexity of solving the linear system than the encoding length of the data, and help identify interesting subclasses of LPs that can be solved in strongly polynomial time (see for example [5]).

Q. Louveaux and M. Skutella (Eds.): IPCO 2016, LNCS 9682, pp. 26–37, 2016.
DOI: 10.1007/978-3-319-33461-5_3

Additionally, our algorithms can be adapted to solve the more general *maximum support* problem: find a solution to $Ax = 0, x \geq 0$ whose support $\{j : x_j > 0\}$ is inclusionwise maximum. Geometrically, this means finding the face of the convex hull of the columns of A that contains 0 in its relative interior. While general LP feasibility (and thus LP optimization) can be reduced to (1) via standard perturbation methods (see for example [18]), this is not desirable for numerical stability. On the other hand, any algorithm for the maximum support problem can be used directly to test feasibility of a system of the form $Ax = b, x \geq 0$ via simple homogenziation. We note that it is an open problem to devise any polynomial method for solving the maximum support problem that does not depend directly on the bit complexity L, but only on purely geometric parameters.

Preliminaries. Throughout the paper, we denote $\mathcal{L} := \{x \in \mathbb{R}^n : Ax = 0\}$, $\mathcal{L}_+ := \mathcal{L} \cap \mathbb{R}^n_+$, $\mathcal{L}_> := \mathcal{L} \cap \mathbb{R}^n_>$. We will also let $\mathcal{L}^\perp$ denote the orthogonal complement of $\mathcal{L}$; clearly, $\mathcal{L}^\perp = \{z \in \mathbb{R}^n : \exists y \in \mathbb{R}^m, z = A^\mathsf{T} y\}$. Let $\mathcal{L}^\perp_+ := \mathcal{L}^\perp \cap \mathbb{R}^n_+$ and $\mathcal{L}^\perp_> := \mathcal{L}^\perp \cap \mathbb{R}^n_>$. Therefore (1) is the problem of finding a point in $\mathcal{L}_>$. By strong duality, (1) is feasible if and only if $\mathcal{L}^\perp_+ = \{0\}$, that is,

$$A^\mathsf{T} y \geq 0, \tag{2}$$

has no solution other than $y = 0$.

Denote by $\mathrm{supp}(\mathcal{L}_+) \subseteq [n]$ the maximum support of a point in $\mathcal{L}_+$. Obviously $\mathrm{supp}(\mathcal{L}_+) \cap \mathrm{supp}(\mathcal{L}^\perp_+) = \emptyset$, whereas the strong duality theorem implies that $\mathrm{supp}(\mathcal{L}_+) \cup \mathrm{supp}(\mathcal{L}^\perp_+) = [n]$.

For any vector $v \in \mathbb{R}^m$ we denote by $\hat{v}$ the normal vector in the direction of v, that is $\hat{v} := v/\|v\|$. We let $\hat{A} := [\hat{a}_1, \ldots, \hat{a}_n]$. Note that, given $v, w \in \mathbb{R}^m$, $\hat{v}^\mathsf{T} \hat{w}$ is the cosine of the angle between them. Let $\mathbb{B}^m = \{x \in \mathbb{R}^m : \|x\| \leq 1\}$ denote the m-dimensional Euclidean ball. Let e_j denote the jth unit vector an e denote the all-ones vector of appropriate dimension (depending on the context).

Coordinate Descent Algorithms. Various coordinate descent methods are known for finding non-zero points in $\mathcal{L}_+$ or $\mathcal{L}^\perp_+$. Most algorithms address either the $\mathrm{supp}(\mathcal{L}_+) = [n]$ or the $\mathrm{supp}(\mathcal{L}^\perp_+) = [n]$ case; here we outline the common update steps.

At every iteration, maintain a non-negative, non-zero vector $x \in \mathbb{R}^n$, and let $y = Ax$. If $y = 0$, then x is a non-zero point in $\mathcal{L}_+$. If $A^\mathsf{T} y > 0$, then $A^\mathsf{T} y \in \mathcal{L}^\perp_>$. Otherwise, choose an index $k \in [n]$ such that $a_k^\mathsf{T} y \leq 0$, and update x and y as follows:

$$y' := \alpha y + \beta \hat{a}_k; \quad x' := \alpha x + \frac{\beta}{\|a_k\|} e_k, \tag{3}$$

where $\alpha, \beta > 0$ depend on the specific algorithm. Below we discuss various possible update choices. These can be seen as coordinate descent methods for minimizing $\|y\|^2$ subject to $y = Ax, x \geq 0$, and some further constraint is added, e.g. $e^\mathsf{T} x = 1$ in the von Neumann algorithm.

An important quantity in the convergence analysis of the algorithms we will describe is (a slight variant) of the condition measure introduced by Goffin [10]:

$$\rho_A := \max_{\|y\|=1, y \in \mathbb{R}^m} \min_{j \in [n]} a_j^\top y \tag{4}$$

We will most often be concerned with the quantity $\rho_{\hat{A}}$, where the columns of A have been scaled to have norm 1, however both will be useful to us. While ρ_A and $\rho_{\hat{A}}$ can be very different, note that they always have the same sign.

Geometrically, $|\rho_A|$ is the distance of the origin from the boundary of $\mathrm{conv}(A)$, where $\rho_A > 0$ if and only if $\mathrm{supp}(\mathcal{L}_+^\perp) = [n]$ (in which case the origin is outside $\mathrm{conv}(A)$), $\rho_A < 0$ if and only if $\mathrm{supp}(\mathcal{L}_+) = [n]$ (in which case the origin is in the interior $\mathrm{conv}(\hat{A})$), and $\rho_A = 0$ otherwise. In particular, if $\rho_A < 0$, then $-\rho_A$ is the radius of the largest ball in $\mathbb{R}^n$ inscribed in $\mathrm{conv}(A)$ and centered at the origin. If $\rho_{\hat{A}} > 0$, then $\rho_{\hat{A}}$ is the width of the *dual cone* $\{y \in \mathbb{R}^m : A^\top y > 0\}$, that is, the radius of the largest ball in $\mathbb{R}^m$ inscribed in the dual cone and centered at a point at distance one from the origin.

von Neumann's algorithm maintains at every iteration the condition that y is a convex combination of $\hat{a}_1, \ldots, \hat{a}_n$. The parameters $\alpha, \beta > 0$ are chosen so that $\alpha + \beta = 1$ and $\|y'\|$ is smallest possible. That is, y' is the point of minimum norm on the line segment joining y and $\hat{a}_k$. If we denote by y^t the vector at iteration t, a simple argument shows that $\|y^t\| \leq 1/\sqrt{t}$ (see Dantzig [7]). If 0 is contained in the interior of the convex hull, that is $\rho_{\hat{A}} < 0$, Epelman and Freund [9] showed that $\|y^t\|$ decreases by a factor of $\sqrt{1 - \rho_{\hat{A}}^2}$ in every iteration. Though the norm of y converges exponentially to 0, we note that this method may not actually terminate in finite time. If 0 is outside the convex hull however, that is, $\rho_{\hat{A}} > 0$, then the algorithm terminates after at most $1/\rho_{\hat{A}}^2$ iterations.

Betke [3] gave a polynomial time algorithm, based on a combinatorial variant of von Neumann's update, for the case $\mathrm{supp}(\mathcal{L}_+^\perp) = [n]$. Chubanov uses von Neumann's update on the columns of the projection matrix to $\mathcal{L}$, and is able to solve the maximum support problem in time $O(n^4 L)$.[1]

Perceptron chooses $\alpha = \beta = 1$ at every iteration. If $\rho_{\hat{A}} > 0$, then, similarly to the von Neumann algorithm, the perceptron algorithm terminates with a solution to the system $A^\top y > 0$ after at most $1/\rho_{\hat{A}}^2$ iterations (see Novikoff [13]). Peña and Soheili gave a smoothed variant of the perceptron update which guarantees termination in time $O(\sqrt{\log n}/\rho_{\hat{A}})$ [14], and showed how this gives rise to a polynomial-time algorithm [15] using the rescaling introduced by Betke in [3]. The same running time $O(\sqrt{\log n}/\rho_{\hat{A}})$ was achieved by Wei Yu et al. [21] by adapting the Mirror-Prox algorithm of Nemirovski [12].

Dunagan-Vempala [8] choose $\alpha = 1$ and $\beta = -(\hat{a}_k^\top y)$. The choice of β is the one that makes $\|y'\|$ the smallest possible when $\alpha = 1$. It can be readily computed that

---

[1] It had been suggested by Prof. C. Roos that Chubanov's algorithm could be further improved to $O(n^{3.5} L)$, but the paper was subsequently withdrawn due to a gap in the argument.

$$\|y'\| = \|y\|\sqrt{1 - (\hat{a}_k^\mathsf{T}\hat{y})^2}. \tag{5}$$

In particular, the norm of y' decreases at every iteration, and the larger is the angle between a_k and y, the larger the decrease. If $\rho_{\hat{A}} < 0$, then $|\hat{a}_k^\mathsf{T}\hat{y}| \geq |\rho_{\hat{A}}|$, therefore this guarantees a decrease in the norm of at least $\sqrt{1 - \rho_{\hat{A}}^2}$.

Our Algorithms. Both our algorithms use Dunagan-Vempala updates: Algorithm 1 on the columns of A, and Algorithm 2 on the orthogonal projection matrix Π to the space $\mathcal{L}^\perp$. These iterations are performed as long as we obtain a substantial decrease in $\|y\|$. Otherwise, a rescaling is performed in order to improve a volumetric potential which serves as a proxy to the condition measure $|\rho_{\hat{A}}|$. The rescaling in Algorithm 1 is the same as in Dunagan-Vempala [8], even though they solve the dual problem of finding a point in $\mathcal{L}^\perp_>$. We will describe the differences after the description of the algorithm.

Our Algorithm 2 is inspired by the work of Chubanov [6], and it uses the same rescaling. Our algorithms are in some sense dual to each other however: Chubanov uses von Neumann updates on the projection matrix to $\mathcal{L}$ whereas we use Dunagan-Vempala on the projection Π to $\mathcal{L}^\perp$. For the same algorithm, we provide two entirely different analyses, one similar to Chubanov's, and another volumetric one, as for Algorithm 1. Thus, while the rescaling is seemingly very different from the one used in Algorithm 1, there is indeed a similar underlying geometry. We compare our algorithm to Chubanov's at the end of Sect. 3.

The running time of our Algorithm 1 is $O((m^3n + n^2m)L)$, where as Algorithm 2 runs in $O(mn^4L)$ time. Although the second running time bound is worse, this algorithm can be extended to solve the full support problem within the same running time estimation. We note that Algorithm 1 could also be extended to solve the maximum support problem, but in a more indirect way, and at the expense of substantially increasing the running time.

2   Algorithm 1

Algorithm 1, described below, solves (1) (that is, finding a point in $\mathcal{L}_>$), using the Dunagan-Vempala (DV) update. It uses the parameters

$$\varepsilon := \frac{1}{11m}, \quad N := 6mL, \quad \delta := \min_{j \in [n]} \frac{1}{\|(AA^\mathsf{T})^{-1}a_j\|}. \tag{6}$$

It follows from (5) that, if in a given iteration there exists $k \in [n]$ such that $\hat{a}_k^\mathsf{T}\hat{y} \leq -\varepsilon$, then we obtain a substantial decrease in the norm, namely

$$\|y'\| \leq \|y\|\sqrt{1 - \varepsilon^2}. \tag{7}$$

On the other hand, if $\hat{a}_j^\mathsf{T}\hat{y} \geq -\varepsilon$ for all $j \in [n]$, then it follows that $|\rho_{\hat{A}}| < \varepsilon$, that is, the condition measure is small. Our aim is to perform a geometric rescaling that improves the condition measure. As a proxy for $|\rho_{\hat{A}}|$, we use the volume of the polytope P_A defined by

$$P_A := \operatorname{conv}(\hat{A}) \cap (-\operatorname{conv}(\hat{A})). \tag{8}$$

Recall that $|\rho_{\hat{A}}|$ is the radius of the largest ball around the origin inscribed in P_A.

Algorithm 1

Input: A matrix $A \in \mathbb{Z}^{m \times n}$ with rank m.
Output: Either a solution to the system (1) or the statement that
    (1) is infeasible.
1: Set $x_j := 1$ for all $j \in [n]$ and $y := Ax$. Set $t := 0$.
2: **while** $\|y\| \geq \delta$ and $t \leq N$ **do**
3:    **if** $A^{\mathsf{T}}y \geq 0$ **then** Terminate, **return** (1) is infeasible.
4:    **else**
5:        Let $k := \arg\min_{j \in [n]} \hat{a}_j^{\mathsf{T}} \hat{y}$;
6:        **if** $\hat{a}_k^{\mathsf{T}} \hat{y} < -\varepsilon$ **then**
7:            **update** $x := x - \dfrac{a_k^{\mathsf{T}} y}{\|a_k\|^2} e_k$; $y := y - (\hat{a}_k^{\mathsf{T}} y)\hat{a}_k$.
8:        **else**
9:            **rescale** $A := \left(I + \hat{y}\hat{y}^{\mathsf{T}}\right) A$; $y := 2y$; $t := t + 1$;
10: **if** $\|y\| < \delta$ **then return** feasible solution $x - A^{\mathsf{T}}(AA^{\mathsf{T}})^{-1}Ax$;
11: **else return** (1) is infeasible.

If $\hat{a}_j^{\mathsf{T}} \hat{y} \geq -\varepsilon$ for all $j \in [n]$, then P_A is contained in a "narrow strip" of width 2ε, namely $P_A \subseteq \{z \in \mathbb{R}^m : -\varepsilon \leq \hat{y}^{\mathsf{T}} \hat{z} \leq \varepsilon\}$. If we replace A with the matrix $A' := (I + \hat{y}\hat{y}^{\mathsf{T}})A$, Lemma 2.2 shows that the volume of $P_{A'}$ is at least $3/2$ times the volume of P_A. Geometrically, A' is obtained by applying to the columns of A the linear transformation that "stretches" them by a factor of two in the direction of $\hat{y}$ (see Fig. 1).

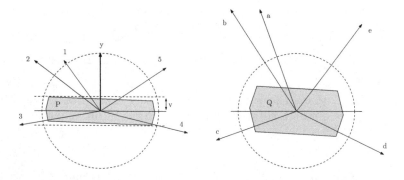

Fig. 1. Effect of rescaling. The dashed circle represent the set of points of norm 1. The shaded areas are P_A and $P_{A'}$.

Thus, at every iteration we either have a substantial decrease in the length of the current y, or we have a constant factor increase in the volume of P_A. Since the volume of P_A is bounded by the volume of the unit ball in $\mathbb{R}^m$, it follows that the algorithm cannot perform too many rescalings, unless (1) is infeasible.

After a polynomial number of iterations we either conclude that (1) is infeasible or we achieve a vector $y = Ax$ of tiny norm. In the latter case, it can be shown that projecting x to the kernel of A yields a positive kernel solution. We now state our main result.

Theorem 2.1. *For any input matrix $A \in \mathbb{Z}^{m \times n}$, Algorithm 1 returns a feasible solution x for (1) if and only if (1) is feasible. The total number of iterations of the while cycle is $O(m^3 L)$, and the total number of arithmetic operations performed is $O\left((m^3 n + mn^2)L\right)$.*

Relation to Previous Work. Even though our update step and rescaling are the same as the one used by Dunagan and Vempala [8], the algorithm and analysis are substantially different. In fact [8] assumes that $\operatorname{supp}(\mathcal{L}_+^\perp) = [n]$, and shows that the dual cone width $\rho_{\hat{A}}$ increases with a high probability. Their algorithm makes use of both perceptron as well as the DV update steps. The latter is always restarted from a random point y in the unit sphere (so in their algorithm y is not a conic combination of the a_i's). In contrast, our algorithm is fully deterministic, and uses the coordinate descent method in a more natural and direct way for the primal full dimensional case $\operatorname{supp}(\mathcal{L}_+) = [n]$.

An earlier volumetric rescaling for the $\operatorname{supp}(\mathcal{L}_+^\perp) = [n]$ case was introduced by Betke [3]. Given any convex combination x, $y = Ax$, $\|y\| \le 1/(\sqrt{mn})$, Betke's rescaling shrinks each column of A in the direction of the a_i that has the largest coefficient x_i, i.e. $a_j \leftarrow a_j - 1/2(\hat{a}_i^\top a_j)\hat{a}_i$. This has the effect of increasing the volume of the intersection of the cone $A^\top z > 0$ with the unit Euclidean ball, which can be interpreted as a smooth proxy for $\rho_{\hat{A}}$. Here, one can view our potential as the natural primal counterpart to Betke's.

Convergent Coordinate Descent. Let us consider a modification of Algorithm 1 without any rescaling: first normalize the columns of A to have unit norm (i.e. set $A = \hat{A}$), initialize δ as above (with the normalized A) and x to e, perform Dunagan-Vempala updates until $\|y\| \le \delta$, and terminate with the orthogonal projection of x onto $\mathcal{L}$. As we explain below, this yields a new "pure" coordinate descent method for (Algorithm 1) (perhaps with the exception of the rounding step) with finite convergence.

If (1) is feasible (that is, $\mathcal{L}_> \ne \emptyset$ and $\rho_{\hat{A}} < 0$), the above will terminate with a feasible solution in at most $O(\log(n/|\rho_{\hat{A}}|)/\rho_{\hat{A}}^2)$ iterations. To understand this, note that after normalizing A, the initial $y = Ae$ has norm at most n by the triangle inequality. Since the rate of norm decrease for the DV updates is still $\sqrt{1 - \rho_{\hat{A}}^2}$, after $O(\log(n/|\rho_{\hat{A}}|)/\rho_{\hat{A}}^2)$ iterations the norm of y is less that $|\rho_{\hat{A}}|$. Termination then follows by $|\rho_{\hat{A}}| = |\rho_A| \le \delta$.

The modified algorithm just described can in fact be seen as the first coordinate descent method with termination bounded in terms of $|\rho_{\hat{A}}|$ and n for the

$\rho_{\hat{A}} < 0$ case. In comparison, perceptron and von Neumann may not even terminate in a finite amount of iterations in this setting. In contrast, for the case $\rho_{\hat{A}} > 0$ these algorithms converge to a feasible solution $\mathcal{L}_>^{\perp}$ in $O(1/\rho_{\hat{A}}^2)$ iterations, whereas our algorithm need not finitely converge. We note that Wolfe's algorithm [22], also implicitly used by Betke [3], is a simple finite method for arbitrary values of $\rho_{\hat{A}}$, including $\rho_{\hat{A}} = 0$. However, finiteness is due to the Simplex-like nature and there is no bound known on the number of iterations in terms of $|\rho_{\hat{A}}|$ if $\rho_{\hat{A}} \leq 0$.

Analysis. The crucial part of the analysis is to bound the volume increase of P_A at every rescaling iteration.

Lemma 2.2. *Assume* (1) *is feasible. For some $0 < \varepsilon < 1/(11m)$, let $v \in \mathbb{R}^m$, $\|v\| = 1$, such that $\hat{a}_j^\top v \geq -\varepsilon \ \forall j \in [n]$. Let $A' = (I + vv^\top)A$. Then $\mathrm{vol}(P_{A'}) \geq \frac{3}{2}\mathrm{vol}(P_A)$.*

We now sketch the proof of Theorem 2.1. It can be shown that if $\mathrm{conv}(A)$ contains the origin in its interior, then at the beginning P_A would contain a ball of radius at least 2^{-3L}. During the entire algorithm, P_A remains inside the unit ball. These facts, together with Lemma 2.2, imply that if the algorithm does not terminate within N rescalings, then $\mathrm{conv}(A)$ cannot contain the origin in its interior. If the algorithm terminates for the condition $\|y\| \leq \delta$, then one can show using the definition of δ that the returned solution is feasible.

To bound the number of iterations, note that by (7), $\|y\|^2$ decreases by a factor of $(1 - \varepsilon^2)$ every time we perform an update. Every time we perform a rescaling, $\|y\|^2$ increases by a factor of 4; however, this may only happen $N = O(mL)$ times. We terminate once $\|y\|^2 \leq \delta^2$. Combining these bounds gives a bound $O(m^3L)$ on the total number of iterations. Every update can be computed in linear time. The number of rescalings is $O(mL)$, and each of them can be performed in $O(n^2)$ arithmetic operations, provided that we had previously computed $A^\top A$. Therefore the total number of arithmetic operations is $O((m^3n + mn^2)L)$.

3   Algorithm 2: A Dual Chubanov Algorithm

Let $\Pi = A^\top(AA^\top)^{-1}A$ denote the orthogonal projection matrix to $\mathcal{L}^\perp$ (i.e., the space spanned by the rows of A), and let $\pi_1, \ldots, \pi_n$ denote the columns of Π and π_{ij} $(i, j \in [n])$ denote the (i, j) entry of Π. We recall the following well known properties of the projection matrix Π.

Proposition 3.1. *Let $A \in \mathbb{R}^{m \times n}$ and let $\Pi = A^\top(AA^\top)^{-1}A$. The following hold (i) For all $x, z \in \mathbb{R}^n$, $\Pi x = 0$ if and only if $x \in \mathcal{L}$, and $\Pi z = z$ if and only if $z \in \mathcal{L}^\perp$; (ii) $\Pi^2 = \Pi$; (iii) For every $w \in \mathbb{R}^n$, $\|\Pi w\| \leq \|w\|$; (iv) For all $j \in [n]$, $\pi_j = \Pi e_j$, thus $\|\pi_j\| \leq 1$; (v) $\pi_{jj} = \|\pi_j\|^2$ for all $j \in [n]$; (vi) $\mathrm{trace}(\Pi) = \sum_{j=1}^n \|\pi_j\|^2 = m$.*

In Algorithm 2 below, we set $\varepsilon := \frac{1}{16n\sqrt{3m}}$. Throughout this section, for every $I \subseteq [n]$ we denote by D_I the diagonal matrix with $d_{jj} = 1/2$ if $j \in I$, $d_{jj} = 1$ if $j \notin I$. Thus $D_I = I - (1/2)\sum_{j \in I} e_j e_j^\mathsf{T}$.

Note that, since $z_j = \pi_j^\mathsf{T} z$ for all $j \in [n]$, the update step is just the Dunagan-Vempala update applied to the matrix Π instead of on A. Thus, at each update the norm of the current z decreases by at least a multiplicative factor $\sqrt{1 - \varepsilon^2}$.

Observe also that at every iteration $w_j \geq 1$ for all $j \in [n]$, so in particular $\|z\| < 1$ immediately implies $w - z > 0$, thus the algorithm terminates with the solution $x := w - z$ if $\|z\| \leq 1$.

We give a proof of correctness of the algorithm. Afterwards, we provide a different analysis, reminiscent of Lemma 2.2, which relates the rescaling step to the change of a certain geometric quantity related to the condition measure of Π.

Algorithm 2

Input: A matrix $A \in \mathbb{Z}^{m \times n}$ with rank m.

Output: Either a solution $x \in \mathcal{L}_>$, or a set $R \subseteq [n]$ disjoint from the support of $\mathcal{L}_+$.

1: Compute $\Pi = A^\mathsf{T}(AA^\mathsf{T})^{-1}A$. Set $D = I_n$.
2: Set $w_j := 1$ for all $j \in [n]$, $z := \Pi w$, $\mathrm{count}_j := 0$ for all $j \in [n]$.
3: **while** $\mathrm{count}_j < L$ for all $j \in [n]$ **do**
4:     **if** $w - z > 0$ **then** Terminate, **return** $x := D^{-1}(w - z)$.
5:     **if** $z \geq 0$ **then** Terminate, **return** $R := \{j \in [n] : z_j \neq 0\}$.
6:     **else**
7:         Let $i := \arg\min_{j \in [n]} \dfrac{z_j}{\|z\|\|\pi_j\|}$;
8:         **if** $\dfrac{z_i}{\|z\|\|\pi_i\|} < -\varepsilon$ **then**
9:             **update** $w := w - \dfrac{z_i e_i}{\|\pi_i\|^2}$; $z := z - \dfrac{z_i \pi_i}{\|\pi_i\|^2}$;
10:        **else rescale**
11:          Let $I := \{j \in [n] : \dfrac{z_j}{\|z\|} > \dfrac{1}{\sqrt{3n}}\}$; $D := DD_I$;
12:          Recompute $\Pi = DA^\mathsf{T}(AD^2A^\mathsf{T})^{-1}AD$;
13:          Set $w_j := 1$ for all $j \in [n]$, $z := \Pi w$;
14:     $\mathrm{count}_j := \mathrm{count}_j + 1$ for all $j \in I$;
    return $R := \{j : \mathrm{count}_j = L\}$.

Correctness of the Algorithm. For any $a \in \mathbb{R}$, we let $a^+ := \max\{0, a\}$ and $a^- = (-a)^+$. The correctness of the algorithm is based on the following simple bound due to Roos [16].

Lemma 3.2 (Roos). *Let $z \in \mathcal{L}^\perp$ and let $k \in [n]$ such that $z_k > 0$. Then, for every $x \in \mathcal{L} \cap [0,1]^n$.*

$$x_k \leq \frac{\sum_{j=1}^n z_j^-}{z_k}. \tag{9}$$

This bound justifies the rescaling in our algorithm, as stated in the following lemma.

Lemma 3.3. *Let A be the current matrix at a given iteration of Algorithm 2. Suppose that the current $z = \Pi w$ satisfies $z_j \geq -\varepsilon \|z\| \|\pi_j\|$. Then the set*

$$I = \left\{ j \in [n] : \frac{z_j}{\|z\|} > \frac{1}{\sqrt{3n}} \right\}$$

is nonempty. Furthermore, every $x \in \mathcal{L} \cap [0,1]^n$ satisfies $x_k \leq \frac{1}{2}$ for all $k \in I$.

Observe that rescaling has the effect of replacing the null space $\mathcal{L}$ of A with $D_I^{-1}\mathcal{L}$, that is, multiplying by 2 the components indexed by I of all vectors in $\mathcal{L}$. Let $\mathcal{L}^0$ be the null space of the input matrix A (i.e. before any rescaling). Lemmas 3.2 and 3.3 show that, at any iteration of the algorithm, $\mathcal{L}^0 \cap [0,1] \subseteq \{x \in \mathbb{R}^n : x_j < 2^{-\text{count}_j}\}$. It is well-known (see for example Schrijver [18]) that, if $j \in [n]$ is in the support $Ax = 0$, $x \geq 0$, then there exists a solution with $x_j \geq 2^{-L}$. This shows that, whenever $\text{count}_j = L$ for some $j \in [n]$, j cannot be in the support.

Running Time. At the beginning of the algorithm and after each rescaling, $z = \Pi e$, therefore $\|z\| \leq \|e\| = \sqrt{n}$. Every Dunagan-Vempala update decreases $\|z\|^2$ by a factor $1 - \varepsilon^2$, and the algorithm terminates with $x := w - z > 0$ when $\|z\| < 1$. This gives the strongly polynomial bound $O(n^2 m \log(n))$ on the number of iterations between any two rescaling. Since the algorithm performs at most L rescaling for every variable, and each update requires $O(n)$ operations, therefore the running-time of the algorithm is $O(n^4 m \log(n)L)$. (It should be noted here that the recomputation of the matrix Π at every rescaling can be performed in $O(|I|n^2)$ arithmetic operations using the Sherman-Morrison formula [19], therefore the total number of arithmetic operations performed during the rescalings is $O(n^3 L)$). Finally, one can improve the running time bound to $O(n^4 m L)$ by slightly modifying the algorithm, choosing the next w after each rescaling more carefully, using the following lemma.

Lemma 3.4. *Let $A \in \mathbb{R}^{m \times n}$, $\Pi = A^\mathsf{T}(AA^\mathsf{T})^{-1}A$. Let $D > 0$ be an $n \times n$ diagonal matrix, and let $\Pi' = DA^\mathsf{T}(AD^2A^\mathsf{T})^{-1}AD$. Given $z = \pi w$ for some $w \in \mathbb{R}^n$, if we let $w' = D^{-1}w$ and $z' = \Pi'w'$, then $\|z'\| \leq \left(\max_{i \in [n]} D_{ii}^{-1}\right)\|z\|$.*

The Maximum Support Problem. Algorithm 2 can be used to identify the support of $Ax = 0$, $x \geq 0$: whenever the algorithm returns a set R of indices not in the support, we set $x_j := 0$ for all $j \in R$, remove the columns of A indexed by R, and repeat. If the algorithm terminates with a feasible solution $x > 0$ for the current system, this defines a maximal support solution for the original problem. In the worst case, we need to run Algorithm 2 n times, giving a naïve running time estimate of $O(n^5 m L)$. However, observe that whenever Algorithm 2 terminates with a set R of indices, at the subsequent call to the algorithm we can initialize count_j, $j \notin R$, to the values computed at the end of the last call. Therefore, the total number of arithmetic operations needed to compute a maximum support solution is $O(n^4 m L)$, the same as the worst-case running time of Algorithm 2.

Analysis Based on a Geometric Potential. An alternative volumetric analysis can be given, similar to the one in Sect. 2. Let $Q_\Pi := \text{conv}(\Pi) \cap \text{conv}(-\Pi)$. Let us denote by $\widehat{\text{vol}}(\cdot)$ the volume with respect to the measure induced on $\mathcal{L}^\perp$. We will consider as a potential $\widehat{\text{vol}}(Q_\Pi)$.

Lemma 3.5. *Let $\varepsilon' = 1/(16\sqrt{3n}m)$. Let $z \in \mathcal{L}^\perp$ such that $z_j \geq -\varepsilon' \|z\| \|\pi_j\|$ for all $j \in [n]$. Let $I = \{j \in [n] : \frac{z_j}{\|z\|} > \frac{1}{\sqrt{3n}}\}$, and $\Pi' = D_I A^\mathsf{T}(A D_I^2 A^\mathsf{T})^{-1} A D_I$. Then*

$$\widehat{\text{vol}}(Q_{\Pi'}) \geq e^{1/8} \widehat{\text{vol}}(Q_\Pi).$$

Since $\varepsilon \leq \varepsilon'$, it follows that when Algorithm 2 performs a rescaling, the current point $z = \Pi w$ satisfies the hypothesis of Lemma 3.5, thus after rescaling, $\widehat{\text{vol}}(Q_\Pi)$ increases by a constant factor. As in Sect. 2, the total number of rescalings can be bounded by $O(mL)$. In particular, in $O(mL)$ rescalings one can either find a solution to $Ax = 0$, $x > 0$, or argue that none exists. Since $m \leq n$, this means that typically we may be able to prove that $Ax = 0$, $x > 0$ has no solution before we are actually able to identify any index j not in the support.

3.1 Refinements

Note that the two analyses we provided are somewhat "loose", in the sense that the parameters in Algorithm 2 have been chosen to ensure that both analyses hold. Here we propose a few refinements and variants.

(a) To optimize the algorithm based on the potential $\widehat{\text{vol}}(Q_\Pi)$, we can use $\varepsilon' = 1/(8\sqrt{n}m)$ instead of $\varepsilon = 1/(8\sqrt{m}n)$. This improves the total running time to $O(n^2 m^3 L)$.

(b) The analysis of the algorithm based on the argument in Sect. 3 can be simplified if we set $\bar{\varepsilon} = 1/(2\sqrt{mn})$, and do an update when the condition $z_i \leq -\bar{\varepsilon} \|\pi_i\|$ is satisfied by some $i \in [n]$ (rather then when $z_i \leq -\varepsilon \|z\| \|\pi_i\|$). This implies that the norm of $z' := z - (z_i/\|\pi_i\|^2)\pi_i$ satisfies $\|z'\|^2 \leq \|z\|^2(1 - (\bar{\varepsilon}/\|z\|)^2) = \|z\|^2 - 1/(4mn)$. Since after each rescaling $\|z\| \leq \sqrt{n}$, this ensures that between every two rescalings there are at most $4mn^2$ updates (without the need of resorting to Lemma 3.4). When $z_j \geq -\bar{\varepsilon}\|\pi_j\|$ for every $j \in [n]$, it follows that there must be at least one $k \in [n]$ such that the bound in (9) is at most $1/2$. Indeed, for any k such that $z_k \geq 1$ (one such k must exist because $w - z \not> 0$ and $w_j \geq 1$ for all $j \in [n]$) we have $(\sum_{j=1}^n z_j^-)/z_k \leq \varepsilon \sum_{j=1}^n \|\pi_j\| \leq \varepsilon\sqrt{nm} = 1/2$.

(c) A variant of the algorithm that gives the same running time but could potentially be more efficient in practice is the following. Define $\tilde{\varepsilon} = 1/(2\sqrt{n})$. At each iteration, let $N(z) := \{j : z_j < 0\}$, and compute $q := \sum_{j \in N(z)} \pi_j$. Note that $\|q\| \leq \sqrt{|N(z)|}$, since q is the projection onto $\mathcal{L}^\perp$ of the incidence vector of $N(z)$.

Instead of checking if there exists $i \in [n]$ such that $z_i \leq -\varepsilon\|z\|\|\pi_i\|$, check if $q^\mathsf{T} z \leq -\tilde{\varepsilon}\|q\|$. If such an index exists, then update as follows

$$z' := z - q\frac{q^\mathsf{T} z}{\|q\|^2}; \quad w' := w - \frac{q^\mathsf{T} z}{\|q\|^2}\sum_{j \in N(z)} e_j.$$

It follows that $\|z'\|^2 \leq \|z\|^2 - 1/(4n)$, hence the maximum number of updates between rescalings is $4n^2$. If instead $q^\mathsf{T} z > -\bar\varepsilon\|q\|$, then for every $k \in [n]$ such that $z_k \geq 1$, we have $(\sum_{j=1}^n z_j^-)/z_k = (-q^\mathsf{T} z)/z_k \leq \bar\varepsilon\|q\| \leq \bar\varepsilon\sqrt{n} = \frac{1}{2}$.

Note that the total number of updates performed by the algorithm is $O(n^3 L)$, which is better than $O(mn^3 L)$ updates performed by Algorithm 2. However, the number of arithmetic operations needed to compute q is, in the worst case, $O(n^2)$, therefore the total number of arithmetic operations is still $O(n^5 L)$. Nevertheless, this variant may be better in practice because it provides faster convergence.

Comparison with Chubanov's Algorithm. Chubanov's algorithm works on the projection matrix $\bar\Pi = [\bar\pi_1, \ldots, \bar\pi_n]$ to the null space $\mathcal{L}$ of A, that is, $\bar\Pi = I - \Pi$. At every iteration, Chubanov maintains a vector $v \in \mathbb{R}^n_+$ such that $e^\mathsf{T} v = 1$, starting from $y = \bar\pi_j$ for some $j \in [n]$, and computes $y = \bar\Pi v$. If $y > 0$, then Chubanov's algorithm terminates with $y \in \mathcal{L}_>$, else it selects an index $i \in [n]$ with $y_i \leq 0$ and performs a von Neumann step $y' = \lambda y + (1 - \lambda)\bar\pi_i$. By Dantzig's analysis of von Neumann's algorithm [7], $\|y'\|^{-2} \geq \|y\|^{-2} + 1$, hence after at most $4n^3$ operations $\|y\| \leq 1/(2n\sqrt{n})$. Now, if $k = \arg\max_{j \in [n]} v_j$, then $v_k \geq 1/n$, therefore we have that for every $x \in \mathcal{L}_+ \cap [0,1]^n$, $x_k \leq (v^\mathsf{T} x)/v_k = (y^\mathsf{T} x)/v_k \leq (\|x\|\|y\|)/v_k \leq \sqrt{n}\|y\|/v_k \leq 1/2$. Thus, after at most $O(n^3)$ steps, Chubanov's algorithm performs the same rescaling as Algorithm 2 using $I := \{j \in [n] : \|y\|/v_k \leq 1/(2\sqrt{n})\}$.

Note that, while the rescaling used by Algorithm 2 and Chubanov's algorithm are the same, and both algorithm ultimately produce a point in $\mathcal{L}_>$ if one exists, the updating steps work in the opposite direction. Indeed, both algorithms maintain a nonnegative vector in $\mathbb{R}^n$, but every von Neumann step in Chubanov's algorithm decreases the norm of the orthogonal projection of the nonnegative vector onto $\mathcal{L}$, whereas every Dunagan-Vempala update of Algorithm 2 decreases the norm of the orthogonal projection z onto $\mathcal{L}^\perp$. Also, Chubanov's iterations guarantee a fixed increase in $\|y\|^{-2}$, and rescaling occurs when $\|y\|$ is small enough, whereas Algorithm 2 terminates when $\|z\|$ is small enough (that is, when $\|z\| \leq 1$), and rescaling occurs when the updating step would not produce a sufficient decrease in $\|z\|$.

We note that Chubanov's algorithm solves the maximum support problem in $O(n^4 L)$, and hence is faster than ours. The full version of the paper will include an enhanced version of Algorithm 2 with running time bound $O(n^3 mL)$.

References

1. Agmon, S.: The relaxation method for linear inequalities. Can. J. Math. **6**, 382–392 (1954)
2. Basu, A., De Loera, J., Junod, M.: On Chubanov's method for linear programming. INFORMS J. Comput. **26**(2), 336–350 (2014)

3. Betke, U.: Relaxation, new combinatorial and polynomial algorithms for the linear feasibility problem. Discrete Comput. Geom. **32**, 317–338 (2004)
4. Chubanov, S.: A strongly polynomial algorithm for linear systems having a binary solution. Math. Prog. **134**, 533–570 (2012)
5. Chubanov, S.: A polynomial algorithm for linear optimization which is strongly polynomial under certain conditions on optimal solutions (2015). http://www.optimization-online.org/DB_FILE/2014/12/4710.pdf
6. Chubanov, S.: A polynomial projection algorithm for linear programming. Math. Prog. **153**, 687–713 (2015)
7. Dantzig, G.B.: An ε-precise feasible solution to a linear program with a convexity constraint in $1/\varepsilon^2$ iterations independent of problem size, Report SOL 92–5, Stanford University (1992)
8. Dunagan, J., Vempala, S.: A simple polynomial-time rescaling algorithm for solving linear programs. Math. Prog. **114**, 101–114 (2006)
9. Epelman, M., Freund, R.M.: Condition number complexity of an elementary algorithm for computing a reliable solution of a conic linear system. Math. Prog. **88**(3), 451–485 (2000)
10. Goffin, J.: The relaxation method for solving systems of linear inequalities. Math. Oper. Res. **5**, 388–414 (1980)
11. Motzkin, T., Schoenberg, I.J.: The relaxation method for linear inequalities. Can. J. Math. **6**, 393–404 (1954)
12. Nemirovski, A.: Prox-method with rate of convergence $o(1/t)$ for variational inequalities with Lipschitz continuous monotone operators and smooth convex-concave saddle point problems. SIAM J. Optim. **15**, 229–251 (2004)
13. Novikoff, A.B.J.: On convergence proofs for perceptrons. In: Proceedings of the Symposium on the Mathematical Theory of Automata XII, pp. 615–622 (1962)
14. Soheili, N., Peña, J.: A smooth perceptron algorithm. SIAM J. Optim. **22**, 728–737 (2012)
15. Peña, J., Soheili, N.: A deterministic rescaled perceptron algorithm. Math. Prog. **155**(1), 497–510 (2016)
16. Roos, K.: On Chubanov's method for solving a homogeneous inequality system. Numer. Anal. Optim. **134**, 319–338 (2015)
17. Rosenblatt, F.: The perceptron: a probabilistic model for information storage and organization in the brain. Psychol. Rev. **65**, 386–408 (1958). Cornell Aeronautical Laboratory
18. Schrijver, A.: Theory of Linear and Integer Programming. Wiley, New York (1986)
19. Sherman, J., Morrison, W.J.: Adjustment of an inverse matrix corresponding to a change in one element of a given matrix. Ann. Math. Stat. **21**, 124–127 (1949)
20. Végh, L.A., Zambelli, G.: A polynomial projection-type algorithm for linear programming. Oper. Res. Lett. **42**, 91–96 (2014)
21. Yu, A.W., Kılınç-Karzan, F., Carbonell, J.: Saddle points and accelerated perceptron algorithms. In: Proceedings of the 31st International Conference on Machine Learning. Journal of Machine Learning Research **32**, 1827–1835 (2014)
22. Wolfe, P.: Finding the nearest point in a polytope. Math. Prog. **11**(1), 128–149 (1976)

Approximating Min-Cost Chain-Constrained Spanning Trees: A Reduction from Weighted to Unweighted Problems

André Linhares$^{(\boxtimes)}$ and Chaitanya Swamy

Combinatorics and Optimization, University of Waterloo,
Waterloo, ON N2L 3G1, Canada
{alinhare,cswamy}@uwaterloo.ca

Abstract. We study the *min-cost chain-constrained spanning-tree* (abbreviated MCCST) problem: find a min-cost spanning tree in a graph subject to degree constraints on a nested family of node sets. We devise the *first* polytime algorithm that finds a spanning tree that (i) violates the degree constraints by at most a constant factor *and* (ii) whose cost is within a constant factor of the optimum. Previously, only an algorithm for *unweighted* CCST was known [13], which satisfied (i) but did not yield any cost bounds. This also yields the first result that obtains an $O(1)$-factor for *both* the cost approximation and violation of degree constraints for any spanning-tree problem with general degree bounds on node sets, where an edge participates in multiple degree constraints.

A notable feature of our algorithm is that we *reduce* MCCST to unweighted CCST (and then utilize [13]) via a novel application of *Lagrangian duality* to simplify the *cost structure* of the underlying problem and obtain a decomposition into certain uniform-cost subproblems.

We show that this Lagrangian-relaxation based idea is in fact applicable more generally and, for any cost-minimization problem with packing side-constraints, yields a reduction from the weighted to the unweighted problem. We believe that this reduction is of independent interest. As another application of our technique, we consider the *k-budgeted matroid basis* problem, where we build upon a recent rounding algorithm of [4] to obtain an improved $n^{O(k^{1.5}/\epsilon)}$-time algorithm that returns a solution that satisfies (any) one of the budget constraints exactly and incurs a $(1 + \epsilon)$-violation of the other budget constraints.

1 Introduction

Constrained spanning-tree problems, where one seeks a minimum-cost spanning tree satisfying additional ({0, 1}-coefficient) packing constraints, constitute an important and widely-studied class of problems. In particular, when the packing

A full version of the paper is available on the CS arXiv.

A. Linhares and C. Swamy—Research supported partially by NSERC grant 327620-09 and the second author's Discovery Accelerator Supplement Award, and Ontario Early Researcher Award.

© Springer International Publishing Switzerland 2016
Q. Louveaux and M. Skutella (Eds.): IPCO 2016, LNCS 9682, pp. 38–49, 2016.
DOI: 10.1007/978-3-319-33461-5_4

constraints correspond to node-degree bounds, we obtain the classical *min-cost bounded-degree spanning tree* (MBDST) problem, which has a rich history of study [5,7,10–12,15] culminating in the work of [15] that yielded an optimal result for MBDST. Such degree-constrained network-design problems arise in diverse areas including VLSI design, vehicle routing and communication networks (see, e.g., the references in [14]), and their study has led to the development of powerful techniques in approximation algorithms.

Whereas the *iterative rounding and relaxation* technique introduced in [15] (which extends the iterative-rounding framework of [9]) yields a versatile technique for handling node-degree constraints (even for more-general network-design problems), we have a rather limited understanding of spanning-tree problems with more-general degree constraints, such as constraints $|T \cap \delta(S)| \leq b_S$ for sets S in some (structured) family $\mathcal{S}$ of node sets.[1] A fundamental impediment here is our inability to leverage the techniques in [10,15]. The few known results yield: (a) (sub-)optimal cost, but a *super-constant* additive- or multiplicative-factor violation of the degree bounds [1–3,6]; or (b) a multiplicative $O(1)$-factor violation of the degree bounds (when $\mathcal{S}$ is a nested family), but *no cost guarantee* [13]. In particular, in stark contrast to the results known for node-degree-bounded network-design problems, there is no known algorithm that yields an $O(1)$-factor cost approximation *and* an (additive or multiplicative) $O(1)$-factor violation of the degree bounds. (Such guarantees are only known when each edge participates in $O(1)$ degree constraints [2]; see however [16] for an exception.)

We consider the *min-cost chain-constrained spanning-tree* (MCCST) problem introduced by [13], which is perhaps the most-basic setting involving general degree bounds where there is a significant gap in our understanding vis-a-vis node-degree bounded problems. In MCCST, we are given an undirected connected graph $G = (V, E)$, nonnegative edge costs $\{c_e\}$, a nested family $\mathcal{S}$ (or *chain*) of node sets $S_1 \subsetneq S_2 \subsetneq \cdots \subsetneq S_\ell \subsetneq V$, and integer degree bounds $\{b_S\}_{S \in \mathcal{S}}$. The goal is to find a minimum-cost spanning tree T such that $|\delta_T(S)| \leq b_S$ for all $S \in \mathcal{S}$, where $\delta_T(S) := T \cap \delta(S)$. Olver and Zenklusen [13] give an algorithm for *unweighted CCST* that returns a tree T such that $|\delta_T(S)| = O(b_S)$ (i.e., there is *no bound on* $c(T)$), and show that, for some $\rho > 0$, it is *NP*-complete to obtain an additive $\rho \cdot \frac{\log |V|}{\log \log |V|}$ violation of the degree bounds. We therefore focus on bicriteria (α, β)-guarantees for MCCST, where the tree T returned satisfies $c(T) \leq \alpha \cdot OPT$ and $|\delta_T(S)| \leq \beta \cdot b_S$ for all $S \in \mathcal{S}$.

Our Contributions. Our main result is the *first* $(O(1), O(1))$-approximation algorithm for MCCST. Given any $\lambda > 1$, our algorithm returns a tree T with $c(T) \leq \frac{\lambda}{\lambda - 1} \cdot OPT$ *and* $|\delta_T(S)| \leq 9\lambda \cdot b_S$ for all $S \in \mathcal{S}$, using the algorithm of [13] for unweighted CCST, denoted $\mathcal{A}_{\text{OZ}}$, as a black box (Theorem 3). As noted above, this is also the *first* algorithm that achieves an $(O(1), O(1))$-approximation for any spanning-tree problem with general degree constraints where an edge belongs to a super-constant number of degree constraints.

---

[1] Such general degree constraints arise in the context of finding *thin trees* [1], where $\mathcal{S}$ consists of all node sets, which turn out to be a very useful tool in devising approximation algorithms for *asymmetric TSP*.

We show in Sect. 4 that our techniques are applicable more generally. We give a *reduction* showing that for *any* cost-minimization problem with packing side-constraints, if we have an algorithm for the *unweighted* problem that returns a solution with an $O(1)$-factor violation of the packing constraints and satisfies a certain property, then one can utilize it to obtain an $(O(1), O(1))$-approximation for the cost-minimization problem. Furthermore, we show that if the algorithm for the unweighted counterpart satisfies a stronger property, then we can utilize it to obtain a $(1, O(1))$-approximation (Theorem 9).

We believe that our reductions are of independent interest and will be useful in other settings as well. Demonstrating this, we show an application to the *k-budgeted matroid basis* problem, wherein we seek to find a basis satisfying k budget constraints. Grandoni et al. [8] devised an $n^{O(k^2/\epsilon)}$-time algorithm that returned a $(1, 1+\epsilon, \ldots, 1+\epsilon)$-solution: i.e., the solution satisfies (any) one budget constraint exactly and violates the other budget constraints by a $(1+\epsilon)$-factor (if the problem is feasible). Very recently, Bansal and Nagarajan [4] improved the running time to $n^{O(k^{1.5}/\epsilon)}$ but return only a $(1+\epsilon, \ldots, 1+\epsilon)$-solution. Applying our reduction (to the algorithm in [4]), we obtain the *best of both worlds*: we return a $(1, 1+\epsilon, \ldots, 1+\epsilon)$-solution in $n^{O(k^{1.5}/\epsilon)}$-time (Theorem 12).

The chief novelty in our algorithm and analysis, and the key underlying idea, is an unorthodox use of *Lagrangian duality*. Whereas typically Lagrangian relaxation is used to drop complicating constraints and thereby simplify the constraint structure of the underlying problem, in contrast, we use Lagrangian duality to simplify the *cost structure* of the underlying problem by equalizing edge costs in certain subproblems. To elaborate (see Sect. 3.1), the algorithm in [13] for unweighted CCST can be viewed as taking a solution x to the natural linear-programming (LP) relaxation for MCCST, converting it to another feasible solution x' satisfying a certain structural property, and exploiting this property to round x' to a spanning tree. The main bottleneck here in handling costs (as also noted in [13]) is that $c^\mathsf{T}x'$ could be much larger than $c^\mathsf{T}x$ since the conversion ignores the c_es and works with an alternate "potential" function.

Our crucial insight is that *we can exploit Lagrangian duality to obtain perturbed edge costs $\{c_e^{y^*}\}$ such that the change in perturbed cost due to the conversion process is bounded*. Loosely speaking, if the conversion process shifts weight from x_f to x_e, then we ensure that $c_e^{y^*} = c_f^{y^*}$ (see Lemma 5); thus, $(c^{y^*})^\mathsf{T}x = (c^{y^*})^\mathsf{T}x'!$. The perturbation also ensures that applying $\mathcal{A}_{OZ}$ to x' yields a tree whose perturbed cost is equal to $(c^{y^*})^\mathsf{T}x' = (c^{y^*})^\mathsf{T}x$. Finally, we show that for an optimal LP solution x^*, the "error" $(c^{y^*} - c)^\mathsf{T}x^*$ incurred in working with the c^{y^*}-cost is $O(OPT)$; this yields the $(O(1), O(1))$-approximation.

We extend the above idea to an arbitrary cost-minimization problem with packing side-constraints as follows. Let x^* be an optimal solution to the LP-relaxation, and $\mathcal{P}$ be the polytope obtained by dropping the packing constraints. We observe that the same Lagrangian-duality based perturbation ensures that all points on the minimal face of $\mathcal{P}$ containing x^* have the same perturbed cost. Therefore, if we have an algorithm for the unweighted problem that rounds x^* to a point $\hat{x}$ on this minimal face, then we again obtain that $(c^{y^*})^\mathsf{T}\hat{x} = (c^{y^*})^\mathsf{T}x^*$, which then leads to an $(O(1), O(1))$-approximation (as in the case of MCCST).

Related Work. Whereas node-degree-bounded spanning-tree problems have been widely studied, relatively few results are known for spanning-tree problems with general degree constraints for a family $\mathcal{S}$ of node-sets. With the exception of the result of [13] for unweighted CCST, these other results [1–3,6] all yield a tree of cost at most the optimum with an $\omega(1)$ additive- or multiplicative- factor violation of the degree bounds. Both [2,3] obtain additive factors via iterative rounding and relaxation. The factor in [3] is $(r-1)$ for an arbitrary $\mathcal{S}$, where r is the maximum number of degree constraints involving an edge (which could be $|V|$ even when $\mathcal{S}$ is a chain), while [2] yields an $O(\log|V|)$ factor when $\mathcal{S}$ is a laminar family (the factor does not improve when $\mathcal{S}$ is a chain). The dependent-rounding techniques in [1,6] yield a tree T satisfying $|\delta_T(S)| \leq \min\{O(\frac{\log|\mathcal{S}|}{\log\log|\mathcal{S}|})b_S,$ $(1+\epsilon)b_S + O(\frac{\log|\mathcal{S}|}{\epsilon})\}$ for all $S \in \mathcal{S}$, for any family $\mathcal{S}$.

For MBDST, Goemans [10] obtained the first $(O(1), O(1))$-approximation; his result yields a tree of cost at most the optimum and at most $+2$ violation of the degree bounds. This was subsequently improved to an (optimal) additive $+1$ violation by [15]. Zenklusen [16] considers an orthogonal generalization of MBDST, where there is a matroid-independence constraint on the edges incident to each node, and obtains a tree of cost at most the optimum and "additive" $O(1)$ violation (defined appropriately) of the matroid constraints. To our knowledge, this is the only prior work that obtains an $O(1)$-approximation to both the cost and packing constraints for a constrained spanning-tree problem where an edge participates in $\omega(1)$ packing constraints (albeit this problem is quite different from spanning tree with general degree constraints).

Finally, we note that our Lagrangian-relaxation based technique is somewhat similar to its use in [11]. However, whereas [11] uses this to reduce uniform-degree MBDST to the problem of finding an MST of minimum maximum degree, which is another *weighted* problem, we utilize Lagrangian relaxation in a more refined fashion to reduce the weighted problem to its *unweighted* counterpart.

2   An LP-Relaxation for MCCST and Preliminaries

We consider the following natural LP-relaxation for MCCST. Throughout, we use e to index the edges of the underlying graph $G = (V, E)$. For a set $S \subseteq V$, let $E(S)$ denote $\{uv \in E : u, v \in S\}$, and $\delta(S)$ denote the edges on the boundary of S. For a vector $z \in \mathbb{R}^E$ and an edge-set F, we use $z(F)$ to denote $\sum_{e \in F} z_e$.

$$\min \quad \sum_e c_e x_e \tag{P}$$

$$\text{s.t.} \quad x\big(E(S)\big) \leq |S| - 1 \quad \forall \emptyset \neq S \subsetneq V \tag{1}$$

$$x(E) = |V| - 1 \tag{2}$$

$$x\big(\delta(S)\big) \leq b_S \quad \forall S \in \mathcal{S} \tag{3}$$

$$x \geq 0. \tag{4}$$

For any $x \in \mathbb{R}_+^E$, let $\text{supp}(x) := \{e : x_e > 0\}$ denote the support of x. It is well known that the polytope, $P_{ST}(G)$, defined by (1), (2), and (4) is the convex hull of spanning trees of G. We call points in $\mathrm{P_{ST}}(G)$ *fractional spanning trees*. We refer to (1), (2) as the *spanning-tree constraints*. We will also utilize (P_λ), the modified version of (P) where we replace (3) with $x\big(\delta(S)\big) \leq \lambda b_S$ for all $S \in \mathcal{S}$, where $\lambda \geq 1$. Let $OPT(\lambda)$ denote the optimal value of (P_λ), and let $OPT := OPT(1)$.

Preliminaries. A family $\mathcal{L} \subseteq 2^V$ of sets is a *laminar family* if for all $A, B \in \mathcal{L}$, we have $A \subseteq B$ or $B \subseteq A$ or $A \cap B = \emptyset$. As is standard, we say that $A \in \mathcal{L}$ is a child of $L \in \mathcal{L}$ if L is the minimal set of $\mathcal{L}$ such that $A \subsetneq L$. For each $L \in \mathcal{L}$, let $G_L^{\mathcal{L}} = (V_L^{\mathcal{L}}, E_L^{\mathcal{L}})$ be the graph obtained from $\big(L, E(L)\big)$ by contracting the children of L in $\mathcal{L}$; we drop the superscript $\mathcal{L}$ when $\mathcal{L}$ is clear from the context.

Given $x \in \mathrm{P_{ST}}(G)$, define a *laminar decomposition* $\mathcal{L}$ of x to be a (inclusion-wise) maximal laminar family of sets whose spanning-tree constraints are tight at x, so $x\big(E(A)\big) = |A| - 1$ for all $A \in \mathcal{L}$. Note that $V \in \mathcal{L}$ and $\{v\} \in \mathcal{L}$ for all $v \in V$. A laminar decomposition can be constructed in polytime (using network-flow techniques). For any $L \in \mathcal{L}$, let $x_L^{\mathcal{L}}$, or simply x_L if $\mathcal{L}$ is clear from context, denote x restricted to E_L. Observe that x_L is a fractional spanning tree of G_L.

3   An LP-Rounding Approximation Algorithm

3.1   An Overview

We first give a high-level overview. Clearly, if (P) is infeasible, there is no spanning tree satisfying the degree constraints, so in the sequel, we assume that (P) is feasible. We seek to obtain a spanning tree T of cost $c(T) = O(OPT)$ such that $|\delta_T(S)| = O(b_S)$ for all $S \in \mathcal{S}$, where $\delta_T(S)$ is the set of edges of T crossing S.

In order to explain the key ideas leading to our algorithm, we first briefly discuss the approach of Olver and Zenklusen [13] for unweighted CCST. Their approach *ignores* the edge costs $\{c_e\}$ and instead starts with a feasible solution x to (P) that minimizes a suitable (linear) potential function. They use this potential function to argue that if $\mathcal{L}$ is a laminar decomposition of x, then $(x, \mathcal{L})$ satisfies a key structural property called *rainbow freeness*. Exploiting this, they give a rounding algorithm, hereby referred to as $\mathcal{A}_{\mathrm{OZ}}$, that for every $L \in \mathcal{L}$, rounds x_L to a spanning tree T_L of G_L such that $|\delta_{T_L}(S)| \approx O(x_L(\delta(S)))$ for all $S \in \mathcal{S}$, so that concatenating the T_Ls yields a spanning tree T of G satisfying $|\delta_T(S)| = O\big(x(\delta(S))\big) = O(b_S)$ for all $S \in \mathcal{S}$ (Theorem 2). However, as already noted in [13], a fundamental obstacle towards generalizing their approach to handle the weighted version (i.e., MCCST) is that in order to achieve rainbow freeness, which is crucial for their rounding algorithm, one needs to *abandon the cost function c and work with an alternate potential function*.

We circumvent this difficulty as follows. First, we note that the algorithm in [13] can be equivalently viewed as rounding an *arbitrary* solution x to (P) as follows. Let $\mathcal{L}$ be a laminar decomposition of x. Using the same potential-function idea, we can convert x to another solution x' to (P) that admits a laminar decomposition $\mathcal{L}'$ refining $\mathcal{L}$ such that $(x', \mathcal{L}')$ satisfies rainbow freeness (see Lemma 1), and then round x' using $\mathcal{A}_{\mathrm{OZ}}$. Of course, the difficulty noted above remains, since the move to rainbow freeness (which again ignores c and uses a potential function) does not yield *any* bounds on the cost $c^\mathsf{T}x'$ relative to $c^\mathsf{T}x$. We observe however that there is one simple property (*) under which one can guarantee that $c^\mathsf{T}x' = c^\mathsf{T}x$, namely, if for every $L \in \mathcal{L}$, all edges in $\mathrm{supp}(x) \cap E_L$ have the same cost. However, it is unclear how to utilize this observation since there is no reason to expect our instance to have this rather special property: for instance, all edges of G could have very different costs!

Now let x^* be an optimal solution to (P) (we will later modify this somewhat) and $\mathcal{L}$ be a laminar decomposition of x^*. The crucial insight that allows us to leverage property (*), and a key notable aspect of our algorithm and analysis, is that *one can leverage Lagrangian duality to suitably perturb the edge costs so that the perturbed costs satisfy property (*)*. More precisely, letting $y^* \in \mathbb{R}_+^{\mathcal{S}}$ denote the optimal values of the dual variables corresponding to constraints (3), if we define the perturbed cost of edge e to be $c_e^{y^*} := c_e + \sum_{S \in \mathcal{S}: e \in \delta(S)} y_S^*$, then *the c^{y^*}-cost of all edges in* $\mathrm{supp}(x^*) \cap E_L$ *are indeed equal, for every* $L \in \mathcal{L}$ (Lemma 5). In essence, this holds because for any $e' \in \mathrm{supp}(x^*)$, by complementary slackness, we have $c_{e'} = $ (dual contribution to e' from (1),(2)) $- \sum_{S \in \mathcal{S}: e' \in \delta(S)} y_S^*$. Since any two edges $e, f \in \mathrm{supp}(x^*) \cap E_L$ appear in the *same sets of* $\mathcal{L}$, one can argue that the dual contributions to e and f from (1), (2) are *equal*, and thus, $c_e^{y^*} = c_f^{y^*}$.

Now since $(x^*, \mathcal{L}^*)$ satisfies (*) with the perturbed costs c^{y^*}, we can convert $(x^*, \mathcal{L}^*)$ to $(x', \mathcal{L}')$ satisfying rainbow freeness without altering the perturbed cost, and then round x' to a spanning tree T using $\mathcal{A}_{\mathrm{OZ}}$. This immediately yields $|\delta_T(S)| = O(b_S)$ for all $S \in \mathcal{S}$. To bound the cost, we argue that $c(T) \leq c^{y^*}(T) = \sum_e c_e^{y^*} x_e^* = c^\mathsf{T}x^* + \sum_{S \in \mathcal{S}} b_S y_S^*$ (Lemma 6), where the last equality follows from complementary slackness. (Note that the perturbed costs are used only in the analysis.) However, $\sum_{S \in \mathcal{S}} b_S y_S^*$ need not be bounded in terms of $c^\mathsf{T}x^*$. To fix this, we modify our starting solution x^*: we solve (P_λ) (which recall is (P) with inflated degree bounds $\{\lambda b_S\}$), where $\lambda > 1$, to obtain x^*, and use this x^* in our algorithm. Now, letting y^* be the optimal dual values corresponding to the inflated degree constraints, a simple duality argument shows that $\sum_{S \in \mathcal{S}} b_S y_S^* \leq \frac{OPT(1) - OPT(\lambda)}{\lambda - 1}$ (Lemma 7), which yields $c(T) = O(OPT)$ (see Theorem 3).

A noteworthy feature of our algorithm is the rather unconventional use of Lagrangian relaxation, where we use duality to simplify the *cost structure* (as opposed to the constraint-structure) of the underlying problem by equalizing edge costs in certain subproblems. This turns out to be the crucial ingredient that allows us to utilize the algorithm of [13] for unweighted CCST *as a black box* without worrying about the difficulties posed by (the move to) rainbow freeness. In fact, as we show in Sect. 4, this Lagrangian-relaxation idea is applicable more generally, and yields a novel reduction from weighted problems to their unweighted counterparts. We believe that this reduction is of independent interest and will find use in other settings as well.

3.2   Algorithm Details and Analysis

To specify our algorithm formally, we first define the rainbow-freeness property and state the main result of [13] (which we utilize as a black box) precisely.

For an edge e, define $\mathcal{S}_e := \{S \in \mathcal{S} : e \in \delta(S)\}$. Note that $\mathcal{S}_e$ could be empty. We say that two edges $e, f \in E$ form a *rainbow* if $\mathcal{S}_e \subseteq \mathcal{S}_f$ or $\mathcal{S}_f \subseteq \mathcal{S}_e$. (This definition is slightly different from the one in [13], in that we allow $\mathcal{S}_e = \mathcal{S}_f$.) We say that $(x, \mathcal{L})$ is a *rainbow-free decomposition* if $\mathcal{L}$ is a laminar decomposition of x and for every set $L \in \mathcal{L}$, no two edges of $\mathrm{supp}(x) \cap E_L$ form a rainbow. (Recall that $G_L = (V_L, E_L)$ denotes the graph obtained from $(L, E(L))$ by contracting the children of L.) The following lemma shows that one can convert an arbitrary decomposition $(x, \mathcal{L})$ to a rainbow-free one; the proof mimics the potential-function argument in [13] and is deferred to the full version.

Lemma 1. *Let $x \in P_{ST}(G)$ and $\mathcal{L}$ be a laminar decomposition of x. We can compute in polytime a fractional spanning tree $x' \in P_{ST}(G)$ and a rainbow-free decomposition $(x', \mathcal{L}')$ such that: (i) $\mathrm{supp}(x') \subseteq \mathrm{supp}(x)$; (ii) $\mathcal{L} \subseteq \mathcal{L}'$; and (iii) $x'(\delta(S)) \leq x(\delta(S))$ for all $S \in \mathcal{S}$.*

Theorem 2 [13]. *There is a polytime algorithm, $\mathcal{A}_{OZ}$, that given a fractional spanning tree $x' \in P_{ST}(G)$ and a rainbow-free decomposition $(x', \mathcal{L}')$, returns a spanning tree $T_L \subseteq \mathrm{supp}(x')$ of G_L for every $L \in \mathcal{L}'$ such that the concatenation T of the T_Ls is a spanning tree of G satisfying $|\delta_T(S)| \leq 9x'(\delta(S))$ for all $S \in \mathcal{S}$.*

We can now describe our algorithm compactly. Let $\lambda > 1$ be a parameter.

1. Compute an optimal solution x^* to (P_λ), a laminar decomposition $\mathcal{L}$ of x^*.
2. Apply Lemma 1 to $(x^*, \mathcal{L})$ to obtain a rainbow-free decomposition $(x', \mathcal{L}')$.
3. Apply $\mathcal{A}_{OZ}$ to the input $(x', \mathcal{L}')$ to obtain spanning trees $T_L^{\mathcal{L}'}$ of $G_L^{\mathcal{L}'}$ for every $L \in \mathcal{L}'$. Return the concatenation T of all the $T_L^{\mathcal{L}'}$s.

Analysis. We show that the above algorithm satisfies the following guarantee.

Theorem 3. *The above algorithm run with parameter $\lambda > 1$ returns a spanning tree T satisfying $c(T) \leq \frac{\lambda}{\lambda-1} \cdot OPT$ and $|\delta_T(S)| \leq 9\lambda b_S$ for all $S \in \mathcal{S}$.*

The bound on $|\delta_T(S)|$ follows immediately from Lemma 1 and Theorem 2 since x^*, and hence x' obtained in step 2, is a feasible solution to (P_λ). So we focus on bounding $c(T)$. This will follow from three things. First, we define the perturbed c^{y^*}-cost precisely. Next, Lemma 5 proves the key result that for every $L \in \mathcal{L}$, all edges in $\mathrm{supp}(x^*) \cap E_L$ have the same perturbed cost. Using this it is easy to show that $c(T) \leq c^{y^*}(T) = \sum_e c_e^{y^*} x_e^* = OPT(\lambda) + \lambda \sum_{S \in \mathcal{S}} b_S y_S^*$ (Lemma 6). Finally, we show that $\sum_{S \in \mathcal{S}} b_S y_S^* \leq \frac{OPT - OPT(\lambda)}{\lambda - 1}$ (Lemma 7), which yields the bound stated in Theorem 3.

To define the perturbed costs, we consider the Lagrangian dual of (P_λ) obtained by dualizing the (inflated) degree constraints $x(\delta(S)) \leq \lambda b_S$ for all $S \in \mathcal{S}$:

$$\max_{y \in \mathbb{R}_+^{\mathcal{S}}} \left(g_\lambda(y) := \min_{x \in P_{ST}(G)} \left(\sum_e c_e x_e + \sum_{S \in \mathcal{S}} (x(\delta(S)) - \lambda b_S) y_S \right) \right). \quad (\mathrm{LD}_\lambda)$$

For $x \in \mathbb{R}^E$ and $y \in \mathbb{R}^{\mathcal{S}}$, let $\mathcal{G}_\lambda(x, y) := \sum_e c_e x_e + \sum_{S \in \mathcal{S}} (x(\delta(S)) - \lambda b_S) y_S = \sum_e c_e^y x_e - \lambda \sum_{S \in \mathcal{S}} b_S y_S$ denote the objective function of $g_\lambda(y)$, where $c_e^y := c_e + \sum_{S \in \mathcal{S}: e \in \delta(S)} y_S$.

Let y^* be an optimal solution to (LD_λ). One can show via LP-duality that this holds iff there exist dual multipliers $\mu^* = (\mu_S^*)_{\emptyset \neq S \subseteq V}$ corresponding to constraints (1), (2) of (P_λ) such that (μ^*, y^*) is an optimal solution to the LP dual of (P_λ). This also implies that $g_\lambda(y^*) = \mathcal{G}_\lambda(x^*, y^*) = OPT(\lambda)$. Our *perturbed costs* are $\{c_e^{y^*}\}$. We defer the proof of Lemma 4 to the full version and use it to show that $c_e^{y^*} = c_f^{y^*}$ for any $L \in \mathcal{L}$ and any edges $e, f \in \mathrm{supp}(x^*) \cap E_L$.

Lemma 4. *We have $g_\lambda(y^*) = \mathcal{G}_\lambda(x^*, y^*) = OPT(\lambda)$. Further, there exists $\mu^* = (\mu_S^*)_{\emptyset \neq S \subseteq V}$ such that (μ^*, y^*) is an optimal solution to the dual (D_λ) of (P_λ).*

Lemma 5. *For any $L \in \mathcal{L}$, all edges of $\mathrm{supp}(x^*) \cap E_L$ have the same c^{y^*}-cost.*

Proof Sketch. Consider any two edges $e, f \in \mathrm{supp}(x^*) \cap E_L$. Suppose for a contradiction that $c_e^{y^*} < c_f^{y^*}$. Obtain $\hat{x}$ from x^* by increasing x_e^* by ϵ and decreasing x_f^* by ϵ (so $\hat{x}_{e'} = x_{e'}^*$ for all $e' \notin \{e, f\}$). We argue that one can choose a sufficiently small $\epsilon > 0$ such that $\hat{x} \in P_{ST}(G)$. This follows since any spanning-tree constraint that is tight at x^* can be expressed as a linear combination of the spanning-tree constraints for the sets in $\mathcal{L}$. Since $c_e^{y^*} < c_f^{y^*}$, we have $g_\lambda(y^*) \leq \mathcal{G}_\lambda(\hat{x}, y^*) < \mathcal{G}_\lambda(x^*, y^*) = g_\lambda(y^*)$, which is a contradiction.   □

Lemma 6. *We have $c(T) \leq \sum_e c_e^{y^*} x_e^* = \sum_e c_e x_e^* + \lambda \sum_{S \in \mathcal{S}} b_S y_S^*$.*

Proof. Observe that $c(T) \leq c^{y^*}(T)$ since $c_e \leq c_e^{y^*}$ for all $e \in E$ as $y^* \geq 0$. We now bound $c^{y^*}(T)$. To keep notation simple, we use $G_L = (V_L, E_L)$ and x_L^* to denote $G_L^{\mathcal{L}}$ and $(x^*)_L^{\mathcal{L}}$ (which recall is x^* restricted to $E_L^{\mathcal{L}}$) respectively, and $G_L' = (V_L', E_L')$ and x_L' to denote $G_L^{\mathcal{L}'}$ and $(x^*)_L^{\mathcal{L}'}$ respectively.

We have $c^{y^*}(T) = \sum_{L \in \mathcal{L}} c^{y^*}(T \cap E_L)$ since the sets $\{E_L\}_{L \in \mathcal{L}}$ partition E. Fix $L \in \mathcal{L}$. Note that x_L^* is a fractional spanning tree of $G_L = (V_L, E_L)$ since for any $\emptyset \neq Q \subseteq V_L$, if R is the subset of V corresponding to Q, and $A_1, \ldots, A_k$ are the children of L whose corresponding contracted nodes lie in Q, we have $x_L^*(E_L(Q)) = x^*(E(R)) - \sum_{i=1}^k x^*(E(A_i)) \leq |R \setminus (A_1 \cup \ldots \cup A_k)| + k - 1 = |Q| - 1$ with equality holding when $Q = V_L$. Note that $T \cap E_L$ is a spanning tree of G_L since T is obtained by concatenating spanning trees for the graphs $\{G_{L'}'\}_{L' \in \mathcal{L}'}$, and $\mathcal{L}'$ refines $\mathcal{L}$. Also, all edges of $\mathrm{supp}(x^*) \cap E_L$ have the same c^{y^*}-cost by Lemma 5. So we have $c^{y^*}(T \cap E_L) = \sum_{e \in E_L} c_e^{y^*} x_e^*$. It follows that

$$c^{y^*}(T) = \sum_e c_e^{y^*} x_e^* = \sum_e \left(c_e x_e^* + \sum_{S \in \mathcal{S}: e \in \delta(S)} y_S^* x_e^* \right)$$

$$= \sum_e c_e x_e^* + \sum_{S \in \mathcal{S}} y_S^* x^*(\delta(S)) = \sum_e c_e x_e^* + \lambda \sum_{S \in \mathcal{S}} b_S y_S^*.$$ □

Lemma 7. *We have $\sum_{S \in \mathcal{S}} b_S y_S^* \leq \frac{OPT(1) - OPT(\lambda)}{\lambda - 1}$.*

Proof. By Lemma 4, there exists μ^* such that (μ^*, y^*) is an optimal solution to the dual (D_λ) of (P_λ). Also, (μ^*, y^*) is a feasible solution to (D_1). Therefore,

$$OPT(1) \geq -\sum_{\emptyset \neq S \subseteq V} (|S| - 1)\mu_S^* - \sum_{S \in \mathcal{S}} b_S y_S^*, \quad OPT(\lambda) = -\sum_{\emptyset \neq S \subseteq V} (|S| - 1)\mu_S^* - \lambda \sum_{S \in \mathcal{S}} b_S y_S^*.$$

Hence $OPT(1) - OPT(\lambda) \geq (\lambda - 1)\sum_{S \in \mathcal{S}} b_S y_S^*$.   □

Proof of Theorem 3. As noted earlier, the bounds on $\delta_T(S)$ follow immediately from Lemma 1 and Theorem 2: for any $S \in \mathcal{S}$, we have $|\delta_T(S)| \leq 9x'(\delta(S)) \leq 9x^*(\delta(S)) \leq 9\lambda b_S$. The bound on $c(T)$ follows from Lemmas 6 and 7 since $\sum_e c_e x_e^* = OPT(\lambda)$. □

4   A Reduction from Weighted to Unweighted Problems

We now show that our ideas are applicable more generally, and yield bicriteria approximation algorithms for any cost-minimization problem with packing side-constraints, provided we have a suitable approximation algorithm for the *unweighted* counterpart. Thus, our technique yields a *reduction* from weighted to unweighted problems, which we believe is of independent interest.

To demonstrate this, we first isolate the key properties of the rounding algorithm $\mathcal{B}$ used above for unweighted CCST that enable us to use it as a black box to obtain our result for MCCST; this will yield an alternate, illuminating explanation of Theorem 3. Note that $\mathcal{B}$ is obtained by combining the procedure in Lemma 1 and $\mathcal{A}_{\mathrm{OZ}}$ (Theorem 2). First, we of course utilize that $\mathcal{B}$ is an approximation algorithm for unweighted CCST, so it returns a spanning tree T such that $|\delta_T(S)| = O(x^*(\delta(S)))$ for all $S \in \mathcal{S}$. Second, we exploit the fact that $\mathcal{B}$ returns a tree T that *preserves tightness of all spanning-tree constraints that are tight at x^*. This is the crucial property that allows us to bound $c(T)$*, since this implies (as we explain below) that $c^{y^*}(T) = \sum_e c_e^{y^*} x_e^*$, which then yields the bound on $c(T)$ as before. The equality follows because the set of optimal solutions to the LP $\min_{x \in \mathrm{P_{ST}}(G)} \mathcal{G}_\lambda(x, y^*)$ is a face of $\mathrm{P_{ST}}(G)$; thus *all* points on the *minimal* face of $\mathrm{P_{ST}}(G)$ containing x^* are optimal solutions to this LP, and the stated property implies that the characteristic vector of T lies on this minimal face. In other words, while $\mathcal{A}_{\mathrm{OZ}}$ proceeds by exploiting the notions of rainbow freeness and laminar decomposition, these notions are not essential to obtaining our result; *any* rounding algorithm for unweighted CCST satisfying the above two properties can be utilized to obtain our guarantee for MCCST.

We now formalize the above two properties for an arbitrary cost-minimization problem with packing side-constraints, and prove that they suffice to yield a bicriteria guarantee. Consider the following abstract problem:

$$\min \quad c^{\mathsf{T}} x \quad \text{s.t.} \quad x \text{ is an extreme point of } \mathcal{P}, \quad Ax \leq b, \qquad (Q^{\mathcal{P}})$$

where $\mathcal{P} \subseteq \mathbb{R}_+^n$ is a fixed polytope, $c, b \geq 0$, and $A \geq 0$. Observe that we can cast MCCST as a special case of $(Q^{\mathcal{P}})$, by taking $\mathcal{P} = \mathrm{P_{ST}}(G)$ (whose extreme points are spanning trees of G), c to be the edge costs, and $Ax \leq b$ to be the degree constraints. Moreover, by taking $\mathcal{P}$ to be the convex hull of a bounded set $\{x \in \mathbb{Z}_+^n : Cx \leq d\}$ we can use $(Q^{\mathcal{P}})$ to encode a discrete-optimization problem.

We say that $\mathcal{B}$ is a *face-preserving rounding algorithm* (FPRA) for unweighted $(Q^{\mathcal{P}})$ if given any point $x \in \mathcal{P}$, $\mathcal{B}$ finds in polytime an extreme point $\hat{x}$ of $\mathcal{P}$ such that:

(P1) $\hat{x}$ belongs to the minimal face of $\mathcal{P}$ that contains x.

We say that $\mathcal{B}$ is a β-*approximation FPRA* (where $\beta \geq 1$) if we *also* have:

(P2) $A\hat{x} \leq \beta Ax$.

Let $(R_\lambda^{\mathcal{P}})$ denote the linear program: $\min\{c^\mathsf{T}x : x \in \mathcal{P}, \ Ax \leq \lambda b\}$; note that $(R_1^{\mathcal{P}})$ is the LP-relaxation of $(Q^{\mathcal{P}})$. Let $\mathsf{opt}(\lambda)$ denote the optimal value of $(R_\lambda^{\mathcal{P}})$, and let $\mathsf{opt} := \mathsf{opt}(1)$. We say that an algorithm is a (ρ_1, ρ_2)-approximation algorithm for $(Q^{\mathcal{P}})$ if it finds in polytime an extreme point $\hat{x}$ of $\mathcal{P}$ such that $c^\mathsf{T}\hat{x} \leq \rho_1\mathsf{opt}$ and $A\hat{x} \leq \rho_2 b$.

Theorem 8. *Let $\mathcal{B}$ be a β-approximation FPRA for unweighted $(Q^{\mathcal{P}})$. Then, given any $\lambda > 1$, one can obtain a $\left(\frac{\lambda}{\lambda-1}, \beta\lambda\right)$-approximation algorithm for $(Q^{\mathcal{P}})$ using a single call to $\mathcal{B}$.*

Proof Sketch. We dovetail the algorithm for MCCST and its analysis. We simply compute an optimal solution x^* to $(R_\lambda^{\mathcal{P}})$ and round it to an extreme point $\hat{x}$ of $\mathcal{P}$ using $\mathcal{B}$. By property (P2), it is clear that $A\hat{x} \leq \beta(Ax^*) \leq \beta\lambda b$.

Let m be the number of rows of A. For $y \in \mathbb{R}_+^m$, define $c^y := c + A^\mathsf{T}y$. To bound the cost, as before, we consider the Lagrangian dual of $(R_\lambda^{\mathcal{P}})$ obtained by dualizing the side-constraints $Ax \leq \lambda b$.

$$\max_{y \in \mathbb{R}_+^m}\left(h_\lambda(y) := \min_{x \in \mathcal{P}}\mathcal{H}_\lambda(x, y)\right), \quad \text{where } \mathcal{H}_\lambda(x, y) := (c^y)^\mathsf{T}x - \lambda y^\mathsf{T}b.$$

Let $y^* = \mathrm{argmax}_{y \in \mathbb{R}_+^m}\, h_\lambda(y)$. We can mimic the proof of Lemma 4 to show that x^* is an optimal solution to $\min_{x \in \mathcal{P}}\mathcal{H}_\lambda(x, y^*)$. The set of optimal solutions to this LP is a face of $\mathcal{P}$. So all points on the minimal face of $\mathcal{P}$ containing x^* are optimal solutions to this LP. By property (P1), $\hat{x}$ belongs to this minimal face and so is an optimal solution to this LP. So $(c^{y^*})^\mathsf{T}\hat{x} = (c^{y^*})^\mathsf{T}x^* = c^\mathsf{T}x^* + (y^*)^\mathsf{T}Ax^* = \mathsf{opt}(\lambda) + \lambda(y^*)^\mathsf{T}b$, where the last equality follows by complementary slackness. Also, by the same arguments as in Lemma 7, we have $(y^*)^\mathsf{T}b \leq \frac{\mathsf{opt}(1) - \mathsf{opt}(\lambda)}{\lambda - 1}$. Since $c \leq c^{y^*}$, we have $c^\mathsf{T}\hat{x} \leq (c^{y^*})^\mathsf{T}\hat{x} \leq \frac{\lambda}{\lambda-1} \cdot \mathsf{opt}$. $\qquad\square$

5 Towards a $(1, O(1))$-Approximation Algorithm for $(Q^{\mathcal{P}})$

A natural question that emerges from Theorems 3 and 8 is whether one can obtain a $(1, O(1))$-approximation, i.e., obtain a solution of *cost at most* opt that violates the packing side-constraints by an (multiplicative) $O(1)$-factor. Such results are known for degree-bounded spanning tree problems with various kinds of degree constraints [3,10,15,16], so, in particular, it is natural to ask whether such a result also holds for MCCST. (Note that for MCCST, the dependent-rounding techniques in [1,6] yield a tree T with $c(T) \leq OPT$ and $|\delta_T(S)| \leq \min\{O(\frac{\log |S|}{\log\log |S|})b_S, (1 + \epsilon)b_S + O(\frac{\log |S|}{\epsilon})\}$ for all $S \in \mathcal{S}$.) We show that our approach is versatile enough to yield such a guarantee provided we assume a stronger property from the rounding algorithm $\mathcal{B}$ for unweighted $(Q^{\mathcal{P}})$.

Let A_i denote the i-th row of A, for $i = 1, \ldots, m$. We say that $\mathcal{B}$ is an (α, β)-*approximation FPRA* for unweighted $(Q^{\mathcal{P}})$ if *in addition* to properties (P1), (P2), it satisfies:

(P3) it rounds a feasible solution x to $(R_\alpha^{\mathcal{P}})$ to an extreme point $\hat{x}$ of $\mathcal{P}$ satisfying $A_i^\mathsf{T}\hat{x} \geq \frac{A_i^\mathsf{T}x}{\alpha}$ for every i such that $A_i^\mathsf{T}x = \alpha b_i$.

(For MCCST, property (P3) requires that $|\delta_T(S)| \geq b_S$ for every set $S \in \mathcal{S}$ whose degree constraint (in (P_α)) is tight at the fractional spanning tree x.)

Theorem 9. *Let $\mathcal{B}$ be an (α, β)-approximation FPRA for unweighted $(Q^{\mathcal{P}})$. Then, one can obtain a $(1, \alpha\beta)$-approximation algorithm for $(Q^{\mathcal{P}})$ using a single call to $\mathcal{B}$.*

Proof. We show that applying Theorem 8 with $\lambda = \alpha$ yields the claimed result. It is clear that the extreme point $\hat{x}$ returned satisfies $A\hat{x} \leq \alpha\beta b$. As in the proof of Theorem 8, let y^* be an optimal solution to $\max_{y \in \mathbb{R}_+^m} h_\lambda(y)$ (where $\lambda = \alpha$). In Lemma 6 and the proof of Theorem 8, we use the weak bound $c^\mathsf{T}\hat{x} \leq (c^{y^*})^\mathsf{T}\hat{x}$. We tighten this to obtain the improved bound on $c^\mathsf{T}\hat{x}$. We have $c^\mathsf{T}\hat{x} = (c^{y^*})^\mathsf{T}\hat{x} - (y^*)^\mathsf{T}A\hat{x}$, and

$$(y^*)^\mathsf{T}A\hat{x} = \sum_{i:A_i^\mathsf{T}x^*=\lambda b_i} y_i^*(A_i^\mathsf{T}\hat{x}) \geq \sum_{i:A_i^\mathsf{T}x^*=\lambda b_i} \frac{y_i^*A_i^\mathsf{T}x^*}{\alpha} = \sum_{i:A_i^\mathsf{T}x^*=\lambda b_i} y_i^*b_i = (y^*)^\mathsf{T}b.$$

The first and last equalities above follow because $y_i^* > 0$ implies that $A_i^\mathsf{T}x^* = \lambda b_i$. The inequality follows from property (P3). Thus, following the rest of the arguments in the proof of Theorem 8, we obtain that

$$c^\mathsf{T}\hat{x} \leq (c^{y^*})^\mathsf{T}\hat{x} - (y^*)^\mathsf{T}b = c^\mathsf{T}x^* + (\lambda - 1)(y^*)^\mathsf{T}b \leq \mathsf{opt}(1). \qquad \square$$

We also obtain the following variant of Theorem 9 with two-sided *additive* guarantees (which can be proved by essentially the same arguments).

Theorem 10. *Let $\mathcal{B}$ be an FPRA for unweighted $(Q^{\mathcal{P}})$ that given $x \in \mathcal{P}$ returns an extreme point $\hat{x}$ of $\mathcal{P}$ such that $Ax - \Delta \leq A\hat{x} \leq Ax + \Delta$. Using a single call to $\mathcal{B}$, we can obtain an extreme point $\tilde{x}$ of $\mathcal{P}$ such that $c^\mathsf{T}\tilde{x} \leq \mathsf{opt}$ and $A\tilde{x} \leq b + 2\Delta$.*

Application to k-budgeted Matroid Basis. Here, we seek to find a basis S of a matroid M satisfying k budget constraints $\{d_i(S) \leq B_i\}_{1 \leq i \leq k}$, where $d_i(S) := \sum_{e \in S} d_i(e)$. Note that this can be cast a special case of $(Q^{\mathcal{P}})$, where $\mathcal{P} = \mathcal{P}(M)$ is the basis polytope of M, the objective function encodes (say) the first budget constraint and $Ax \leq b$ encodes the remaining budget constraints. Applying Theorem 10 to a recent randomized algorithm of [4], we obtain a (randomized) algorithm that, for any $\epsilon > 0$, returns in $n^{O(k^{1.5}/\epsilon)}$ time a basis that (exactly) satisfies a chosen budget constraint, and violates the other budget constraints by (at most) a $(1 + \epsilon)$-factor, where n is the size of the ground-set of M. This *matches* the current-best approximation guarantee of [8] (who give a deterministic algorithm) *and* the current-best running time of [4].

Theorem 11 [4]. *There exists a randomized FPRA, $\mathcal{B}_{BN}$, for unweighted $(Q^{\mathcal{P}(M)})$ that rounds any $x \in \mathcal{P}(M)$ to an extreme point $\hat{x}$ of $\mathcal{P}(M)$ such that $Ax - O(\sqrt{k})\Delta \leq A\hat{x} \leq Ax + O(\sqrt{k})\Delta$, where $\Delta = (\max_{1 \leq j \leq n} a_{ij})_{1 \leq i \leq k-1} = (\max_e d_{i+1}(e))_{1 \leq i \leq k-1}$.*

Applying Theorem 10 with $\mathcal{B}=\mathcal{B}_{\mathrm{BN}}$, we obtain a basis S of M such that $d_1(S) \leq B_1$, and $d_i(S) \leq B_i + O(\sqrt{k}) \max_e d_i(e)$ for all $2 \leq i \leq k$. We combine this with a partial-enumeration step, where we "guess" all elements e of an optimal solution having $d_i(e) = \Omega\left(\frac{\epsilon}{\sqrt{k}}\right) \cdot B_i$ for at least one index $i \in \{2, \ldots, k\}$, update M and the budget constraints, and then apply Theorem 10. This yields the following result.

Theorem 12. *There exists a randomized algorithm that, given any $\epsilon > 0$, finds in $n^{O(k^{1.5}/\epsilon)}$ time a basis S of M such that $d_1(S) \leq B_1$ and $d_i(S) \leq (1 + \epsilon)B_i$ for all $2 \leq i \leq k$.*

References

1. Asadpour, A., Goemans, M., Madry, A., Oveis Gharan, S., Saberi, A.: An $O(\log n/\log\log n)$-approximation algorithm for the asymmetric traveling salesman problem. In: Proceedings of the 20th SODA, pp. 379–389 (2010)
2. Bansal, N., Khandekar, R., Könemann, J., Nagarajan, V., Peis, B.: On generalizations of network design problems with degree bounds. Math. Program. **141**(1–2), 479–506 (2013)
3. Bansal, N., Khandekar, R., Nagarajan, V.: Additive guarantees for degree-bounded directed network design. SICOMP **39**(4), 1413–1431 (2009)
4. Bansal, N., Nagarajan, V.: Approximation-friendly discrepancy rounding. In: Louveaux. Q., Skutella, M. (eds.) IPCO 2016. LNCS, vol. 9682, pp. 375–386. Springer, Heidelberg (2016). Also appears arXiv:1512.02254 (2015)
5. Chaudhuri, K., Rao, S., Riesenfeld, S., Talwar, K.: What would Edmonds do? Augmenting paths and witnesses for degree-bounded MSTs. Algorithmica **55**, 157–189 (2009)
6. Chekuri, C., Vondrak, J., Zenklusen, R.: Dependent randomized rounding via exchange properties of combinatorial structures. In: 51st FOCS (2010)
7. Fürer, M., Raghavachari, B.: Approximating the minimum-degree Steiner tree to within one of optimal. J. Algorithms **17**(3), 409–423 (1994)
8. Grandoni, F., Ravi, R., Singh, M., Zenklusen, R.: New approaches to multiobjective optimization. Math. Program. **146**(1–2), 525–554 (2014)
9. Jain, K.: A factor 2 approximation algorithm for the generalized Steiner network problem. Combinatorica **21**, 39–60 (2001)
10. Goemans, M.: Minimum bounded degree spanning trees. In: 47th FOCS (2006)
11. Könemann, J., Ravi, R.: A matter of degree: improved approximation algorithms for degree-bounded minimum spanning trees. SICOMP **31**, 1783–1793 (2002)
12. Könemann, J., Ravi, R.: Primal-dual meets local search: approximating MST's with nonuniform degree bounds. In: Proceedings of the 35th STOC, pp. 389–395 (2003)
13. Olver, N., Zenklusen, R.: Chain-constrained spanning trees. In: Goemans, M., Correa, J. (eds.) IPCO 2013. LNCS, vol. 7801, pp. 324–335. Springer, Heidelberg (2013)
14. Ravi, R., Marathe, M., Ravi, S., Rosenkrantz, D., Hunt III, H.: Approximation algorithms for degree-constrained minimum-cost network-design problems. Algorithmica **31**(1), 58–78 (2001)
15. Singh, M., Lau, L.: Approximating minimum bounded degree spanning trees to within one of optimal. In: Proceedings of the 39th STOC, pp. 661–670 (2007)
16. Zenklusen, R.: Matroidal degree-bounded minimum spanning trees. In: Proceedings of the 23rd SODA, pp. 1512–1521 (2012)

Max-Cut Under Graph Constraints

Jon Lee, Viswanath Nagarajan, and Xiangkun Shen[✉]

Department of IOE, University of Michigan, Ann Arbor, MI 48109, USA
xkshen@umich.edu

Abstract. An instance of the graph-constrained max-cut (GCMC) problem consists of (i) an undirected graph $G = (V, E)$ and (ii) edge-weights $c : \binom{V}{2} \to \mathbb{R}_+$ on a complete undirected graph. The objective is to find a subset $S \subseteq V$ of vertices satisfying some graph-based constraint in G that maximizes the weight $\sum_{u \in S, v \notin S} c_{uv}$ of edges in the cut $(S, V \backslash S)$. The types of graph constraints we can handle include independent set, vertex cover, dominating set and connectivity.

Our main results are for the case when G is a graph with bounded treewidth, where we obtain a $\frac{1}{2}$-approximation algorithm. Our algorithm uses an LP relaxation based on the Sherali-Adams hierarchy. It can handle any graph constraint for which there is a (certain type of) dynamic program that exactly optimizes linear objectives.

Using known decomposition results, these imply essentially the same approximation ratio for GCMC under constraints such as independent set, dominating set and connectivity on a planar graph G (more generally for bounded-genus or excluded-minor graphs).

1 Introduction

The max-cut problem is an extensively studied combinatorial-optimization problem. Given an undirected edge-weighted graph, the goal is to find a subset $S \subseteq V$ of vertices that maximizes the weight of edges in the cut $(S, V \backslash S)$. Max-cut has a 0.878-approximation algorithm [13] which is known to be best-possible assuming the "unique games conjecture" [16]. It also has a number of practical applications, e.g., in circuit layout, statistical physics and clustering.

In some applications, one needs to solve the max-cut problem under additional constraints on the subset S. Consider for example, the following clustering problem. The input is an undirected graph $G = (V, E)$ representing, say, a social network (vertices V denote users and edges E denote connections between users), and a weight function $c : \binom{V}{2} \to \mathbb{R}_+$ representing, a dissimilarity measure between pairs of users. The goal is to find a subset $S \subseteq V$ of users that are connected in G while maximizing the weight of edges in the cut $(S, V \backslash S)$. This corresponds to finding a cluster of connected users that is as different as possible

Research of J. Lee was partially supported by NSF grant CMMI–1160915 and ONR grant N00014-14-1-0315.

Research of V. Nagarajan supported in part by a faculty award from Bloomberg Labs.

© Springer International Publishing Switzerland 2016
Q. Louveaux and M. Skutella (Eds.): IPCO 2016, LNCS 9682, pp. 50–62, 2016.
DOI: 10.1007/978-3-319-33461-5_5

from its complement set. This "connected max-cut" problem also arises in image segmentation applications [15,21].

Designing algorithms for constrained versions of max-cut is also interesting from a theoretical standpoint. For max-cut under certain types of constraints (such as cardinality or matroid constraints) good approximation algorithms are known, e.g., [1,2]. In fact, many of these results have since been extended to the more general setting of submodular objectives [8,11]. However, not much is known for max-cut under "graph-based" constraints as in the example above.

In this paper, we study a large class of graph-constrained max-cut problems and present unified approximation algorithms for them. Our results require that the constraint be defined on a graph G of bounded treewidth. (Treewidth is a measure of how similar a graph is to a tree structure — see Sect. 2 for definitions.) We note however that for a number of constraints (including the connectivity example above), we can combine our algorithm with known decomposition results [9,10] to obtain essentially the same approximation ratios when the constraint graph G is planar/bounded-genus/excluded-minor.

Problem Definition. The input to the *graph-constrained max-cut* (GCMC) problem consists of (i) an n-vertex graph $G = (V, E)$ with a graph-property which implicitly specifies a collection $\mathcal{C}_G \subseteq 2^V$ of feasible vertex subsets, and (ii) (symmetric) edge-weights $c : \binom{V}{2} \to \mathbb{R}_+$. The GCMC problem is then as follows:

$$\max_{S \in \mathcal{C}_G} \sum_{u \in S, v \notin S} c(u, v). \tag{1}$$

In this paper, we assume that the constraint graph G has bounded treewidth. We also assume that the graph constraint $\mathcal{C}_G$ admits an exact dynamic program (of a specific form) for optimizing a linear objective, i.e. for:

$$\max_{S \in \mathcal{C}_G} \sum_{u \in S} f(u), \qquad \text{where } f : V \to \mathbb{R} \text{ is any given vertex weights.} \tag{2}$$

Note that the GCMC objective (1) is a quadratic function of the solution S, whereas our assumption (2) involves a *linear* function of the solution S. See Sect. 2 for more precise definitions/assumptions.

1.1 Our Results and Techniques

Our main result can be stated informally as follows.

Theorem 1 (GCMC result — informal). *Consider any instance of the* GCMC *problem on a bounded-treewidth graph $G = (V, E)$. Suppose there is an exact dynamic program for optimizing any linear function subject to constraint $\mathcal{C}_G$. Then we obtain a $\frac{1}{2}$-approximation algorithm for* GCMC.

This algorithm uses a linear-programming relaxation for GCMC based on the dynamic program (for linear objectives) which is further strengthened via the

Sherali-Adams LP hierarchy. The resulting LP has polynomial size whenever the number of dynamic program states associated with a single tree-decomposition node is constant (see Sect. 2 for the formal definition).[1] The rounding algorithm is a natural top-down procedure that randomly chooses a "state" for each tree-decomposition node using the LP's probability distribution conditional on the choices at its ancestor nodes. The final solution is obtained by combining the chosen states at each tree-decomposition node, which is guaranteed to satisfy constraint $\mathcal{C}_G$ due properties of the dynamic program. We note that the choice of variables in the Sherali-Adams LP as well as the rounding algorithm are similar to those used in [14] for the sparsest cut problem on bounded-treewidth graphs. An important difference in our result is that we apply the Sherali-Adams hierarchy to a non-standard LP that is defined using the dynamic program for linear objectives. (If we were to apply Sherali-Adams to the standard LP, then it is unclear how to enforce the constraint $\mathcal{C}_G$ during the rounding algorithm.) Another difference is that our rounding algorithm needs to make a correlated choice in selecting the states of sibling nodes in order to satisfy constraint $\mathcal{C}_G$ — this causes the number of variables in the Sherali-Adams LP to increase, but it still remains polynomial because the tree decomposition has constant degree.

The requirements in Theorem 1 on the graph constraint $\mathcal{C}_G$ are satisfied by several interesting constraints and thus we obtain approximation algorithms for all these GCMC problems.

Theorem 2 (Applications). *There is a $\frac{1}{2}$-approximation algorithm for GCMC under the following constraints in a bounded-treewidth graph: independent set, vertex cover, dominating set, connectivity.*

We note that many other constraints such as precedence, connected dominating set, and triangle matching also satisfy our requirement. We note that for some of these constraints (e.g., independent set), there are problem-specific algorithms where the approximation ratio depends on the treewidth k. Our result is stronger: the algorithm is more general, and the ratio is independent of k.

For many of the constraints above, we can use known decomposition results [9, 10] to obtain approximation algorithms for GCMC when the constraint graph has bounded genus or excludes some fixed minor (e.g., planar graphs).

Corollary 1. *There is a $\left(\frac{1}{2} - \epsilon\right)$-approximation algorithm for GCMC under the following constraints in an excluded-minor graph: independent set, vertex cover, dominating set. Here $\epsilon > 0$ is a fixed constant.*

Corollary 2. *There is a $\left(\frac{1}{2} - \epsilon\right)$-approximation algorithm for connected max-cut in a bounded-genus graph. Here $\epsilon > 0$ is a fixed constant.*

1.2   Related Work

For the basic undirected max-cut problem, there is an elegant 0.878-approximation algorithm [13] via semidefinite programming. This is also the best one can hope for, assuming the *unique games conjecture* [16].

---

[1] For other polynomial-time dynamic programs, the LP has *quasi-polynomial* size.

Most of the prior work on constrained max-cut has focused on cardinality, matroid and knapsack constraints [1,2,8,11,17,18]. Constant-factor approximation algorithms are known for max-cut under the intersection of any constant number of such constraints — these results hold in the substantially more general setting of non-negative submodular functions. The main techniques used here are local search and the multilinear extension [7] of submodular functions. These results made crucial use of certain exchange properties of the underlying constraints, which are not true for graph-based constraints that we consider.

Closer to our setting, a version of the connected max-cut problem was studied recently in [15], where the connectivity constraint as well as the weight function were defined on the *same* graph G. The authors obtained an $O(\log n)$-approximation algorithm for general graphs, and an exact algorithm on bounded-treewidth graphs (which implied a PTAS for bounded-genus graphs); their algorithms relied heavily on the uniformity of the constraint/weight graphs. In contrast, we consider the connected max-cut problem where the connectivity constraint and the weight function are *unrelated*; in particular, our problem generalizes max-cut even when G is a trivial graph (e.g., a star). Moreover, our algorithms work for a much wider class of constraints. We note however that our results require graph G to have bounded treewidth — this is also necessary because some of the constraints we consider (e.g., independent set) are inapproximable in general graphs. (For connected max-cut itself, obtaining a non-trivial approximation ratio when G is a general graph remains an open question.)

In terms of techniques, the closest work to ours is [14]. We use ideas from [14] in formulating the (polynomial size) Sherali-Adams LP as well as in the rounding algorithm. There are important differences too, as discussed in Sect. 1.1.

Finally, our result adds to a somewhat small list [3–5,12,14,19] of algorithmic results based on the Sherali-Adams [20] LP hierarchy. We are not aware of a more direct approach to obtain a constant-factor approximation algorithm even for connected max-cut when the constraint graph G is a tree.

2   Preliminaries

Basic Definitions. For an undirected complete graph on vertices V and subset $S \subseteq V$, let δS be the set of edges with exactly one end-point in S. For any weight function $c : \binom{V}{2} \to \mathbb{R}_+$ and subset $F \subseteq \binom{V}{2}$, we use $c(F) := \sum_{e \in F} c_e$.

Tree Decomposition. Given an undirected graph $G = (V, E)$, a tree decomposition consists of a tree $\mathcal{T} = (I, F)$ and a collection of vertex subsets $\{X_i \subseteq V\}_{i \in I}$ such that:

– for each $v \in V$, the nodes $\{i \in I : v \in X_i\}$ are connected in $\mathcal{T}$, and
– for each edge $(u, v) \in E$, there is some node $i \in I$ with $u, v \in X_i$.

The width of such a tree decomposition is $\max_{i \in I}(|X_i| - 1)$, and the treewidth of G is the smallest width of any tree decomposition for G.

We will work with "rooted" tree decompositions that also specify a root node $r \in I$. The depth d of such a tree decomposition is the length of the longest root-leaf path in $\mathcal{T}$. The depth of any node $i \in I$ is the length of the $r - i$ path in $\mathcal{T}$.

For any $i \in I$, the set V_i denotes all the vertices at or below node i, that is

$$V_i := \cup_{k \in T_i} X_k, \quad \text{where } T_i = \{k \in I : k \text{ in subtree of } T \text{ rooted at } i\}.$$

The following known result provides a convenient representation of T.

Theorem 3 (Balanced Tree Decomposition [6]). *Let $G = (V, E)$ be a graph of treewidth k. Then G has a rooted tree decomposition $(T = (I, F), \{X_i | i \in I\})$ where T is a binary tree of depth $2\lceil \log_{\frac{5}{4}}(2|V|) \rceil$ and treewidth at most $3k + 2$. This tree decomposition can be found in $O(|V|)$ time.*

Dynamic Program for Linear Objectives. We assume that the constraint $\mathcal{C}_G$ admits an exact dynamic programming (DP) algorithm for optimizing linear objectives, i.e. for the problem (2). There is some additional notation that is needed to formally describe the DP: this is necessary due to the generality of our results.

Definition 1 (DP). *With any tree decomposition $(T = (I, F), \{X_i | i \in I\})$, we associate the following:*

1. *For each node $i \in I$, there is a state space Σ_i.*
2. *For each node $i \in I$ and $\sigma \in \Sigma_i$, there is a collection $\mathcal{H}_{i,\sigma} \subseteq 2^{V_i}$ of subsets.*
3. *For each node $i \in I$, its children nodes $\{j, j'\}$ and $\sigma \in \Sigma_i$, there is a collection $\mathcal{F}_{i,\sigma} \subseteq \Sigma_j \times \Sigma_{j'}$ of valid combinations of children states.*

Assumption 1 (Linear Objective DP for $\mathcal{C}_G$). *Let $(T = (I, F), \{X_i | i \in I\})$ be any tree decomposition. Then there exist Σ_i, $\mathcal{F}_{i,\sigma}$ and $\mathcal{H}_{i,\sigma}$ (see Definition 1) that satisfy the following:*

1. *(bounded state space) Σ_i and $\mathcal{F}_{i,\sigma}$ are all bounded by constant, that is, $\max_i |\Sigma_i| = t$ and $\max_{i,\sigma} |\mathcal{F}_{i,\sigma}| = p$, where $t, p = O(1)$.*
2. *(required state) For each $i \in I$ and $\sigma \in \Sigma_i$, the intersection with X_i of every set in $\mathcal{H}_{i,\sigma}$ is the same, denoted $X_{i,\sigma}$, that is $S \cap X_i = X_{i,\sigma}$ for all $S \in \mathcal{H}_{i,\sigma}$.*
3. *(subproblem) For each non-leaf node $i \in I$ with children $\{j, j'\}$ and $\sigma \in \Sigma_i$,*

$$\mathcal{H}_{i,\sigma} = \left\{ X_{i,\sigma} \cup S_j \cup S_{j'} : S_j \in \mathcal{H}_{j,w_j}, S_{j'} \in \mathcal{H}_{j',w_{j'}}, (w_j, w_{j'}) \in \mathcal{F}_{i,\sigma} \right\}.$$

By condition 2, for any leaf $\ell \in I$ and $\sigma \in \Sigma_\ell$, we have $\mathcal{H}_{\ell,\sigma} = \{X_{\ell,\sigma}\}$ or $\emptyset$.
4. *(cover all constraints) At the root node r, we have $\mathcal{C}_G = \bigcup_{\sigma \in \Sigma_r} \mathcal{H}_{r,\sigma}$.*

We note that Assumption 1 implies a polynomial-time dynamic program for optimizing linear objectives, i.e. problem (2). The converse is not necessarily true. However, to the best of our knowledge, all natural constraints that admit a dynamic program on bounded-treewidth graphs satisfy a weaker version of Assumption 1 where conditions 2–4 are satisfied exactly and a weaker version of condition 1 is satisfied (with t and p being polynomial in n instead of constant). Our approach works even in such cases and gives a *quasi-polynomial* time $\frac{1}{2}$-approximation algorithm.

We also note that our algorithm just relies on the structure given by Assumption 1, and does not explicitly solve the linear-objective problem (2).

Example: Here we outline how independent set satisfies these requirements.

- The state space of each node $i \in I$ consists of all independent subsets of X_i.
- The subsets $\mathcal{H}_{i,\sigma}$ consist of all independent subsets $S \subseteq V_i$ with $S \cap X_i = \sigma$.
- The valid combinations $\mathcal{F}_{i,\sigma}$ consist of all tuples $(w_j, w_{j'})$ where the child states w_j and $w_{j'}$ are "consistent" with state σ at node i. Here "consistent" means both w_j and $w_{j'}$ make the same choice as σ on the vertices of X_i, formally, $w_j \cap X_i = \sigma \cap X_j$ and $w_{j'} \cap X_i = \sigma \cap X_{j'}$.

A formal proof appears in the full version. There, we also discuss a number of other graph constraints satisfying our assumption.

The following result follows from Assumption 1.

Claim 1. *For any $S \in \mathcal{C}_G$, there is a collection $\{b(i) \in \Sigma_i\}_{i \in I}$ such that:*

- *for each node $i \in I$ with children j and j', $(b(j), b(j')) \in \mathcal{F}_{i,b(i)}$,*
- *for each leaf ℓ we have $\mathcal{H}_{\ell,b(\ell)} \neq \emptyset$, and*
- *$S = \bigcup_{i \in I} X_{i,b(i)}$.*

Moreover, for any vertex $u \in V$, if $\bar{u} \in I$ denotes the highest node containing u then $u \in S \iff u \in X_{\bar{u},b(\bar{u})}$.

Sherali-Adams LP Hierarchy. This is one of the several "lift-and-project" procedures that, given a $\{0, 1\}$ integer program, produces systematically a sequence of increasingly tighter convex relaxations. The Sherali-Adams procedure [20] involves generating stronger *LP relaxations* by adding new variables and constraints. The rth round of this procedure has a variable $y(S)$ for every subset S of at most $r + 1$ variables in the original integer program — the new variable $y(S)$ corresponds to the joint event that all the original variables in S are one.

3   Approximation Algorithm for GCMC

In this section, we prove:

Theorem 4. *Consider any instance of the* GCMC *problem on a bounded-treewidth graph $G = (V, E)$. If the graph constraint $\mathcal{C}_G$ satisfies Assumption 1 then we obtain a $\frac{1}{2}$-approximation algorithm.*

Algorithm Outline. We start with a balanced tree decomposition $\mathcal{T}$ of graph G, as given in Theorem 3; recall the associated definitions from Sect. 2. Then we formulate an LP relaxation of the problem using Assumption 1 (i.e. the dynamic program for linear objectives) and further strengthened by applying the Sherali-Adams operator. Finally we use a natural top-down rounding that relies on Assumption 1 and the Sherali-Adams constraints.

All missing proofs will appear in the upcoming full version.

3.1   Linear Program

We start with some additional notation related to the tree decomposition $\mathcal{T}$ (from Theorem 3) and our dynamic program assumption (Assumption 1).

- For any node $i \in I$, T_i is the set of nodes on the $r - i$ path along with the children of all nodes except the children of i. See also Fig. 1.
- $\mathcal{P}$ is the collection of all node subsets J such that $J \subseteq T_{\ell_1} \cup T_{\ell_2}$ for some pair of leaf-nodes ℓ_1, ℓ_2. See also Fig. 1.
- $s(i) \in \Sigma_i$ denotes a state at node i. Moreover, for any subset of nodes $N \subseteq I$, we use the shorthand $s(N) := \{s(k) : k \in N\}$.
- $\bar{u} \in I$ denotes the highest tree-decomposition node containing vertex u.

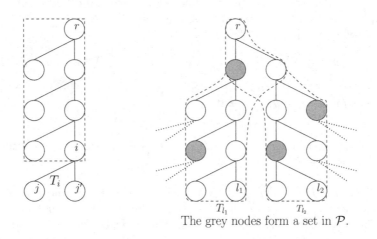

The grey nodes form a set in $\mathcal{P}$.

Fig. 1. Examples of (i) a set T_i and (ii) a set in $\mathcal{P}$.

The variables in our LP are $y(s(N))$ for all $\{s(k) \in \Sigma_k\}_{k \in N}$ and $N \in \mathcal{P}$. Variable $y(s(N))$ corresponds to the joint event that the solution (in $\mathcal{C}_G$) "induces" state $s(k)$ (in terms of Assumption 1) at each node $k \in N$.

We also use variables z_{uv} defined in constraint (4) that measure the probability of an edge (u, v) being cut. Constraints (5) are the Sherali-Adams constraints that enforce consistency among the y variables. Our LP uses a subset of variables and constraints from $O(\log n)$ rounds of the Sherali-Adams hierarchy. Constraints (6)–(8) are from the dynamic program (Assumption 1) and require valid state selections.

Claim 2. *Let y be a feasible solution to LP. For any node $i \in I$ with children j, j' and $s(k) \in \Sigma_k$ for all $k \in T_i$,*

$$y(s(T_i)) = \sum_{s(j) \in \Sigma_j} \sum_{s(j') \in \Sigma_{j'}} y(s(T_i \cup \{j, j'\})). \tag{3}$$

$$\text{maximize} \quad \sum_{\{u,v\} \in \binom{V}{2}} c_{uv} z_{uv} \tag{LP}$$

$$z_{uv} = \sum_{\substack{s(\bar{u})\in\Sigma_{\bar{u}}\\u\in X_{\bar{u},s(\bar{u})}}} \sum_{\substack{s(\bar{v})\in\Sigma_{\bar{v}}\\v\notin X_{\bar{v},s(\bar{v})}}} y(s(\{\bar{u},\bar{v}\})) + \sum_{\substack{s(\bar{u})\in\Sigma_{\bar{u}}\\u\notin X_{\bar{u},s(\bar{u})}}} \sum_{\substack{s(\bar{v})\in\Sigma_{\bar{v}}\\v\in X_{\bar{v},s(\bar{v})}}} y(s(\{\bar{u},\bar{v}\})),$$

$$\forall \{u,v\} \in \binom{V}{2}; \qquad (4)$$

$$y(s(N)) = \sum_{s(i)\in\Sigma_i} y(s(N\cup\{i\})), \qquad \forall N\in\mathcal{P}, i\notin N : N\cup\{i\}\in\mathcal{P}; \qquad (5)$$

$$\sum_{s(r)\in\Sigma_r} y(s(r)) = 1; \qquad (6)$$

$$y(s(\{i,j,j'\})) = 0, \qquad \forall i\in I, s(i)\in\Sigma_i, (s(j),s(j'))\notin\mathcal{F}_{i,s(i)}; \qquad (7)$$

$$y(s(\ell)) = 0, \qquad \forall \ell\in I, s(\ell)\in\Sigma_\ell : \mathcal{H}_{\ell,s(\ell)}=\emptyset; \qquad (8)$$

$$0 \le y(s(N)) \le 1, \qquad \forall N\in\mathcal{P}, \{s(k)\in\Sigma_k\}_{k\in N}. \qquad (9)$$

In constraint (7), we use j and j' to denote the two children of node $i\in I$.

3.2 The Rounding Algorithm

We start with the root node $r\in I$. Here $\{y(s(r)) : s(r)\in\Sigma_r\}$ defines a probability distribution over the states of r. We sample a state $a(r)\in\Sigma_r$ from this distribution. Then we continue top-down: given the chosen state $a(i)$ of any node i, we sample states for both children of i *simultaneously* from their joint distribution given at node i.

Input : Optimal solution of LP.
Output: A vertex set in $\mathcal{C}_G$.
1 Sample a state $a(r)$ at the root node by distribution $y(s(r))$;
2 **Do** *process all nodes i in $\mathcal{T}$ in order of increasing depth* :
3   Sample states $a(j), a(j')$ for the children of node i by joint distribution

$$\Pr[a(j)=s(j) \text{ and } a(j')=s(j')] = \frac{y(s(T_i\cup\{j,j'\}))}{y(s(T_i))}, \qquad (10)$$

    where $s(T_i)=a(T_i)$.
4 **end**
5 **Do** *process all nodes i in $\mathcal{T}$ in order of decreasing depth* :
6     $R_i = X_{i,a(i)}\cup R_j\cup R_{j'}$ where j,j' are the children of i.
7 **end**
8 $R = R_r$;
9 **return** R.

Algorithm 1. Rounding Algorithm for LP

3.3   Algorithm Analysis

Lemma 1. *(LP) is a valid relaxation of* GCMC.

Proof. Let $S \in \mathcal{C}_G$ be any feasible solution to the GCMC instance. Let $\{b(i)\}_{i \in I}$ denote the states given by Claim 1 corresponding to S. For any subset $N \in \mathcal{P}$ of nodes, and for all $\{s(i) \in \Sigma_i\}_{i \in N}$, set

$$y(s(N)) = \begin{cases} 1, & \text{if } s(i) = b(i) \text{ for all } i \in N; \\ 0, & \text{otherwise.} \end{cases}$$

Clearly that constraints (5) and (9) are satisfied. By the first two properties in Claim 1, it follows that constraints (7) and (8) are also satisfied. The last property in Claim 1 implies that $u \in S \iff u \in X_{\bar{u},b(\bar{u})}$ for any vertex $u \in V$. So any edge $\{u, v\}$ is cut exactly when one of the following occurs:

- $u \in X_{\bar{u},b(\bar{u})}$ and $v \notin X_{\bar{v},b(\bar{v})}$;
- $u \notin X_{\bar{u},b(\bar{u})}$ and $v \in X_{\bar{v},b(\bar{v})}$.

Using the setting of variable z_{uv} in (4) it follows that z_{uv} is exactly the indicator of edge $\{u, v\}$ being cut by S. Thus the objective value in (LP) is $c(\delta S)$.   ∎

Lemma 2. *(LP) has a polynomial number of variables and constraints. Hence the overall algorithm runs in polynomial time.*

Proof. There are $\binom{n}{2} = O(n^2)$ variables z_{uv}. Because the tree is binary, we have $|T_i| \leq 2d$ for any node i, where $d = O(\log n)$ is the depth of the tree decomposition. Moreover there are only $O(n^2)$ pairs of leaves as there are $O(n)$ leaf nodes. For each pair ℓ_1, ℓ_2 of leaves, we have $|T_{\ell_1} \cup T_{\ell_2}| \leq 4d$. Thus $|\mathcal{P}| \leq O(n^2) \cdot 2^{4d} = poly(n)$. By Assumption 1, we have $\max |\mathcal{H}_{i,\sigma}| = t = O(1)$, so the number of y-variables is at most $|\mathcal{P}| \cdot t^{4d} = poly(n)$. This shows that (LP) has polynomial size and can be solved optimally in polynomial time. Finally, it is clearly that the rounding algorithm runs in polynomial time.   ∎

Lemma 3. *The algorithm's solution R is always feasible.*

Proof. Note that the distributions used in Steps 1 and 2 are valid due to Claim 2; so the states $a(i)$s are well-defined. Moreover, by the choice of these distributions, for each node i, $y(a(T_i)) > 0$.

We now show that for any node $i \in I$ with children j, j' we have $(a(j), a(j')) \in \mathcal{F}_{i,a(i)}$. Indeed, at the iteration for node i (when $a(j)$ and $a(j')$ are set) using the conditional probability distribution in (10) and by constraint (7), we obtain that $(a(j), a(j')) \in \mathcal{F}_{i,a(i)}$ with probability one.

We show by induction that for each node $i \in I$, the subset $R_i \in \mathcal{H}_{i,a(i)}$. The base case is when i is a leaf. In this case, due to constraint (8) and the fact that $y(a(T_i)) > 0$ we know that $\mathcal{H}_{i,a(i)} \neq \emptyset$. So $R_i = X_{i,a(i)} \in \mathcal{H}_{i,a(i)}$ by Assumption 1(3). For the inductive step, consider node $i \in I$ with children j, j' where $R_j \in \mathcal{H}_{j,a(j)}$ and $R_{j'} \in \mathcal{H}_{j',a(j')}$. Moreover, from the property above, $(a(j), a(j')) \in \mathcal{F}_{i,a(i)}$. Now using Assumption 1(3) we have $R_i = X_{i,a(i)} \cup R_j \cup R_{j'} \in \mathcal{H}_{i,a(i)}$. Thus the final solution $R \in \mathcal{C}_G$.   ∎

Claim 3. *A vertex u is contained in solution R if and only if $u \in X_{\bar{u},a(\bar{u})}$.*

In the rest of this section, we show that every edge (u, v) is cut by solution R with probability at least $z_{uv}/2$, which would prove the algorithm's approximation ratio. Lemma 4 handles the case when $\bar{u} \in T_{\bar{v}}$ (the case $\bar{v} \in T_{\bar{u}}$ is identical). And Lemma 5 handles the (harder) case when $\bar{u} \notin T_{\bar{v}}$ and $\bar{v} \notin T_{\bar{u}}$.

We first state some useful claims before proving the lemmas.

Observation 1 (see [14] for a similar use of this principle). *Let X, Y be two jointly distributed $\{0, 1\}$ random variables. Then $\Pr(X = 1)\Pr(Y = 0) + \Pr(X = 0)\Pr(Y = 1) \geq \frac{1}{2}[\Pr(X = 0, Y = 1) + \Pr(X = 1, Y = 0)]$.*

Claim 4. *For any node i and state $s(k) \in \Sigma_k$ for all $k \in T_i$, the rounding algorithm satisfies $\Pr[a(T_i) = s(T_i)] = y(s(T_i))$.*

Proof. We proceed by induction on the depth of node i. It is clearly true when $i = r$, i.e. $T_i = \{r\}$. Assuming the statement is true for node i, we will prove it for i's children. Let j, j' be the children nodes of i; note that $T_j = T_{j'} = T_i \cup \{j, j'\}$. Then using (10), we have

$$\Pr[a(T_j) = s(T_j) \mid a(T_i) = s(T_i)] = \frac{y(s(T_i \cup \{j, j'\}))}{y(s(T_i))}.$$

Combined with $\Pr[a(T_i) = s(T_i)] = y(s(T_i))$ we obtain $\Pr[a(T_j) = s(T_j)] = y(s(T_j))$ as desired. ∎

Claim 5. *For any $u, v \in V$, $s(\bar{u}) \in \Sigma_{\bar{u}}$ and $s(\bar{v}) \in \Sigma_{\bar{v}}$, we have*

$$y(s(\{\bar{u}, \bar{v}\})) = \sum_{\substack{s(k) \in \Sigma_k \\ k \in T_i \setminus \bar{u} \setminus \bar{v}}} y(s(T_i \cup \{\bar{u}, \bar{v}\})),$$

where i is the least common ancestor of $\bar{u}$ and $\bar{v}$.

Proof. Because i is the least common ancestor of $\bar{u}$ and $\bar{v}$, we have $T_i \cup \{\bar{u}, \bar{v}\} \in \mathcal{P}$. Then the claim follows by repeatedly applying constraint (5). ∎

Lemma 4. *Consider any $u, v \in V$ such that $\bar{u} \in T_{\bar{v}}$. Then the probability that edge (u, v) is cut by solution R is z_{uv}.*

Lemma 5. *Consider any $u, v \in V$ such that $\bar{u} \notin T_{\bar{v}}$ and $\bar{v} \notin T_{\bar{u}}$. Then the probability that edge (u, v) is cut by solution R is at least $z_{uv}/2$.*

Proof. In order to simplify notation, we define:

$$z_{uv}^+ = \sum_{\substack{s(\bar{u}) \in \Sigma_{\bar{u}} \\ u \in X_{\bar{u}, s(\bar{u})}}} \sum_{\substack{s(\bar{v}) \in \Sigma_{\bar{v}} \\ v \notin X_{\bar{v}, s(\bar{v})}}} y(s(\{\bar{u}, \bar{v}\})), \quad z_{uv}^- = \sum_{\substack{s(\bar{u}) \in \Sigma_{\bar{u}} \\ u \notin X_{\bar{u}, s(\bar{u})}}} \sum_{\substack{s(\bar{v}) \in \Sigma_{\bar{v}} \\ v \in X_{\bar{v}, s(\bar{v})}}} y(s(\{\bar{u}, \bar{v}\})).$$

Note that $z_{uv} = z_{uv}^+ + z_{uv}^-$.

Let $D_u = \{s(\bar{u}) \in \Sigma_{\bar{u}} | u \in s(\bar{u})\}$ and $D_v = \{s(\bar{v}) \in \Sigma_{\bar{v}} | v \in s(\bar{v})\}$. Let i denote the least common ancestor of nodes $\bar{u}$ and $\bar{v}$. For any choice of states $\{s(k) \in \Sigma_k\}_{k \in T_i}$ define:

$$z_{uv}^+(s(T_i)) = \sum_{s(\bar{u}) \in D_u} \sum_{s(\bar{v}) \notin D_v} \frac{y(s(T_i \cup \{\bar{u}, \bar{v}\}))}{y(s(T_i))},$$

and similarly $z_{uv}^-(s(T_i))$.

In the rest of the proof we fix states $\{s(k) \in \Sigma_k\}_{k \in T_i}$ and condition on the event $\mathcal{E}$ that $a(T_i) = s(T_i)$. We will show:

$$\Pr[|\{u, v\} \cap R| = 1 \,|\, \mathcal{E}] \geq \frac{1}{2}\left(z_{uv}^+(s(T_i)) + z_{uv}^-(s(T_i))\right). \tag{11}$$

By taking expectation over the conditioning $s(T_i)$, this would imply Lemma 5. We now define the following indicator random variables (conditioned on $\mathcal{E}$).

$$I_u = \begin{cases} 0 & \text{if } a(\bar{u}) \notin D_u \\ 1 & \text{if } a(\bar{u}) \in D_u \end{cases} \quad \text{and} \quad I_v = \begin{cases} 0 & \text{if } a(\bar{v}) \notin D_v \\ 1 & \text{if } a(\bar{v}) \in D_v. \end{cases}$$

Observe that I_u and I_v (*conditioned* on $\mathcal{E}$) are independent because $\bar{u} \notin T_{\bar{v}}$ and $\bar{v} \notin T_{\bar{u}}$. So,

$$\Pr[|\{u, v\} \cap R| = 1 \,|\, \mathcal{E}] = \Pr[I_u = 1] \cdot \Pr[I_v = 0] + \Pr[I_u = 0] \cdot \Pr[I_v = 1] \tag{12}$$

For any $s(k) \in \Sigma_k$ for $k \in T_{\bar{u}} \setminus T_i$, we have by Claim 4 and $T_i \subseteq T_{\bar{u}}$ that

$$\Pr[a(T_{\bar{u}}) = s(T_{\bar{u}}) \,|\, a(T_i) = s(T_i)] = \frac{\Pr[a(T_{\bar{u}}) = s(T_{\bar{u}})]}{\Pr[a(T_i) = s(T_i)]} = \frac{y(s(T_{\bar{u}}))}{y(s(T_i))}.$$

Therefore

$$\Pr[I_u = 1] = \sum_{s(\bar{u}) \in D_u} \sum_{k \in T_{\bar{u}} \setminus T_i \setminus \{\bar{u}\} s(k) \in \Sigma_k} \frac{y(s(T_{\bar{u}}))}{y(s(T_i))} = \sum_{s(\bar{u}) \in D_u} \frac{y(s(T_i \cup \{\bar{u}\}))}{y(s(T_i))}.$$

The last equality follows from the (5) constraint. Similarly,

$$\Pr[I_v = 1] = \sum_{s(\bar{v}) \in D_v} \frac{y(s(T_i \cup \{\bar{v}\}))}{y(s(T_i))}.$$

Now define $\{0, 1\}$ random variables X and Y jointly distributed as:

	$Y = 0$	$Y = 1$
$X = 0$	$\Pr[I_v = 1] - z_{uv}^-(s(T_i))$	$z_{uv}^-(s(T_i))$
$X = 1$	$z_{uv}^+(s(T_i))$	$\Pr[I_u = 1] - z_{uv}^+(s(T_i))$

Note that $\Pr[X = 1] = \Pr[I_u = 1]$ and $\Pr[Y = 1] = \Pr[I_v = 1]$. So, applying Observation 1 and using (12) we have:

$$\Pr[|\{u, v\} \cap R| = 1 \,|\, \mathcal{E}] \geq \frac{1}{2}\left(\Pr[X = 0, Y = 1] + \Pr[X = 1, Y = 0]\right),$$

which implies (11). ∎

References

1. Ageev, A.A., Hassin, R., Sviridenko, M.: A 0.5-approximation algorithm for MAX DICUT with given sizes of parts. SIAM J. Discrete Math. **14**(2), 246–255 (2001)
2. Ageev, A.A., Sviridenko, M.I.: Approximation algorithms for maximum coverage and max cut with given sizes of parts. In: Cornuéjols, G., Burkard, R.E., Woeginger, G.J. (eds.) IPCO 1999. LNCS, vol. 1610, pp. 17–30. Springer, Heidelberg (1999)
3. Bansal, N., Lee, K.W., Nagarajan, V., Zafer, M.: Minimum congestion mapping in a cloud. SIAM J. Comput. **44**(3), 819–843 (2015)
4. Bateni, M., Charikar, M., Guruswami, V.: Maxmin allocation via degree lower-bounded arborescences. In: STOC, pp. 543–552 (2009)
5. Bienstock, D., Özbay, N.: Tree-width and the Sherali-Adams operator. Discrete Optim. **1**(1), 13–21 (2004)
6. Bodlaender, H.L.: NC-algorithms for graphs with small treewidth. In: van Leeuwen, J. (ed.) WG 1988. LNCS, vol. 344, pp. 1–10. Springer, Heidelberg (1989)
7. Călinescu, G., Chekuri, C., Pál, M., Vondrák, J.: Maximizing a monotone submodular function subject to a matroid constraint. SIAM J. Comput. **40**(6), 1740–1766 (2011)
8. Chekuri, C., Vondrák, J., Zenklusen, R.: Submodular function maximization via the multilinear relaxation and contention resolution schemes. SIAM J. Comput. **43**(6), 1831–1879 (2014)
9. Demaine, E.D., Hajiaghayi, M.T., Kawarabayashi, K.: Algorithmic graph minor theory: decomposition, approximation, and coloring. In: FOCS, pp. 637–646 (2005)
10. Demaine, E.D., Hajiaghayi, M., Kawarabayashi, K.: Contraction decomposition in h-minor-free graphs and algorithmic applications. In: STOC, pp. 441–450 (2011)
11. Feldman, M., Naor, J., Schwartz, R.: A unified continuous greedy algorithm for submodular maximization. In: FOCS, pp. 570–579 (2011)
12. Friggstad, Z., Könemann, J., Kun-Ko, Y., Louis, A., Shadravan, M., Tulsiani, M.: Linear programming hierarchies suffice for directed Steiner tree. In: Lee, J., Vygen, J. (eds.) IPCO 2014. LNCS, vol. 8494, pp. 285–296. Springer, Heidelberg (2014)
13. Goemans, M.X., Williamson, D.P.: Improved approximation algorithms for maximum cut and satisfiability problems using semidefinite programming. J. Assoc. Comput. Mach. **42**(6), 1115–1145 (1995)
14. Gupta, A., Talwar, K., Witmer, D.: Sparsest cut on bounded treewidth graphs: algorithms and hardness results. In: STOC, pp. 281–290 (2013)
15. Hajiaghayi, M.T., Kortsarz, G., MacDavid, R., Purohit, M., Sarpatwar, K.: Approximation algorithms for connected maximum cut and related problems. In: Bansal, N., Finocchi, I. (eds.) ESA 2015. LNCS, vol. 9294, pp. 693–704. Springer, Heidelberg (2015)
16. Khot, S., Kindler, G., Mossel, E., O'Donnell, R.: Optimal inapproximability results for MAX-CUT and other 2-variable csps? SIAM J. Comput. **37**(1), 319–357 (2007)
17. Lee, J., Mirrokni, V.S., Nagarajan, V., Sviridenko, M.: Maximizing nonmonotone submodular functions under matroid or knapsack constraints. SIAM J. Discrete Math. **23**(4), 2053–2078 (2010)
18. Lee, J., Sviridenko, M., Vondrák, J.: Submodular maximization over multiple matroids via generalized exchange properties. Math. Oper. Res. **35**(4), 795–806 (2010)
19. Magen, A., Moharrami, M.: Robust algorithms for MAX INDEPENDENT SET on minor-free graphs based on the Sherali-Adams hierarchy. In: Dinur, I., Jansen, K., Naor, J., Rolim, J. (eds.) Approximation, Randomization, and Combinatorial Optimization. LNCS, vol. 5687, pp. 258–271. Springer, Heidelberg (2009)

20. Sherali, H.D., Adams, W.P.: A hierarchy of relaxations between the continuous and convex hull representations for zero-one programming problems. SIAM J. Discrete Math. **3**(3), 411–430 (1990)
21. Vicente, S., Kolmogorov, V., Rother, C.: Graph cut based image segmentation with connectivity priors. In: CVPR (2008)

Sparsest Cut in Planar Graphs, Maximum Concurrent Flows and Their Connections with the Max-Cut Problem

Mourad Baïou[1]([✉]) and Francisco Barahona[2]

[1] CNRS, Université Clermont II,
Campus des Cézeaux BP 125, 63173 Aubière Cedex, France
baiou@isima.fr
[2] IBM T. J. Watson Research Center, Yorktown Heights, NY 10589, USA

Abstract. We study the sparsest cut problem when the "capacity-demand" graph is planar, and give a combinatorial algorithm. In this type of graphs there is an edge for each positive capacity and also an edge for each positive demand. We extend this result to graphs with no K_5 minor. We also show how to find a maximum concurrent flow in these two cases. We use ideas that had been developed for the max-cut problem, and show how to exploit the connections among these problems.

Keywords: Sparsest cut · Maximum concurrent flow · Planar graphs · Max-cut

1 Introduction

Given a graph $C = (V, E_C)$ with a *capacity* function $c : E_C \to \Re_+$, and a graph $D = (V, E_D)$ with a *demand* function $d : E_D \to \Re_+$, the *sparsest cut problem* consists of finding a node-set $S \subset V$ with $d(\delta_D(S)) > 0$ that minimizes

$$\frac{c(\delta_C(S))}{d(\delta_D(S))}.$$

For $S \subseteq V$ we denote by $\delta_C(S)$ (resp. $\delta_D(S)$) the set of edges in E_C (resp. E_D) with exactly one endnode in S. The set $\delta_C(S)$ (resp. $\delta_D(S)$) is called a *cut*. We use $c(\delta_C(S))$ and $d(\delta_D(S))$ to denote $\sum_{e \in \delta_C(S)} c(e)$ and $\sum_{e \in \delta_D(S)} d(e)$ respectively.

The sparsest cut problem is used to find approximate solutions of problems in VLSI layout and network routing, see e.g. [6,20,31]. The sparsest cut problem is NP-hard, see [22]. Approximation algorithms have been given in [1,2,20] and many others. Approximation algorithms for C being planar and for C having bounded treewidth were given in [13,27] respectively. If C is planar (or even series-parallel), a lower bound of 17/16 on the approximability of the problem (unless P = NP) was given in [13]. If C is planar, an algorithm to find the minimum of $\frac{|\delta(S)|}{\min(|S|,|V \setminus S|)}$ was given in [23]. This is called the *minimum quotient cut problem*.

Q. Louveaux and M. Skutella (Eds.): IPCO 2016, LNCS 9682, pp. 63–76, 2016.
DOI: 10.1007/978-3-319-33461-5_6

This was generalized to graphs of bounded genus in [24]. Also in [24] an $O(m^{4g+7})$ algorithm for finding the minimum of $\frac{|\delta(S)|}{|S|(|V|-|S|)}$, for graphs of genus g was given. This is called the *uniform sparsest cut problem*. Here $m = |E_C|$.

Exact algorithms for sparsest cut have been given for the case when C is a tree [22], C is planar and all endnodes of the edges in E_D are in one face of C [22], and for the case when all edges in E_D have one endnode in a node-set of fixed size [15]. When the demand edge-set E_D consists of exactly one edge, we have the well known minimum cut problem [10].

For each edge $e \in E_D$ (the demand graph), let $\mathcal{P}_e$ be the set of paths between the endnodes of e, in E_C (the capacity graph). Let $\mathcal{P} = \cup_{e \in E_D} \mathcal{P}_e$. A *concurrent flow of throughput* λ is a function $f : \mathcal{P} \to \Re_+$ such that

$$\sum_{p \in \mathcal{P}_e} f(p) = \lambda d(e), \text{ for all } e \in E_D,$$

$$\sum_{p \,:\, e \in p} f(p) \leq c(e), \text{ for all } e \in E_C.$$

The first set of equations says that the flow sends $\lambda d(e)$ units between the endnodes of $e \in E_D$. The second set of inequalities says that the sum of the flows going through each edge $e \in E_C$, should satisfy the capacity $c(e)$. A *maximum concurrent flow* is a flow that maximizes the throughput.

For a cut $\delta_D(S)$ with $d(\delta_D(S)) > 0$, considering the flow across $\delta_C(S)$, we obtain the inequality $\lambda d(\delta_D(S)) \leq c(\delta_C(S))$. Thus the value of a sparsest cut is an upper bound for the value of a maximum concurrent flow, [22]. In general there is a gap between the two values, and many authors have given bounds for this gap, see e.g. [12,17,18,20,25].

Here we consider the case when the graph $H = (V, E_C \cup E_D)$ is planar, and we give an exact polynomial combinatorial algorithm to find a sparsest cut. We extend this algorithm to graphs with no K_5 minor. For this class of graphs, we also give a polynomial combinatorial algorithm that starting from the value of a sparsest cut, finds a maximum concurrent flow. We use ideas that had been developed for the max-cut problem, and we exploit the connections among these problems. We review all these techniques to try to make this paper self-contained.

This paper is organized as follows. In Sect. 2 we give some notation and cite some preliminary results. In Sect. 3 we deal with planar graphs. In Sect. 4 we study graphs with no K_5 minor. Section 5 is devoted to maximum concurrent flows.

2   Preliminaries

We start with some notation. Given a graph $G = (V, E)$, we use n to denote $|V|$. For an edge $e \in E$ with endnodes u and v we also use uv to denote the edge e. For $S \subseteq V$ we denote by $\delta(S)$ the set of edges with exactly one endnode in S. The set $\delta(S)$ is called a *cut*. Notice that $\delta(V) = \emptyset$ is also a cut. We use $\delta(v)$ instead of $\delta(\{v\})$. Sometimes we use δ_G to indicate that this is a cut of the graph G. We denote by $E(S)$ the set of edges with both endnodes in S. The subgraphs $(S, E(S))$ and $(V \setminus S, E(V \setminus S))$ are called the *shores* of the cut $\delta(S)$. If $F \subseteq E$, the graph $H = (V, F)$ is called a spanning graph of G.

For an edge set $S \subseteq E$, its *incidence vector* x^S is defined by $x^S(e) = 1$ if $e \in S$, and $x^S(e) = 0$ otherwise, for each edge $e \in E$. The *cycle space* of a graph is obtained by taking sums (mod 2) of incidence vectors of cycles. An element of this space is the incidence vector of a spanning subgraph so that every node has even degree. A *cycle basis* is a basis of this vector space. For a planar graph, its faces minus one, form a cycle basis.

Let $\langle \cdot, \cdot \rangle$ denote the inner product between two vectors. If x^C is the incidence vector of a cycle C, and y^K and is the incidence vector of a cut K, then $\langle x^C, y^K \rangle \equiv 0 \pmod 2$. This is because the intersection between a cut and a cycle has even cardinality. Moreover, if $\{C_1, \ldots, C_p\}$ is a cycle basis, then y is the incidence vector of a cut if and only if $\langle x^{C_i}, y \rangle \equiv 0 \pmod 2$, for $i = 1, \ldots, p$. Notice that the vector $y = 0$ is the incidence vector of the cut $\delta(V) = \emptyset$.

Now we review a classic result in combinatorial optimization.

2.1 The Chinese Postman Problem and Minimum T-joins

Given an undirected connected graph $G = (V, E)$ with nonnegative edge weights $w(e)$ for each edge e, this problem consists of finding a tour of the graph of minimum weight, so that every edge is visited at least once. Edmonds & Johnson [9] gave a polynomial algorithm for this. Since some edges might have to be visited more than once, the problem reduces to find a set of edges of minimum weight that should be visited twice. This can be formulated as follows.

$$\text{minimize} \sum_{e \in E} w(e)x(e) \tag{1}$$

$$\sum_{e \in \delta(v)} x(e) \equiv \begin{cases} 1 & \pmod 2 \text{ if } v \in T, \\ 0 & \pmod 2 \text{ if } v \in V \setminus T, \end{cases} \tag{2}$$

$$x(e) \in \{0, 1\} \text{ for all } e \in E. \tag{3}$$

Here T denotes the set of nodes of odd degree. A solution of this corresponds to a set of paths matching the nodes in T. For this Edmonds & Johnson gave a combinatorial algorithm that solves the following linear program.

$$\text{minimize} \sum_{e \in E} w(e)x(e) \tag{4}$$

$$\sum_{e \in \delta(S)} x(e) \geq 1 \quad \text{for each node-set } S \text{ with } |S \cap T| \text{ odd,} \tag{5}$$

$$x(e) \geq 0 \quad \text{for all } e \in E. \tag{6}$$

Their algorithm shows that this linear program always has an optimal solution that is integer valued.

If T is an arbitrary set of nodes with $|T|$ even, the same results hold, and this is called the *Minimum T-join problem*. We are going to use this in Sects. 3 and 5.

If G is a complete graph, this problem can be solved in $O(n^3)$ time, see [11,19]. If the graph is planar, one can use the planar separator theorem of [21] to solve this in $O(n^{3/2} \log n)$ time, see [4].

3   Sparsest Cut in Planar Graphs

The max-cut problem with positive edge-weights, in planar graphs, can be solved in polynomial time using planar duality, see [14]. This is easy to extend to graphs with positive and negative edge-weights, see [4]. Here we use this extension for the sparsest cut problem. To simplify notation, we assume that $G = (V, E)$ is a planar graph with a capacity function $c : E \to \Re_+$ and a demand function $d : E \to \Re_+$. Some edges might have capacity (or demand) equal to zero. We have to find a set $S \subset V$ with $d(\delta(S)) > 0$ and that minimizes

$$\frac{c(\delta(S))}{d(\delta(S))}.$$

For fractional optimization problems, Newton's method, also known as Dinkelbach method [8], is frequently used. In our case it is as follows.

Newton's method

Step 0. Pick any set $\hat{S} \subset V$ with $d(\delta(\hat{S})) > 0$. Set $\lambda = \frac{c(\delta(\hat{S}))}{d(\delta(\hat{S}))}$.
Step 1. Find $\bar{S} \subset V$ such that $c(\delta(\bar{S})) - \lambda d(\delta(\bar{S})) = \min c(\delta(S)) - \lambda d(\delta(S))$.
Step 2. If $c(\delta(\bar{S})) - \lambda d(\delta(\bar{S})) < 0$, then

$$\frac{c(\delta(\bar{S}))}{d(\delta(\bar{S}))} < \frac{c(\delta(\hat{S}))}{d(\delta(\hat{S}))}.$$

In this case set $\lambda = \frac{c(\delta(\bar{S}))}{d(\delta(\bar{S}))}$, $\hat{S} \leftarrow \bar{S}$, and go to Step 1.

If $c(\delta(\bar{S})) - \lambda d(\delta(\bar{S})) = 0$, stop, $\hat{S}$ is an optimal solution.

Notice that in Step 1, it is possible to have $\bar{S} = \hat{S}$, thus the minimum is always less than or equal to zero. The value λ decreases at each iteration, thus the number of iterations is finite. Radzik [26] analysed Newton's method, his results imply that if all capacities and demands are bounded by U, then Newton's method takes $O(\log(nU))$ iterations, and for general capacities and demands, the number of iterations is $O(n^2 \log^2 n)$. This last bound was improved to $O(n^2 \log n)$ by Wang et al. [33]. Their analysis was for maximization problems, so here we could maximize $d(\delta(S))/c(\delta(S))$. Then in Step 1 we would have to look for $\max d(\delta(S)) - \mu c(\delta(S))$, where $\mu = d(\delta(\hat{S}))/c(\delta(\hat{S}))$, and notice that this second algorithm will produce exactly the same set of intermediate solutions as the first algorithm. Moreover if $\{\lambda_i\}$ are the different values of λ produced by the first algorithm, and $\{\mu_i\}$ are the different values of μ given by the second algorithm, then $\lambda_i = 1/\mu_i$ for all i. This also shows that we can maximize this ratio and solve what we can call the *Densest cut problem*.

Now we have to discuss how to find the minimum in Step 1. We are looking for a cut of minimum weight. Since the edge-weights can be positive or negative, this cannot be done with network flow techniques. We define weights $w(e) = c(e) - \lambda d(e)$ for each edge $e \in E$. Since the intersection between a cut and a cycle has even cardinality, we define 0–1 variables $x(e)$ for $e \in E$, and solve

$$\text{minimize} \sum_{e \in E} w(e)x(e) \tag{7}$$

$$\sum_{e \in C} x(e) \equiv 0 \quad (\text{mod } 2), \text{for each cycle } C \text{ of } G. \tag{8}$$

Let $\mathcal{F}$ be the set of faces of G. Since $\mathcal{F}$ contains a cycle basis, we just need to impose Eq. (8) for the elements of $\mathcal{F}$. Then we define

$$x'(e) = \begin{cases} x(e) & \text{if } w(e) \geq 0, \\ 1 - x(e) & \text{if } w(e) < 0. \end{cases}$$

Also we call *odd* a face with an odd number of edges e with $w(e) < 0$, and *even* a face with an even number of edges e with $w(e) < 0$. Then problem (7)–(8) is equivalent to

$$\text{minimize} \sum_{e \in E} |w(e)|x'(e) \tag{9}$$

$$\sum_{e \in F} x'(e) \equiv \begin{cases} 1 \,(\text{mod } 2) \text{ if } F \text{ is an odd face,} \\ 0 \,(\text{mod } 2) \text{ if } F \text{ is an even face.} \end{cases} \tag{10}$$

Now consider the dual graph $\tilde{G} = (\mathcal{F}, E)$. Let T be the set of odd faces, then problem (9)–(10) is equivalent to

$$\text{minimize} \sum_{e \in E} |w(e)|x'(e) \tag{11}$$

$$\sum_{e \in \delta(u)} x'(e) \equiv \begin{cases} 1 \,(\text{mod } 2) \text{ if } u \in T, \\ 0 \,(\text{mod } 2) \text{ if } u \in \mathcal{F} \setminus T. \end{cases} \tag{12}$$

This is a minimum T-join problem [9], discussed in Sect. 2.

For planar graphs one can use planar separators [21], and a primal version of the T-join algorithm [4], to find a minimum T-join in $O(n^{3/2} \log n)$ time. Since the number of iterations of Newton's algorithm is $O(n^2 \log n)$, we can state the following.

Theorem 1. *The sparsest cut problem when $H = (V, E_C \cup E_D)$ is planar can be solved in $O(n^{3.5} \log^2 n)$ time.*

4   Graphs with no K_5 Minor

In this section we extend the algorithm from the previous section to graphs with no K_5 minor. Wagner [32] showed that these graphs can be decomposed into planar graphs and a graph with eight nodes, we describe that decomposition procedure below.

Let $G = (V, E)$ be a connected graph, and let $Y \subset V$ be a minimal articulation set (that is, the deletion of Y produces a disconnected graph, but no proper subset of Y has this property). Choose nonemtpy subsets T_1, T_2 of V, such that (T_1, Y, T_2) is a partition of V, and no edge joins a node in T_1 to a node in T_2. Add a set Z of new edges joining each pair of nonadjacent nodes in Y. Let $G_1 = (V_1, E_1)$, $G_2 = (V_2, E_2)$ be subgraphs so that $V_i = T_i \cup Y$, $E_i = E(V_i) \cup Z$, $i = 1, 2$. Then if $|Y| = k$, $1 \le k \le 3$, G is called a k-sum of G_1 and G_2, see Fig. 1. Wagner [32] showed that a graph has no K_5 minor if and only if it can be obtained by means of k-sums starting from planar graphs and copies of V_8, which is the graph in Fig. 1. This decomposition can be found in linear time, see [28]. To extend the algorithm of the previous section, we have to show how to find the minimum in Step 1 of Newton's Algorithm. So we assume that $G = (V, E)$ is a graph with no K_5 minor, with a weight function $w : E \to \Re$, and we have to find a cut of minimum weight. Again notice that some of the weights can be negative, so this cannot be solved with network flow techniques. The algorithm is as follows.

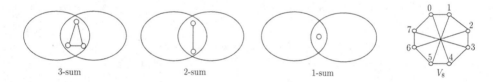

3-sum   2-sum   1-sum V_8

Fig. 1. Different k-sums, and the graph V_8.

If G is planar we treat it as in the previous section. If G is V_8 we use enumeration. Now we assume that G is a k-sum of G_1 and G_2, where G_2 is a planar graph or a copy of V_8. We need a way to decompose. Denote by $\alpha(S, T, H)$ the minimum weight of a cut of the graph H, containing the edge set S and having empty intersection with the edge set T. We write $\alpha(H)$ instead of $\alpha(\emptyset, \emptyset, H)$. We have three cases.

- If $k = 3$, let e, f and g be the edges in $G_1 \cap G_2$. The edge weights in G_2 are taken to be the same as for G. Then, a minimum weight cut problem is solved in G_1 where all the edge weights are taken to be the same as for G, except for e, f, g, which are redefined as the solution of the following system of linear equations:

$$w'(e) + w'(f) = \alpha(\{e, f\}, \emptyset, G_2) - \alpha(\emptyset, \{e, f, g\}, G_2)$$
$$w'(f) + w'(g) = \alpha(\{f, g\}, \emptyset, G_2) - \alpha(\emptyset, \{e, f, g\}, G_2)$$
$$w'(e) + w'(g) = \alpha(\{e, g\}, \emptyset, G_2) - \alpha(\emptyset, \{e, f, g\}, G_2).$$

The above system reflects the fact that a cut contains zero or two of these edges. We have that

$$\alpha(G) = \alpha(G_1) + \alpha(\emptyset, \{e, f, g\}, G_2).$$

The new weights for e, f, g contain all the information about G_2 needed to reduce the problem to a minimum weighted cut in G_1. So we solve four problems in G_2, to compute the different α values, and then continue working recursively in G_1.

- If $k = 2$, let e be the edge in $G_1 \cap G_2$. We take in G_2 the same weights as for G. In G_1 we redefine only the weight of e as

$$w'(e) = \alpha(\{e\}, \emptyset, G_2) - \alpha(\emptyset, \{e\}, G_2).$$

Then $\alpha(G) = \alpha(G_1) + \alpha(\emptyset, \{e\}, G_2)$. Thus we have to solve two problems in G_2 to obtain the α values, and then continue working recursively in G_1.
- If $k = 1$, then the problem is solved independently in G_1 and G_2.

This type of decomposition was used in [7] for the Travelling Salesman Problem, and in [3] for the Max-cut Problem in graphs with no K_5 minor. Since the decomposition of the graph can be found in linear time, each planar graph with p nodes can be treated in $O(p^{3/2} \log p)$ time, and each copy of V_8 can be treated in constant time, then a minimum weighted cut in G can be found in $O(n^{3/2} \log n)$ time. Since Newton's algorithm takes $O(n^2 \log n)$ iterations, we derive the following.

Theorem 2. If $H = (V, E_C \cup E_D)$ is a graph with no K_5 minor, then the sparsest cut problem can be solved in $O(n^{3.5} \log^2 n)$ time.

5   Maximum Concurrent Flow

Again we assume that we have a graph $G = (V, E)$ with a *capacity* function $c : E \to \Re_+$, and a *demand* function $d : E \to \Re_+$. For $\lambda > 0$ we define $w(e) = c(e) - \lambda d(e)$. Denote by $\mathscr{C}$ the set of cycles containing exactly one edge e with $w(e) < 0$. A flow is a function $f : \mathscr{C} \to \Re_+$ such that

$$\sum \{f(C) \mid e \in C\} \begin{cases} \leq w(e) & \text{if } w(e) \geq 0 \\ = -w(e) & \text{if } w(e) < 0, \end{cases} \tag{13}$$

for each edge e. An edge e with $w(e) < 0$ represents a demand of value $|w(e)|$ between the endnodes of e. Given a flow, it is easy to derive a concurrent flow of throughput λ.

If there is a flow, and $K = \delta(S)$ is a cut, by considering the flow across the cut we obtain

$$w(K) \geq 0. \tag{14}$$

This inequality says that the sum of the capacities across the cut K should be at least the sum of the demands across the cut. So condition (14) is necessary for the existence of a flow, and it is called the *cut condition*. A cut that violates (14) will be called a *negative cut*. For a concurrent flow of throughput λ this condition translates into $\lambda \leq \dfrac{c(\delta(S))}{d(\delta(S))}$, for any node set $S \subset V$ with $d(\delta(S)) > 0$.

Let $\hat{\lambda}$ be the value given by the maximum concurrent flow problem. The sparsest cut value is an upper bound for $\hat{\lambda}$ and in general this bound is not tight. Seymour [29, 30], proved that for planar graphs and for graphs with no K_5 minor, the cut condition (14) is sufficient for the existence of a flow. This implies that for these classes of graphs, the maximum throughput is exactly the value of a sparsest cut. In what follows we show how to find a maximum concurrent flow starting from the value of a sparsest cut, for the graphs studied here. To obtain the flow we use the dual variables associated with a linear programming formulation of the max-cut problem, see also [5]. We assume that we have a graph with no negative cut and give an algorithm that produces a flow, thus this is an algorithmic proof of Seymour's Theorem.

5.1 Planar Graphs

Let $\hat{\lambda}$ be the value of a sparsest cut, obtained by Newton's algorithm. If we set $w(e) = c(e) - \hat{\lambda}d(e)$, then the value given by problem (7)–(8) is zero, this is the termination criterion. Therefore the value given by (9)–(10) (and (11)–(12)) is $\sum\{|w(e)| : w(e) < 0\}$. Edmonds & Johnson [9] proved that (11)–(12) can be formulated as

$$\text{minimize} \sum |w(e)|x'(e) \tag{15}$$

$$\sum_{e \in \delta(S)} x'(e) \geq 1, \text{ for all sets } S \subset \mathcal{F}, \text{ with } |S \cap T| \text{ odd}, \tag{16}$$

$$x' \geq 0. \tag{17}$$

Here $\mathcal{F}$ is the set of faces (the nodes of the dual graph), and T is the set of faces with an odd number of edges e with $w(e) < 0$. We need the following lemma whose proof is omitted.

Lemma 3. *Consider an inequality* (16), *if one of the shores of* $\delta(S)$ *is not connected, then the inequality is redundant.*

Thus we concentrate on sets S such that both shores of $\delta(S)$ are connected. Since cuts in the dual graph correspond to cycles in the original graph, here we obtain the set of simple cycles in the original graph G, containing an odd number of edges e with $w(e) < 0$. Let us denote by $\mathbb{C}$ this set of cycles. We have the linear program below.

$$\text{minimize} \sum |w(e)| x'(e) \tag{18}$$

$$\sum_{e \in C} x'(e) \geq 1, \text{for all cycles } C \text{ of } \mathbb{C}. \tag{19}$$

$$x' \geq 0. \tag{20}$$

Its dual is

$$\text{maximize} \sum_{C \in \mathbb{C}} y_C \tag{21}$$

$$\sum_{C : e \in C} y_C \leq |w(e)|, \text{ for each edge } e, \tag{22}$$

$$y \geq 0. \tag{23}$$

Let $\bar{x}(e) = 1$ if $w(e) < 0$, and $\bar{x}(e) = 0$ otherwise. We have that $\bar{x}$ is an optimal solution of (18)–(20), (because $\bar{x}$ is a solution of (11)–(12) of value $\sum\{|w(e)| : w(e) < 0\}$). Let $\bar{y}$ be an optimal solution of (21)–(23), then the complementary slackness conditions imply that if $\bar{y}_C > 0$, then C contains exactly one edge e with $w(e) < 0$. It also follows from complementary slackness that

$$\sum_{C : e \in C} \bar{y}_C = |w(e)|$$

for each edge e with $w(e) < 0$. Thus $\bar{y}$ is the required flow. For planar graphs this dual vector can be obtained in $O(n^{3/2} \log n)$ time, see [4].

5.2   Graphs with no K_5 Minor

Here we have to consider the following three cases: (1) G is planar. (2) G is a copy of V_8. (3) G is a k-sum of two graphs where one of them is planar or a copy of V_8. If G is planar we treat it is as in the previous subsection, so now we treat the two remaining cases.

The graph V_8. For this graph we assume that we have a set of edge-weights w so that the cut condition (14) is satisfied, then we have to show how to find a flow. We use the same notation as in the previous subsection, in particular $\mathbb{C}$ denotes the set of cycles with an odd number of edges e with $w(e) < 0$. As before, we plan to obtain the flow from the linear program (21)–(23).

We need some definitions related to two-commodity flows. A *two-commodity flow* is a flow with two demands. A set of *two-commodity paths* is a set of paths in the capacity graph, each of these paths is either between the endnodes of the first demand, or between the endnodes of the second demand. A two-commodity cut is a minimal edge set of the capacity graph that intersects all two-commodity paths. A *maximum two-commodity flow* is a two-commodity flow that maximizes the total flow. Now we need the following three lemmas. Their proofs are omitted for space reasons.

Lemma 4. *There are two nodes that cover all cycles in* $\mathbb{C}$.

Lemma 5. *The cycles in* $\mathbb{C}$ *are a set of two-commodity paths.*

This construction implies that problem (21)–(23) (that is the dual of (18)–(20)) can be seen as a maximum two-commodity flow problem. Hu [16] showed that the value of a maximum two-commodity flow is equal to the value of a minimum two-commodity cut. Lemma 5 shows that the incidence vector of a two-commodity cut satisfies (19)–(20), thus the linear program (18)–(20) has an integer optimum.

Define $\bar{x}(e) = 1$ if $w(e) < 0$ and $\bar{x}(e) = 0$ otherwise. If $\bar{x}$ is an optimal solution of (18)–(20) then complementary slackness implies that an optimal solution of (21)–(23) gives a flow of value $\sum\{|w(e)| : w(e) < 0\}$, this is similar to what we did in Subsect. 5.1. The lemma below shows that $\bar{x}$ is optimal.

Lemma 6. *If* $\bar{x}$ *is not optimal, then there is a cut that violates condition* (14).

A k-sum. The remaining case is when G is a k-sum of G_1 and G_2, and G_2 is either a planar graph or a copy of V_8. We treat this below.

We have that G with the weights w has no negative cut, so the value obtained in Step 1 of Newton's algorithm is zero. We proceed as in Sect. 4 and compute the weights w' to be used in G_1. First we assume that $k = 3$. We need the following lemma.

Lemma 7. *We should have* $\alpha(\emptyset, \{e, f, g\}, G_2) = 0$.

Proof. The empty set is also considered as a cut, thus $\alpha(\emptyset, \{e, f, g\}, G_2) \leq 0$. Assume now that there is a cut $K = \delta(S)$ of G_2 that gives $\alpha(\emptyset, \{e, f, g\}, G_2) < 0$, and such that S contains the three nodes in $G_1 \cap G_2$. If we add to S all nodes in G_1 we obtain a cut of G with weight $\alpha(\emptyset, \{e, f, g\}, G_2) < 0$. □

Thus the procedure of Sect. 4 gives $\alpha(G) = \alpha(G_1) = 0$. Now for each edge $a \in \{e, f, g\} = G_1 \cap G_2$, we redefine its weight to $w''(a) = w(a) - w'(a)$ and we keep the original weights for the other edges in G_2. We need the lemma below.

Lemma 8. *The graph* G_2 *with the weights* w'' *has no negative cut.*

Proof. Let $K = \delta_{G_2}(S)$ be a cut of minimum weight with respect to w'', and assume that $w''(K) < 0$. Consider first the case when the nodes in $G_1 \cap G_2$ are in S. Then by adding to S all nodes in G_1, we would have a negative cut in G.

Now assume that $e, f \in K$. Then $w''(K) = w(K) - w'(e) - w'(f)$, from the definition of w''. If K is a cut of minimum weight with respect to w'', it is also of minimum weight with the constraint that e and f should be in the cut. Then $-w'(e) - w'(f)$ is a constant for the cuts containing e and f. Therefore K minimizes $w(C)$ among the cuts C that contain e and f. Thus $w(K) = \alpha(\{e, f\}, \emptyset, G_2)$, from the definition of α. On the other hand, from the definition of w' and Lemma 7 we have $w'(e) + w'(f) = \alpha(\{e, f\}, \emptyset, G_2)$. Then we have $w''(K) = \alpha(\{e, f\}, \emptyset, G_2) - \alpha(\{e, f\}, \emptyset, G_2) = 0$, a contradiction. □

Then we obtain a flow in G_2 with the weights w'' that can be combined with a flow in G_1 with the weights w', to produce a flow in G, as shown below.

Lemma 9. *Let f' be a flow in $G_1 = (V_1, E_1)$ with the weights w', and f'' a flow in $G_2 = (V_2, E_2)$ with the weights w'', then these two flows can be combined to produce a flow $\bar{f}$ in G with the weights w.*

Proof. Let $\{e, f, g\} = E_1 \cap E_2$. Let $\{e^1, f^1, g^1\}$ be the copies of $\{e, f, g\}$ in G_1, and $\{e^2, f^2, g^2\}$ be the copies of $\{e, f, g\}$ in G_2. Let $\bar{G}$ be the graph obtained from G_1 and G_2 by identifying the nodes in $V_1 \cap V_2$ and keeping the edges $\{e^1, f^1, g^1\}$ and $\{e^2, f^2, g^2\}$. We define $\bar{f}(C) = f'(C)$ if C is a cycle in G_1, $\bar{f}(C) = f''(C)$ if C is a cycle in G_2, and $\bar{f}(C) = 0$ for any other cycle $C \in \mathscr{C}$. Then $\bar{f}$ satisfies (13) and it is a flow in $\bar{G}$. Next we discuss how to replace the parallel edges e^1 and e^2 with e. We have four cases.

Case 1. If $w'(e^1) \geq 0$ and $w''(e^2) \geq 0$, then $w(e) = w'(e^1) + w''(e^2) \geq 0$. For each cycle C in G_1 with $\bar{f}(C) > 0$, containing e^1, we replace e^1 with e in C, and keep the same value for $\bar{f}(C)$. We proceed similarly with cycles in G_2 containing e^2. Finally we remove e^1 and e^2 from $\bar{G}$ and keep e. Then $\sum\{\bar{f}(C) \mid e \in C\} = \sum\{f'(C) \mid e^1 \in C\} + \sum\{f''(C) \mid e^2 \in C\} \leq w'(e^1) + w''(e^2) = w(e)$. Therefore this new flow satisfies (13).

Case 2. If $w'(e^1) < 0$ and $w''(e^2) < 0$, then $w(e) = w'(e^1) + w''(e^2) < 0$. We proceed as in the preceding case, then $\sum\{\bar{f}(C) \mid e \in C\} = \sum\{f'(C) \mid e^1 \in C\} + \sum\{f''(C) \mid e^2 \in C\} = |w'(e^1)| + |w''(e^2)| = |w(e)|$. So the new flow satisfies (13).

Case 3. If $w'(e^1) \geq 0$ and $w''(e^2) < 0$. Let $\{C_1, \ldots, C_k\}$ be the set of cycles in G_1 containing e^1 and with $\bar{f}(C_i) > 0$. Let $\{D_1, \ldots, D_l\}$ be the set of cycles in G_2 containing e^2 and with $\bar{f}(D_j) > 0$. Let $\mu = \sum_i \bar{f}(C_i)$, we have $\mu \leq w'(e^1)$, and $\sum_j \bar{f}(D_j) = |w''(e^2)|$.

Pick C_i with $\bar{f}(C_i) > 0$ and D_j with $\bar{f}(D_j) > 0$, and set $\epsilon = \min\{\bar{f}(C_i), \bar{f}(D_j)\}$. Let $C = (C_i \setminus \{e^1\}) \cup (D_j \setminus \{e^2\})$. If C_i or D_j contains exactly two nodes in $G_1 \cap G_2$, then C is a simple cycle and we set $\bar{f}(C) \leftarrow \bar{f}(C) + \epsilon$, $\bar{f}(C_i) \leftarrow \bar{f}(C_i) - \epsilon$, $\bar{f}(D_j) \leftarrow \bar{f}(D_j) - \epsilon$. If C_i and D_j contain the three nodes in $G_1 \cap G_2$, then C as defined above is not be a simple cycle, but the union of two simple cycles C' and C''. Only one of them, C' say, contains an edge with negative weight. Then we keep C', remove C'', and proceed as above.

Notice that after this either $\bar{f}(C_i) = 0$ or $\bar{f}(D_j) = 0$. Continue doing this until either (a) $\bar{f}(C_i) = 0$ for all i, or (b) $\bar{f}(D_j) = 0$ for all j. We treat these two sub-cases below.

Sub-case (a). All flow going through e^1 has been re-routed, and there are still $|w''(e^2)| - \mu$ units of flow associated with e^2. If $w(e) = w'(e^1) + w''(e^2) \geq 0$, we just set $\bar{f}(D_j) = 0$ for all remaining cycles D_j with $\bar{f}(D_j) > 0$. Then there is no flow going through the edge e. If $w(e) = w'(e^1) + w''(e^2) < 0$, since $|w''(e^2)| - \mu \geq |w(e)|$, we can apply the procedure below.

Re-route

Step 0. Set $T = |w(e)|$.

Step 1. Pick a cycle D_j with $\bar{f}(D_j) > 0$. Set $\epsilon = \min\{f(D_j), T\}$.

Step 2. Let $D = D_j \setminus \{e^2\} \cup \{e\}$. Set $\bar{f}(D) = \epsilon$. Set $T \leftarrow T - \epsilon$, $\bar{f}(D_j) \leftarrow \bar{f}(D_j) - \epsilon$. If $T > 0$ go to Step 1, otherwise stop.

After this procedure for each cycle D_j that still has $\bar{f}(D_j) > 0$, we set $\bar{f}(D_j) = 0$. Then we have

$$\sum \{\bar{f}(C) \,|\, e \in C\} = |w(e)|.$$

Sub-Case (b). All the demand associated with e^2 has been re-routed, and we have re-routed $|w''(e^2)|$ units of flow going through e^1. There are $\mu - |w''(e^2)|$ units going through e^1 remaining to re-route. Then for every remaining cycle C_i with $\bar{f}(C_i) > 0$, we set $C = C_i \setminus \{e^1\} \cup \{e\}$, and set $\bar{f}(C) = \bar{f}(C_i)$. Then
$$\mu - |w''(e^2)| = \sum \{\bar{f}(C) \,|\, e \in C\} \leq w'(e^1) - |w''(e^2)| = w(e).$$

Then the edges e^1 and e^2 are removed and the edge e is kept.

Case 4. If $w'(e^1) < 0$ and $w''(e^2) \geq 0$, this is treated as in Case 3.

Then the pairs of parallel edges $\{f^1, f^2\}$ and $\{g^1, g^2\}$ are treated in a similar way. $\square$

Then we continue working recursively in G_1. The cases $k = 2$ and $k = 1$ are similar.

Let $n_2 = |V_2|$. If G_2 is planar, then computing the weights w' requires treating G_2 three times, this takes $O(n_2^{3/2} \log n_2)$ time, and then the flow in G_2 with weights w'' can be obtained in $O(n_2^{3/2} \log n_2)$ time. If G_2 is a copy of V_8 then the flow can be obtained in constant time. This leads to the result below.

Theorem 10. *For a graph $H = (V, E_C \cup E_D)$ that has no K_5 minor, once the value of a sparsest cut has been computed, a maximum concurrent flow can be obtained in $O(n^{3/2} \log n)$ time.*

References

1. Arora, S., Lee, J., Naor, A.: Euclidean distortion and the sparsest cut. J. Am. Math. Soc. **21**, 1–21 (2008)
2. Arora, S., Rao, S., Vazirani, U.: Expander flows, geometric embeddings and graph partitioning. J. ACM (JACM) **56**, 5 (2009)
3. Barahona, F.: The max-cut problem on graphs not contractible to K_5. Oper. Res. Lett. **2**, 107–111 (1983)
4. Barahona, F.: Planar multicommodity flows, max-cut, and the Chinese postman problem. In: Cook, W., Seymour, P.D. (eds.) Polyhedral Combinatorics. DIMACS Series in Discrete Mathematics and Theoretical Computer Science, vol. 1, pp. 189–202. American Mathematical Society, Providence (1990)
5. Barahona, F.: On cuts and matchings in planar graphs. Math. Program. **60**, 53–68 (1993)
6. Bhatt, S.N., Leighton, F.T.: A framework for solving VLSI graph layout problems. J. Comput. Syst. Sci. **28**, 300–343 (1984)
7. Cornuejols, G., Naddef, D., Pulleyblank, W.: The traveling salesman problem in graphs with 3-edge cutsets. J. ACM (JACM) **32**, 383–410 (1985)

8. Dinkelbach, W.: On nonlinear fractional programming. Manage. Sci. **13**, 492–498 (1967)
9. Edmonds, J., Johnson, E.L.: Matching, euler tours and the Chinese postman. Math. Program. **5**, 88–124 (1973)
10. Ford, L.R., Fulkerson, D.R.: Maximal flow through a network. Can. J. Math. **8**, 399–404 (1956)
11. Gabow, H.N.: An efficient implementation of Edmonds' algorithm for maximum matching on graphs. J. ACM (JACM) **23**, 221–234 (1976)
12. Günlük, O.: A new min-cut max-flow ratio for multicommodity flows. SIAM J. Discrete Math. **21**, 1–15 (2007)
13. Gupta, A., Talwar, K., Witmer, D.: Sparsest cut on bounded treewidth graphs: algorithms and hardness results. In: Proceedings of the Forty-Fifth Annual ACM Symposium on Theory of Computing, pp. 281–290. ACM (2013)
14. Hadlock, F.: Finding a maximum cut of a planar graph in polynomial time. SIAM J. Comput. **4**, 221–225 (1975)
15. Hong, S.-P., Choi, B.-C.: Polynomiality of sparsest cuts with fixed number of sources. Oper. Res. Lett. **35**, 739–742 (2007)
16. Hu, T.C.: Multi-commodity network flows. Oper. Res. **11**, 344–360 (1963)
17. Klein, P., Plotkin, S.A., Rao, S.: Excluded minors, network decomposition, and multicommodity flow. In: Proceedings of the Twenty-Fifth Annual ACM Symposium on Theory of Computing, pp. 682–690. ACM (1993)
18. Klein, P., Rao, S., Agrawal, A., Ravi, R.: An approximate max-flow min-cut relation for undirected multicommodity flow, with applications. Combinatorica **15**, 187–202 (1995)
19. Lawler, E.L.: Combinatorial Optimization: Networks and Matroids. Courier Corporation, New York (1976)
20. Leighton, T., Rao, S.: Multicommodity max-flow min-cut theorems and their use in designing approximation algorithms. J. ACM (JACM) **46**, 787–832 (1999)
21. Lipton, R.J., Tarjan, R.E.: A separator theorem for planar graphs. SIAM J. Appl. Math. **36**, 177–189 (1979)
22. Matula, D.W., Shahrokhi, F.: Sparsest cuts and bottlenecks in graphs. Discrete Appl. Math. **27**, 113–123 (1990)
23. Park, J.K., Phillips, C.A.: Finding minimum-quotient cuts in planar graphs. In: Proceedings of the Twenty-Fifth Annual ACM Symposium on Theory of Computing, pp. 766–775. ACM (1993)
24. Patel, V.: Determining edge expansion and other connectivity measures of graphs of bounded genus. SIAM J. Comput. **42**, 1113–1131 (2013)
25. Plotkin, S., Tardos, É.: Improved bounds on the max-flow min-cut ratio for multicommodity flows. Combinatorica **15**, 425–434 (1995)
26. Radzik, T.: Fractional combinatorial optimization. In: Du, D.-Z., Pardalos, P.M. (eds.) Handbook of Combinatorial Optimization, pp. 429–478. Springer, USA (1999)
27. Rao, S.: Small distortion and volume preserving embeddings for planar and euclidean metrics. In: Proceedings of the Fifteenth Annual Symposium on Computational Geometry, pp. 300–306. ACM (1999)
28. Reed, B., Li, Z.: Optimization and recognition for K_5-minor free graphs in linear time. In: Laber, E.S., Bornstein, C., Nogueira, L.T., Faria, L. (eds.) LATIN 2008. LNCS, vol. 4957, pp. 206–215. Springer, Heidelberg (2008)
29. Seymour, P.: Matroids and multicommodity flows. Eur. J. Comb. **2**, 257–290 (1981)
30. Seymour, P.D.: On odd cuts and plane multicommodity flows. Proc. London Math. Soc. **3**, 178–192 (1981)

31. Shmoys, D.B.: Cut problems and their application to divide-and-conquer. In: Hochbaum, D.S. (ed.) Approximation Algorithms for NP-Hard Problems, pp. 192–235. PWS Publishing, Boston (1997)
32. Wagner, K.: Über eine eigenschaft der ebenen komplexe. Math. Ann. **114**, 570–590 (1937)
33. Wang, Q., Yang, X., Zhang, J.: A class of inverse dominant problems under weighted l_∞ norm and an improved complexity bound for Radzik's algorithm. J. Global Optim. **34**, 551–567 (2006)

Intersection Cuts for Bilevel Optimization

Matteo Fischetti[1(✉)], Ivana Ljubić[2], Michele Monaci[1], and Markus Sinnl[3]

[1] DEI, University of Padua, Padua, Italy
{matteo.fischetti,michele.monaci}@unipd.it
[2] ESSEC Business School of Paris, Cergy-Pontoise, France
ivana.ljubic@essec.edu
[3] ISOR, University of Vienna, Vienna, Austria
markus.sinnl@univie.ac.at

Abstract. The exact solution of bilevel optimization problems is a very challenging task that received more and more attention in recent years, as witnessed by the flourishing recent literature on this topic. In this paper we present ideas and algorithms to solve to proven optimality generic Mixed-Integer Bilevel Linear Programs (MIBLP's) where all constraints are linear, and some/all variables are required to take integer values. In doing so, we look for a general-purpose approach applicable to any MIBLP (under mild conditions), rather than ad-hoc methods for specific cases. Our approach concentrates on minimal additions required to convert an effective branch-and-cut MILP exact code into a valid MIBLP solver, thus inheriting the wide arsenal of MILP tools (cuts, branching rules, heuristics) available in modern solvers.

1 Introduction

A general bilevel optimization problem is defined as

$$\min_{x \in \mathbb{R}^{n_1}, y \in \mathbb{R}^{n_2}} F(x,y) \tag{1}$$

$$G(x,y) \leq 0 \tag{2}$$

$$y \in \arg \min_{y' \in \mathbb{R}^{n_2}} \{ f(x,y') : g(x,y') \leq 0 \}, \tag{3}$$

where $F, f : \mathbb{R}^{n_1+n_2} \to \mathbb{R}$, $G : \mathbb{R}^{n_1+n_2} \to \mathbb{R}^{m_1}$, and $g : \mathbb{R}^{n_1+n_2} \to \mathbb{R}^{m_2}$. Let $n = n_1 + n_2$ denote the total number of decision variables.

We will refer to $F(x,y)$ and $G(x,y) \leq 0$ as the *leader* objective function and constraints, respectively, and to (3) as the *follower* subproblem. In case the follower subproblem has multiple optimal solutions, we assume that one with minimum leader cost among those with $G(x,y) \leq 0$ is chosen—i.e. we consider the *optimistic* version of bilevel optimization.

By defining the follower value function for a given $x \in \mathbb{R}^{n_1}$

$$\Phi(x) = \min_{y \in \mathbb{R}^{n_2}} \{ f(x,y) : g(x,y) \leq 0 \}, \tag{4}$$

© Springer International Publishing Switzerland 2016
Q. Louveaux and M. Skutella (Eds.): IPCO 2016, LNCS 9682, pp. 77–88, 2016.
DOI: 10.1007/978-3-319-33461-5_7

one can restate the bilevel optimization problem as follows:

$$\min F(x, y) \tag{5}$$
$$G(x, y) \leq 0 \tag{6}$$
$$g(x, y) \leq 0 \tag{7}$$
$$(x, y) \in \mathbb{R}^n \tag{8}$$
$$f(x, y) \leq \Phi(x). \tag{9}$$

Note that the above optimization problem would be hard (both theoretically and in practice) even if one would assume convexity of F, G, f and g (which would imply that of Φ), due to the intrinsic nonconvexity of (9).

Dropping condition (9) leads the so-called *High Point Relaxation* (HPR). As customary in the bilevel context, we assume that HPR is feasible and bounded, and that the minimization problem in (4) is bounded for each feasible solution of HPR—while its feasibility follows directly from the definition of HPR. As HPR contains all the follower constraints, any HPR solution (x, y) satisfies $f(x, y) \geq \Phi(x)$, hence (9) actually implies $f(x, y) = \Phi(x, y)$.

A point $(x, y) \in \mathbb{R}^n$ will be called *bilevel infeasible* if it violates (9). A point $(x, y) \in \mathbb{R}^n$ will be called *bilevel feasible* if it is satisfies all constraints (6)–(9).

2   Literature Overview

In this paper we will mainly focus on Mixed-Integer Bilevel Linear Programs (MIBLP's) where some/all variables are required to be integer, and all HPR constraints (plus objective function) are linear.

The first generic branch-and-bound approach to the MIBLP's has been given in [7], where the authors propose to solve HPR embedded into a branch-and-bound scheme and basically enumerate bilevel feasible solutions. Recently, [4,5] proposed a sound branch-and-cut approach that builds upon the ideas from [7] and cuts off integer bilevel infeasible solutions, by adding cuts that exploit the integrality property of the leader and the follower variables. The authors provide an open-source MIBLP solver MibS [8]. More recently, [3] again propose to embed HPR into a branch-and-bound tree, bilevel infeasible solutions being cut off by adding a continuous follower subproblem into HPR, each time a new bilevel infeasible solution is detected. Continuous follower subproblems are then reformulated using KKT conditions and linearized in a standard way. Another generic approach for MIBLP's is a branch-and-sandwich method in [6], where the authors propose novel ideas for deriving lower and upper bounds of the follower's value function.

As this is usually the case with intersection cuts for MILPs, our IC's for MIBLP's also use disjunctive arguments. Disjunctive cuts in connection to bilevel linear programming have been investigated in [1], where the continuous follower subproblem is reformulated using KKT conditions, and disjunctive cuts are used to enforce complementary slackness conditions.

3   Bilevel-Free Sets

The following result is valid for generic bilevel problems and was implicit in some early references (including [9]) where it was only used as a guide for branching.

Lemma 1. *For any $\hat{y} \in \mathbb{R}^{n_2}$, the set*

$$S(\hat{y}) = \{(x,y) \in \mathbb{R}^n : f(x,y) \geq f(x,\hat{y}), \ g(x,\hat{y}) \leq 0\} \tag{10}$$

does not contain any bilevel feasible point in its interior.

Proof. It is enough to prove that no bilevel feasible (x,y) exists such that $f(x,y) > f(x,\hat{y})$ and $g(x,\hat{y}) < 0$. We will in fact prove a tighter result where the latter condition is replaced by $g(x,\hat{y}) \leq 0$, as this will be required in the proof of the next theorem. Indeed, for any bilevel feasible solution (x,y) with $g(x,\hat{y}) \leq 0$, one has $f(x,y) \leq \Phi(x) = \min_{y'}\{f(x,y') : g(x,y') \leq 0\} \leq f(x,\hat{y})$.

In some relevant settings, the above result can be strengthened to obtain the following enlarged bilevel-free set.

Theorem 1. *Assume that $g(x,y)$ is integer for all HPR solutions (x,y). Then, for any $\hat{y} \in \mathbb{R}^{n_2}$, the extended set*

$$S^+(\hat{y}) = \{(x,y) \in \mathbb{R}^n : f(x,y) \geq f(x,\hat{y}), \ g(x,\hat{y}) \leq 1\} \tag{11}$$

does not contain any bilevel feasible point in its interior, where 1 denotes a vector of all ones.

Proof. To be in the interior of $S^+(\hat{y})$, a bilevel feasible (x,y) should satisfy $f(x,y) > f(x,\hat{y})$ and $g(x,\hat{y}) < 1$. By assumption, the latter condition can be replaced by $g(x,\hat{y}) \leq 0$, hence the claim follows from the proof of previous lemma.

As far as we know, the above result is new. In spite of its simplicity, it will play a fundamental role in our solution method.

4   Mixed-Integer Bilevel Linear Programming

In the remaining part of the paper we will focus on the case where some/all variables are required to be integer, and all HPR constraints (plus objective function) are linear. This leads to the following Mixed-Integer Bilevel Linear Program (MIBLP):

$$\min F(x,y) \tag{12}$$
$$G(x,y) \leq 0 \tag{13}$$
$$g(x,y) \leq 0 \tag{14}$$
$$(x,y) \in \mathbb{R}^n \tag{15}$$
$$f(x,y) \leq \Phi(x) \tag{16}$$
$$x_j \text{ integer}, \quad \forall j \in J_1 \tag{17}$$
$$y_j \text{ integer}, \quad \forall j \in J_2, \tag{18}$$

where F, G, f, g are now assumed to be affine functions, sets $J_1 \subseteq \{1, \cdots, n_1\}$ and $J_2 \subseteq \{1, \cdots, n_2\}$ identify the (possibly empty) indices of the integer-constrained variables in x and y, respectively, and the value function reads

$$\Phi(x) = \min_{y \in \mathbb{R}^{n_2}} \{f(x,y) : g(x,y) \leq 0, \; y_j \in \mathbb{Z} \; \forall j \in J_2\}. \tag{19}$$

Dropping (16) leads to the HPR, which is a MILP in this setting. Dropping integrality conditions as well leads to the LP relaxation of HPR, namely (12)–(15), an LP which will be denoted by $\overline{\mathrm{HPR}}$.

Our main goal is to solve the above MIBLP by using a standard simplex-based branch-and-cut algorithm where the hard constraint (16) is enforced, on the fly, by adding cutting planes. The minimal requisite for the correctness of such an approach is the ability of cutting any *vertex*, say (x^*, y^*), of $\overline{\mathrm{HPR}}$ which satisfies the integrality requirements (17) and (18) but is bilevel infeasible because

$$f(x^*, y^*) > \Phi(x^*), \tag{20}$$

thus preventing a wrong update of the incumbent. To this end, we will propose a novel application of Balas' intersection cuts [2] in the MIBLP context.

5   A New Family of Cuts for MIBLP

Intersection cuts (IC's) for a given (x^*, y^*) require the definition of two sets: (1) a cone pointed at (x^*, y^*) that contains all the bilevel feasible solutions, and (2) a convex set S^* that contains (x^*, y^*) but no bilevel feasible solutions in its interior. The reader is referred to [2] for technical details.

As customary in mixed-integer programming, IC's are generated for vertices (x^*, y^*) of an LP relaxation of the problem to be solved, so a suitable cone is just the corner polyhedron associated with the corresponding optimal basis. All relevant information in this cone is readily available in the "optimal tableau" and requires no additional computational effort.

As to the convex set S^*, we propose to use the set defined in Lemma 1 (or, better, in Theorem 1 if applicable) by choosing

$$\hat{y} = \arg\min_{y} \{f(x^*, y) : g(x^*, y) \leq 0, \; y_j \in \mathbb{Z} \; \forall j \in J_2\} \tag{21}$$

(assuming this problem is not unbounded). Indeed, such a set S^* does not contain any bilevel feasible point in its interior, as required, while $(x^*, y^*) \in S^*$ because of (20) and $\Phi(x^*) = f(x^*, \hat{y})$ by definition. Note that $\hat{y}$ is well defined when (x^*, y^*) is a solution of HPR, and that S^* is a convex polyhedron in the MIBLP case.

However, an important property is still missing, namely, (x^*, y^*) must belong to the *interior* of S^* if we want to generate a violated intersection cut. This is always the case for MILBP's for which S^* is the *extended* set defined as in Theorem 1. This includes problems with *all-integer follower* where $J_2 = \{1, \cdots, n_2\}$, all g-coefficients are integer, and $j \in J_1$ for all x_j's appearing with nonzero coefficient in some follower constraint.

A relevant consequence of the above discussion is that, at least in the all-integer follower case, an exact branch-and-cut MIBLP solver can be obtained from a MILP solver by just adding a separation function for IC's based on the extended set $S^+(\hat{y})$ defined by (11) and (21). Indeed, observe that an exact MIBLP solver can be obtained by applying a general-purpose simplex-based MILP solver to HPR. To avoid the incumbent be updated with bilevel infeasible solutions, it is enough to cut any HPR solution (x^*, y^*) with $f(x^*, y^*) > \Phi(x^*)$. Without loss of generality, by disabling internal MILP heuristics, we can assume that (x^*, y^*) is a *vertex* of the current $\overline{\mathrm{HPR}}$ so we can always cut it by an (locally-valid) IC as, by definition, (x^*, y^*) is in the interior of the extended $S^+(\hat{y})$ when $\hat{y}$ is defined as in (21). In addition, assuming that all leader's variables x are integer and bounded, the number of HPR solutions to cut is finite, so a finite number of branching nodes (and hence of IC's) will be generated, i.e., the method converges in a finite number of iterations.

In the heuristic attempt of producing violated IC's for a generic vertex (x^*, y^*) of the $\overline{\mathrm{HPR}}$ polyhedron, one could also consider the following alternative definition of the point $\hat{y}$ that defines the bilevel-free set $S^+(\hat{y})$:

$$(\hat{y}, \hat{d}) = \arg\max_{y,d}\{d : f(x^*, y) + \varphi d \leq f(x^*, y^*),$$

$$g(x^*, y) + \gamma d \leq 1, \ y_j \in \mathbb{Z} \ \forall j \in J_2\}, \tag{22}$$

where $\varphi \in \mathbb{R}_+$ and $\gamma \in \mathbb{R}_+^{m_2}$ are suitable normalization factors, e.g., the Euclidean norm of the corresponding left-hand-side coefficient vectors. The rationale of this definition is that one wants to detect a bilevel-free set $S(\hat{y})$ whose closest face to (x^*, y^*) has a maximum distance from it.

Example. Figure 1 illustrates the application of IC's on an example given in [7], which is frequently used in the literature:

$$\min_{x \in \mathbb{Z}} -x - 10y \tag{23}$$

$$y \in \arg\min_{y' \in \mathbb{Z}}\{ \ y' : \tag{24}$$

$$-25x + 20y' \leq 30 \tag{25}$$

$$x + 2y' \leq 10 \tag{26}$$

$$2x - y' \leq 15 \tag{27}$$

$$2x + 10y' \geq 15 \ \}. \tag{28}$$

In this all-integer example, there are 8 bilevel feasible points (depicted as crossed squares in Fig. 1), and the optimal bilevel solution is $(2, 2)$. The drawn polytope corresponds to the $\overline{\mathrm{HPR}}$ feasible set.

We first apply the definition of the bilevel-free set from Lemma 1 with $\hat{y}$ defined as in (21). After solving the first $\overline{\mathrm{HPR}}$, the point $A = (2, 4)$ is found. This point is bilevel infeasible, as for $x^* = 2$ we have $f(x^*, y^*) = y^* = 4$ while

$\Phi(x^*) = 2$. From (21) we compute $\hat{y} = 2$ and the intersection cut derived from the associated $S(\hat{y})$ is depicted in Fig. 1(a). In the next iteration, the optimal $\overline{\text{HPR}}$ solution moves to $B = (6, 2)$. Again, for $x^* = 6$, $f(x^*, y^*) = y^* = 2$ while $\Phi(x^*) = 1$. So we compute $\hat{y} = 1$ and generate the IC induced by the associated $S(\hat{y})$, namely $2x + 11y \leq 27$ (cf. Fig. 1(b)). In the next iteration, the fractional point $C = (5/2, 2)$ is found and $\hat{y} = 1$ is again computed. In this case, C is not in the interior of $S(\hat{y})$ so we cannot generate an IC cut from C but we should proceed and optimize HPR to integrality by using standard MILP tools such as MILP cuts or branching. This produces the optimal HPR solution $(2, 2)$ which is bilevel feasible and hence optimal.

We next apply the definition of the enlarged bilevel-free set from Theorem 1 (whose assumption is fulfilled) with $\hat{y}$ defined again as in (21); see Fig. 1(c) and (d). After the first iteration, the point $A = (2, 4)$ is cut off by a slightly larger $S^+(\hat{y} = 2)$, but with the same IC as before ($y \leq 2$). After the second iteration, from the bilevel infeasible point $B = (6, 2)$ we derive a larger set $S^+(\hat{y} = 1)$ and a stronger IC ($x + 6y \leq 14$). In the third iteration, solution $D = (2, 2)$ is found which is the optimal bilevel solution, so no branching at all is required in this example.

6   Informed No-Good Cuts

A known drawback of IC's is their dependency on the LP basis associated with the point to cut, which can create cut accumulation in the LP relaxation and hence shallow cuts and numerical issues. Moreover, IC's are not directly applicable if the point to cut is not a vertex of a certain LP relaxation of the problem at hand, as it happens e.g. when it is computed by the internal MILP heuristics.

We next describe a general-purpose variant of IC's whose derivation does not require any LP basis and is based on the well-known interpretation of IC's as disjunctive cuts. It turns out that the resulting inequality is valid and violated by any bilevel infeasible solution of HPR in the relevant special case where all x and y variables are binary.

Suppose we are given a point $\xi^* = (x^*, y^*) \in \mathbb{R}^n$ and a polyhedron $S^* = \{\xi \in \mathbb{R}^n : \alpha_i^T \xi \leq \alpha_{i0}, i = 1, \cdots, k\}$ whose interior contains ξ^* but no feasible points. Assume that variable-bound constraints $l \leq \xi \leq u$ are present, where some entries of l or u can be $-\infty$ or $+\infty$, respectively. Given ξ^*, define $L := \{j : \xi_j^* - l_j \leq u_j - \xi_j^*\}$ and $U := \{1, \cdots, n\} \setminus L$ and the corresponding linear mapping $\xi \mapsto \overline{\xi} \in \mathbb{R}^n$ with $\overline{\xi}_j := \xi_j - l_j$ for $j \in L$, and $\overline{\xi}_j := u_j - \xi_j$ for $j \in U$ (variable shift and complement).

By assumption, any feasible point ξ must satisfy the disjunction

$$\bigvee_{i=1}^{k} \{\xi \in \mathbb{R}^n : \sum_{j=1}^{n} \alpha_{ij} \xi_j \geq \alpha_{i0}\}, \tag{29}$$

whereas ξ^* violates all the above inequalities. Now, each term of (29) can be rewritten in terms of $\overline{\xi}$ as

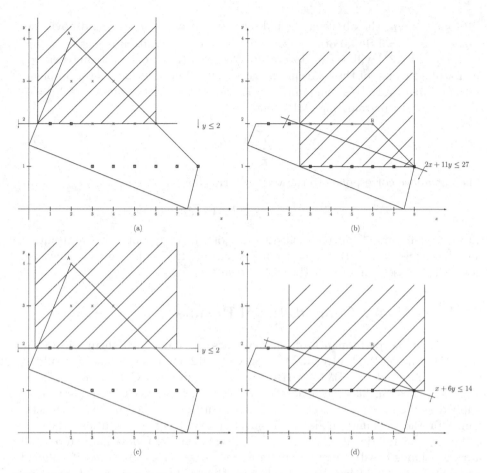

Fig. 1. Illustration of the effect of alternative intersection cuts for a notorious example from [7]. Shaded regions correspond to the bilevel-free sets for which the cut is derived.

$$\sum_{j=1}^{n} \overline{\alpha}_{ij}\,\overline{\xi}_j \geq \overline{\beta}_i := \alpha_{i0} - \sum_{j \in L} \alpha_{ij} l_j - \sum_{j \in U} \alpha_{ij} u_j, \qquad (30)$$

with $\overline{\alpha}_{ij} := \alpha_{ij}$ if $j \in L$, $\overline{\alpha}_{ij} = -\alpha_{ij}$ otherwise. If $\overline{\beta}_i > 0$ for all $i = 1, \cdots, k$, one can normalize the above inequalities to get $\sum_{j=1}^{n} (\overline{\alpha}_{ij}/\overline{\beta}_i)\,\overline{\xi}_j \geq 1$ and derive the valid disjunctive cut in the $\overline{\xi}$ space

$$\sum_{j=1}^{n} \overline{\gamma}_j \overline{\xi}_j \geq 1, \qquad (31)$$

where $\overline{\gamma}_j := \max\{\overline{\alpha}_{ij}/\overline{\beta}_i : i = 1, \cdots, k\}$, and then one can transform it back to the ξ space in the obvious way. It is easy to see that, in case $\xi_j^* \in \{l_j, u_j\}$ for all $j = 1, \cdots, n$, the above cut is indeed valid (because $\overline{\beta} > 0$) and obviously violated as $\overline{\xi}^* = 0$. In all other cases, the above cut separation is just heuristic.

Inequality (31) will be called *Informed No-Good* (ING) cut as it can be viewed as a strengthening of the following no-good cut often used for bilevel problems with all-binary variables—and in many other Constraint Programming (CP) and Mathematical Programming (MP) contexts:

$$\sum_{j \in L} \xi_j + \sum_{j \in U}(1 - \xi_j) \geq 1. \tag{32}$$

The cut above corresponds to the very generic choice

$$S^* = \{\xi \in \mathbb{R}^n : \xi_j \leq 1 \,\forall j \in L, \ 1 - \xi_j \leq 1 \,\forall j \in U\}$$

and is violated by ξ^* but is satisfied by any other binary point, hence resulting into a very weak cut. To the best our knowledge, ING cuts are new; they will hopefully be useful in other CP and MP contexts.

7   Preliminary Computational Results

To evaluate the performance of our new cuts, we embedded them within the general-purpose MILP solver IBM ILOG Cplex 12.6.2 using callbacks, resulting into a branch-and-cut (B&C) MIBLP approach. Internal Cplex's heuristics as well preprocessing have been deactivated in all experiments. IC separation is applied at the root node on all LP solutions (in the so-called usercut callback), while for the remaining nodes it is only applied to integer solutions (lazycut callback). For fractional solutions, IC's whose normalized violation is too small are just skipped. All generated cuts are treated as local cuts (even if no-good and ING cuts would be globally valid) as this reduces the node LP size and significantly improves node throughput. To improve the quality of IC cuts, the bilevel-free set is enlarged by removing all its defining inequalities $\alpha^T(x, y) \leq \alpha_0$ (say) such that imposing the reverse condition $\alpha^T(x, y) \geq \alpha_0$ would trivially lead to an infeasible HPR relaxation due to the current bounds on the x and y variables (this step turns out to be very important for the success of our method). More implementation details will be given in the full paper.

We first compared our code with the one in [3] on the testbed proposed therein. All such instances turned out to be very easy, both for our approach and for MibS. More precisely, each instance could be solved in less than a second by our code and in at most 3 s by MibS, i.e., both codes were 2–3 orders of magnitude faster than the one in [3]. Therefore we addressed more difficult instances, obtained according to the following procedure.

We took a familiar testbed (MILPLIB 3.0) that contains instances that are easily solvable by modern MILP solvers (except instance seymour which is very hard even as a MILP). As we planned to also run the open-source MIBLP solver MibS [8] to check our code, we skipped all instances involving equations or continuous variables, as well as those involving noninteger coefficients—all the above cases being not supported by the current release of MibS. This produced a set of 10 basic 0–1 MILP instances, that we converted into bilevel problems by labeling the first $Y\%$ (rounded up) variables as y's, and the remaining ones as x's. In our test, we considered the three cases with $Y \in \{10, 50, 90\}$ leading to instances named name-0.1.mps, name-0.5.mps, and name-0.9.mps, respectively. All constraints in the resulting model belong to the follower subproblem, as MibS cannot handle leader-specific constraints $G(x, y) \le 0$, while the objective function is used as the leader's objective $F(x, y)$. Finally, the follower's objective is defined as $f(x, y) = -F(x, y)$.

In Table 1, we use MibS to assess the computational difficulty of the instances we generated. The table also reports results for our basic B&C code (with IC's but not ING cuts) when run in single-thread mode and with internal Cplex cuts disabled. Note that the two solvers cannot be compared directly, as they are based on a different underlying MILP code, namely: Cplex for our code, and COIN-OR (BLIS) plus Cplex for MibS. For both codes, we report in Table 1 the following values: the best obtained upper bound (UB), the best obtained lower bound (LB), the final percentage gap (%gap) calculated as (UB - LB) / UB × 100. Computing times (t.[s]) are wall-clock seconds on an Intel Xeon E5-2670v2 @ 2.5 Ghz computer with 12 GB ram. The timelimit was set to 600 s as larger values produced memory issues for some instances where the number of tree nodes is very large. If the time-limit was reached, this is notified as "TL" in the time column. These results clearly indicate that we managed to generate a testbed which is sufficiently challenging for state-of-the-art MIBLP solvers.

Table 2 compares four settings for our code: (1) only no-good cuts are generated, (2) only ING cuts are generated, (3) only IC's are generated, and (4) IC's are generated for fractional solutions at root node, while only ING cuts generated for integer ones. Note that all settings lead to an exact method as all instances in our testbed are pure binary. All versions were run in 4-thread opportunistic mode, without disabling internal Cplex cuts, on a Intel Xeon E3-1220V2 quadcore PC @ 3.10 GHz with 16 GB of RAM. Setting (1) is intended to assess the difficulty of the created data set for a method built on top of Cplex, but using the most basic MIBLP cuts (no-good). Setting (2) is intended to measure the performance improvement obtained by replacing generic no-good cuts with bilevel-specific ING cuts, while the impact of IC's is addressed in setting (3). Finally, setting (4) combines IC's and ING cuts to limit the negative effect of cut accumulation in the LP basis.

For each of the four setting and for each instance, in Table 2 we report the same information as in Table 1, plus the overall number of branch-and-bound nodes (#nodes).

The influence of IC's to the performance of the B&C can be measured by comparing the quality of lower bounds of the setting (3), with the settings (1) and (2). In 14, respectively 11 cases, the LBs obtained by IC's are strictly stronger than those obtained by pure no-good and ING cuts, respectively. The quality of lower bounds when IC's are combined with ING cuts remains roughly the same across all instances. As expected, the setting (1) exhibits the worst performance with 22 instances remaining unsolved within the given time-limit. ING cuts perform better (in particular considering the quality of lower bounds), but still

Table 1. Instance difficulty when using two different MIBLP solvers

name	Mibs				B&C with IC's			
	UB	LB	%gap	t.[s]	UB	LB	%gap	t.[s]
fast0507-0.1	–	173	100.00	TL	12553	173	98.62	TL
fast0507-0.5	–	173	100.00	TL	61503	174	99.72	TL
fast0507-0.9	–	173	100.00	TL	109916	109916	0.00	7
lseu-0.1	1120	1120	0.00	4	1120	1120	0.00	2
lseu-0.5	2400	1205	49.79	TL	2263	1235	45.43	TL
lseu-0.9	5838	1171	79.94	TL	5838	1275	78.75	TL
p0033-0.1	3089	3089	0.00	0	3089	3089	0.00	0
p0033-0.5	3095	3095	0.00	0	3095	3095	0.00	0
p0033-0.9	4679	4679	0.00	90	4679	4679	0.00	3
p0201-0.1	12615	7859	37.70	TL	12465	7931	36.37	TL
p0201-0.5	14220	7832	44.92	TL	13910	7925	43.03	TL
p0201-0.9	15025	7809	48.03	TL	15025	7925	47.25	TL
p0282-0.1	261188	258435	1.05	TL	260781	260067	0.27	TL
p0282-0.5	276338	258432	6.48	TL	272659	259331	4.89	TL
p0282-0.9	724572	258427	64.33	TL	636846	284519	55.32	TL
p0548-0.1	–	317	100.00	TL	10982	8691	20.86	TL
p0548-0.5	–	317	100.00	TL	22450	8620	61.60	TL
p0548-0.9	–	317	100.00	TL	48959	8694	82.24	TL
p2756-0.1	–	2691	100.00	TL	12765	2734	78.58	TL
p2756-0.5	–	2691	100.00	TL	23976	2723	88.64	TL
p2756-0.9	–	2691	100.00	TL	35867	2733	92.38	TL
seymour-0.1	–	407	100.00	TL	480	407	15.21	TL
seymour-0.5	–	407	100.00	TL	823	408	50.43	TL
seymour-0.9	–	407	100.00	TL	1251	1251	0.00	2
stein27-0.1	18	18	0.00	0	18	18	0.00	1
stein27-0.5	19	19	0.00	7	19	19	0.00	3
stein27-0.9	24	20	16.67	TL	24	24	0.00	0
stein45-0.1	30	30	0.00	103	30	30	0.00	32
stein45-0.5	33	31	6.06	TL	32	32	0.00	205
stein45-0.9	40	31	22.50	TL	40	40	0.00	0

with 20 instances remaining unsolved. Both settings with IC's and IC's with ING cuts manage to solve 12 instances to optimality. The number of enumerated branch-and-bound nodes varies strongly between the instances, even between those being derived from the same MIPLIB source. This indicates that, despite the fact that some instances are derived from the identical HPR formulation, the difficulty is mainly determined by the structure of the follower subproblem.

Table 2. Comparison of different settings of our B&C approach.

name	No-good cuts only					ING cuts only					IC's only					IC's and ING cuts				
	UB	LB	%gap	t.[s]	#nodes	UB	LB	%gap	t.[s]	#nodes	UB	LB	%gap	t.[s]	#nodes	UB	LB	%gap	t.[s]	#nodes
fast0507-0.1	12547	173	98.62	TL	2766	12548	173	98.62	TL	11k	12550	173	98.62	TL	4451	12552	173	98.62	TL	5371
fast0507-0.5	61485	173	99.72	TL	2699	61485	173	99.72	TL	5215	-	5440	100.00	TL	33k	-	5440	100.00	TL	33k
fast0507-0.9	109928	173	99.84	TL	2697	109928	173	99.84	TL	864	109916	109916	0.00	4	2	109916	109916	0.00	4	2
lseu-0.1	1120	1120	0.00	0	38	1120	1120	0.00	0	40	1120	1120	0.00	0	39	1120	1120	0.00	0	40
lseu-0.5	2314	1219	47.32	TL	141k	2263	1324	41.49	TL	1M	2263	1318	41.76	TL	2M	2274	1323	41.82	TL	1M
lseu-0.9	5838	1213	79.22	TL	128k	5838	1355	76.79	TL	2M	5838	1384	76.29	TL	2M	5838	1385	76.28	TL	2M
p0033-0.1	3089	3089	0.00	0	2	3089	3089	0.00	0	2	3089	3089	0.00	0	2	3089	3089	0.00	0	2
p0033-0.5	3095	3095	0.00	0	42	3095	3095	0.00	0	45	3095	3095	0.00	0	41	3095	3095	0.00	0	43
p0033-0.9	4679	4679	0.00	9	11k	4679	4679	0.00	1	4646	4679	4679	0.00	1	4071	4679	4679	0.00	1	3355
p0201-0.1	12610	7802	38.13	TL	126k	12495	7915	36.65	TL	794k	12345	7945	35.64	TL	944k	12345	7922	35.83	TL	738k
p0201-0.5	13925	7803	43.96	TL	117k	13910	7932	42.98	TL	922k	13920	7944	42.93	TL	1M	13850	7945	42.64	TL	965k
p0201-0.9	15025	7804	48.06	TL	115k	15025	7925	47.25	TL	718k	15025	7933	47.20	TL	722k	15025	7927	47.24	TL	716k
p0282-0.1	260781	258431	0.90	TL	102k	260781	258448	0.89	TL	2M	260781	258449	0.89	TL	3M	260781	258448	0.89	TL	2M
p0282-0.5	274422	258432	5.83	TL	120k	274422	258447	5.82	TL	2M	274422	258448	5.82	TL	3M	274422	258447	5.82	TL	2M
p0282-0.9	685640	258432	62.31	TL	124k	638243	258446	59.51	TL	2M	639964	271734	57.54	TL	15M	644113	271734	57.81	TL	15M
p0548-0.1	11100	8691	21.70	TL	123k	11100	8691	21.70	TL	365k	11348	8691	23.41	TL	1M	11348	8691	23.41	TL	494k
p0548-0.5	22083	8691	60.64	TL	64k	22078	8691	60.64	TL	76k	22083	8691	60.64	TL	74k	22083	8691	60.64	TL	70k
p0548-0.9	50162	8691	82.67	TL	103k	50162	8691	82.67	TL	220k	50253	9147	81.80	TL	42k	50253	9147	81.80	TL	44k
p2756-0.1	14540	3124	78.51	TL	20k	14430	3124	78.35	TL	38k	13936	3124	77.58	TL	65k	14500	3124	78.46	TL	32k
p2756-0.5	25654	3124	87.82	TL	19k	25654	3124	87.82	TL	44k	23931	3124	86.95	TL	66k	24181	3124	87.08	TL	49k
p2756-0.9	36449	3124	91.43	TL	17k	35242	3124	91.14	TL	182k	34092	3124	90.84	TL	95k	34703	3124	91.00	TL	175k
seymour-0.1	477	415	13.00	TL	29k	477	415	13.00	TL	25k	477	415	13.00	TL	27k	478	415	13.18	TL	25k
seymour-0.5	823	415	49.57	TL	31k	816	415	49.14	TL	37k	823	415	49.57	TL	34k	814	415	49.02	TL	39k
seymour-0.9	1252	415	66.85	TL	31k	1252	415	66.85	TL	23k	1251	1251	0.00	5	2	1251	1251	0.00	5	2
stein27-0.1	18	18	0.00	0	1202	18	18	0.00	0	1244	18	18	0.00	0	1209	18	18	0.00	0	1247
stein27-0.5	19	19	0.00	1	7377	19	19	0.00	1	7060	19	19	0.00	1	6465	19	19	0.00	1	7001
stein27-0.9	24	19	20.83	TL	110k	24	24	0.00	2	13k	24	24	0.00	0	2	24	24	0.00	0	2
stein45-0.1	30	30	0.00	4	14k	30	30	0.00	4	14k	30	30	0.00	5	13k	30	30	0.00	5	14k
stein45-0.5	32	32	0.00	176	211k	32	32	0.00	31	133k	32	32	0.00	37	161k	32	32	0.00	47	202k
stein45-0.9	40	30	25.00	TL	158k	40	40	0.00	234	1M	40	40	0.00	0	2	40	40	0.00	0	2

Acknowledgment. This research was funded by the Vienna Science and Technology Fund (WWTF) through project ICT15-014. The work of M. Fischetti and M. Monaci was also supported by the University of Padova (Progetto di Ateneo "Exploiting randomness in Mixed Integer Linear Programming"), and by MiUR, Italy (PRIN project "Mixed-Integer Nonlinear Optimization: Approaches and Applications"). The work of I. Ljubić and M. Sinnl was also supported by the Austrian Research Fund (FWF, Project P 26755-N19). The authors thank Ted Ralphs for his technical support and instructions regarding MibS, and Massimiliano Caramia for providing the instances used in [3].

References

1. Audet, C., Haddad, J., Savard, G.: Disjunctive cuts for continuous linear bilevel programming. Optimization Letters **1**(3), 259–267 (2007)
2. Balas, E.: Intersection cuts-a new type of cutting planes for integer programming. Oper. Res. **19**(1), 19–39 (1971)
3. Caramia, M., Mari, R.: Enhanced exact algorithms for discrete bilevel linear problems. Optimization Letters **9**(7), 1447–1468 (2015)
4. DeNegre, S.: Interdiction and Discrete Bilevel Linear Programming. Ph.D. thesis, Lehigh University (2011)
5. DeNegre, S., Ralphs, T.K.: A branch-and-cut algorithm for integer bilevel linear programs. In: Chinneck, J.W., Kristjansson, B., Saltzman, M.J. (eds.) Operations Research and Cyber-Infrastructure, vol. 47, pp. 65–78. Springer, New York (2009)
6. Kleniati, P.-M., Adjiman, C.S.: A generalization of the branch-and-sandwich algorithm: from continuous to mixed-integer nonlinear bilevel problems. Comput. Chem. Eng. **72**, 373–386 (2015)
7. Moore, J., Bard, J.: The mixed integer linear bilevel programming problem. Oper. Res. **38**(5), 911–921 (1990)
8. Ralphs, T.K.: MibS. https://github.com/tkralphs/MibS
9. Xu, P., Wang, L.: An exact algorithm for the bilevel mixed integer linear programming problem under three simplifying assumptions. Comput. Oper. Res. **41**, 309–318 (2014)

Exact Algorithms for the Chance-Constrained Vehicle Routing Problem

Thai Dinh[1], Ricardo Fukasawa[2]($\boxtimes$), and James Luedtke[1]($\boxtimes$)

[1] Department of Industrial and Systems Engineering,
University of Wisconsin-Madison, Madison, WI 53706, USA
{tndinh,jim.luedtke}@wisc.edu
[2] Department of Combinatorics and Optimization, Faculty of Mathematics,
University of Waterloo, Waterloo, ON N2L 3G1, Canada
rfukasawa@uwaterloo.ca

Abstract. We study the chance-constrained vehicle routing problem (CCVRP), a version of the vehicle routing problem (VRP) with stochastic demands, where a limit is imposed on the probability that each vehicle's capacity is exceeded. A distinguishing feature of our proposed methodologies is that they allow correlation between random demands, whereas nearly all existing methods for the stochastic VRP require independent demands. We first study an edge-based formulation for the CCVRP, in particular addressing the challenge of how to determine a lower bound on the number of trucks required to serve a subset of customers. We then investigate the use of a branch-and-cut-and-price (BCP) algorithm. While BCP algorithms have been considered the state of the art in solving the deterministic VRP, few attempts have been made to extend this framework to the stochastic VRP.

1 Introduction

The deterministic vehicle routing problem (VRP) [8] is the problem of finding routes for a fleet of identical, fixed capacity vehicles that collect known amounts of goods from customers. When demands of customers are random variables, the problem is referred to as the vehicle routing problem with stochastic demands (VRPSD). In an optimization model for the VRPSD, one must determine how to model the possibility that the demands on a planned route might exceed the capacity of a truck. One approach, taken, e.g., in [3,9,18,20,21], is to consider a recourse model, in which a recourse action must be taken if a truck's capacity is exceeded. This leads to a two-stage stochastic programming formulation, in which routes are determined in advance of knowing the random demands (first-stage decisions), and then, when the routes are implemented and demands are observed, recourse actions (second-stage decisions) are taken if a vehicle's capacity is exceeded. The objective is to minimize the expected travel cost, including travel taken in the recourse stage. In order to make the evaluation of the expected recourse costs tractable, restrictive assumptions are usually placed on the form of the recourse taken (e.g., that it consists of a trip to/from the depot) and on

© Springer International Publishing Switzerland 2016
Q. Louveaux and M. Skutella (Eds.): IPCO 2016, LNCS 9682, pp. 89–101, 2016.
DOI: 10.1007/978-3-319-33461-5_8

the random demands. In particular, nearly all existing work assumes the random demands are independent of each other.

We study an alternative approach, the chance-constrained VRP (CCVRP), which does not explicitly model the recourse actions to be taken when a truck's capacity is exceeded, and instead requires that such an event happens with low probability. This type of model leads to operational benefits like more consistent service and less need for complex recourse actions to be taken. Indeed, the CCVRP model is not dependent on the particular choice of recourse actions which can be either oversimplified or too complex to be computationally tractable. The first attempt to solve the CCVRP was proposed in [20], where they derive conditions in which it can be reduced to a deterministic VRP. These conditions are restrictive, as they require customer demands to be independent and have identical coefficients of variation. In [13], the first exact solution technique for CCVRP was proposed using a branch-and-cut framework, but their implementation requires random demands to be independent and normally distributed. Though other works exist on the CCVRP, to the best of our knowledge, these methods, which require restrictive assumptions on the distributions of customer demands, are the only proposed exact methods for the problem.

In addition to the shortage of exact algorithms for the CCVRP, the methodology applied to VRPSD problems also has some deficiencies. The current best known algorithms for solving the deterministic VRP are based on a Dantzig-Wolfe reformulation, strengthened by valid inequalities, solved using a branch-and-cut-and-price (BCP) algorithm [1,2,7,10,17]. On the other hand, very little work has attempted to apply the BCP framework to solve either a recourse-based or chance-constrained VRPSD model. Most of the exact solution techniques for the recourse-based VRPSD models rely on variants of the integer L-Shaped method proposed by [15] or the branch-and-price algorithm proposed by [4], while most of the attempts to solve the CCVRP rely on variants of the branch-and-cut algorithms proposed in [13,14]. To the best of our knowledge, the only work that considers solving a VRPSD model using a BCP framework has been proposed by [11], but is again restricted to the assumption that random demands are independent.

Main Contributions. We present exact solution methods for the CCVRP which do not require the customer demands to be independent, and are able to solve to optimality, or near-optimality, instances with more than 50 customers. The only assumption we require on the customer demands is that we can compute a quantile of the random variable defined by the sum of customer demands in any subset of customers. This assumption holds for customer demands having joint normal distribution and for a scenario model of customer demands. Using sample average approximation, the scenario model can be used to approximate a problem in which customer demands follow any distribution from which samples can be taken [16]. First, we derive strong and computationally tractable bounds on the number of trucks required to serve a subset of customers and remain chance constraint feasible, leading to improved capacity inequalities. This allows us to extend the formulation of [13] to more general cases. Second, we find that a direct extension of the pricing routine used in BCP for the deterministic VRP is challenging since the associated pricing problem is strongly $\mathcal{NP}$-hard (as opposed to pseudopolynomially solvable in the deterministic case). We thus propose a relaxed pricing scheme to overcome this challenge. We also empirically compare the solutions obtained with the CCVRP model to those obtained with a recourse-based model of the VRPSD, and find that the CCVRP solutions provide high quality solutions to the recourse-based model, whereas the reverse is often not true.

2 Problem Definition and an Edge-Based Formulation

Let $G = (V, E)$ be an undirected graph with vertices $V = \{0, 1, ..., n\}$. Vertex 0 represents the depot and the vertices $V_+ = \{1, ..., n\}$ represent the customers. Each customer $i \in V_+$ has a random demand D_i. The set of demands D is a random vector defined in a probability space $(\Omega, \mathcal{F}, \mathbb{P})$. The expected value and variance of demand for customer $i \in V_+$ are denoted by d_i and σ_i^2, respectively. The length of edge $e \in E$ is denoted by $\ell_e \geq 0$. There are K available vehicles and each vehicle has a capacity of b. A route is a simple cycle C going through 0 (or an edge $0v$ twice, representing the route $0 - v - 0$ for $v \in V_+$). We say a route serves S if $V(C) \setminus \{0\} = S$. A *chance-constraint feasible route* is a route for which the set of customers $S \subseteq V_+$ that it serves satisfies $\mathbb{P}\{D(S) \leq b\} \geq 1 - \epsilon$, where $\epsilon \in (0, 1)$ is a given, typically small, parameter. In this expression, the following standard notation is used (and will be used throughout the paper): given values w_t for a ground set T, define $w(Q) := \sum_{t \in Q} w_t, \forall Q \subseteq T$. The objective is to find a minimum length set of K chance constraint feasible routes such that every customer is visited exactly once.

2.1 Edge-Based Formulation

Let x_e represent the number of times edge e is used in a solution. For a subset of customers $S \subseteq V_+$, we let $\delta(S)$ be the cut-set defined by S and let $r_\epsilon(S)$ be the minimum number of vehicles needed to serve S with chance-constraint feasible routes. We call $r_\epsilon(S)$ the *minimum vehicle requirements*. The edge-based formulation is then [13]:

$$\min_x \quad \sum_{e \in E} \ell_e x_e \tag{1a}$$

$$\text{s.t.} \quad \sum_{e \in \delta(\{i\})} x_e = 2 \qquad i \in V_+ \tag{1b}$$

$$\sum_{e \in \delta(\{0\})} x_e = 2K \tag{1c}$$

$$\sum_{e \in \delta(S)} x_e \geq 2r_\epsilon(S) \qquad S \subseteq V_+ \tag{1d}$$

$$x_e \leq 1 \qquad e \in E \setminus \delta(\{0\}) \tag{1e}$$

$$x_e \in \mathbb{Z}_+ \qquad e \in E. \tag{1f}$$

Constraints (1b) require that each customer is visited exactly once by some vehicle, whereas (1c) states that K vehicles must leave and enter at the depot. Constraints (1d) are the capacity inequalities, which enforce that enough vehicles are assigned to any subset of customers.

A similar model has been used for the deterministic VRP with customer demands $d_i, i \in V_+$ where $r_\epsilon(S)$ in (1d) is replaced with the minimum number of trucks $r_0(S)$ required to serve the customers in the set S. Calculating this quantity exactly requires solving the strongly $\mathcal{NP}$-hard bin-packing problem. Fortunately, for the deterministic VRP, the easily computed lower bound $k(S) := \lceil d(S)/b \rceil$ yields a valid formulation, and the resulting cuts have been shown empirically to be effective. A key challenge for the CCVRP is to determine how to compute a lower bound for $r_\epsilon(S)$ that is at least sufficient to provide a valid formulation, and that is as close to $r_\epsilon(S)$ as possible in order to yield strong inequalities. When the formulation (1) was studied in [13], they proposed to use the value

$$k_\epsilon^I(S) = \left\lceil \left(d(S) + \Phi^{-1}(1 - \epsilon)\sqrt{\sigma^2(S)} \right)/b \right\rceil, \tag{2}$$

as an approximation of $r_\epsilon(S)$ which is a valid lower bound when demands are independent normal, but is not necessarily valid in other cases. In (2), Φ^{-1} is the inverse of the cumulative density function of the standard normal distribution.

2.2   Vehicle Requirements in the Capacity Inequalities

We now discuss how to obtain $k_\epsilon(S) \leq r_\epsilon(S)$, such that formulation (1) is still valid for the CCVRP if we replace $r_\epsilon(S)$ with $k_\epsilon(S)$. We refer to such lower bounds on the minimum vehicle requirements as *valid* lower bounds. We begin with a simple valid lower bound, $k_\epsilon(S)$, defined as:

$$k_\epsilon(S) = \begin{cases} 1, & \text{if } \mathbb{P}\{D(S) \leq b\} \geq 1 - \epsilon \\ 2, & \text{otherwise.} \end{cases}$$

which just states that at least two trucks are needed to serve the set of customers S if the probability that the sum of customer demands in the set S exceeding a single truck's capacity is too high.

Proposition 1. $k_\epsilon(S)$ *is a valid lower bound for* $r_\epsilon(S)$.

Note that computing $k_\epsilon(S)$ only requires computing the probability $\mathbb{P}\{D(S) \leq b\}$. Moreover, any value that is between $k_\epsilon(S)$ and $r_\epsilon(S)$ will yield a valid lower bound. To improve on $k_\epsilon(S)$, given a random variable X, we use $Q_p(X)$ to denote the p-th quantile of X, that is, $Q_p(X) = \inf\{\alpha : \mathbb{P}\{X \leq \alpha\} \geq p\}$.[1] One initial lower bound on $r_\epsilon(S)$ is given as follows. The proof of this lemma (and any results without proof) will appear in the full version of this paper.

Lemma 1. *Given* $p := 1 - r_\epsilon(S)\epsilon$, *then for all* $S \subseteq V_+$, *we have* $r_\epsilon(S) \geq \lceil Q_p(D(S))/b \rceil$.

---

[1] Note that $\mathbb{P}\{D(S) \leq b\} \geq 1 - \epsilon \iff Q_{1-\epsilon}(D(S)) \leq b$.

The lower bound given by Lemma 1 cannot be directly used because its calculation uses the value of $r_\epsilon(S)$ itself. However, one can use Lemma 1 to derive a computable lower bound. For $S \subseteq V_+$, define $a(S, 1) = 1$ and for $k = 2, \ldots, K$ define

$$a(S, k) := \min\left\{k, \lceil Q_{1-(k-1)\epsilon}(D(S))/b \rceil\right\}.$$

Using $a(S, k)$ we get to our second valid lower bound:

$$k_\epsilon^*(S) = \max\left\{a(S, k) : k = 1, 2, ..., K\right\}.$$

Theorem 1. *For all $S \subseteq V_+$, $k_\epsilon(S) \leq k_\epsilon^*(S) \leq r_\epsilon(S)$.*

Proof. For brevity, we skip the (simple) proof of $k_\epsilon(S) \leq k_\epsilon^*(S)$.

We will show that for any k, $a(S, k)$ is a lower bound on $r_\epsilon(S)$. Indeed, this is trivially true if $r_\epsilon(S) \geq k$ or if $k = 1$. Otherwise, $r_\epsilon(S) \leq k - 1$, in which case Lemma 1 shows that $r_\epsilon(S) \geq \lceil Q_p(D(S))/b \rceil$, where $p = 1 - r_\epsilon(S)\epsilon$. Then $a(S, k) \leq r_\epsilon(S)$ since $Q_p(D(S))$ is nondecreasing in p. Therefore, the maximum of these lower bounds over $k = 1, \ldots, K$ is also a lower bound.   □

We next discuss how alternative valid lower bounds can be obtained in the special case when the customer random demands are joint normally distributed.

2.3   Joint Normal Random Demands

In this section we assume customer demands follow a joint normal distribution with covariance matrix $\Sigma \succeq 0$. For a given subset of customers $S \subseteq V_+$, let $\underline{\lambda}_S$ denote the smallest eigenvalue of the submatrix of Σ associated with rows and columns corresponding to customers in the set S. In addition, for each customer $i \in S$, define

$$\bar{d}_i(S) = d_i + \Phi^{-1}(1 - \epsilon)\sqrt{\underline{\lambda}_S/|S|}. \tag{3}$$

Then, for $S \subseteq V_+$ we define the following bound on $r_\epsilon(S)$:

$$k_\epsilon^J(S) = \begin{cases} 1, & \text{if } \mathbb{P}\{D(S) \leq b\} \geq 1 - \epsilon \\ \max\left\{\lceil \sum_{i \in S} \bar{d}_i(S)/b \rceil, 2\right\}, & \text{otherwise.} \end{cases}$$

Theorem 2. *If the demands follow a joint normal distribution with mean vector d and covariance matrix Σ and $\epsilon \leq 0.5$, then*

$$k_\epsilon(S) \leq k_\epsilon^J(S) \leq r_\epsilon(S), \forall S \subseteq V_+.$$

Proof. The proof is omitted due to space limitations, but is based on deriving a deterministic bin packing problem that provides a lower bound on $r_\epsilon(S)$.   □

The assumption that $\epsilon \leq 0.5$ is not restrictive as the typical use of this model is for settings in which ϵ is smaller than 0.5. Note that neither $k_\epsilon^*(S)$ nor $k_\epsilon^J(S)$ dominates the other. Thus, we use $\max\{k_\epsilon^*(S), k_\epsilon^J(S)\}$ as the valid lower bound when solving CCVRP instances with joint normal random demands. We note that when demands follow an independent normal distribution, then $k_\epsilon^J(S) = k_\epsilon^I(S)$, so this result generalizes the bound from [13].

3   Dantzig-Wolfe Formulation

Set partitioning formulations for routing problems are based on enumerating *elementary routes* or relaxations of them. We start by describing such an approach for the deterministic VRP. For that case, an *elementary route* is a closed walk $v_0, v_1, \ldots, v_k, v_{k+1} = v_0$, for some $k \geq 1$ such that (i) $v_0 = 0$, $v_i \in V_+, \forall i = 1, \ldots, k$ and $v_{i-1}, v_i \in E, \forall i = 1, \ldots, k+1$; (ii) $v_i \neq v_j, \forall 0 < i < j < k+1$ and (iii) $d^\top y \leq b$, where $y_v := \sum_{i=1}^{k} \mathbb{1}_{\{v=v_i\}}$ is the number of times v appears in the route and d is the vector of deterministic customer demands.

Let $\mathcal{Q}$ be the set of elementary routes and $\lambda_j \in \{0,1\}$ represent if elementary route j is used. Let $q_j^e := \sum_{i=0}^{k} \mathbb{1}_{\{e=v_i v_{i+1}\}}$, that is, the number of times edge e appears in route j. By using the relationship $x_e = \sum_{j \in \mathcal{Q}} q_j^e \lambda_j, \ \forall e \in E$, we obtain from (1) a set-partitioning based formulation for the deterministic VRP [10]:

$$\min_{\lambda} \quad \sum_{j \in \mathcal{Q}} \sum_{e \in E} \ell_e q_j^e \lambda_j \tag{4a}$$

$$\text{s.t.} \quad \sum_{j \in \mathcal{Q}} \sum_{e \in \delta(\{i\})} q_j^e \lambda_j = 2 \qquad i \in V_+ \tag{4b}$$

$$\sum_{j \in \mathcal{Q}} \sum_{e \in \delta(\{0\})} q_j^e \lambda_j = 2K \tag{4c}$$

$$\sum_{j \in \mathcal{Q}} \sum_{e \in \delta(S)} q_j^e \lambda_j \geq 2r_0(S) \qquad S \subseteq V_+ \tag{4d}$$

$$\sum_{j \in \mathcal{Q}} q_j^e \lambda_j \leq 1 \qquad e \in E \setminus \delta(\{0\}) \tag{4e}$$

$$\lambda_j \in \{0,1\} \qquad j \in \mathcal{Q}. \tag{4f}$$

In order to use (4) in a BCP approach it must be possible to solve the pricing subproblem of λ variables efficiently. The pricing subproblem consists of finding elementary routes of minimum reduced cost, which is strongly $\mathcal{NP}$-hard. In [10] condition (ii) was relaxed, leading to what is called a q-route [5]. Pricing q-routes is still $\mathcal{NP}$-hard due to the knapsack-type condition (iii), but it can be solved in pseudo-polynomial time [5] if the demands are integer. We note that more complex column-generation schemes have also been proposed, strengthening (4) by forbidding some (or all) cycles in q-routes. Our approach can be adapted to those cases, but we choose to only present it based on q-routes.

To adapt (4) for the CCVRP, all we need to do is replace $r_0(S)$ in (4d) by $r_\epsilon(S)$ or any of its valid lower bounds derived in Sect. 2 and consider $\mathcal{Q}$ as the set of *chance-constraint feasible q-routes (CCq-routes)*, where a CCq-route is a closed walk satisfying (i) and replacing condition (iii) by

$$\mathbb{P}\{D^\top y \leq b\} \geq 1 - \epsilon \tag{5}$$

We note that chance-constraint feasible routes are CCq-routes satisfying (ii).

Unfortunately, in contrast to the deterministic VRP, the pricing of CCq-routes is *strongly* $\mathcal{NP}$-hard in general. This is proved by showing the problem of pricing elementary routes can be reduced to it.

Theorem 3. *Suppose the distribution of demands is specified by N scenarios d^k, $k = 1, \ldots, N$, where $\mathbb{P}\{D = d^k\} = 1/N$ for $k = 1, \ldots, N$. Then the pricing of CCq-routes is strongly $\mathcal{NP}$-hard.*

To overcome the difficulty in pricing CCq-routes, we propose to further relax the capacity constraints defining CCq-routes used in the set partitioning formulation. In the following subsections we present two approaches for doing this: one that applies to any distribution for which we can evaluate (5) and the other that uses distribution-specific arguments.

3.1   Relaxed Pricing

The key advantage of using q-routes instead of elementary routes in (4) was to go from a strongly $\mathcal{NP}$-hard pricing to a pseudo-polynomially solvable one. The approach is valid since the set of q-routes contains the set of elementary routes and, in any $\{0, 1\}$ solution to (4), constraints (4b) ensure that only elementary routes are chosen. We build upon that idea to further relax constraint (5) so that any chance constraint feasible route is still feasible to the pricing subproblem, and then use constraints in the master problem to ensure that in a $\{0, 1\}$ solution only chance constraint feasible routes are chosen.

Since the original condition (iii) can be handled by dynamic programming, we choose to relax the condition (5) in CCq-routes to a similar knapsack-type constraint $\pi^\top y \le b_\pi$. To make sure that chance constraint feasible routes are still feasible, we must have that

$$b_\pi \ge b_\pi^* := \max\left\{\pi^\top y : \mathbb{P}\{D^\top y \le b\} \ge 1 - \epsilon, y_i \in \{0, 1\}, i \in V_+\right\}. \qquad (6)$$

It is then clear that the following proposition holds.

Proposition 2. *Let y be a binary vector satisfying (5). If (6) holds, then y satisfies:*

$$\pi^\top y \le b_\pi. \qquad (7)$$

We call *relaxed chance constraint feasible q-routes (rCCq-routes)* the closed walks satisfying conditions (i) in the definition of elementary routes and replacing (iii) in that definition by (7).

For any integral and nonnegative choice of the π coefficients, once b_π^* (or an upper bound) is determined, we can proceed with the pricing exactly the same way as is done in the deterministic VRP, using (7) as the knapsack constraint.

In the deterministic case VRP, the constraints (4b) impose that, in a $\{0, 1\}$ solution to (4) only elementary routes will have their corresponding variable equal to 1. Since each q-route satisfies the capacity constraints, the constraints (4d) are not required for validity of the formulation (4), and so a pure branch-and-price algorithm could be applied. In contrast, in our proposed BCP approach for the CCVRP, even if a $rCCq$-route is an elementary route, it may not be a chance constraint feasible route. The formulation thus requires the constraints (4d), which impose that in a $\{0, 1\}$ solution to (4) only chance constraint feasible routes will have their corresponding variable equal to 1.

We next discuss the choice of the coefficients π and calculation of the value b_π^*. Any integer values of π can be used, but it is natural to choose values that correlate with the size of the items, and so we chose to use $\pi_i = d_i$ (these may be scaled and rounded to obtain acceptably small integers). Calculating b_π^* requires solving the chance-constrained knapsack problem (6). This can be a computationally challenging problem, although it only needs to be solved once as a preprocessing step. When the demands assume a discrete distribution having finitely many demand scenarios, the preprocessing problem (6) can be solved using specialized techniques [19]. With joint normal random demands with mean vector d and covariance matrix $\Sigma \succeq 0$, the capacity chance constraint in (6) can be modeled as

$$d^\top y + \Phi^{-1}(1 - \epsilon)\sqrt{y^\top \Sigma y} \leq b. \tag{8}$$

which can be reformulated as a second-order cone constraint when $\epsilon \leq 0.5$. In this case, (6) becomes a binary second-order cone programming program. In addition, if desired, any upper bound on b_π^* that is computationally cheaper to compute can be used. In the next section, we discuss an alternative relaxation scheme that can be used when customer demands are joint normally distributed that avoids solving (6) altogether.

3.2   Relaxed Pricing for Joint Normal Demands

In the case that demands are joint normally distributed with covariance matrix $\Sigma \succeq 0$, we will proceed as follows. Define η^* as the optimal objective value of the following semidefinite program

$$\eta^* = \max_{\eta, p, Q} \eta \tag{9a}$$

$$\text{s.t.} \quad d_i \eta \leq p_i \qquad\qquad i \in V_+ \tag{9b}$$

$$\Sigma = \text{diag}(p_1, ..., p_n) + Q \tag{9c}$$

$$Q \succeq 0, \tag{9d}$$

and derive the following constraint based on η^*

$$d^\top y + \Phi^{-1}(1 - \epsilon)\sqrt{\eta^* d^\top y} \leq b. \tag{10}$$

Proposition 3. *Let y be a binary vector satisfying (8), then y satisfies (10).*

Proof. Let (η^*, p^*, Q^*) be an optimal solution of (9) and let $P^* = \text{diag}(p_1^*, ..., p_n^*)$. It suffices to show that $\eta^* d^\top y \leq y^\top \Sigma y$. It follows that $y^\top \Sigma y = y^\top P^* y + y^\top Q^* y \geq y^\top P^* y$, since $Q^* \succeq 0$. Since y is a binary vector, $y^\top P^* y = \sum_{i \in V_+} p_i^* y_i^2 = (p^*)^\top y$. Since p^* is feasible to (9), $p_i^* \geq d_i \eta^*, \forall i \in V_+$. Therefore, it follows that $y^\top \Sigma y \geq (p^*)^\top y \geq \eta^* d^\top y$.   □

Proposition 3 shows that (10) is a relaxation of (5) which is cheap to compute as it only requires solving a semidefinite program. Though it is not linear in terms of y, the left-hand-side of (10) is a monotone increasing function of $d^\top y$, making it possible to represent this constraint as

$$d^\top y \leq \hat{b}_d(\eta^*) \tag{11}$$

where

$$\hat{b}_d(\eta^*) = \left(\sqrt{b + I_\epsilon \eta^*} - \sqrt{I_\epsilon \eta^*}\right)^2 \text{ and } I_\epsilon = \left(\frac{1}{2}\Phi^{-1}(1 - \epsilon)\right)^2.$$

Thus, the linear constraint (11) is a relaxation of the exact capacity chance constraint (8) and so it can be used to generate $rCCq$-routes. We conclude by noting that for *independent* normal random variables, the value of η^* has the closed-form solution $\eta^* = \min\{\sigma_i^2/d_i : i \in V_+\}$.

4   Computational Experiments

Our BCP implementation is based on the code of [10], with similar ideas with respect to computational choices which, for the sake of brevity, will not be discussed. One point worth mentioning is that we also adapt the Clarke and Wright heuristic [6] for the CCVRP to obtain initial primal bounds.

The instances used for testing are adapted from the deterministic VRP instances available at http://vrp.atd-lab.inf.puc-rio.br/. Independent normal instances were generated letting D_i be normal with mean d_i equal to the deterministic demand of the instance. Low variance instances were generated by setting standard deviations σ_i uniformly at random in the interval $[0.07*d_i, 0.13*d_i]$. High variance instances had σ_i uniformly distributed in $[0.14 * d_i, 0.26 * d_i]$.

Joint normal distributions were generated from independent normal ones by setting the correlation ρ_{ij} to be inversely proportional to the distance ℓ_{ij} times a uniform random variable. The idea is that customers that are closer together are likely to be impacted by similar factors and so should have higher correlation. For the scenario model, we generated 200 scenarios samples from the low variance joint normal instances. High variance instances were generated in a similar way, except that each customer also had a probability of having zero demand. For such a distribution, it is difficult to exactly calculate $\mathbb{P}\{D(S) \leq b\}$ for a subset of customers S, motivating the use of the scenario approximation. For both of these cases, the sampled demands were rounded to the nearest integer.

The formulations were tested on 10 instances with 32–55 vertices. In all our tests, we chose $\epsilon = 0.05$. Tables 1, 2 and 3 present aggregate statistics for the independent normal, joint normal and scenario models, respectively.

We classified the instances as: *small* (7 instances with < 50 vertices) and *large* (3 instances with ≥ 50 vertices). This grouping highlights the behavior of the algorithm on instances of different size. For each set of instances, each variance setting (low and high) and each formulation, the tables present the geometric mean of time to solve (**AvT**), the number of instances solved to optimality (**NumSolv**) and the arithmetic mean of the gap left after the time limit of

7200 s (**AvGap**). Instances solved within the time limit have final gap equal to
0 %, while instances not solved within the time limit have time equal to 7200 s.

The formulations tested were as follows: Formulation (1) with valid lower
bounds $k_\epsilon(S)$, $k_\epsilon^*(S)$ and $\max\{k_\epsilon^*(S), k_\epsilon^J(S)\}$ for $r_\epsilon(S)$ (named BC, BC^*, BC^J
respectively). In addition, formulation (4) was tested with the valid lower bound
$\max\{k_\epsilon^*(S), k_\epsilon^J(S)\}$ for joint and independent normal distributions and $k_\epsilon^*(S)$
for the scenario model, and with the relaxed pricing proposed in Sect. 3 for the
corresponding distribution (named BCP_r). For independent normally distributed customer demands, we also tested exact pricing of CCq-routes based on
discretizing both the means and the sum of the variances and considering the
sum of the variances as a resource in the pricing [12]. This was labeled BCP_e.
All experiments were run on a Dell R510 machine with 128 G memory, and using
one core of a 2.66 G X5650 Xeon Chip.

The results from the tables show that the improved valid lower bounds
on $r_\epsilon(S)$ have a very significant impact in all instances, with the use of the
distribution-specific valid bound being usually the best option, although the
generic bound $k_\epsilon^*(S)$ typically performs well too. As for the use of (4) versus
(1), it seems that for smaller instances, the overhead of solving (4) is too much
to overcome. However, for larger instances, the tighter bound provided by (4)
pays off and we are able to solve more of the larger instances using BCP. In
addition, in the independent normal instances, exact pricing seems to be better
than relaxed pricing – in the small instances, the loss is not big and sometimes
even a gain is observed, while in the large instances, there is a clear advantage.

4.1   Comparison of the CCVRP with Recourse Models

One additional experiment performed aimed to compare the solution obtained
by the CCVRP with the one obtained by a recourse model for VRPSD, where
the assumed recourse is that a return trip must be made to the depot whenever
a truck's capacity is exceeded. This experiment was done for instances having
independent normal customer demands, since these are the only instances that
can be solved for the recourse model. For such instances, the optimal solution for
a recourse-based model and for the CCVRP were obtained, denoted as x^r and
x^{cc}, respectively. Both solutions were then compared according to two quantities:

Table 1. Summary of computational results for independent normal random demands.

		Low variance					High variance				
		BC	BC^*	BC^J	BCP_r	BCP_e	BC	BC^*	BC^J	BCP_r	BCP_e
Small	AvT (s)	2646	71	40	153	125	3436	354	144	939	1722
	NumSolv	3	7	7	6	7	1	4	6	4	4
	AvGap (%)	4.73	0	0	0.76	0	10.40	3.17	1.34	1.70	1.44
Large	AvT (s)	7200	7200	7200	3311	257	7200	7200	7200	7200	2379
	NumSolv	0	0	0	3	3	0	0	0	0	2
	AvGap (%)	14.52	8.59	7.27	0	0	14.74	11.12	11.45	5.72	1.10

Table 2. Summary of computational results for joint normal random demands.

		Low variance				High variance			
		BC	BC^*	BC^J	BCP_r	BC	BC^*	BC^J	BCP_r
Small	AvT (s)	3238	107	109	585	4628	414	393	2335
	NumSolv	3	6	6	5	1	4	4	4
	AvGap (%)	4.67	1.10	1.12	1.29	11.72	3.41	3.41	2.76
Large	AvT (s)	7200	7200	7200	5743	7200	7200	7200	891
	NumSolv	0	0	0	1	0	0	0	2
	AvGap (%)	14.04	8.73	8.73	2.40	14.58	10.74	10.74	3.06

$z^r(x)$ (the objective value of the recourse-based two-stage stochastic program) and $\eta(x)$, which represents the largest probability that a vehicle will have its capacity violated. Note that by design, $z^r(x^r) \leq z^r(x^{cc})$ and that $\eta(x^{cc}) \leq 0.05$ – though it is possible that $\eta(x^r) < \eta(x^{cc})$.

We find that when evaluating the CCVRP solution in the recourse model, the value $z^r(x^{cc})$ was, on average, only about 1 % more than the optimal value $z^r(x^r)$, and the largest increase was 3.4 %. On the other hand, while $\eta(x^{cc})$ was always (by design) less than 0.05, $\eta(x^r)$ was greater than 0.15 in three of the instances, and as high as 0.5, meaning that in the solution there was a truck whose capacity would be exceeded on average 50 % of the time. We thus conclude that the CCVRP model tends to yield solutions that are high quality for the recourse model, whereas the reverse is not true. In addition, the CCVRP model is not dependent on a particular assumption of the recourse taken, and can be solved also when customer demands are not independent.

Table 3. Summary of computational results for scenario model of random demands.

	Small-Low var			Small-High var			Large-Low var			Large-High var		
	BC	BC^*	BCP_r	BC	BC^*	BCP_r	BC	BC^*	BCP_r	BC	BC^*	BCP_r
AvT (s)	1947	217	803	4328	499	4207	7200	7200	7200	7200	7200	7200
NumSolv	3	6	7	1	5	2	0	0	0	0	0	0
AvGap (%)	5.54	0.79	0	11.76	2.64	5.09	13.68	9.07	4.89	19.26	15.78	11.16

5   Conclusion

We propose two exact approaches for the CCVRP with very mild assumptions on distribution of customer demands. In particular, we allow for correlations between random customer demands. Significant contributions were made both to cutting and to pricing, thus allowing for the first time a successful BCP approach for the CCVRP. Our results show that the formulations proposed are promising and can be used to solve instances of up to 55 vertices.

Several improvements can still be made to the implementation of the pricing routines, as well as the investigation of further valid inequalities for the problem. Those remain the topic of further research.

Acknowledgments. Fukasawa was supported by NSERC Discovery Grant RGPIN-05623. Luedtke was supported by NSF grants CMMI-0952907 and CMMI-1130266, and ONR award N00014-15-1-2268.

References

1. Baldacci, R., Mingozzi, A.: A unified exact method for solving different classes of vehicle routing problems. Math. Program. **120**(2), 347–380 (2009)
2. Baldacci, R., Mingozzi, A., Roberti, R.: New route relaxation and pricing strategies for the vehicle routing problem. Oper. Res. **59**(5), 1269–1283 (2011)
3. Bertsimas, D.J.: A vehicle routing problem with stochastic demand. Oper. Res. **40**(3), 574–585 (1992)
4. Christiansen, C.H., Lysgaard, J., Wøhlk, S.: A branch-and-price algorithm for the capacitated arc routing problem with stochastic demands. Oper. Res. Lett. **37**(6), 392–398 (2009)
5. Christofides, N., Mingozzi, A., Toth, P.: Exact algorithms for the vehicle routing problem, based on spanning tree and shortest path relaxations. Math. Program. **20**(1), 255–282 (1981)
6. Clarke, G., Wright, J.W.: Scheduling of vehicles from a central depot to a number of delivery points. Oper. Res. **12**(4), 568–581 (1964)
7. Contardo, C., Martinelli, R.: A new exact algorithm for the multi-depot vehicle routing problem under capacity and route length constraints. Discr. Optim. **12**, 129–146 (2014)
8. Dantzig, G.B., Ramser, J.H.: The truck dispatching problem. Man. Sci. **6**(1), 80–91 (1959)
9. Dror, M., Laporte, G., Louveaux, F.V.: Vehicle routing with stochastic demands and restricted failures. Zeitschrift für Oper. Res. **37**(3), 273–283 (1993)
10. Fukasawa, R., Longo, H., Lysgaard, J., de Aragão, M.P., Reis, M., Uchoa, E., Werneck, R.F.: Robust branch-and-cut-and-price for the capacitated vehicle routing problem. Math. Program. **106**(3), 491–511 (2006)
11. Gauvin, C., Desaulniers, G., Gendreau, M.: A branch-cut-and-price algorithm for the vehicle routing problem with stochastic demands. Comput. & Oper. Res. **50**, 141–153 (2014)
12. Irnich, S., Desaulniers, G.: Shortest path problems with resource constraints. In: Desaulniers, G., Desrosiers, J., Solomon, M. (eds.) Column Generation. Springer, New York (2005)
13. Laporte, G., Louveaux, F., Mercure, H.: Models and exact solutions for a class of stochastic location-routing problems. European J. Oper. Res. **39**(1), 71–78 (1989)
14. Laporte, G., Louveaux, F., Mercure, H.: The vehicle routing problem with stochastic travel times. Trans. Sci. **26**(3), 161–170 (1992)
15. Laporte, G., Louveaux, F.V., Van Hamme, L.: An integer L-shaped algorithm for the capacitated vehicle routing problem with stochastic demands. Oper. Res. **50**(3), 415–423 (2002)
16. Luedtke, J., Ahmed, S.: A sample approximation approach for optimization with probabilistic constraints. SIAM J. Optim. **19**, 674–699 (2008)

17. Pecin, D., Pessoa, A., Poggi, M., Uchoa, E.: Improved branch-cut-and-price for capacitated vehicle routing. In: Lee, J., Vygen, J. (eds.) IPCO 2014. LNCS, vol. 8494, pp. 393–403. Springer, Heidelberg (2014)
18. Secomandi, N., Margot, F.: Reoptimization approaches for the vehicle-routing problem with stochastic demands. Oper. Res. **57**(1), 214–230 (2009)
19. Song, Y., Luedtke, J.R., Küçükyavuz, S.: Chance-constrained binary packing problems. INFORMS J. Comput. **26**(4), 735–747 (2014)
20. Stewart, W.R., Golden, B.L.: Stochastic vehicle routing: a comprehensive approach. Euro. J. Oper. Res. **14**(4), 371–385 (1983)
21. Yang, W.H., Mathur, K., Ballou, R.H.: Stochastic vehicle routing problem with restocking. Trans. Sci. **34**(1), 99–112 (2000)

Extended Formulations in Mixed-Integer Convex Programming

Miles Lubin[1(✉)], Emre Yamangil[2], Russell Bent[2], and Juan Pablo Vielma[1]

[1] Massachusetts Institute of Technology, Cambridge, MA, USA
mlubin@mit.edu
[2] Los Alamos National Laboratory, Los Alamos, NM, USA

Abstract. We present a unifying framework for generating extended formulations for the polyhedral outer approximations used in algorithms for mixed-integer convex programming (MICP). Extended formulations lead to fewer iterations of outer approximation algorithms and generally faster solution times. First, we observe that all MICP instances from the MINLPLIB2 benchmark library are conic representable with standard symmetric and nonsymmetric cones. Conic reformulations are shown to be effective extended formulations themselves because they encode separability structure. For mixed-integer conic-representable problems, we provide the first outer approximation algorithm with finite-time convergence guarantees, opening a path for the use of conic solvers for continuous relaxations. We then connect the popular modeling framework of disciplined convex programming (DCP) to the existence of extended formulations independent of conic representability. We present evidence that our approach can yield significant gains in practice, with the solution of a number of open instances from the MINLPLIB2 benchmark library.

1 Introduction

Mixed-Integer convex programming (MICP) is the class of problems where one seeks to minimize a convex objective function subject to convex constraints and integrality restrictions on the variables. MICP is less general than mixed-integer nonlinear programming (MINLP), where the objective and constraints may be nonconvex, but unlike the latter, one can often develop finite-time algorithms to find a global solution. These finite-time algorithms depend on convex nonlinear programming (NLP) solvers to solve continuous subproblems. MICP, also called *convex MINLP*, has broad applications and is supported in various forms by both academic solvers like Bonmin [7] and SCIP [3] and commercial solvers like KNITRO [9]; see Bonami et al. [5,8] for a review.

The most straightforward approach for MICP is NLP-based branch and bound, an extension of the branch and bound algorithm for mixed-integer linear programming (MILP) where a convex NLP relaxation is solved at each node of the branch and bound tree [16]. However, driven by the availability of effective solvers for linear programming (LP) and MILP, it was observed in the early

© Springer International Publishing Switzerland 2016
Q. Louveaux and M. Skutella (Eds.): IPCO 2016, LNCS 9682, pp. 102–113, 2016.
DOI: 10.1007/978-3-319-33461-5_9

1990s by Leyffer and others [21] that it is often more effective to avoid solving NLP relaxations when possible in favor of solving polyhedral relaxations using MILP. Polyhedral relaxations form the basis of the majority of the existing solvers recently reviewed and benchmarked by Bonami et al. [8].

While traditional MICP approaches construct polyhedral approximations in the original space of variables, a number of authors have considered introducing auxiliary variables and forming a polyhedral approximation in a higher dimensional space [19,20,28,30]. Such constructions are called *extended formulations* or *lifted formulations*, the motivation for which is the fact that the projection of these polyhedra onto the original space can provide a higher quality approximation than one built from scratch in the original space. Tawarmalani and Sahinidis [28] propose, in the context of nonconvex MINLP, extended formulations for compositions of functions. For MICP, Hijazi et al. [19] demonstrate the effectiveness of extended formulations in the special case where all nonlinear functions can be written as a sum of *univariate* convex functions. Their method obtains promising speed-ups over Bonmin on the instances which exhibit this structure. Hijazi et al. generated these extended formulations by hand, and no subsequent work has proposed techniques for off-the-shelf MICP solvers to detect and exploit separability. Building on these results, Vielma et al. [30] propose extended formulations for second-order cones. These extended formulations improved solution times for mixed-integer second-order cone programming (MISOCP) over state of the art commercial solvers CPLEX and Gurobi quite significantly; both solvers adopted the technique within a few months after its publication.

A major contribution of this work is to propose a new, unifying framework for generating extended formulations for the polyhedral outer approximations used in MICP algorithms. This framework generalizes the work of Hijazi et al. [19] which was specialized for separable problems to include all MICPs whose objective and constraints can be expressed in closed algebraic form. We begin in Sect. 2 by considering conic representability. While many MICP instances are representable by using MISOCP, reformulation to MISOCP has not been widely adopted, and MICP is still considered a significantly more general form. We demonstrate that with the introduction of the nonsymmetric exponential and power cones, surprisingly, all convex instances in the MINLPLIB2 benchmark library [1] are representable by a combination of these nonsymmetric cones and the second-order cone. We discuss how the conic-form representation of a problem is itself a strong extended formulation. Hence, the guideline to "just solve the conic form problem" is surprisingly effective.

We note that conic-form problems have modeling strength beyond that of smooth MICP, in particular for handling of nonsmooth perspective functions useful in disjunctive convex programming [15]. With the recent development of conic solvers supporting nonsymmetric cones [26,27], it may be advantageous to use these solvers over derivative-based NLP solvers, in which case the standard convergence theory for outer approximation algorithms no longer applies. In Sect. 3, we present the first finite-time outer approximation algorithm applicable to mixed-integer conic programming with any closed, convex cones (symmetric and nonsymmetric), so long as conic duality holds in strong form. This algorithm extends the work of Drewes and Ulbrich [11] for MISOCP with a much simpler and more general proof.

In Sect. 4, we generalize the idea of extended formulations through conic representability by considering the modeling framework of *disciplined convex programming* (DCP) [14], a popular modeling paradigm for convex optimization which has so far received little notice in the MICP realm. In DCP, convex expressions are specified in an algebraic form such that convexity can be verified by simple composition rules. We establish a 1-1 connection between these rules for verifying convexity and the existence of extended formulations. Hence, all MICPs expressed in mixed-integer disciplined convex programming (MIDCP) form have natural extended formulations regardless of conic representability. This view has connections with techniques for nonconvex MINLP, where it is already common practice to construct extended outer approximations based on the algebraic representation of the problem [5].

In our computational experiments, we translate MICP problems from the MINLPLIB2 benchmark library into MIDCP form and demonstrate significant gains from the use of extended formulations, including the solution of a number of open instances. Our open-source solver, *Pajarito*, is the first solver specialized for MIDCP and is accessible through existing DCP modeling languages.

2   Extended Formulations and Conic Representability

We state a generic mixed-integer convex programming problem as

$$\text{minimize}_{x,y} \; f(x,y)$$
$$\text{subject to } g_j(x,y) \leq 0 \quad \forall j \in J, \qquad\qquad \text{(MICONV)}$$
$$L \leq x \leq U, \quad x \in \mathbb{Z}^n, y \in \mathbb{R}^p_+,$$

where the set J indexes the nonlinear constraints, the functions $f, g_j : \mathbb{R}^n \times \mathbb{R}^p \to \mathbb{R} \cup \{\infty\}$ are convex, and the vectors L and U are finite bounds on x. Without loss of generality, when convenient, we may assume that the objective function f is linear (via epigraph reformulation [8]).

Vielma et al. [30] discuss the motivation for extended formulations in MICP: many successful MICP algorithms use polyhedral outer approximations of nonlinear constraints, and polyhedral outer approximations in a higher dimensional space can often be much stronger than approximations in the original space. Hijazi et al. [19] give an example of an approximation of an ℓ_2 ball in $\mathbb{R}^n$ which requires 2^n tangent hyperplanes in the original space to prove that the intersection of the ball with the integer lattice is in fact empty. By exploiting the summation structure in the definition of the ℓ_2 ball, [19] demonstrate that an extended formulation requires only $2n$ hyperplanes to prove an empty intersection. More generally, [19,28] propose to reformulate constraints with separable structure $\sum_{k=1}^{q} g_k(x_k) \leq 0$, where $g_k : \mathbb{R} \to \mathbb{R}$ are *univariate* convex functions by introducing auxiliary variables t_k and imposing the constraints

$$\sum_{k=1}^{q} t_k \leq 0, g_k(x_k) \leq t_k \forall k. \qquad\qquad (1)$$

A consistent theme in this paper is the *representation* of the convex functions f and g_j, $\forall j \in J$ in (MICONV). Current MICP solvers require continuous differentiability of the nonlinear functions and access to black-box oracles for querying the values and derivatives of each at any given point (x, y). The difficulty in the reformulation (1) is that the standard representation (MICONV) does not encode the necessary information, since separability is an algebraic property which is not detectable given only oracles to evaluate function values and derivatives. As such, we are not aware of any off-the-shelf MICP solver which exploits this special-case structure, despite the promising experimental results of [19].

In this section, we consider the equally general, yet potentially more useful, representation of (MICONV) as a mixed-integer conic programming problem:

$$\min_{x,z} \quad c^T z$$
$$\text{s.t. } A_x x + A_z z = b \qquad \qquad \text{(MICONE)}$$
$$L \le x \le U, x \in \mathbb{Z}^n, z \in \mathcal{K},$$

where $\mathcal{K} \subseteq \mathbb{R}^k$ is a closed convex cone. Without loss of generality, we assume integer variables are not restricted to cones, since we may introduce corresponding continuous variables by equality constraints. The representation of (MICONV) as (MICONE) is equally as general in the sense that given a convex function f, we can define a closed convex cone $\mathcal{K}_f = \text{cl}\{(x, y, \gamma, t) : \gamma f(x/\gamma, y/\gamma) \le t, \gamma > 0\}$ where $\text{cl } S$ is defined as the closure of a set S. Using this, we can reformulate (MICONV) to the equivalent optimization problem

$$\min \quad t_f$$
$$\text{s.t. } t_j + s_j = 0 \quad \forall j \in J,$$
$$\gamma_f = 1, x = x_f, y = y_f,$$
$$\gamma_j = 1, x = x_j, y = y_j, \forall j \in J, \qquad \qquad (2)$$
$$L \le x \le U, x \in \mathbb{Z}^n, y \in \mathbb{R}_+^p,$$
$$(x_f, y_f, \gamma_f, t_f) \in \mathcal{K}_f,$$
$$(x_j, y_j, \gamma_j, t_j) \in \mathcal{K}_{g_j}, s_j \in \mathbb{R}_+ \quad \forall j \in J.$$

The problem (2) is in the form of (MICONE) with $\mathcal{K} = \mathbb{R}_+^{n+|J|} \times \mathcal{K}_f \times \mathcal{K}_{g_1} \times \cdots \times \mathcal{K}_{g_{|J|}}$. Such a tautological reformulation is not particularly useful, however. What *is* useful is a reformulation of (MICONV) into (MICONE) where the cone $\mathcal{K}$ is a product $\mathcal{K}_1 \times \mathcal{K}_2 \times \cdots \times \mathcal{K}_r$, where each $\mathcal{K}_i$ is one of a small number of recognized cones, such as the positive orthant $\mathbb{R}_+^n$, the second-order cone $SOC_n = \{(t, x) \in \mathbb{R}^n : ||x|| \le t\}$, the exponential cone, $EXP = \text{cl}\{(x, y, z) \in \mathbb{R}^3 : y \exp(x/y) \le z, y > 0\}$, and the power cone (given $0 < \alpha < 1$), $POW_\alpha = \{(x, y, z) \in \mathbb{R}^3 : |z| \le x^\alpha y^{1-\alpha}, x \ge 0, y \ge 0\}$.

The question of which functions can be represented by second-order cones has been well studied [6, 22]. More recently, a number of authors have considered nonsymmetric cones, in particular the exponential cone, which can be used to model logarithms, entropy, logistic regression, and geometric programming [27], and the power cone, which can be used to model p-norms and powers [18].

The folklore within the conic optimization community is that almost all convex constraints which arise in practice are representable by using these cones[1], in addition to the positive semidefinite cone which we do not consider here. To substantiate this claim, we classified the 333 MICP instances in MINLPLIB2 according to their conic representability and found that *all* of the instances are conic representable; see Table 1.

Table 1. A categorization of the 333 MICP instances in the MINLPLIB2 library according to conic representability. Over two thirds are pure MISOCP problems and nearly one third is representable by using the exponential (EXP) cone alone. All instances are representable by using standard cones.

SOC only	EXP only	SOC and EXP	POW only	Not representable	Total
217	107	7	2	0	333

While solvers for SOC-constrained problems (SOCPs) are mature and commercially supported, the development of effective and reliable algorithms for handling exponential cones and power cones is an emerging, active research area [26,27]. Nevertheless, we claim that the conic view of (MICONV) is useful *even lacking* reliable solvers for continuous conic relaxations.

As a motivating example, we consider the trimloss [17] (tls) instances from MINLPLIB2, a convex formulation of the cutting stock problem. These instances are notable as some of the few unsolved instances in the benchmark library and also because they exhibit a separability structure more general than what can be handled by Hizaji et al. [19].

The trimloss instances have constraints of the form

$$\sum_{k=1}^{q} -\sqrt{x_k y_k} \leq c^T z + b, \tag{3}$$

where x, y, z are arbitrary variables unrelated to the previous notation in this section. Harjunkoski et al. [17] obtain these constraints from a clever reformulation of nonconvex bilinear terms. The function $-\sqrt{xy}$ is the negative of the geometric mean of x and y. It is convex for nonnegative x and y and its epigraph $E = \{(t, x, y) : -\sqrt{xy} \leq t, x \geq 0, y \geq 0\}$ is representable as an affine transformation of the three-dimensional second-order cone SOC_3 [6]. A conic formulation for (3) is constructed by introducing an auxiliary variable for each term in the sum plus a slack variable, resulting in the following constraints:

$$\sum_{k=1}^{q} t_k + s = c^T z + b, (t_k, x_k, y_k) \in E \; \forall k, \text{ and } s \in \mathbb{R}_+. \tag{4}$$

---

[1] http://erlingdandersen.blogspot.com/2010/11/which-cones-are-needed-to-represent.html.

Equation (4) provides an *extended* formulation of the constraint (3), that is, an equivalent formulation using additional variables.

If we take the MINLPLIB2 library to be representative, then conic structure using standard cones exists in the overwhelming majority of MICP problems in practice. This observation calls for considering (MICONE) as a standard form of MICP, one which is perhaps more useful for computation than (MICONV) precisely because it is an extended formulation which encodes separability structure in a natural and general way. There is a large body of work and computational infrastructure for automatically generating the conic-form representation given an algebraic representation, a discussion we defer to Sect. 4.

The benefits of reformulation from (MICONV) to (MICONE) are quite tangible in practice. By direct reformulation from MICP to MISOCP, we were able to solve to global optimality the trimloss tls5 and tls6 instances from MINLPLIB2 by using Gurobi 6.0[2]. These instances from this public benchmark library had been unsolved since 2001, perhaps indicating that the value of conic formulations is not widely known.

3 An Outer-Approximation Algorithm for Mixed-Integer Conic Programming

Although the conic representation (MICONE) does not preclude the use of derivative-based solvers for continuous relaxations, derivative-based nonlinear solvers are typically not appropriate for conic problems because the nonlinear constraints which define the standard cones have points of nondifferentiability [13]. Sometimes the nondifferentiability is an artifact of the conic reformulation (e.g., of smooth functions x^2 and $\exp(x)$), but in a number of important cases the nondifferentiability is intrinsic to the model and provides additional modeling power. Nonsmooth perspective functions, for example, which are used in disjunctive convex programming, have been particularly challenging for derivative-based MICP solvers and have motivated smooth approximations [15]. On the other hand, conic form can handle these nonsmooth functions in a natural way, so long as there is a solver capable of solving the continuous conic relaxations.

There is a growing body of work as well as some (so far) experimental solvers supporting mixed second-order and exponential cone problems [26,27], which opens the door for considering conic solvers in place of derivative-based solvers. To the best of our knowledge, however, no outer-approximation algorithm or finite-time convergence theory has been proposed for general mixed-integer conic programming problems of the form (MICONE).

In this section, we present the first such algorithm for (MICONE) with arbitrary closed, convex cones. This algorithm generalizes the work of Drewes and Ulbrich [11] for MISOCP with a much simpler proof based on conic duality. In stating this algorithm, we hope to motivate further development of conic solvers for cones beyond the second-order and positive semidefinite cones.

---

[2] Solutions reported to Stefan Vigerske, October 5, 2015.

We begin with the definition of dual cones.

Definition 1. *Given a cone* $\mathcal{K}$, *we define* $\mathcal{K}^* := \{\beta \in \mathbb{R}^k : \beta^T z \geq 0 \; \forall z \in \mathcal{K}\}$ *as the dual cone of* $\mathcal{K}$.

Dual cones provide an equivalent outer description of any closed, convex cone, as the following lemma states. We refer readers to [6] for the proof.

Lemma 1. *Let* $\mathcal{K}$ *be a closed, convex cone. Then* $z \in \mathcal{K}$ *iff* $z^T \beta \geq 0 \; \forall \beta \in \mathcal{K}^*$.

Based on the above lemma, we will consider an outer approximation of (MICONE):

$$
\begin{aligned}
\min_{x,z} \quad & c^T z \\
\text{s.t.} \quad & A_x x + A_z z = b \qquad\qquad\qquad\qquad \text{(MIOA(T))} \\
& L \leq x \leq U, x \in \mathbb{Z}^n, \\
& \beta^T z \geq 0 \; \forall \beta \in T.
\end{aligned}
$$

Note that if $T = \mathcal{K}^*$, MIOA(T) is an equivalent semi-infinite representation of (MICONE). If $T \subset \mathcal{K}^*$ and $|T| < \infty$ then MIOA(T) is an MILP outer approximation of (MICONE) whose objective value is a lower bound on the optimal value of (MICONE).

The outer approximation (OA) algorithm is based on iteratively building up T until convergence in a finite number of steps to the optimal solution. First, we define the continuous subproblem for fixed integer value $\hat{x}$ which plays a key role in the OA algorithm:

$$
\begin{aligned}
v_{\hat{x}} = \min_z \quad & c^T z \\
\text{s.t.} \quad & A_z z = b - A_x \hat{x}, \qquad\qquad\qquad\qquad \text{(CP($\hat{x}$))} \\
& z \in \mathcal{K}.
\end{aligned}
$$

The dual of $(CP(\hat{x}))$ is

$$
\begin{aligned}
\max_{\beta,\lambda} \quad & \lambda^T (b - A_x \hat{x}) \\
\text{s.t.} \quad & \beta = c - A_z^T \lambda \qquad\qquad\qquad\qquad (5) \\
& \beta \in \mathcal{K}^*.
\end{aligned}
$$

The following lemmas demonstrate, essentially, that the dual solutions to $(CP(\hat{x}))$ provide the only elements of $\mathcal{K}^*$ that we need to consider.

Lemma 2. *Given* $\hat{x}$, *assume* $CP(\hat{x})$ *is feasible and strong duality holds at the optimal primal-dual solution* $(z_{\hat{x}}, \beta_{\hat{x}}, \lambda_{\hat{x}})$. *Then for any* z *with* $A_z z = b - A_x \hat{x}$ *and* $\beta_{\hat{x}}^T z \geq 0$, *we have* $c^T z \geq v_{\hat{x}}$.

Proof.

$$\beta_{\hat{x}}^T z = (c - A_z^T \lambda_{\hat{x}})^T z = c^T z - \lambda_{\hat{x}}^T (b - A_x \hat{x}) = c^T z - v_{\hat{x}} \geq 0. \tag{6}$$

Lemma 3. *Given $\hat{x}$, assume $CP(\hat{x})$ is infeasible and (5) is unbounded, such that we have a ray $(\beta_{\hat{x}}, \lambda_{\hat{x}})$ satisfying $\beta_{\hat{x}} \in \mathcal{K}^*$, $\beta_{\hat{x}} = -A_z^T \lambda_{\hat{x}}$, and $\lambda_{\hat{x}}^T (b - A_x \hat{x}) > 0$. Then for any z satisfying $A_z z = b - A_x \hat{x}$ we have $\beta_{\hat{x}}^T z < 0$.*

Proof.

$$\beta_{\hat{x}}^T z = -\lambda_{\hat{x}}^T A_z z = -\lambda_{\hat{x}}^T (b - A_x \hat{x}) < 0. \tag{7}$$

Algorithm 1. The conic outer approximation (OA) algorithm

> **Initialize:** $z_U \leftarrow \infty, z_L \leftarrow -\infty, T \leftarrow \emptyset$. Fix convergence tolerance ϵ.
> **while** $z_U - z_L \geq \epsilon$ **do**
>   Solve MIOA(T).
>   **if** MIOA(T) is infeasible **then**
>   (MICONE) is infeasible, so terminate.
>   **end if**
>   Let $(\hat{x}, \hat{z})$ be the optimal solution of MIOA(T) with objective value w_T.
>   Update lower bound $z_L \leftarrow w_T$.
>   Solve $CP(\hat{x})$.
>   **if** $CP(\hat{x})$ is feasible **then**
>   Let $(z_{\hat{x}}, \beta_{\hat{x}}, \lambda_{\hat{x}})$ be an optimal primal-dual solution with objective value $v_{\hat{x}}$.
>   **if** $v_{\hat{x}} < z_U$ **then**
> $z_U \leftarrow v_{\hat{x}}$
>   Record $(\hat{x}, z_{\hat{x}})$ as the best known solution.
>   **end if**
>   **else if** $CP(\hat{x})$ is infeasible **then**
>   Let $(\beta_{\hat{x}}, \lambda_{\hat{x}})$ be a ray of (5).
>   **end if**
> $T \leftarrow T \cup \{\beta_{\hat{x}}\}$
> **end while**

Finite termination of the algorithm is guaranteed because integer solutions $\hat{x}$ cannot repeat, and only a finite number of integer solutions is possible.

This algorithm is arguably incomplete because the assumptions of Lemmas 2 and 3 need not always hold. The assumption of strong duality at the solution is analogous to the constraint qualification assumption of the NLP OA algorithm [12]. Drewes and Ulbrich [11] describe a procedure in the case of MISOCP to ensure finite termination if this assumption does not hold. The assumption that a ray of the dual exists if the primal problem is infeasible is also not always true in the conic case, though [6] provide a characterization of when this can occur. These cases will receive full treatment in future work.

A notable difference between the conic OA algorithm and the standard NLP OA algorithm is that there is no need to solve a second subproblem in the case of infeasibility, although some specialized NLP solvers may also obviate this need [2]. In contrast, Drewes and Ulbrich [11] propose a second subproblem in the case of MISOCP even when dual rays would suffice.

Finally, the algorithm is presented in terms of a single cone $\mathcal{K}$ for simplicity. When $\mathcal{K}$ is a product of cones, our implementation disaggregates the elements of $\mathcal{K}^*$ per individual cone, adding one OA cut per cone per iteration.

4   Extended Formulations and Disciplined Convex Programming

While many problems are representable in conic form, the transformation from the user's algebraic representation of the problem often requires expert knowledge. Disciplined convex programming (DCP) is an algebraic modeling concept proposed by Grant et al. [14], one of whose original motivations was to provide a means to make these transformations automatic and transparent to users. DCP is not intrinsically tied to conic representations, however. In this section, we present the basic concepts of DCP from the viewpoint of extended formulations. This perspective both provides insight into how conic formulations are generated and enables further generalization of the technique to problems which are not conic representable using standard cones.

Detection of convexity of arbitrary nonlinear expressions is NP-Hard [4], and since a conic-form representation is a proof of convexity, it is unreasonable to expect a modeling system to be able to reliably generate these representations from arbitrary input. Instead, DCP requires users to construct expressions whose convexity can be proven by simple composition rules. A DCP implementation (e.g., the MATLAB package CVX) provides a library of basic operations like addition, subtraction, norms, square root, square, geometric mean, logarithms, exponential, entropy $x \log(x)$, powers, absolute value, $\max\{x, y\}$, $\min\{x, y\}$, etc. whose curvature (convex, concave, or affine) and monotonicity properties are known. These basic operations are called *atoms*.

All expressions representing the objective function and constraints are built up via compositions of these atoms in such a way that guarantees convexity. For example, the expression $\exp(x^2 + y^2)$ is convex and *DCP compliant* because $\exp(\cdot)$ is convex and monotone increasing and $x^2 + y^2$ is convex because it is a convex composition (through addition) of two convex atoms. The expression $\sqrt{xy}$ is concave when $x, y \geq 0$ as we noted previously, but not *DCP compliant* because the inner term xy has indefinite curvature. In this case, users must reformulate their expression using a different atom like $geomean(x, y)$. We refer readers to [10,14] for further introduction to DCP.

An important yet not well-known aspect of DCP is that the composition rules for DCP have a 1-1 correspondence with the existence of extended formulations of epigraphs. For example, suppose g is convex and monotone increasing and f is convex. Then the function $h(x) := g(f(x))$ is convex and recognized as such by DCP. If $E_h := \{(x, t) : h(x) \leq t\}$ is the epigraph of h, then we can represent E_h through an extended formulation using the epigraphs E_g and E_f of g and f, respectively. That is, $(x, t) \in E_h$ iff $\exists s$ such that $(x, s) \in E_f$ and $(s, t) \in E_g$. The validity of this extended formulation follows directly from monotonicty of g. Furthermore, if E_f and E_g are conic representable, then so is E_h, which is precisely how DCP automatically generates conic formulations. The conic form representation is not necessary, however; one may instead represent E_f and E_g using smooth nonlinear constraints if f and g are smooth.

This correspondence between composition of functions and extended formulations was considered by Tawarmalani and Sahinidis [28], although in the context of nonconvex MINLP. Composition generalizes the notion of separability far beyond summations of univariate functions as proposed by Hijazi et al. [19].

DCP, based on the philosophy that users should be "disciplined" in their modeling of convex functions, describes a simple set of rules for verifying convexity and rejects any expressions not satisfying them; it is not based on ad-hoc detection of convexity which is common among nonconvex MINLP solvers. DCP is well established within the convex optimization community as a practical modeling technique, and many would agree that it is reasonable to ask users to formulate convex optimization problems in DCP form. By doing so they unknowingly provide all of the information needed to generate powerful extended formulations.

5   Computational Results

In this section we present preliminary computational results implementing the extended formulations proposed in this work. We have implemented a solver, *Pajarito*, which currently accepts input as mixed-integer conic programming problems with a mix of second-order and exponential cones. We have translated 194 convex problems from MINLPLIB2 representable using these cones into Convex.jl [29], a DCP algebraic modeling language in Julia which performs automatic transformation into conic form. We exclude instances without integer constraints, some which are pure quadratic, and some which Bonmin is unable to solve within time limits. Pajarito currently implements traditional OA using derivative-based NLP solvers [7] applied to the conic extended formulation, as the conic solvers we tested were not sufficiently reliable. Pajarito itself relies on JuMP [23], and the implementation of the core algorithm spans less than 1000 lines of code. Pajarito will be released as open source in the upcoming months.

Our main comparison is with Bonmin's OA algorithm, which in 2014 benchmarks by H. Mittelmann was found to be the overall fastest MICP solver when using CPLEX as the inner MILP solver [25]. For the MISOCP instances, we also compare with CPLEX. Due to space limitations, we refer readers to [24] for the appendix and figures containing our preliminary results. Their highlights are:

1. We observe that the extended formulation helps significantly reduce the number of OA iterations. This can be seen as a sign of scalability provided by the extended formulation.
2. Pajarito is much faster on many of the challenging problems (slay,netmod, portfol_classical), although these problems are MISOCPs where CPLEX

dominates (note that CPLEX 12.6.2 already applies extended formulations for MISOCPs [30]). Pajarito has not been optimized for performance, leading Bonmin to be faster on the relatively easy instances.

3. Perhaps the strongest demonstration of Pajarito's strength is the gams01 instance, which was previously unsolved and whose conic representation requires a mix of SOC and EXP cones. The best known bound was 1735.06 and the best known solution was 21516.83. Pajarito solved the instance to optimality with an objective value of 21380.20 in 6 iterations. Unfortunately, the origin of the instance is unknown and confidential.

Acknowledgements. We thank the anonymous referees for their comments. They greatly improved the clarity of the manuscript. We also thank one of the anonymous referees for pointing out the SOC-representability of the sssd family of instances originally derived in [15]. M. Lubin was supported by the DOE Computational Science Graduate Fellowship, which is provided under grant number DE-FG02-97ER25308. The work at LANL was funded by the Center for Nonlinear Studies (CNLS) and was carried out under the auspices of the NNSA of the U.S. DOE at LANL under Contract No. DE-AC52-06NA25396. J.P. Vielma was funded by NSF grant CMMI-1351619.

References

1. MINLPLIB2 library. http://www.gamsworld.org/minlp/minlplib2/html/
2. Abhishek, K., Leyffer, S., Linderoth, J.: FilMINT: An outer approximation-based solver for convex mixed-integer nonlinear programs. INFORMS J. Comput. **22**, 555–567 (2010)
3. Achterberg, T.: SCIP: solving constraint integer programs. Math. Program. Comput. **1**, 1–41 (2009)
4. Ahmadi, A., Olshevsky, A., Parrilo, P., Tsitsiklis, J.: NP-hardness of deciding convexity of quartic polynomials and related problems. Math. Program. **137**, 453–476 (2013)
5. Belotti, P., Kirches, C., Leyffer, S., Linderoth, J., Luedtke, J., Mahajan, A.: Mixed-Integer nonlinear optimization. Acta Numerica **22**, 1–131 (2013)
6. Ben-Tal, A., Nemirovski, A.: Lectures on Modern Convex Optimization. Society for Industrial and Applied Mathematics, Philadelphia (2001)
7. Bonami, P., Biegler, L.T., Conn, A.R., Cornuéjols, G., Grossmann, I.E., Laird, C.D., Lee, J., Lodi, A., Margot, F., Sawaya, N., Wächter, A.: An algorithmic framework for convex mixed integer nonlinear programs. Discrete Optim. **5**, 186–204 (2008)
8. Bonami, P., Kilinç, M., Linderoth, J.: Algorithms and software for convex mixed integer nonlinear programs. In: Lee, J., Leyffer, S. (eds.) Mixed Integer Nonlinear Programming. The IMA Volumes in Mathematics and its Applications, vol. 154, pp. 1–39. Springer, New York (2012)
9. Byrd, R.H., Nocedal, J., Waltz, R.: KNITRO: An integrated package for nonlinear optimization. In: di Pillo, G., Roma, M. (eds.) Large-Scale Nonlinear Optimization. Nonconvex Optimization and its Applications, vol. 83, pp. 35–59. Springer, Berlin (2006)
10. Diamond, S., Chu, E., Boyd, S.: Disciplined convex programming. http://dcp.stanford.edu/

11. Drewes, S., Ulbrich, S.: Subgradient based outer approximation for mixed integer second order cone programming. In: Lee, J., Leyffer, S. (eds.) Mixed Integer Nonlinear Programming. The IMA Volumes in Mathematics and its Applications, vol. 154, pp. 41–59. Springer, New York (2012)
12. Fletcher, R., Leyffer, S.: Solving mixed integer nonlinear programs by outer approximation. Math. Program. **66**, 327–349 (1994)
13. Goldberg, N., Leyffer, S.: An active-set method for second-order conic-constrained quadratic programming. SIAM J. Optim. **25**, 1455–1477 (2015)
14. Grant, M., Boyd, S., Ye, Y.: Disciplined convex programming. In: Liberti, L., Maculan, N. (eds.) Global Optimization. Nonconvex Optimization and its Applications, vol. 84, pp. 155–210. Springer, US (2006)
15. Günlük, O., Linderoth, J.: Perspective reformulation and applications. In: Lee, J., Leyffer, S. (eds.) Mixed Integer Nonlinear Programming. The IMA Volumes in Mathematics and its Applications, vol. 154, pp. 61–89. Springer, New York (2012)
16. Gupta, O.K., Ravindran, A.: Branch and bound experiments in convex nonlinear integer programming. Manag. Sci. **31**, 1533–1546 (1985)
17. Harjunkoski, I., Westerlund, T., Pörn, R., Skrifvars, H.: Different transformations for solving non-convex trim-loss problems by MINLP. Eur. J. Oper. Res. **105**, 594–603 (1998)
18. Hien, L.: Differential properties of euclidean projection onto power cone. Math. Methods Oper. Res. **83**(3), 265–284 (2015)
19. Hijazi, H., Bonami, P., Ouorou, A.: An outer-inner approximation for separable mixed-integer nonlinear programs. INFORMS J. Comput. **26**, 31–44 (2014)
20. Kılınç, M.R.: Disjunctive cutting planes and algorithms for convex mixed integer nonlinear programming. Ph.D. thesis, University of Wisconsin-Madison (2011)
21. Leyffer, S.: Deterministic methods for mixed integer nonlinear programming. Ph.D. thesis, University of Dundee, December 1993
22. Lobo, M.S., Vandenberghe, L., Boyd, S., Lebret, H.: Applications of second-order cone programming. Linear Algebra Appl. **284**, 193–228 (1998). International Linear Algebra Society (ILAS) Symposium on Fast Algorithms for Control, Signals and Image Processing
23. Lubin, M., Dunning, I.: Computing in operations research using Julia. INFORMS J. Comput. **27**, 238–248 (2015)
24. Lubin, M., Yamangil, E., Bent, R., Vielma, J.P.: Extended formulations in mixed-integer convex programming, ArXiv e-prints (2015)
25. Mittelmann, H.: MINLP benchmark. http://plato.asu.edu/ftp/minlp_old.html
26. O'Donoghue, B., Chu, E., Parikh, N., Boyd, S.: Operator splitting for conic optimization via homogeneous self-dual embedding, ArXiv e-prints (2013)
27. Serrano, S.A.: Algorithms for unsymmetric cone optimization and an implementation for problems with the exponential cone. Ph.D. thesis, Stanford University, Stanford, CA, March 2015
28. Tawarmalani, M., Sahinidis, N.V.: A polyhedral branch-and-cut approach to global optimization. Math. Program. **103**, 225–249 (2005)
29. Udell, M., Mohan, K., Zeng, D., Hong, J., Diamond, S., Boyd, S.: Convex optimization in Julia. In: Proceedings of HPTCDL 2014, Piscataway, NJ, USA, pp. 18–28. IEEE Press (2014)
30. Vielma, J.P., Dunning, I., Huchette, J., Lubin, M.: Extended formulations in mixed integer conic quadratic programming, ArXiv e-prints (2015)

k-Trails: Recognition, Complexity, and Approximations

Mohit Singh[1] and Rico Zenklusen[2]([✉])

[1] Microsoft Research, Redmond, USA
mohits@microsoft.com
[2] ETH Zurich, Zurich, Switzerland
ricoz@math.ethz.ch

Abstract. The notion of degree-constrained spanning hierarchies, also called k-trails, was recently introduced in the context of network routing problems. They describe graphs that are homomorphic images of connected graphs of degree at most k. First results highlight several interesting advantages of k-trails compared to previous routing approaches. However, so far, only little is known regarding computational aspects of k-trails.

In this work we aim to fill this gap by presenting how k-trails can be analyzed using techniques from algorithmic matroid theory. Exploiting this connection, we resolve several open questions about k-trails. In particular, we show that one can recognize efficiently whether a graph is a k-trail, and every graph containing a k-trail is a $(k + 1)$-trail. Moreover, further leveraging the connection to matroids, we consider the problem of finding a minimum weight k-trail contained in a graph G. We show that one can efficiently find a $(2k - 1)$-trail contained in G whose weight is no more than the cheapest k-trail contained in G, even when allowing negative weights.

The above results settle several open questions raised by Molnár, Newman, and Sebő.

1 Introduction

Motivated by applications in network routing, the notion of *degree-constrained spanning hierarchies* was introduced as a way to obtain lower-degree routing structures as what could be obtained with low-degree spanning subgraphs [5,6]. These hierarchies, which, for brevity, have also simply been called *k-trails* [7,9], describe how a given graph can be described as the homomorphic image of another low-degree graph. First results have been reported that show advantages of k-trails compared to traditional methods in network routing contexts [6]. However, many basic questions around k-trails remained open, including whether one can efficiently decide if a graph is a k-trail; we refer the interested reader to [9] for a nice overview of some open problems around k-trails. The goal of this work is to fill this gap by answering several basic open questions about k-trails, by revealing and exploiting a connection to matroids.

© Springer International Publishing Switzerland 2016
Q. Louveaux and M. Skutella (Eds.): IPCO 2016, LNCS 9682, pp. 114–125, 2016.
DOI: 10.1007/978-3-319-33461-5_10

Before giving a summary of our main results, we start by formally defining k-trails as well as some closely related notions. In particular, the notion of homomorphic images introduced below is the basis for defining k-trails. Throughout this paper, we focus on undirected connected graphs with possibly loops and parallel edges, and with at least 2 vertices to avoid trivial special cases.

Definition 1 (Homomorphic Image). *A graph $G = (V, E)$ is the homomorphic image of a graph $H = (W, F)$ if there is an onto function $\phi : W \to V$ such that for any two vertices $u, v \in V$ (with possibly $u = v$), the number of edges in G between u and v is equal to the number of edges in H whose endpoints get mapped by ϕ to $\{u, v\}$.*

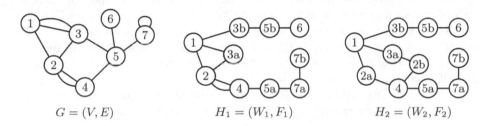

$$G = (V, E) \qquad\qquad H_1 = (W_1, F_1) \qquad\qquad H_2 = (W_2, F_2)$$

Fig. 1. The graph G is the homomorphic image of H_1 as well as H_2. The naming of the vertices has been chosen to highlight the corresponding homomorphisms, e.g., the nodes 3a and 3b in H_1 are both mapped to node 3 in G. G is a 3-trail, because the homomorphic preimage H_2 has maximum degree 3.

Figure 1 shows an example graph and 2 homomorphic preimages of it.

Hence, a preimage $H = (W, F)$ of G corresponds to a graph obtained from G by splitting each of its vertices $v \in V$ into $|\phi^{-1}(v)|$ many copies. We therefore call $|\phi^{-1}(v)|$ the ϕ-*multiplicity*, or simply *multiplicity*, of v.

Definition 2 (k-Trail). *A graph $G = (V, E)$ is a k-trail if it is the homomorphic image of a connected graph $H = (W, F)$ with maximum degree at most k.*

The graph G shown in Fig. 1 is a 3-trail since it is the homomorphic image of H_2, which has maximum degree 3. It is not hard to see that k-trails can equivalently be defined as preimages of trees of degree at most k.

Definition 3 (k-Tree). *A graph is a k-tree if it is a spanning tree of maximum degree at most k.*

Lemma 1. *If G is a k-trail then it is a homomorphic image of a k-tree. More precisely, given a connected graph $H = (W, F)$ and onto function $\phi : W \to V$ such that G is the homomorphic image of H by ϕ, we can construct efficiently a tree $H' = (W', F')$ and onto function $\phi' : W' \to W$ such that H is the homomorphic image of H' by ϕ'. Thus, G is the $\phi' \circ \phi$-homomorphic image of H', and for $v \in V$, the $\phi' \circ \phi$-multiplicity of v is at least the ϕ-multiplicity of v.*

Many basic questions on k-trails remained open. This includes the complexity status of deciding whether a graph is a k-trail for a given k, an open question raised in [9]. A nice result shown in [7,9] is that every 2-edge-connected graph is a 3-trail. Further interesting open questions on k-trails that are motivated by routing applications are linked to the notion of whether a graph *contains* a k-trail, which is defined as follows.

Definition 4 (Containing a k-Trail). *We say that a graph $G = (V, E)$ contains a k-trail if there is a set $U \subseteq E$ such that $G' = (V, U)$ is a k-trail.*

Notice that all k-trails are connected graphs, since they are homomorphic images of connected graphs. Hence, candidate edge sets $U \subseteq E$, for $G = (V, E)$ to contain a k-trail (V, U), must be such that (V, U) is connected.

Also many questions linked to the containment of k-trails are open [9], including questions related to approximation algorithms for finding minimum weight k-trails. In particular, it was conjectured in [9] that for any nonnegative edge weights $w : E \to \mathbb{Z}_{\geq 0}$ and any $k \geq 3$, there exists a polynomial algorithm returning a $(2k-2)$-trail in G whose cost is not larger than the minimum weight k-trail in G, if G contains a k-trail.

In this paper we are able to settle most of the above-mentioned open questions, by presenting a new viewpoint on k-trails in terms of matroids.

1.1   Our Results

One of our main results, whose derivation will also be used to highlight a strong link between k-trails and matroids, is the fact that k-trails can be recognized efficiently.

Theorem 1. *Given a graph $G = (V, E)$ and $k \in \mathbb{Z}_{>0}$, it can be checked efficiently whether G is a k-trail, and if so, obtain a connected graph $H = (W, F)$ with degrees bounded by k and onto function $\phi : W \to V$ such that G is the homomorphic image of H by ϕ.*

Contrary to the recognition problem, the containment problem is hard. Our hardness proof can be interpreted as a natural extension of a hardness proof shown in [5], for a weighted version of the problem. For completeness, we provide a full proof in the long version of the paper.

Theorem 2. *For any $k \in \mathbb{Z}_{\geq 2}$, the problem of deciding whether a graph contains a k-trail is NP-complete.*

Despite the different complexity status of the containment and recognition question, the following theorem shows that they are closely related.

Theorem 3. *If G contains a k-trail then it is a $(k+1)$-trail.*

Using that recognition is polynomial time solvable, we can thus find the smallest k for which a given graph G is a k-trail, which then implies by Theorem 3 that the smallest k' for which G contains a k'-trail is either k or $k - 1$. Finally, we obtain the following result on the containment of weighted k-trails.

Theorem 4. *There exists a polynomial time algorithm that, given a graph $G = (V, E)$ with weight function $w : E \to \mathbb{Z}$ and an integer $k \geq 2$, either shows that there is no k-trail contained in G or returns a $(2k-1)$-trail contained in G whose total weight is at most the weight of any k-trail contained in G.*

Theorem 4 almost resolves a conjecture in [9], claiming that one can efficiently find a $(2k - 2)$-trail in G of weight no more than the weight of any k-trail contained in G, assuming $k \geq 3$ and nonnegativity of the weights. Our result only implies the existence of a cheap $(2k-1)$-trail; however, it holds for arbitrary weights. Furthermore, we can show that, for arbitrary weights, the factor $2k - 1$ is optimal when comparing to a natural LP relaxation.

Organization of Paper. Section 2 proves Theorem 1 and shows how k-trails can be studied using tools from algorithmic matroid theory. Section 3 discusses our algorithm for Theorem 4. Due to space constraints, the proofs of Lemmas 1–5, Theorems 1, 2, and the correctness proof of the algorithm in Sect. 3, which completes the proof of Theorem 4, have been omitted from this extended abstract.

Basic Terminology. The degree of a vertex $v \in V$ in a graph $G = (V, E)$ is denoted by $\deg_G(v)$, or simply $\deg(v)$ if there is no danger of confusion. If the graph is clear from context, we will also use the notation $\deg_U(v) := |\delta(v) \cap U|$ for a set $U \subseteq E$ and $v \in V$.

2   Recognition of *k*-Trails

Let $G = (V, E)$ be an undirected graph and assume we want to show that G is a k-trail for k as small as possible. Consider some connected graph $H = (W, F)$ such that G is the homomorphic image of H by some onto function $\phi : W \to V$. Let $v \in V$ and consider all vertices of H that get mapped to v, i.e., $\phi^{-1}(v) = \{w_1, \ldots, w_\ell\}$, where $\ell = |\phi^{-1}(v)|$ is the multiplicity of v. Clearly, we have

$$\deg_G(v) = \sum_{i=1}^{\ell} \deg_H(w_i).$$

Knowing that v gets split into ℓ vertices by ϕ, for degrees to be low in H it would be best if all w_i for $i \in [\ell]$ have about the same degree. It turns out that starting with any H and corresponding ϕ, we can balance out the degrees of vertices in H that correspond to the same vertex in G, using a simple modification algorithm. The following lemma formalizes this statement.

Lemma 2. *Given two connected graphs $G = (V, E)$ and $H = (W, F)$, and an onto function $\phi : W \to V$ such that G is the homomorphic image of H, one can determine in polynomial time a connected graph $H' = (W, F')$ such that*

(i) G is the homomorphic image of H' by ϕ.

(ii) For any $v \in V$ and $w \in W$ such that $\phi(w) = v$, the degree of w in H' is either $\left\lfloor \frac{\deg_G(v)}{|\phi^{-1}(v)|} \right\rfloor$ or $\left\lceil \frac{\deg_G(v)}{|\phi^{-1}(v)|} \right\rceil$.

We leverage the above lemma to rephrase the problem of whether a graph is a k-trail in terms of multiplicities.

Definition 5 (Feasible Multiplicity Vector). *Let $G = (V, E)$ be a graph. A vector $\lambda \in \mathbb{Z}_{>0}^V$ is a feasible multiplicity vector (for G) if G is the homomorphic image of a connected graph with multiplicities given by λ; more formally, if there is a connected graph $H = (W, F)$ such that G is the homomorphic image of H by some onto function $\phi : W \to V$, and $|\phi^{-1}(v)| = \lambda(v) \ \forall v \in V$.*

Feasible multiplicity vectors fulfill the following down-monotonicity property.

Lemma 3. *Let G be a graph and $\lambda \in \mathbb{Z}_{>0}^V$ be a feasible multiplicity vector. Then any vector $\lambda' \in \mathbb{Z}_{>0}^V$ with $\lambda' \leq \lambda$ (component-wise) is also a feasible multiplicity vector.*

Furthermore, this result is constructive: Given a connected graph H and homomorphism ϕ such that G is the ϕ-homomorphic image of H and λ is the multiplicity vector corresponding to ϕ, we can efficiently construct for any $\lambda' \leq \lambda$ a connected graph H' and homomorphism ϕ' such that G is the ϕ'-homomorphic image of H' with corresponding multiplicity vector λ'.

Lemmas 2 and 3 easily imply that the question of whether G is a k-trail for some given k can be reduced to the problem of deciding whether some multiplicity vector $\lambda \in \mathbb{Z}_{>0}^V$ is feasible.

Lemma 4. *A graph $G = (V, E)$ is a k-trail if and only if the following multiplicity vector $\lambda \in \mathbb{Z}_{>0}^V$ is feasible:*

$$\lambda(v) = \left\lceil \frac{\deg_G(v)}{k} \right\rceil \qquad \forall v \in V.$$

To finally provide an efficient recognition algorithm to decide whether a graph is a k-trail, we show that feasible multiplicity vectors are highly structured.

Notice that a feasible multiplicity vector is at least 1 in each coordinate. For simplicity, we introduce a shifted version of feasible multiplicity vectors, called *feasible split vector*; a vector $\mu \in \mathbb{Z}_{\geq 0}^V$ is a feasible split vector if $\mu + \mathbf{1}$ is a feasible multiplicity vector, where $\mathbf{1} \in \mathbb{Z}^V$ is the all-ones vector. Hence, a split vector tells us how many times a vertex is split.

Theorem 5. *Let G be an undirected graph. The set of feasible split vectors correspond to the integral points of a polymatroid[1], i.e.,*

$$P_G := \mathrm{conv}(\{\mu \in \mathbb{Z}_{\geq 0}^V \mid \mu \text{ is a feasible split vector}\}),$$

---

[1] A polymatroid over a finite set N is a polytope $P \subseteq \mathbb{R}_{\geq 0}^N$ described by $P = \{x \in \mathbb{R}_{\geq 0}^N \mid x(S) \leq f(S) \ \forall S \subseteq N\}$, where $f : 2^N \to \mathbb{Z}_{\geq 0}$ is a submodular function, and $x(S) = \sum_{v \in S} x_v$. We refer the interested reader to [8, vol. B] for more information on polymatroids.

is a polymatroid. Furthermore, we can efficiently optimize over P_G, and for any feasible split vector $\mu \in \mathbb{Z}_{\geq 0}^V$ we can efficiently find a connected graph $H = (W, F)$ and an onto function $\phi : W \to V$ such that G is the homomorphic image of H, and $|\phi^{-1}(v)| = \mu(v) \; \forall v \in V$.

Before proving the theorem, we start with a few observations and show that Theorem 5 implies Theorem 1. It is well-known that P_G being a polymatroid implies that P_G is given by

$$P_G = \{x \in \mathbb{R}_{\geq 0}^V \mid x(S) \leq f(S) \; \forall S \subseteq V\},$$

where $f : 2^V \to \mathbb{Z}_{\geq 0}$ is the submodular function defined by

$$f(S) = \max\{x(S) \mid x \in P_G\} \qquad \forall S \subseteq V.$$

Many results on polymatroids typically assume that a polymatroid is given through a value oracle for the function f. Clearly, if we can efficiently optimize over P_G, we can also evaluate efficiently the submodular function f, which, as described above, corresponds to maximizing a $\{0, 1\}$-objective over P_G.

We are particularly interested in checking whether some split vector $\mu \in \mathbb{Z}_{\geq 0}^V$ is feasible. Having an efficient evaluation oracle for f allows for checking whether $\mu \in P_G$ by standard techniques: It suffices to solve the submodular function minimization problem $\min\{f(S) - x(S) \mid S \subseteq V\}$; if the optimal value is negative then $\mu \notin P_G$, otherwise $\mu \in P_G$. We will later see that the link between k-trails and matroids that we establish implies an easy way to check whether $\mu \in P_G$ using a simple matroid intersection problem (without relying on submodular function minimization).

Combining the above results and observations, Theorem 1 easily follows.

Proof (of Theorem 1). Let $\lambda \in \mathbb{Z}_{>0}^V$ be defined as in Lemma 4, and let $\mu = \lambda - 1$. Lemma 4—rephrased in terms of μ—states that G is a k-trail if and only if μ is a feasible split vector, which can be checked efficiently by Theorem 5. Furthermore, if μ is feasible, then Theorem 5 also shows that we can efficiently obtain a connected graph $\bar{H} = (W, \bar{F})$ and onto function $\phi : W \to V$ such that

(i) G is the homomorphic image of $\bar{H}$ by ϕ, and
(ii) $|\phi^{-1}(v)| = \mu(v) + 1 \quad \forall v \in V$.

By Lemma 2 we can balance the degrees of $\bar{H}$ for each vertex set $\phi^{-1}(v)$ efficiently, to obtain a connected graph $H = (W, F)$ with balanced degrees as stated in Lemma 2. It remains to observe that all degrees in H are bounded by k. This indeed holds: let $w \in W$ and $v = \phi(w)$; we thus obtain

$$\deg_H(w) \leq \left\lceil \frac{\deg_G(v)}{|\phi^{-1}(v)|} \right\rceil = \left\lceil \frac{\deg_G(v)}{\mu(v) + 1} \right\rceil = \left\lceil \frac{\deg_G(v)}{\left\lceil \frac{\deg_G(v)}{k} \right\rceil} \right\rceil \leq \left\lceil \frac{\deg_G(v)}{\frac{\deg_G(v)}{k}} \right\rceil = k,$$

where the first inequality follows by Lemma 2 and the second equality by $\mu(v) + 1 = \lambda(v) = \lceil \deg_G(v)/k \rceil$. $\qquad \square$

Matroidal Description of k-Trails and Proof of Theorem 5. We start by introducing an auxiliary graph $G' = (V', E')$ such that spanning trees in G' can be interpreted as graphs H such that G is a homomorphic image of H. Using this connection, we then derive that P_G is a polymatroid over which we can optimize efficiently, and show how to construct a homomorphic preimage of G corresponding to some split vector μ, as claimed by Theorem 5.

Hence, let $G = (V, E)$ be an undirected graph. The graph $G' = (V', E')$ contains a vertex for each of the two endpoints of each edge in E. More formally, for each $v \in V$, the Graph G' contains $\deg_G(v)$ many vertices $V'_v := \{v_e\}_{e \in \delta(v)}$; hence,

$$V' = \bigcup_{v \in V} V'_v.$$

We note that describing V'_v by $\{v_e\}_{e \in \delta(v)}$ is a slight abuse of notation, since for each loop at v we include two vertices in V'_v and not just one. Furthermore,

$$E' = \bar{E} \cup K, \quad \text{where} \quad \bar{E} = \{\{v_e, u_e\} \mid e = \{u, v\} \in E\}, \quad \text{and}$$

$$K = \bigcup_{v \in V} K_v, \quad \text{where} \quad K_v = \{\{v_e, v_f\} \mid v_e, v_f \in V'_v, v_e \neq v_f\} \quad \forall v \in V.$$

See Fig. 2 for an example of the above construction.

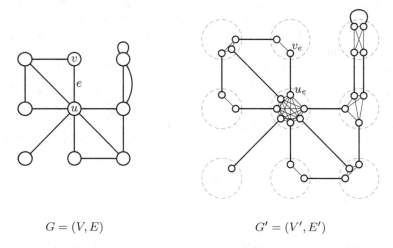

$$G = (V, E) \qquad\qquad\qquad G' = (V', E')$$

Fig. 2. An example for the construction of the auxiliary graph $G' = (V', E')$ from $G = (V, E)$. In G' the thick edges correspond to edges in $\bar{E}$ and the thin ones to edges in K. The dashed gray circles correspond to the cliques (V'_v, K_v) and highlight the link between the vertices $v \in V$ in G and vertex sets V'_v in G' which correspond to v.

For any spanning tree $T \subseteq E'$ in G' that contains $\bar{E}$, we define a graph $H_T = (W_T, F_T)$ and an onto function $\phi_T : W_T \to V$, such that G is the homomorphic image of H_T by ϕ, as follows. Let $K_T = T \cap K = T \backslash \bar{E}$. For each $v \in V$ consider the connected components of (V'_v, K_v). Let q_v be the number of these connected components and let

$$V'_v = C^1_v \cup C^2_v \cup \cdots \cup C^{q_v}_v$$

be the partition of V'_v into vertex sets of the q_v connected components in (V'_v, K_v).

We now define $H_T = (W_T, F_T)$ as the graph obtained from G' by contracting all C^j_v for $v \in V$ and $j \in [q_v] := \{1, \ldots, q_v\}$. For clarity, we call the vertices in H_T *nodes*. Contracting C^j_v corresponds to replacing C^j_v with a single node, which we identify with the set C^j_v for simplicity, thus leading to the following set of nodes for H_T:

$$W_T = \{C^j_v \mid v \in V, j \in [q_v]\}.$$

Furthermore, two nodes $C^j_v, C^\ell_w \in W_T$ are adjacent if and only if there is a pair of vertices, one in C^j_v and one in C^ℓ_w, that are connected by an edge in $T \cap \bar{E}$; formally, this corresponds to the existence of $e \in E$ such that edge in $\bar{E}$ that corresponds to e is in T, and $v_e \in C^j_v$ and $w_e \in C^\ell_w$. Moreover, $\phi_T : W_T \to V$ is defined by

$$\phi_T(C^j_v) = v \quad \forall \, C^j_v \in W_T.$$

Figure 3 shows an example construction of H_T from a spanning tree T that contains $\bar{E}$. We start with some observations that follow immediately from the above construction:

- G is the homomorphic image of H_T by ϕ_T.
- The multiplicity of $v \in V$ is q_v.
- G' can be constructed efficiently from G.
- Several spanning trees T can lead to the same graph H_T and homomorphism ϕ_T; indeed, for any component C^j_v, the edges of T with both endpoints in C^j_v form a spanning tree over C^j_v, which can be replaced by any other spanning tree without changing H_T or ϕ_T.

Conversely, every description of G as a homomorphic image of a tree can be obtained by the above construction:

Claim. Let $H = (W, F)$ be a tree and $\phi : W \to V$ be an onto function such that G is the homomorphic image of H by ϕ. Then there exists a spanning tree T in G' such that $H = H_T$ and $\phi = \phi_T$.

Proof (of claim). Indeed, the spanning tree T can be chosen as follows. Starting with $T = \emptyset$, we first add to T all edges in $\bar{E}$. For each $w \in W$, we do the following: Let $v = \phi(w)$, and let $e_1, \ldots, e_h \in E$ be the ϕ-images of the edges $\delta_H(w)$; in particular, $e_1, \ldots, e_h \in \delta_G(v)$. We add to T an arbitrary set of $h - 1$ edges of G' that form a spanning tree over the vertices $v_{e_1}, \ldots, v_{e_h}$. One can now easily observe that the constructed tree T has the desired properties. $\qquad\square$

The equivalence between (ii) and (iii) in the following statement summarizes the above discussion. The equivalence between (i) and (ii) follows from Lemma 1 (implying (i) $\Rightarrow$ (ii)) and Lemma 3 (implying (ii) $\Rightarrow$ (i)).

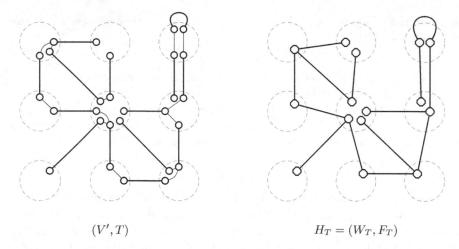

$$(V', T) \qquad\qquad\qquad H_T = (W_T, F_T)$$

Fig. 3. On the left-hand side, a spanning tree T in the auxiliary graph $G' = (V', E')$, that corresponds to the graph G shown in Fig. 2, is highlighted. On the right-hand side, the corresponding graph H_T is shown. The homomorphism $\phi_T : W_T \to V$ maps all vertices of H_T within the same dashed circle to the same vertex of G. For $v \in V$, the number q_v is equal to the number of vertices on the right-hand side lying inside the dashed circle corresponding to v.

Property 1. Let $G = (V, E)$ be an undirected graph and $\mu \in \mathbb{Z}_{\geq 0}^V$. The following three statements are equivalent:

(i) G is the homomorphic image of a connected graph $H = (W, F)$ by an onto function $\phi : W \to V$ such that $|\phi^{-1}(v)| = \mu(v) + 1 \ \forall v \in V$.

(ii) G is the homomorphic image of a tree $H = (V, W)$ by an onto function $\phi : W \to V$ such that $|\phi^{-1}(v)| \geq \mu(v) + 1 \ \forall v \in V$.

(iii) There is a spanning tree T in the auxiliary graph $G' = (V', E')$ such that $\bar{E} \subseteq T$ and $|T \cap K_v| \leq |\delta(v)| - 1 - \mu(v) \ \forall v \in V$.

Furthermore, we highlight that the equivalences in Property 1 are all constructive.

Using the above connection between the auxiliary graph G' and homomorphic preimages of G, Theorem 5 can now be derived as follows. For brevity, we use the following notation. For any spanning tree T in G' such that $\bar{E} \subseteq T$, we define $\alpha_T \in \mathbb{Z}_{\geq 0}^V$ by $\alpha_T(v) := |T \cap K_v| \ \forall v \in V$. Furthermore, let $\deg \in \mathbb{Z}_{\geq 0}^V$ be the degree vector of G, i.e., $\deg(v)$ is the degree of v as usual. The equivalence between point (i) and point (iii) of Property 1 can thus be rephrased as follows.

$$\mu \in \mathbb{Z}_{\geq 0}^V \text{ is a feasible split vector} \quad \Leftrightarrow \quad \begin{array}{l} \exists \text{ spanning tree } T \text{ in } G' \text{ with} \\ \bar{E} \subseteq T \text{ and } \alpha_T \leq \deg -\mathbf{1} - \mu, \end{array} \quad (1)$$

where $\mathbf{1} \in \mathbb{Z}^V$ is the all-ones vector.

The equivalence highlighted by (1) directly leads to an efficient way to check whether a given vector $\mu \in \mathbb{Z}_{\geq 0}^V$ is a feasible split vector, and if so, obtain a homomorphic preimage of G that certifies it. Indeed, finding a spanning tree T in G' that contains $\bar{E}$ and satisfies $\alpha_T \leq \deg{-1} - \mu$ is a matroid intersection problem. More precisely, the task is to find a spanning tree in $G'/\bar{E}$ (the graph G' after contracting $\bar{E}$)—such spanning trees are the bases of the graphic matroid on $G'/\bar{E}$—whose edges are simultaneously an independent set in the partition matroid on the partition $K = \cup_{v \in V} K_v$, requiring that no more than $\deg(v) - 1 - \mu(v)$ edges are selected within K_v for each $v \in V$. If this matroid intersection problem has a solution, then we get the desired spanning tree T fulfilling the conditions of point (iii) in Property 1, which is equivalent to point (i), and this equivalence is constructive, thus leading to the desired homomorphic preimage H of G that corresponds to the split vector μ.

We finish by proving the claims about P_G in Theorem 5. For this, we start by observing that the vectors α_T are integral base vectors of a polymatroid.

Lemma 5. *The polytope*

$$\bar{B}_G = \operatorname{conv}(\{\alpha_T \mid T \text{ is a spanning tree in } G' \text{ with } \bar{E} \subseteq T\})$$

is the base polytope of a polymatroid. Furthermore, we can optimize efficiently over $\bar{B}_G$.

Consider the polymatroid $\bar{P}_G$ that corresponds to the base polytope $\bar{B}_G$, i.e.,

$$\bar{P}_G = \{x \in \mathbb{R}_{\geq 0}^V \mid \exists \alpha \in \bar{B}_G \text{ with } x \leq \alpha\}.$$

We finish the proof of Theorem 5 by showing that P_G is the (polymatroidal) dual of $\bar{P}_G$. More precisely, McDiarmid [4] (see also [8, vol. B, Sect. 44.6f]) introduced the following notion of a dual of a polymatroid, say $\bar{P}_G \subseteq \mathbb{R}^V$. Consider a vector $y \in \mathbb{Z}^V$ such that $\bar{P}_G$ is contained in the box $[0, y]$; we choose $y = \deg{-1}$. Then the set of all points $y - \alpha$ for $\alpha \in \bar{P}_G$ correspond to the bases of a polymatroid, which is called the *dual* of $\bar{P}_G$ with respect to y. By (1), the vectors $\mu \in \mathbb{Z}^V$ obtained by taking any integral point $\alpha \in \bar{B}_G$ and setting $\mu = \deg{-1} - \alpha$ correspond precisely to the maximal vertices of P_G as defined in Theorem 5. Hence, P_G is the dual of $\bar{P}_G$ with respect to y, and thus a polymatroid. Moreover, analogous to matroid duality, we can efficiently optimize over P_G because we can efficiently optimize over $\bar{P}_G$ (see [8, vol, B] for details).

3 Containment of Minimum Weight k-Trails

Now we consider the problem of finding the minimum weight k-trail contained in $G = (V, E)$ and prove Theorem 4. Our goal is to use the auxiliary graph $G' = (V', E')$ described in the proof of Theorem 1 for the recognition algorithm. Recall that edges in E are in one-to-one correspondence with $\bar{E} \subseteq E'$. We extend the weight function $w : E \to \mathbb{Z}$ to all edges in E', where $e \in \bar{E}$ gets the same weight as the corresponding edge in E. The rest of the edges in $E' \backslash \bar{E}$ are assigned weight 0. Recall, V'_v denotes the set of vertices introduced for vertex v and K_v denote the complete graph on V'_v. Identical to Property 1, we state the following property. With a slight abuse of notation, for a subgraph $\hat{G} = (V, \hat{E})$ of G, we will also denote as $\hat{E}$ the set of edges in $\bar{E}$ that correspond to $\hat{E}$.

Property 2. Let $\hat{G} = (V, \hat{E})$ denote a subgraph of G. Let $\mu \in \mathbb{Z}_{\geq 0}^V$. The following two statements are equivalent:

(i) $\hat{G}$ is the homomorphic image of a connected graph $H = (W, F)$ of maximum degree k by an onto function $\phi : W \to V$ with $|\phi^{-1}(v)| = \mu(v) + 1 \; \forall v \in V$.

(ii) There is a spanning tree T in the auxiliary graph $G' = (V', E')$ such that $\hat{E} \subseteq T$, $|T \cap K_v| \leq |\delta_E(v)| - 1 - \mu(v) \; \forall v \in V$, and finally, $\frac{|\delta_T(v)|}{|\delta_E(v)| - |T \cap K_v|} \leq k$.

We give a general linear programming relaxation for the problem. For any set $S \subseteq V'$ and set of edges $F \subseteq E'$, we use the notation $F(S) = \{\{u, v\} \in F : u, v \in S\}$. We introduce a variable x_e for each edge $e \in E'$. The first set of constraints enforce that x is in the convex hull of spanning trees of G'. We place *degree constraints* on $\bar{E}$-edges incident with V'_v for a well chosen subset of vertices $v \in Q$. We will initialize $Q = V$ but remove these constraints successively in later iterations. We also write the linear program with the edge set $\hat{E} \subseteq E'$. Again we initialize $F = E'$. Property 2 implies that the following linear program is a relaxation, where $E^* = E$ is a set to be updated in later steps of the algorithm.

$$
\begin{aligned}
\min \; & \sum_{e \in E^*} w_e x_e \\[4pt]
x(E^*) \quad &= |V'| - 1 \qquad\qquad\qquad\qquad\quad \text{(LPA)}\\
x(E^*(S)) \quad &\leq |S| - 1 \qquad \forall S \subseteq V', S \neq \emptyset \\
x(\delta_{E^*}(V'_v)) + k x(E^*(V'_v)) \quad &\leq k \cdot \deg_E(v) \quad v \in Q \\
x_e \quad &\geq 0 \qquad\qquad\;\; \forall e \in E
\end{aligned}
$$

We now give an algorithm based on the iterative relaxation paradigm. Iterative relaxation and related techniques have previously been applied to degree-constrained spanning tree problems [1,2,10,11], and we refer the reader to [3] for further details and examples related to this technique.

1. Initialize $Q \leftarrow V$, $E^* \leftarrow E$.
2. While $Q \neq \emptyset$
   (a) Let x denote the optimal extreme point solution to LPA.
   (b) If there exists an edge $e \in E^*$ such that $x_e = 0$, then $E^* \leftarrow E^* \backslash \{e\}$.
   (c) If there exists a vertex $v \in V$ such that one of the following is satisfied

$$
\deg_{E^*}(v) + (2k - 1) \cdot |E^*(V'_v)| \leq (2k - 1) \cdot \deg_E(v) \;, \text{ or}
$$
$$
\deg_{E^*}(v) \leq 2k - 1 \;,
$$

   then $Q \leftarrow Q \backslash \{v\}$.
3. Return the optimal extreme point solution x^* to LPA.

In the long version of the paper, we prove that the above algorithm indeed returns a $(2k - 1)$-trail. Clearly, the returned $(2k - 1)$-trail will have weight no more than the value of LPA, since we only drop constraints during the algorithm, and resolve the LP. Furthermore, we also obtained an integrality gap example, showing that when comparing against the LP, the term $2k - 1$ is best possible, assuming arbitrary weight vectors which can also take negative values.

Acknowledgements. We are grateful to Michel Goemans, Anupam Gupta, Neil Olver, and András Sebő for inspiring discussions, and to the anonymous referees for many helpful comments. This research project started while both authors were guests at the Hausdorff Research Institute for Mathematics (HIM) during the 2015 Trimester on Combinatorial Optimization. Both authors are very thankful to the generous support and inspiring environment provided by the HIM and the organizers of the trimester program.

References

1. Bansal, N., Khandekar, R., Könemann, J., Nagarajan, V., Peis, B.: On generalizations of network design problems with degree bounds. Math. Program. Ser. A **141**, 479–506 (2013)
2. Bansal, N., Khandekar, R., Nagarajan, V.: Additive guarantees for degree-bounded directed network design. SIAM J. Comput. **39**(4), 1413–1431 (2009)
3. Lau, L.C., Ravi, R., Singh, M.: Iterative Methods in Combinatorial Optimization, vol. 46. Cambridge University Press, Cambridge (2011)
4. McDiarmid, C.J.H.: Rado's theorem for polymatroids. Math. Proc. Cambridge Philos. Soc. **78**, 263–281 (1975)
5. Molnár, M., Durand, S., Merabet, M.: Approximation of the degree-constrained minimum spanning hierarchies. In: Halldórsson, M.M. (ed.) SIROCCO 2014. LNCS, vol. 8576, pp. 96–107. Springer, Heidelberg (2014)
6. Molnár, M., Durand, S., Merabet, M.: A new formulation of degree-constrained spanning problems. In: Proceedings of 9th International Colloquium on Graph Theory and Combinatorics (ICGT) (2014). http://oc.inpg.fr/conf/icgt2014/
7. Molnár, M., Newman, A., Sebő, A.: Travelling salesmen on bounded degree trails, Hausdorff report (in preparation) (2015)
8. Schrijver, A.: Combinatorial Optimization: Polyhedra and Efficiency. Algorithms and Combinatorics, vol. 24. Springer, Heidelberg (2003)
9. Sebő, A.: Travelling salesmen on bounded degree trails. In: Presentation at HIM Connectivity Workshop in Bonn, Presentation (2015). https://www.youtube.com/watch?v=5Do2JMhgrCM
10. Singh, M., Lau, L.C.: Approximating minimum bounded degree spanning trees to within one of optimal. In: Proceedings of the 39th Annual ACM Symposium on Theory of Computing (STOC), pp. 661–670 (2007)
11. Zenklusen, R.: Matroidal degree-bounded minimum spanning trees. In: Proceedings of the 23rd Annual ACM-SIAM Symposium on Discrete Algorithms (SODA), pp. 1512–1521 (2012)

Better s-t-Tours by Gao Trees

Corinna Gottschalk[1]([✉]) and Jens Vygen[2]

[1] RWTH Aachen University, Aachen, Germany
corinna.gottschalk@oms.rwth-aachen.de
[2] University of Bonn, Bonn, Germany
vygen@or.uni-bonn.de

Abstract. We consider the s-t-path TSP: given a finite metric space with two elements s and t, we look for a path from s to t that contains all the elements and has minimum total distance. We improve the approximation ratio for this problem from 1.599 to 1.566. Like previous algorithms, we solve the natural LP relaxation and represent an optimum solution x^* as a convex combination of spanning trees. Gao showed that there exists a spanning tree in the support of x^* that has only one edge in each narrow cut (i.e., each cut C with $x^*(C) < 2$). Our main theorem says that the spanning trees in the convex combination can be chosen such that many of them are such "Gao trees" simultaneously at all sufficiently narrow cuts.

1 Introduction

The traveling salesman problem (TSP) is one of the best-known NP-hard problems in combinatorial optimization. In this paper, we consider the s-t-path variant: given a finite metric space (V, c) and two elements $s, t \in V$, the goal is to find a sequence $v_1, \ldots, v_n$ containing every element exactly once and with $v_1 = s$ and $v_n = t$, minimizing $\sum_{i=1}^{n-1} c(v_i, v_{i+1})$. For $s = t$, this is the well-known metric TSP; but in this paper we assume $s \neq t$.

The classical algorithm by Christofides (1976) computes a minimum-cost spanning tree (V, S) and then does *parity correction* by adding a minimum-cost matching on the vertices whose degree in S has the wrong parity.

While Christofides' algorithm is still the best known approximation algorithm for metric TSP (with ratio $\frac{3}{2}$), there have recently been improvements for special cases and variants (see e.g. Vygen's 2012 survey), including the s-t-path TSP.

1.1 Previous Work

For the s-t-path TSP, Christofides' algorithm has only an approximation ratio of $\frac{5}{3}$ as shown by Hoogeveen (1991). An, Kleinberg and Shmoys (2015) were the first to improve on this and obtained an approximation ratio of $\frac{1+\sqrt{5}}{2} \approx 1.618$.

This work was done during the trimester program on combinatorial optimization at the Hausdorff Institute for Mathematics in Bonn.

Q. Louveaux and M. Skutella (Eds.): IPCO 2016, LNCS 9682, pp. 126–137, 2016.
DOI: 10.1007/978-3-319-33461-5_11

They first solve the natural LP relaxation and represent an optimum solution x^* as a convex combination of spanning trees. This idea, first proposed by Held and Karp (1970), was exploited earlier for different TSP variants by Asadpour et al. (2010) and Oveis Gharan et al. (2011). Given this convex combination, An, Kleinberg and Shmoys (2015) do parity correction for each of the contributing trees and output the best of these solutions. Sebő (2013) improved the analysis of this *best-of-many Christofides algorithm* and obtained the approximation ratio $\frac{8}{5}$. Gao (2015) gave a unified analysis. Vygen (2015) suggested to "reassemble" the trees: starting with an arbitrary convex combination of spanning trees, he computed a different one, still representing x^*, that avoids certain bad local configurations. This led to the slightly better approximation ratio of 1.599.

In this paper, we will reassemble the trees more systematically to obtain a convex combination with strong global properties. This will enable us to control the cost of parity correction much better, leading to an approximation ratio of 1.566.

This also proves an upper bound of 1.566 on the integrality ratio of the natural LP. The only known lower bound is 1.5. Sebő and Vygen (2014) proved that the integrality ratio is indeed 1.5 for the graph s-t-path TSP, i.e., the special case of *graph metrics* (where $c(v, w)$ is the distance from v to w in a given unweighted graph on vertex set V). Gao (2013) gave a simpler proof of this result, which inspired our work: see Sect. 1.4.

1.2 Notation and Preliminaries

Throughout this paper, (V, c) is the given metric space, $n := |V|$, and E denotes the set of edges of the complete graph on V. For any $U \subseteq V$ we write $E[U]$ for the set of edges with both endpoints in U and $\delta(U)$ for the set of edges with exactly one endpoint in U; moreover, $\delta(v) := \delta(\{v\})$ for $v \in V$. If $F \subseteq E$ and $U \subseteq V$, we denote by $(V, F)[U]$ the subgraph $(U, F \cap E[U])$ of (V, F) induced by U. For $x \in \mathbb{R}_{\geq 0}^E$ we write $c(x) := \sum_{e=\{v,w\} \in E} c(v, w) x_e$ and $x(F) := \sum_{e \in F} x_e$ for $F \subseteq E$. Furthermore, $\chi^F \in \{0, 1\}^E$ denotes the characteristic vector of a subset $F \subseteq E$, and $c(F) := c(\chi^F)$ the cost of F. For $F \subset E$ and $f \in E$, we write $F + f$ and $F - f$ for $F \cup \{f\}$ and $F \backslash \{f\}$, respectively.

For $T \subseteq V$ with $|T|$ even, a *T-join* is a set $J \subseteq E$ for which $|\delta(v) \cap J|$ is odd if and only if $v \in T$. Edmonds (1965) proved that a minimum cost T-join can be computed in polynomial time. Moreover, the minimum cost of a T-join is the minimum over $c(y)$ for y in the *T-join polyhedron* $\{y \in \mathbb{R}_{\geq 0}^E : y(\delta(U)) \geq 1 \ \forall U \subset V$ with $|U \cap T|$ odd$\}$ as proved by Edmonds and Johnson (1973).

To obtain a solution for the s-t-path TSP, it is sufficient to compute a connected multigraph with vertex set V in which exactly s and t have odd degree. We call such a graph an *$\{s, t\}$-tour*. As an $\{s, t\}$-tour contains an Eulerian walk from s to t, we can obtain a *Hamiltonian s-t-path* (i.e. an s-t-path with vertex set V) by traversing the Eulerian walk and shortcutting when the walk encounters a vertex that has been visited already. Since c is a metric, it obeys the triangle inequality. Thus, the resulting path is not more expensive than the $\{s, t\}$-tour.

By $\mathcal{S}$ we denote the set of edge sets of spanning trees in (V, E). For $S \in \mathcal{S}$, T_S denotes the set of vertices whose degree has the wrong parity in (V, S), i.e., even for s or t and odd for $v \in V \backslash \{s, t\}$. Christofides' algorithm computes an $S \in \mathcal{S}$ with minimum $c(S)$ and adds a T_S-join J with minimum $c(J)$; this yields an $\{s, t\}$-tour.

1.3 Best-of-Many Christofides

Like An, Kleinberg and Shmoys (2015), we begin by solving the natural LP relaxation:

$$
\begin{aligned}
\min \ & c(x) \\
\text{subject to } \ & x(\delta(U)) \geq 2 \quad (\emptyset \neq U \subset V, \ |U \cap \{s, t\}| \ \text{even}) \\
& x(\delta(U)) \geq 1 \quad (\emptyset \neq U \subset V, \ |U \cap \{s, t\}| \ \text{odd}) \\
& x(\delta(v)) = 2 \quad (v \in V \backslash \{s, t\}) \\
& x(\delta(v)) = 1 \quad (v \in \{s, t\}) \\
& x_e \geq 0 \quad (e \in E)
\end{aligned}
\tag{1}
$$

whose integral solutions are precisely the incidence vectors of the edge sets of the Hamiltonian s-t-paths in (V, E). This LP can be solved in polynomial time (either by the ellipsoid method Grötschel et al. (1981) or an extended formulation). Let x^* be an optimum basic solution; then x^* has at most $2n - 3$ positive variables (Goemans 2006). Moreover, x^* (in fact, every feasible solution) satisfies $x^*(E) = n - 1$ and $x^*(E[U]) \leq |U| - 1$ for all $\emptyset \neq U \subset V$. Therefore x^* can be written as convex combination of spanning trees, i.e. as $x^* = \sum_{S \in \mathcal{S}} p_S \chi^S$, where p is a *distribution* on $\mathcal{S}$, i.e., $p_S \geq 0$ for all $S \in \mathcal{S}$ and $\sum_{S \in \mathcal{S}} p_S = 1$.

As x^* has at most $2n - 3$ positive variables, we can assume that $p_S > 0$ for at most $2n - 2$ spanning trees (V, S) by Carathéodory's theorem. Such spanning trees and numbers p_S can be computed in polynomial time, using either the ellipsoid method or the splitting-off technique (cf. Genova and Williamson 2015).

The best-of-many Christofides algorithm does the following. Compute an optimum solution x^* for (1) and obtain a distribution p with $x^* = \sum_{S \in \mathcal{S}} p_S \chi^S$ as above. For each tree (V, S) with $p_S > 0$, compute a minimum weight T_S-join J_S. Then, the multigraph $(V, S \ \dot{\cup} \ J_S)$ is an $\{s, t\}$-tour. Output the best of these.

We will fix x^* henceforth. An important concept in An, Kleinberg and Shmoys (2015) and the subsequent works are the so-called *narrow cuts*, i.e., the cuts $C = \delta(U)$ with $\emptyset \neq U \subset V$ and $x^*(C) < 2$. We denote by $\mathcal{C}$ the set of all narrow cuts. We are going to exploit their structure as well.

Lemma 1 (An, Kleinberg and Shmoys 2015)**.** *The narrow cuts form a chain: there are sets* $\{s\} = U_0 \subset U_1 \subset \cdots \subset U_{\ell-1} \subset U_\ell = V \backslash \{t\}$ *so that* $\mathcal{C} = \{\delta(U_i) : i = 0, \ldots, \ell\}$. *These sets can be computed in polynomial time.*

We number the narrow cuts $\mathcal{C} = \{C_0, C_1 \ldots, C_\ell\}$ with $C_i = \delta(U_i)$ $(i = 0, \ldots, \ell)$.

1.4 Gao Trees

Our work was inspired by the following idea of Gao (2013):

Theorem 1 (Gao 2013). *There exists a spanning tree $S \in \mathcal{S}$ with $x_e^* > 0$ for all $e \in S$ and $|C \cap S| = 1$ for all $C \in \mathcal{C}$.*

In fact, Gao (2013) showed this for any vector $x \in \mathbb{R}_{\geq 0}^E$ with $x(\delta(U)) \geq 1$ for all $\emptyset \neq U \subset V$ and $x(\delta(U)) \geq 2$ for all $\emptyset \neq U \subset V$ with $|U \cap \{s, t\}|$ even. For graph s-t-path TSP, one uses only variables corresponding to edges of the given graph. Then every spanning tree has cost $n - 1$. The approximation guarantee of $\frac{3}{2}$ then follows from the fact that for a tree (V, S) with $|S \cap C| = 1$ for all $C \in \mathcal{C}$, the vector $\frac{1}{2}x^*$ is in the T_S-join polyhedron. But, as shown by Gao (2015), for the general s-t-path TSP there may be no tree as in Theorem 1 whose cost is bounded by the LP value.

Let us call a tree $S \in \mathcal{S}$ a *local Gao tree at C* if $|C \cap S| = 1$. We call S a *global Gao tree* if it is a local Gao tree at every narrow cut.

An, Kleinberg and Shmoys (2015) and Sebő (2013) observed that for every distribution p with $x^* = \sum_{S \in \mathcal{S}} p_S \chi^S$ and every narrow cut $C \in \mathcal{C}$, at least a $2 - x^*(C)$ fraction of the trees will be local Gao trees at C. However, in general none of these trees will be a global Gao tree.

1.5 Our Contribution

Our main contribution is a new structural result: Starting from an arbitrary distribution of trees representing x^*, we can compute a new distribution in which a sufficient number of trees are local Gao trees simultaneously for all sufficiently narrow cuts. For example, if $x^*(C) = \frac{3}{2}$ for all $C \in \mathcal{C}$, at least half of our new distribution will be made of global Gao trees. Here is our main structure theorem:

Theorem 2. *For every feasible solution x^* of (1), there are $S_1, \ldots, S_r \in \mathcal{S}$ and $p_1, \ldots, p_r > 0$ with $\sum_{j=1}^r p_j = 1$ such that $x^* = \sum_{j=1}^r p_j \chi^{S_j}$ and for every $C \in \mathcal{C}$ there exists a $k \in \{1, \ldots, r\}$ with $\sum_{j=1}^k p_j \geq 2 - x^*(C)$ and $|C \cap S_j| = 1$ for all $j = 1, \ldots, k$.*

Note that this result immediately implies Theorem 1, simply by taking S_1.

In the next section we will prove Theorem 2. In Sect. 3 we show how to obtain such a distribution in polynomial time. Finally, in Sect. 4, we explain how this leads to an improved approximation guarantee of the best-of-many Christofides algorithm.

The intuition is as follows. For each tree S in our list, we find a vector y^S in the T_S-join polyhedron and use it to bound the cost of parity correction. We aim to bound the average cost of these vectors by an as small as possible multiple of $c(x^*)$. Following Sebő (2013), we design y^S as follows. If $S = I_S \,\dot\cup\, J_S$, where I_S is the s-t-path in $S \in \mathcal{S}$, then χ^{J_S} is in the T_S-join polyhedron. Moreover, $\frac{1}{2}x^*$ satisfies all constraints of the T_S-join polyhedron except those of narrow cuts C with $|S \cap C|$ even. We choose y^S as a convex combination of χ^{J_S} and $\frac{1}{2}x^*$ but need to add a correction term for even narrow cuts. For this we can use some edges in $I_{S'}$, possibly of a different tree S'. The s-t-paths of the early trees in our list, which are global Gao trees and thus do not need correction at narrow cuts, can thus help pay for the late trees.

2 Proof of the Structure Theorem

We will start with an arbitrary convex combination $x^* = \sum_{S \in \mathcal{S}} p_S \chi^S$ where the p_S ($S \in \mathcal{S}$) are rational. If r is a common denominator, then we can take $r p_S$ copies of S and write $x^* = \frac{1}{r} \sum_{j=1}^{r} \chi^{S_j}$.

Starting from this list $S_1, \ldots, S_r \in \mathcal{S}$, we will successively exchange a pair of edges in two of the trees. We will first satisfy the properties for the first tree S_1, then for S_2, and so on. For each S_j, we will work on the narrow cuts $C_1, \ldots, C_{\ell-1}$ in this order; note that $|C_0 \cap S_j| = |C_\ell \cap S_j| = 1$ always holds for all $j = 1, \ldots, r$ due to $x^*(\delta(s)) = x^*(\delta(t)) = 1$. In the following we write

$$\theta_i := \lceil r(2 - x^*(C_i)) \rceil$$

for $i = 0, \ldots, \ell$. Note that $\theta_0 = r \geq \theta_i$ for all $i = 1, \ldots, \ell$ because $x^*(C_i) \geq 1 = x^*(C_0) = x^*(\delta(s))$. Our goal is to obtain $|S_j \cap C_i| = 1$ whenever $j \leq \theta_i$.

Note that $|S_j \cap C_i| = 1$ implies that $(V, S_j)[U_i]$ is connected, and we will first obtain this weaker property by Lemma 6, before obtaining $|S_j \cap C_i| = 1$ by Lemma 7.

We need a few preparations. Throughout this section, we use indices g, h, i, i' for cuts and j, j', k for trees. We first observe that a sufficient number of trees is connected in every section between two narrow cuts.

Lemma 2. *Let* $S_1, \ldots, S_r \in \mathcal{S}$ *such that* $x^* = \frac{1}{r} \sum_{j'=1}^{r} \chi^{S_{j'}}$. *Let* $0 \leq h < i \leq \ell$ *and* $1 \leq j \leq r$ *with* $j \leq \theta_h$ *and* $j \leq \theta_i$. *Let* $M = U_i \setminus U_h$ *or* $M = U_i$. *Then there exists an index* $k \geq j$ *such that* $(V, S_k)[M]$ *is connected.*

Proof. Assume the above is not true. Then $|E[M] \cap S_{j'}| \leq |M| - 1$ for all $j' < j$, and $|E[M] \cap S_{j'}| \leq |M| - 2$ for $j' \geq j$. Therefore,

$$x^*(E[M]) = \frac{1}{r} \sum_{j'=1}^{r} |E[M] \cap S_{j'}|$$

$$\leq \frac{1}{r}\big((j-1)(|M|-1) + (r-j+1)(|M|-2)\big) = |M| - 2 + \frac{j-1}{r}.$$

On the other hand, $x^*(E[M]) = \frac{1}{2}\big(\sum_{v \in M} x^*(\delta(v)) - x^*(\delta(M))\big)$.

If $M = U_i \setminus U_h$, we therefore have $x^*(E[M]) = |M| - \frac{1}{2}(x^*(C_h) + x^*(C_i)) + x^*(C_h \cap C_i) \geq |M| - \frac{1}{2}(x^*(C_h) + x^*(C_i))$.

If $M = U_i$, we have $x^*(E[M]) = |M| - \frac{1}{2} - \frac{1}{2}x^*(C_i) = |M| - \frac{1}{2}(1 + x^*(C_i))$.

Now, $j \leq \theta_h$ and $j \leq \theta_i$ implies $1 \leq x^*(C_h) < 2 - \frac{j-1}{r}$ and $x^*(C_i) < 2 - \frac{j-1}{r}$. Thus, in both cases,

$$x^*(E[M]) > |M| - \frac{1}{2}\left(2 - \frac{j-1}{r} + 2 - \frac{j-1}{r}\right) = |M| - 2 + \frac{j-1}{r},$$

a contradiction.   □

Next, a similar argument shows that, for any pair of narrow cuts, sufficiently many trees have no edge in their intersection:

Lemma 3. *Let* $S_1, \ldots, S_r \in \mathcal{S}$ *such that* $x^* = \frac{1}{r} \sum_{j'=1}^{r} \chi^{S_{j'}}$. *Let* $0 \leq h < i \leq \ell$ *and* $1 \leq j \leq r$ *with* $j \leq \theta_h$ *and* $j \leq \theta_i$. *Then there exists an index* $k \geq j$ *such that* $S_k \cap C_h \cap C_i = \emptyset$.

Proof. Using $x^*(C_{i'}) < 2 - \frac{i-1}{r}$ for $i' \in \{h, i\}$ and $x^*(\delta(U)) \geq 2$ for $|U \cap \{s, t\}|$ even, we obtain

$$x^*(C_h \cap C_i) = \tfrac{1}{2}\big(x^*(C_h) + x^*(C_i) - x^*(\delta(U_i \backslash U_h))\big) < 2 - \tfrac{i-1}{r} - 1 = \tfrac{r-j+1}{r}.$$

Therefore, $\frac{1}{r} \sum_{j'=1}^{r} |S_{j'} \cap C_h \cap C_i| = x^*(C_h \cap C_i) < \frac{r-j+1}{r}$, i.e. at most $r - j$ trees can contain an edge in $C_h \cap C_i$. $\qquad \square$

Finally, as mentioned already in Sect. 1.4, many trees are local Gao trees at a narrow cut:

Lemma 4. *Let* $S_1, \ldots, S_r \in \mathcal{S}$ *such that* $x^* = \frac{1}{r} \sum_{j'=1}^{r} \chi^{S_{j'}}$. *Let* $1 \leq i \leq \ell - 1$ *and* $j \leq \theta_i$ *with* $|C_i \cap S_j| \geq 2$. *Then there exists an index* $k > \theta_i$ *with* $|C_i \cap S_k| = 1$.

Proof. Suppose there exists no such k. Then we get $rx^*(C_i) = \sum_{j'=1}^{r} |C_i \cap S_{j'}| \geq (r - \theta_i + 1)2 + \theta_i - 1 = 2r - \lceil r(2 - x^*(C_i)) \rceil + 1 > 2r - r(2 - x^*(C_i)) = rx^*(C_i)$, which is a contradiction. $\qquad \square$

Now we proceed to the main components of the proof of Theorem 2. Recall that, in order to obtain $|S_j \cap C_i| = 1$, we plan to first get the weaker condition that $(V, S_j)[U_i]$ is connected. While it is not, we exchange a pair of edges with a later tree. We split the proof into two parts, beginning with the following lemma:

Lemma 5. *Let* $1 \leq i \leq \ell - 1$ *and* $M \subseteq U_i$ *and* $S_j, S_k \in \mathcal{S}$ *such that* $(V, S_j)[M]$ *is disconnected and* $(V, S_k)[M]$ *is connected and* $|S_j \cap \delta(U_i \backslash M)| \leq 1$. *Then there exist edges* $e \in S_j$ *and* $f \in S_k$ *such that* $S_j - e + f \in \mathcal{S}$ *and* $S_k + e - f \in \mathcal{S}$ *and* $e \notin E[U_i]$ *and* $f \in E[M]$.

Proof. Let $A_1, \ldots, A_q$ be the vertex sets of the connected components of $(V, S_j)[M]$; note that $q \geq 2$ (For illustrations, see Fig. 1).

Let $F := S_k \cap \bigcup_{p=1}^{q}(\delta(A_p) \backslash \delta(M))$ be the set of edges of S_k between the sets $A_1, \ldots, A_q$. Note that $F \subseteq E[M]$. For $p = 1, \ldots, q$ let B_p be the set of vertices reachable from A_p in $(V, S_k \backslash F)$. Trivially, $A_p \subseteq B_p$ for all p, and $\{B_1, \ldots, B_q\}$ is a partition of V because $(V, S_k)[M]$ is connected and (V, S_k) is a tree.

Let Y be the union of the edge sets of the unique v-w-paths in S_j for all $v, w \in M$. Note that $Y \subseteq E[V \backslash (U_i \backslash M)]$ because $|S_j \cap \delta(U_i \backslash M)| \leq 1$.

Claim: There exists an index $p \in \{1, \ldots, q\}$ and an edge $e \in Y \cap \delta(B_p)$ such that for every $p' \in \{1, \ldots, q\} \backslash \{p\}$ and $v' \in A_{p'}$, $v'' \in A_p$, the v'-v''-path in S_j contains e.

To prove the Claim, observe that (V, Y) consists of a tree and possibly isolated vertices. Choose an arbitrary root z in this tree and take an edge $e \in Y \cap \bigcup_{p'=1}^{q} \delta(B_{p'})$ with maximum distance from z.

Let $D := \{v \in V : v$ is reachable from z in (V, Y) but not in $(V, Y - e)\}$. We will show that $D \cap M = A_p$ for some $p \in \{1, \dots, q\}$. This will immediately imply that p and e satisfy the properties of the Claim.

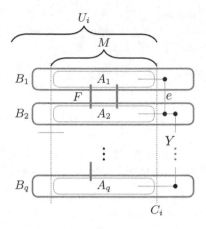

As $e \in Y$, there are $v, w \in M$ such that the v-w-path in S_j contains e. Then exactly one of these two vertices belongs to D, so w.l.o.g. $v \in D \cap M$. Let p be the index with $v \in A_p$.

Since $(V, S_j)[A_p]$ is connected, this implies $A_p \subseteq D$. Next, $(D, Y \cap E[D])$ is a tree and by the choice of e, it contains no edge from $\bigcup_{p'=1}^{q} \delta(B_{p'})$. Therefore, $D \subseteq B_p$ and hence, $D \cap M = A_p$. The Claim is proved.

Fig. 1. Edges of tree S_j are green/thin, edges of S_k are red/bold. Note that e could belong to C_i (but not to $\delta(U_i \setminus M)$). (Color figure online)

Now, take an index p and an edge e as in the Claim. Consider the path P in S_k that connects the endpoints of e. Since $e \in \delta(B_p)$, P has an edge $f \in \delta(B_p) \cap S_k = \delta(A_p) \cap F$. Thus, $(V, S_k + e - f)$ is a tree. The path in S_j that connects the endpoints of f contains e by the Claim. Thus, $(V, S_j - e + f)$ is a tree. We have $f \in E[M]$ since $f \in F$, and $e \notin E[U_i]$ as $e \in Y \cap \delta(B_p)$.   □

Now we make $(V, S_j)[U_i]$ connected without destroying previously obtained properties:

Lemma 6. *Let $S_1, \dots, S_r \in \mathcal{S}$ such that $x^* = \frac{1}{r} \sum_{j'=1}^{r} \chi^{S_{j'}}$. Let $1 \leq j \leq r$ and $1 \leq i \leq \ell - 1$ such that $j \leq \theta_i$ and $|S_j \cap C_h| = 1$ for all $h < i$ with $j \leq \theta_h$. Then we can find $\hat{S}_1, \dots, \hat{S}_r \in \mathcal{S}$ in polynomial time such that $x^* = \frac{1}{r} \sum_{j'=1}^{r} \chi^{\hat{S}_{j'}}$, and $\hat{S}_{j'} = S_{j'}$ for all $j' < j$, and $|\hat{S}_j \cap C_h| = 1$ for all $h < i$ with $j \leq \theta_h$, and $(V, \hat{S}_j)[U_i]$ is connected.*

Proof. Assume that $(V, S_j)[U_i]$ is disconnected, i.e., $|S_j \cap E[U_i]| < |U_i| - 1$. Let h be the largest index smaller than i with $j \leq \theta_h$. Such an index must exist because $\theta_0 \geq \theta_i \geq j$.

Case 1: $|S_j \cap E[U_i \setminus U_h]| < |U_i \setminus U_h| - 1$.

Let $M := U_i \setminus U_h$. Note that $|S_j \cap \delta(U_i \setminus M)| = |S_j \cap C_h| = 1$. Since $(V, S_j)[M]$ is not connected, by Lemma 2 there exists an index $k > j$ such that $(V, S_k)[M]$ is connected.

Now we apply Lemma 5 and obtain two trees $\hat{S}_j := S_j - e + f$ and $\hat{S}_k := S_k + e - f$ with $e \notin E[U_i]$ and $f \in E[M]$.

We have $|\hat{S}_j \cap E[U_i]| = |S_j \cap E[U_i]| + 1$ and $|\hat{S}_j \cap C_{h'}| \leq |S_j \cap C_{h'}|$ for all $h' \leq h$ and hence $|\hat{S}_j \cap C_{h'}| = 1$ for $h' \leq h$ with $j \leq \theta_{h'}$. Note that $j > \theta_{i'}$ for all $h < i' < i$, so a new edge f in such cuts $C_{i'}$ does no harm.

We replace S_j and S_k by $\hat{S}_j$ and $\hat{S}_k$ and leave the other trees unchanged. If $(V, \hat{S}_j)[U_i]$ is still not connected, we iterate.

Case 2: $|S_j \cap E[U_i\backslash U_h]| = |U_i\backslash U_h| - 1$. $(V, S_j)[U_h]$ is connected since $|S_j \cap C_h| = 1$. Moreover, $(V, S_j)[U_i\backslash U_h]$ is connected, but $(V, S_j)[U_i]$ is disconnected. Therefore, S_j must contain an edge in $C_h \cap C_i$ and $S_j \cap C_h \subset S_j \cap C_i$ and $(V, S_j)[U_i]$ has exactly two connected components: U_h and $U_i\backslash U_h$ (For illustrations, see Fig. 2).

Case 2a: $h > 0$. Let g be the largest index smaller than h with $j \le \theta_g$. Set $M := U_i\backslash U_g$. Note that $|S_j \cap \delta(U_i\backslash M)| = |S_j \cap C_g| = 1$. By Lemma 2 there exists an index $k \ge j$ with $(V, S_k)[M]$ connected. As $(V, S_j)[M]$ is not connected, we have $k > j$.

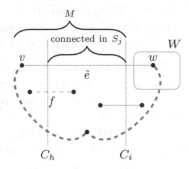

Fig. 2. Tree S_j green/thin, $\hat{S}_{k'}$ red/bold. The dashed edge f is added to S_j by applying Lemma 5. (Color figure online)

Case 2b: $h = 0$. Set $M := U_i$. Note that $|S_j \cap \delta(U_i\backslash M)| = |S_j \cap \delta(\emptyset)| = 0$. By Lemma 2, there exists an index $k \ge j$ such that $(V, S_k)[M]$ is connected. Since $(V, S_j)[M]$ is not connected, $k > j$.

Note that in both cases 2a and 2b, $(V, S_j)[M]$ is disconnected. Now we apply Lemma 5 to S_j and S_k and obtain two trees $\hat{S}_j := S_j - e + f$ and $\hat{S}_k := S_k + e - f$ with $e \notin E[U_i]$ and $f \in E[M]$. We replace S_j and S_k by $\hat{S}_j$ and $\hat{S}_k$ and leave the other trees unchanged. Then $(V, \hat{S}_j)[U_i]$ is connected. We have $|\hat{S}_j \cap C_{h'}| = 1$ for all $h' < h$ with $j \le \theta_{h'}$, but we may have $|\hat{S}_j \cap C_h| = 2$ since $E[M] \cap C_h \ne \emptyset$.

Assume $|\hat{S}_j \cap C_h| = 2$ (otherwise we are done). Then $\hat{S}_j \cap C_h \cap C_i = S_j \cap C_h \cap C_i = S_j \cap C_h$, and this set contains precisely one edge $\hat{e} = \{v, w\}$ (where $v \in U_h$ and $w \in V\backslash U_i$). By Lemma 3 there exists an index $k' > j$ with $\hat{S}_{k'} \cap C_h \cap C_i = \emptyset$.

Let W be the set of vertices reachable from w in $(V, \hat{S}_j\backslash C_i)$. Since $(V, \hat{S}_j)[U_i]$ is connected, $\hat{S}_j \cap \delta(W) = \{\hat{e}\}$. The unique path in $(V, \hat{S}_{k'})$ from v to w contains at least one edge $\hat{f} \in \delta(W)$. Note that $\hat{f} \notin C_h$ by the choice of $\hat{S}_{k'}$. We replace $\hat{S}_j$ and $\hat{S}_{k'}$ by $\hat{\hat{S}}_j := \hat{S}_j - \hat{e} + \hat{f}$ and $\hat{\hat{S}}_{k'} := \hat{S}_{k'} + \hat{e} - \hat{f}$. Then $(V, \hat{\hat{S}}_j)[U_i]$ is still connected and $|\hat{\hat{S}}_j \cap C_{h'}| = 1$ for all $h' < i$ with $j \le \theta_{h'}$. $\square$

Lemma 7. *Let $S_1, \ldots, S_r \in \mathcal{S}$ such that $x^* = \frac{1}{r}\sum_{j'=1}^r \chi^{S_{j'}}$. Let $1 \le i \le \ell - 1$ and $j \le \theta_i$ such that $(V, S_j)[U_i]$ is connected and $|S_j \cap C_h| = 1$ for all $h < i$ with $j \le \theta_h$. Then we can find $\hat{S}_1, \ldots, \hat{S}_r \in \mathcal{S}$ in polynomial time such that $x^* = \frac{1}{r}\sum_{j'=1}^r \chi^{\hat{S}_{j'}}$ and $\hat{S}_{j'} = S_{j'}$ for all $j' < j$ and $|\hat{S}_j \cap C_h| = 1$ for all $h \le i$ with $j \le \theta_h$.*

Proof. Suppose $|S_j \cap C_i| \ge 2$. Then by Lemma 4 there exists an index $k > \theta_i$ with $|S_k \cap C_i| = 1$. We will swap a pair of edges, reducing $|S_j \cap C_i|$ and increasing $|S_k \cap C_i|$ while maintaining the other properties (For illustrations, see Fig. 3). Let $S_k \cap C_i = \{\{x, y\}\}$ with $x \in U_i$ and $y \in V\backslash U_i$. Let A be the set of vertices reachable from y in $(V, S_j\backslash C_i)$. Note that $A \cap U_i = \emptyset$. We have $|\delta(A) \cap S_j \cap C_i| = 1$ because $(V, S_j)[U_i]$ is connected. So let $e = \{v, w\} \in (S_j \cap C_i)\backslash \delta(A)$, with $v \in U_i$ and $w \in V\backslash U_i$. Let B be the set of vertices reachable from w in $(V, S_j\backslash C_i)$. We have $w \in B$, $y \in A$, and $A \cap B = \emptyset$ by the choice of e. Consider the path P in S_k from w to y. Note that P does not contain any vertex in U_i because $|S_k \cap C_i| = 1$. But P contains at least one edge $f \in \delta(B)$.

We will swap e and f. Since $S_k \cap C_i = \{\{x, y\}\}$, the w-v-path in S_k contains P. Therefore, $\hat{S}_k := S_k + e - f$ is a tree. On the other hand, the path in S_j connecting the endpoints of f must use an edge in $\delta(B)$. Since $S_j \cap E[U_i]$ is connected and $S_j \cap (\delta(B) \backslash C_i) = \emptyset$, e is the only edge in $\delta(B) \cap S_j$ and thus, $\hat{S}_j := S_j + f - e$ is a tree.

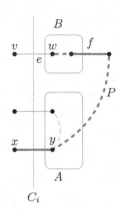

Since $f \in E[V \backslash U_i]$ and $e \in C_i$, we have $|\hat{S}_j \cap C_i| = |S_j \cap C_i| - 1$ and $|\hat{S}_j \cap C_h| = |S_j \cap C_h|$ for all $h < i$ with $j \leq \theta_h$. Moreover, $(V, \hat{S}_j)[U_i]$ is still connected. As before, we replace S_j and S_k by $\hat{S}_j$ and $\hat{S}_k$ and leave the other trees unchanged. If $|\hat{S}_j \cap C_i| > 1$, we iterate. □

Fig. 3. Tree S_j green/ thin, S_k red/bold (Color figure online)

Now the proof of Theorem 2 is a simple induction. We scan the indices of the trees $j = 1, \ldots, r$ in this order. For each j, we consider all narrow cuts C_i with $j \leq \theta_i$. Since $x^*(\delta(s)) = x^*(\delta(t)) = 1$, we always have $|S_j \cap C_0| = 1$ and $|S_j \cap C_\ell| = 1$ for all $j = 1, \ldots, r$. Now let $i \in \{1, \ldots, \ell - 1\}$ with $j \leq \theta_i$. Assuming $|S_j \cap C_h| = 1$ for all $h < i$ with $j \leq \theta_h$, we first apply Lemma 6 and then Lemma 7. The new tree then satisfies $|S_j \cap C_h| = 1$ for all $h \leq i$ with $j \leq \theta_h$, and $S_1, \ldots S_{j-1}$ remain unchanged.

3   Obtaining the Distribution in Polynomial Time

So far it was not clear whether the number r of trees in our distribution can be polynomially bounded. In this section we show two solutions to this question.

First, one can start with an arbitrary distribution p with at most $2n - 2$ trees S with $p_S > 0$ and round the coefficients down to integral multiples of $\frac{\epsilon}{2n^2}$ for a sufficiently small constant $\epsilon > 0$, and then scale up all coefficients so that their sum is 1 again. This way we will get a vector x close to x^* that we can write as $x = \sum_{j=1}^r \frac{1}{r} \chi^{S_j}$, where $r \leq \frac{2n^2}{\epsilon}$ and $S_j \in \mathcal{S}$ for $j = 1, \ldots, r$. It is not difficult to show that $x \in \mathbb{R}_{\geq 0}^E$ satisfies the properties

$$x(\delta(s)) = x(\delta(t)) = 1 \quad \text{and} \quad x^*(F) - \epsilon \leq x(F) \leq x^*(F) + \epsilon \text{ for all } F \subseteq E. \quad (2)$$

In the full version of this paper [arXiv:1511.05514] we show that the proof in Sect. 2 also works in this case. More precisely:

Theorem 3. *Given $S_1, \ldots, S_r \in \mathcal{S}$, a feasible solution x^* of (1) and $\epsilon \geq 0$ such that $x = \frac{1}{r} \sum_{j=1}^r \chi^{S_j}$ satisfies (2), we can find $\hat{S}_1, \ldots, \hat{S}_r \in \mathcal{S}$ in polynomial time such that $x = \frac{1}{r} \sum_{j=1}^r \chi^{\hat{S}_j}$, and for every $C \in \mathcal{C}$ there exists a $k \in \{1, \ldots, r\}$ with $\frac{k}{r} \geq 2 - x^*(C) - \epsilon$ and $|C \cap \hat{S}_j| = 1$ for all $j = 1, \ldots, k$.*

This is sufficient to obtain the claimed approximation ratio if ϵ is chosen small enough. However, Kanstantsin Pashkovich [private communication, 2015] suggested a more elegant solution: Theorem 2 implies a stronger version in which r can be chosen to be less than $2n^2$ and the trees and p can be found in polynomial time. We now explain how.

As before, fix an optimum basic solution x^* of (1). Let $\xi^1 > \cdots > \xi^k = 1$ be the distinct values among $\{x^*(C) : C \in \mathcal{C}\}$ and $\xi^0 := 2$. Note that $k \le \ell \le n - 2$ by Lemma 1. Then Theorem 2 implies that the polytope defined by

$$\sum_{h=1}^{k} (\xi^{h-1} - \xi^h) x^h = x^*$$

$$x^h(C_i) = 1 \qquad\qquad (1 \le h \le k,\ 0 \le i \le \ell,\ x^*(C_i) \le \xi^h)$$
$$x^h(E) = n - 1 \qquad\qquad (1 \le h \le k)$$
$$x^h(E[U]) \le |U| - 1 \qquad\qquad (1 \le h \le k,\ \emptyset \ne U \subset V)$$
$$x_e^h \ge 0 \qquad\qquad (1 \le h \le k,\ e \in E)$$

is nonempty: if $S_1, \ldots, S_r$ and $p_1, \ldots, p_r$ are as in Theorem 2, then

$$x^h = \sum_{j=1}^{r} \frac{\max\left\{0,\ \min\left\{2 - \xi^h, \sum_{j' \le j} p_{j'}\right\} - \max\left\{2 - \xi^{h-1}, \sum_{j' < j} p_{j'}\right\}\right\}}{\xi^{h-1} - \xi^h} \chi^{S_j}$$

defines a feasible solution. We can find a vector x in this polytope in polynomial time by the ellipsoid method because the separation problem can be solved in polynomial time. Then, writing each x^h as convex combination $x^h = \sum_{j=1}^{2n-2} p_j^h \chi^{S_j^h}$ of $2n - 2$ spanning trees, and setting $S_{(h-1)(2n-2)+j} := S_j^h$ and $p_{(h-1)(2n-2)+j} := (\xi^{h-1} - \xi^h) p_j^h$ for $h = 1, \ldots, k$ and $j = 1, \ldots, 2n - 2$, yields a decomposition $x^* = \sum_{j=1}^{k(2n-2)} p_j \chi^{S_j}$ with at most $(2n-2)(n-2)$ spanning trees and the properties of Theorem 2.

4   Analysis of the Approximation Ratio

In this section, we will analyze the best-of-many Christofides algorithm on a distribution as in Theorem 2. We follow the framework from Vygen (2015) (based on An, Kleinberg and Shmoys 2015 and Sebő 2013); see the end of Sect. 1 for an intuition. In particular, we use the following definition and lemma from Vygen (2015):

Definition 1. *Given numbers $0 \le \gamma_S \le 1$ for $S \in \mathcal{S}$ and $\beta < \frac{1}{2}$, we define the benefit of $(S, C) \in \mathcal{S} \times \mathcal{C}$ to be $b_{S,C} := \min\left\{\frac{\beta(2 - x^*(C))}{1 - 2\beta}, \gamma_S\right\}$ if $|S \cap C|$ is even, $b_{S,C} := 1 - \gamma_S$ if $|S \cap C| = 1$, and $b_{S,C} = 0$ otherwise.*

Lemma 8. *Let $0 \leq \beta < \frac{1}{2}$ and $0 \leq \gamma_S \leq 1$ for $S \in \mathcal{S}$. Let p be a distribution on $\mathcal{S}$ with $x^* = \sum_{S \in \mathcal{S}} p_S \chi^S$. If*

$$\sum_{S \in \mathcal{S}} p_S b_{S,C} \geq \frac{\beta}{1-2\beta}(2 - x^*(C)) \sum_{S \in \mathcal{S}:|S \cap C| \text{ even}} p_S \tag{3}$$

for all $C \in \mathcal{C}$, then the best-of-many Christofides algorithm run on the trees $S \in \mathcal{S}$ with $p_S > 0$ returns a solution of cost at most $(2 - \beta)c(x^*)$

We now show how to set the γ-constants in order to maximize the benefits, with the ultimate goal to choose β as large as possible.

Lemma 9. *Let $S_1, \ldots, S_r \in \mathcal{S}$ such that $x^* = \frac{1}{r} \sum_{j=1}^{r} \chi^{S_j}$, r is even, and for every $C \in \mathcal{C}$ there exists a $k \in \{1, \ldots, r\}$ with $\frac{k}{r} \geq 2 - x^*(C)$ and $|C \cap S_j| = 1$ for all $j = 1, \ldots, k$.*

We set $\delta := 0.126$, and $\gamma_{S_j} = \delta$ if $j \leq \frac{r}{2}$ and $\gamma_{S_j} = 1 - \delta$ otherwise. Choose β such that $\frac{\beta}{1-2\beta} = 3.327$. Then

$$\frac{1}{r} \sum_{j=1}^{r} b_{S_j, C_i} \geq 3.327 \, (2 - x^*(C_i)) \frac{1}{r} |\{j : |S_j \cap C_i| \text{ even}\}| \tag{4}$$

for all $i = 0, \ldots, \ell$.

For the proof of this lemma, see the full version of this paper [arXiv:1511.05514]. Now we obtain our approximation guarantee:

Theorem 4. *There is a 1.566-approximation algorithm for the s-t-path TSP.*

Proof. Let x^* be an optimal solution to (1). Let $S_1, \ldots, S_{r'} \in \mathcal{S}$ and rational $p_1, \ldots, p_{r'} > 0$ as in Theorem 2. We showed in Sect. 3 how to obtain this in polynomial time. Therefore, we can apply Lemma 9 to the trees $S_1, \ldots, S_{r'}$ with appropriate multiplicities and obtain inequality (3) for $\frac{\beta}{1-2\beta} = 3.327$. Equivalently, $\beta = \frac{3.327}{7.654} > 0.434$.

Thus, the conditions in Lemma 8 are met and best-of-many Christofides yields an approximation ratio of at most $2 - \beta < 1.566$. $\qquad\square$

5   Conclusion

The approximation ratio can probably be improved slightly by choosing the γ_{S_j} differently, but still depending only on $\frac{j}{r}$. However, using an analysis based on Lemma 8, one cannot obtain a better approximation ratio than $\frac{14}{9}$ because the benefit can never be more than one and there can be cuts C with $x^*(C) = \frac{3}{2}$ and $\sum_{S \in \mathcal{S}:|S \cap C| \text{ even}} p_S = \frac{1}{2}$; therefore $\frac{\beta}{1-2\beta} \leq 4$. Our ratio is already close to this threshold.

On the other hand, it is not impossible that the best-of-many Christofides algorithm on a distribution like the one obtained in Theorem 2, or even on an arbitrary distribution, has a better approximation ratio, maybe even $\frac{3}{2}$.

Acknowledgement. We thank Kanstantsin Pashkovich for allowing us to include his idea described in the second half of Sect. 3.

References

An, H.-C., Kleinberg, R., Shmoys, D.B.: Improving Christofides' algorithm for the s-t path TSP. J. ACM **62**, Article 34, 34:1–34:28 (2015)

Asadpour, A., Goemans, M.X., Mądry, A., Oveis Gharan, S., Saberi, A.: An $O(\log n/\log\log n)$-approximation algorithm for the asymmetric traveling salesman problem. In: Proceedings of the 21st Annual ACM-SIAM Symposium on Discrete Algorithms (SODA 2010), pp. 379–389 (2010)

Christofides, N.: Worst-case analysis of a new heuristic for the traveling salesman problem. Technical report 388, Graduate School of Industrial Administration, Carnegie-Mellon University, Pittsburgh (1976)

Edmonds, J.: The Chinese postman's problem. Bull. Oper. Res. Soc. Am. **13**, B-73 (1965)

Edmonds, J., Johnson, E.L.: Matching, Euler tours and the Chinese postman. Math. Program. **5**, 88–124 (1973)

Gao, Z.: An LP-based $\frac{3}{2}$-approximation algorithm for the s-t path graph traveling salesman problem. Oper. Res. Lett. **41**, 615–617 (2013)

Gao, Z.: On the metric s-t path traveling salesman problem. SIAM J. Discrete Math. **29**, 1133–1149 (2015)

Genova, K., Williamson, D.P.: An experimental evaluation of the best-of-many Christofides' algorithm for the traveling salesman problem. In: Bansal, N., Finocchi, I. (eds.) Algorithms – ESA 2015. LNCS, pp. 570–581. Springer, Heidelberg (2015)

Goemans, M.X.: Minimum bounded-degree spanning trees. In: Proceedings of the 47th Annual IEEE Symposium on Foundations of Computer Science (FOCS 2006), pp. 273–282 (2006)

Grötschel, M., Lovász, L., Schrijver, A.: The ellipsoid method and its consequences in combinatorial optimization. Combinatorica **1**, 169–197 (1981)

Held, M., Karp, R.M.: The traveling-salesman problem and minimum spanning trees. Oper. Res. **18**, 1138–1162 (1970)

Hoogeveen, J.A.: Analysis of Christofides' heuristic: some paths are more difficult than cycles. Oper. Res. Lett. **10**, 291–295 (1991)

Oveis Gharan, S., Saberi, A., Singh, M.: A randomized rounding approach to the traveling salesman problem. In: Proceedings of the 52nd Annual IEEE Symposium on Foundations of Computer Science (FOCS 2011), pp. 550–559 (2011)

Sebő, A.: Eight-fifth approximation for the path TSP. In: Goemans, M., Correa, J. (eds.) IPCO 2013. LNCS, vol. 7801, pp. 362–374. Springer, Heidelberg (2013)

Sebő, A., Vygen, J.: Shorter tours by nicer ears: 7/5-approximation for graph-TSP, 3/2 for the path version, and 4/3 for two-edge-connected subgraphs. Combinatorica **34**, 597–629 (2014)

Vygen, J.: New approximation algorithms for the TSP. OPTIMA **90**, 1–12 (2012)

Vygen, J.: Reassembling trees for the traveling salesman. SIAM J. Discrete Math. To appear. arXiv:1502.03715

Popular Edges and Dominant Matchings

Ágnes Cseh[1]([✉]) and Telikepalli Kavitha[2]

[1] Reykjavík University, Reykjavik, Iceland
cseh@ru.is
[2] Tata Institute of Fundamental Research, Mumbai, India
kavitha@tcs.tifr.res.in

Abstract. Given a bipartite graph $G = (A \cup B, E)$ with strict preference lists and given an edge $e^* \in E$, we ask if there exists a popular matching in G that contains e^*. We call this the *popular edge* problem. A matching M is popular if there is no matching M' such that the vertices that prefer M' to M outnumber those that prefer M to M'. It is known that every stable matching is popular; however G may have no stable matching with the edge e^*. In this paper we identify another natural subclass of popular matchings called "dominant matchings" and show that if there is a popular matching that contains the edge e^*, then there is either a stable matching that contains e^* or a dominant matching that contains e^*. This allows us to design a linear time algorithm for the popular edge problem. When preference lists are complete, we show an $O(n^3)$ algorithm to find a popular matching containing a given set of edges or report that none exists, where $n = |A| + |B|$.

1 Introduction

Our input is an instance $G = (A \cup B, E)$ of the stable marriage problem with strict and possibly incomplete preference lists, along with an edge $e^* \in E$. A matching M is stable if there is no *blocking pair* with respect to M, in other words, there is no pair (a, b) such that a is either unmatched or prefers b to $M(a)$ (a's partner in M) and similarly, b is either unmatched or prefers a to $M(b)$. The problem of deciding if there exists a stable matching that contains the edge e^* is an old and well-studied problem – this was first considered by Knuth [13] in 1976 who showed that a modified version of the Gale-Shapley algorithm solves this problem. Here we consider a related problem that we call the "popular edge" problem: is there a *popular matching* in G that contains the edge e^*?

The notion of popularity, introduced by Gärdenfors [8] in 1975, is a relaxation of stability. A popular matching allows blocking edges with respect to it, however there is *global* acceptance for this matching. We make this formal now.

A vertex $u \in A \cup B$ prefers matching M to matching M' if either u is matched in M and unmatched in M' or u is matched in both and it prefers $M(u)$ to $M'(u)$. For matchings M and M' in G, let $\phi(M, M')$ be the number of vertices that prefer M to M'. If $\phi(M', M) > \phi(M, M')$ then M' is *more popular than* M.

© Springer International Publishing Switzerland 2016
Q. Louveaux and M. Skutella (Eds.): IPCO 2016, LNCS 9682, pp. 138–151, 2016.
DOI: 10.1007/978-3-319-33461-5_12

Definition 1. *A matching M is* popular *if there is no matching that is more popular than M; in other words, $\phi(M, M') \geq \phi(M', M)$ for all matchings M' in G.*

Thus in an election between any pair of matchings, where each vertex casts a vote for the matching that it prefers, a popular matching never loses. Popular matchings always exist in G since every stable matching is popular [8]. It is also known that every stable matching is a minimum size popular matching [10]. As stability is stricter than popularity, it may be the case that there is no stable matching that contains the given edge e^* while there is a popular matching that contains e^*. Figure 1 has such an example.

$$a_1 \; : \; b_1 \; b_2 \qquad b_1 : a_1 \; a_2$$
$$a_2 \; : \; b_1 \qquad b_2 : a_1$$

Fig. 1. The top-choice of both a_1 and of a_2 is b_1; the second choice of a_1 is b_2. The preference lists of the b_i's are symmetric. There is no edge between a_2 and b_2. The matching $S = \{(a_1, b_1)\}$ is the only stable matching here, while there is another popular matching $M = \{(a_1, b_2), (a_2, b_1)\}$. Thus every edge is a popular edge here, while there is only one stable edge, namely (a_1, b_1).

Stability is a very strong condition and there are several problems, for instance, in allocating projects to students or in assigning applicants to training posts, where the total absence of blocking edges may not be necessary. However the popularity of a matching is required, otherwise the vertices could vote to replace the current matching with a more popular one. The popular edge problem has applications in such a setting where the central authority seeks to pair $a \in A$ and $b \in B$ with each other and desires a matching M such that M is popular and $(a, b) \in M$.

A first attempt to solve this problem may be to ask for a stable matching S in the subgraph obtained by deleting the endpoints of e^* from G and add e^* to S. However $S \cup \{e^*\}$ need not be popular. Figure 2 has a simple example where $e^* = (a_2, b_2)$ and the subgraph induced by a_1, b_1, a_3, b_3 has a unique stable matching $\{(a_1, b_1)\}$. However $\{(a_1, b_1), (a_2, b_2)\}$ is not popular in G as $\{(a_1, b_3), (a_2, b_1)\}$ is more popular. Note that there *is* a popular matching $M^* = \{(a_1, b_3), (a_2, b_2), (a_3, b_1)\}$ that contains e^*.

It would indeed be surprising if it was the rule that for every edge e^*, there is always a popular matching that can be decomposed as $e^* +$ a stable matching on the remaining vertices, as popularity is a far more flexible notion than stability; for instance, the set of vertices matched in every stable matching in G is the same [7] while there can be a large variation (up to a factor of 2) in the sizes of popular matchings in G. We need a larger palette than the set of stable matchings to solve the popular edge problem. We now identify another natural subclass of popular matchings called *dominant* popular matchings or dominant matchings, in short.

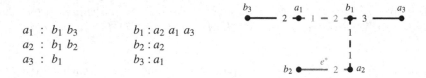

a_1 : b_1 b_3 b_1 : a_2 a_1 a_3
a_2 : b_1 b_2 b_2 : a_2
a_3 : b_1 b_3 : a_1

Fig. 2. Here we have $e^* = (a_2, b_2)$. The top choice for a_1 and a_2 is b_1, while b_3 is a_1's second choice and b_2 is a_2's second choice; b_1's top choice is a_2, second choice is a_1, and third choice is a_3. The vertices b_2, b_3, and a_3 have a_2, a_1, and b_1 as their only neighbors.

Definition 2. *Matching M is* dominant *if M is popular and moreover, for any matching M' we have: if $|M'| > |M|$, then M is more popular than M'.*

When M and M' gather the same number of votes in the election between M and M', instead of declaring these matchings as incomparable, it seems natural to regard the larger of M and M' as the *superior* matching. Dominant matchings are those popular matchings that have no superior matchings. That is, a dominant matching M gets at least as many votes as any other matching M' in an election between them and if $|M'| > |M|$, then M gets more votes than M'.

Note that a dominant matching has to be a maximum size popular matching. However not every maximum size popular matching is a dominant matching, as the example (from [10]) in Fig. 3 demonstrates.

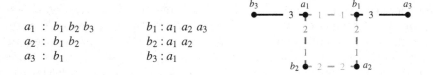

a_1 : b_1 b_2 b_3 b_1 : a_1 a_2 a_3
a_2 : b_1 b_2 b_2 : a_1 a_2
a_3 : b_1 b_3 : a_1

Fig. 3. The vertex b_1 is the top choice for all a_i's and b_2 is the second choice for a_1 and a_2 while b_3 is the third choice for a_1. The preference lists of the b_i's are symmetric. There are 2 maximum size popular matchings here: $M_1 = \{(a_1, b_1), (a_2, b_2)\}$ and $M_2 = \{(a_1, b_2), (a_2, b_1)\}$. The matching M_1 is not dominant since it is *not* more popular than the larger matching $M_3 = \{(a_1, b_3), (a_2, b_2), (a_3, b_1)\}$. The matching M_2 is dominant since M_2 is more popular than M_3.

Our Contribution. Theorem 1 is our main result here. This enables us to solve the popular edge problem in linear time.

Theorem 1. *There exists a popular matching in $G = (A \cup B, E)$ that contains the edge e^*, if and only if there exists either a stable matching in G that contains e^* or a dominant matching in G that contains e^*.*

Techniques. To show Theorem 1, we show that any popular matching M can be decomposed as $M_0 \uplus M_1$, where M_0 is dominant in the subgraph induced by the vertices matched in M_0, and in the subgraph induced by the remaining vertices,

M_1 is stable. If M contains e^*, then e^* is either in M_0 or in M_1. In the former case, we show a dominant matching in G that contains e^* and in the latter case, we show a stable matching in G that contains e^*.

We also show that every dominant matching in G can be realized as an image (under a simple and natural mapping) of a stable matching in a new graph G'. This allows us to determine in linear time if there is a dominant matching with the edge e^*. The above mapping between stable matchings in G' and dominant matchings in G can also be used to efficiently find a max-weight dominant matching in G, where each edge has a weight associated with it.

When every vertex in $G = (A \cup B, E)$ has a complete preference list, then every popular matching is dominant. Thus in such instances, a max-weight popular matching can be efficiently computed and we use this to solve the "popular set" problem. In the popular set problem, we are given a set $\{e_1, e_2, \ldots, e_k\}$ of edges and we want to find a popular matching with *all* these edges, if one exists. We show an $O(n^3)$ algorithm for this problem (via max-weight popular matching) when preference lists are complete, where $n = |A| + |B|$. When preference lists are incomplete, the complexity of the popular set problem is open, for $k \geq 2$.

Related Results. Stable matchings were defined by Gale and Shapley in their landmark paper [6]. The attention of the community was drawn very early to the characterization of *stable edges*: edges and sets of edges that can appear in a stable matching. In the seminal book of Knuth [13], stable edges first appeared under the term "arranged marriages". Knuth presented a linear time algorithm to find a stable matching with a given stable set of edges or report that none exists. Gusfield and Irving [9] provided a similar, simple method for the stable edge problem with the same running time.

The stable edge problem is a highly restricted case of the *max-weight stable matching problem*, where a stable matching that has the maximum edge weight among all stable matchings is sought. With the help of edge weights, various stable matching problems can be modeled, such as stable matchings with restricted edges [3] or egalitarian stable matchings [11]. A simple and elegant formulation of the stable matching polytope of $G = (A \cup B, E)$ is known [15] and using this, a max-weight stable matching can be computed in polynomial time via linear programming. When edge weights are non-negative integers, Feder [4,5] showed a max-weight stable matching algorithm with running time $O(n^2 \cdot \log(\frac{C}{n^2} + 2) \cdot \min\{n, \sqrt{C}\})$, where n is the number of vertices and C is the optimal weight computed based on the weight function represented as the sum of U-shaped weight functions at each vertex.

The popular matching problem is to decide if a given instance $G = (A \cup B, E)$ admits a popular matching or not. When ties are allowed in preference lists, this problem is NP-complete [1,2]. With strict preference lists, the popular matching problem becomes easy since every stable matching is popular [8]. The size of a stable matching in G can be as small as $|M_{\max}|/2$, where $M_{\max}$ is a maximum size matching in G. Relaxing stability to popularity yields larger matchings and it is easy to show that a largest popular matching has size at least $2|M_{\max}|/3$ in $G = (A \cup B, E)$ with strict preference lists. Efficient algorithms for computing a popular matching of maximum size were shown in [10,12]; in fact, both these algorithms compute dominant matchings. The popular edge problem was solved by McDermid and Irving [14] for bipartite instances, where only one side has preferences and is allowed to vote.

Organization of the Paper. A characterization of dominant matchings is given in Sect. 2. In Sect. 3 we show a surjective mapping between stable matchings in a larger graph G' and dominant matchings in G. Section 4 has our algorithm for the popular edge problem. Due to space constraints, some proofs have been omitted from this version of the paper. These proofs will be included in the full version of the paper.

2   A Characterization of Dominant Matchings

Let M be any matching in $G = (A \cup B, E)$. Recall that each $u \in A \cup B$ has a strict and possibly incomplete preference list and let $M(u)$ denote u's partner in M.

Definition 3. *For any $u \in A \cup B$ and distinct neighbors x and y of u, define u's vote between x and y as $+$ if u prefers x to y and $-$ if u prefers y to x.*

If a vertex u is unmatched in M, then $M(u)$ is undefined and this is the least preferred state for u, so $\mathsf{vote}_u(v, M(u)) = +$ for any neighbor v of u. Label each edge $e = (a, b)$ in $E \setminus M$ by the pair (α_e, β_e), where $\alpha_e = \mathsf{vote}_a(b, M(a))$ and $\beta_e = \mathsf{vote}_b(a, M(b))$, i.e., α_e is a's vote for b vs. $M(a)$ and β_e is b's vote for a vs. $M(b)$.

For any edge $(a, b) \notin M$, there are 4 possibilities for the label of edge (a, b):

- it is $(+, +)$ if (a, b) blocks M in the stable matching sense;
- it is $(+, -)$ if a prefers b to $M(a)$ while b prefers $M(b)$ to a;
- it is $(-, +)$ if a prefers $M(a)$ to b while b prefers a to $M(b)$;
- it is $(-, -)$ if both a and b prefer their respective partners in M to each other.

Let G_M be the subgraph of G obtained by deleting edges that are labeled $(-, -)$. The following theorem characterizes popular matchings.

Theorem 2 (from [10]). *A matching M is popular if and only if the following three conditions are satisfied in the subgraph G_M:*

(i) There is no alternating cycle wrt M that contains a $(+, +)$ edge.
(ii) There is no alternating path starting from an unmatched vertex wrt M that contains a $(+, +)$ edge.
(iii) There is no alternating path wrt M that contains two or more $(+, +)$ edges.

Lemma 1 characterizes those popular matchings that are dominant. The "if" side of Lemma 1 was shown in [12]: it was shown that if there is no augmenting path with respect to a popular matching M in G_M then M is more popular than all larger matchings.

Here we show that the converse holds as well, i.e., if M is a popular matching such that M is more popular than all larger matchings, in other words, if M is a dominant matching, then there is no augmenting path with respect to M in G_M.

Lemma 1. *A popular matching M is dominant if and only if there is no augmenting path with respect to M in G_M.*

Corollary 1 is a characterization of dominant matchings. This follows immediately from Lemma 1 and Theorem 2.

Corollary 1. *Matching M is a dominant matching if and only if M satisfies conditions (i)–(iii) of Theorem 2 and condition (iv): there is no augmenting path wrt M in G_M.*

3   The Set of Dominant Matchings

In this section we show a surjective mapping from the set of stable matchings in a new instance $G' = (A' \cup B', E')$ to the set of dominant matchings in $G = (A \cup B, E)$. It will be convenient to refer to vertices in A and A' as *men* and vertices in B and B' as *women*. The construction of $G' = (A' \cup B', E')$ is as follows.

- Corresponding to every man $a \in A$, there will be two men a_0 and a_1 in A' and one woman $d(a)$ in B'. The vertex $d(a)$ will be referred to as the dummy woman corresponding to a. Corresponding to every woman $b \in B$, there will be exactly one woman in B' – for the sake of simplicity, we will use b to refer to this woman as well. Thus $B' = B \cup d(A)$, where $d(A) = \{d(a) : a \in A\}$ is the set of dummy women.
- Regarding the other side of the graph, $A' = A_0 \cup A_1$, where $A_i = \{a_i : a \in A\}$ for $i = 0, 1$, and vertices in A_0 are called level 0 vertices, while vertices in A_1 are called level 1 vertices.

We now describe the edge set E' of G'. For each $a \in A$, the vertex $d(a)$ has exactly two neighbors: these are a_0 and a_1; $d(a)$'s preference order is a_0 followed by a_1. The dummy woman $d(a)$ is a_1's most preferred neighbor and a_0's least preferred neighbor. The preference list of a_0 is all the neighbors of a (in a's preference order) followed by $d(a)$. The preference list of a_1 is $d(a)$ followed by the neighbors of a (in a's preference order) in G.

For any $b \in B$, its preference list in G' is level 1 neighbors in the same order of preference as in G followed by level 0 neighbors in the same order of preference as in G. For instance, if b's preference list in G is a followed by a', then b's preference list in G' is top-choice a_1, then a'_1, and then a_0, and the last-choice is a'_0.

We now define the mapping $T :$ {stable matchings in G'} $\rightarrow$ {dominant matchings in G}. Let M' be any stable matching in G.

– $T(M')$ is the set of edges obtained by deleting all edges involving vertices in $d(A)$ (i.e., dummy women) from M' and replacing every edge $(a_i, b) \in M'$, where $b \in B$ and $i \in \{0, 1\}$, by the edge (a, b).

It is easy to see that $T(M')$ is a valid matching in G. This is because M' has to match $d(a)$, for every $a \in A$, since $d(a)$ is the top-choice for a_1. Thus for each $a \in A$, one of a_0, a_1 has to be matched to $d(a)$. Hence at most one of a_0, a_1 is matched to a non-dummy woman b and thus $M = T(M')$ is a matching in G. It can be shown that M satisfies properties (i)–(iv) that characterize dominant matchings (see Corollary 1). Thus M is a dominant matching in G.

T is Surjective. We now show that corresponding to any dominant matching M in G, there is a stable matching M' in G' such that $T(M') = M$. We will work in G_M, the subgraph of G obtained by deleting all edges labeled $(-, -)$. We now construct sets $A_0, A_1 \subseteq A$ and $B_0, B_1 \subseteq B$ as described in the algorithm below. These sets will be useful in constructing the matching M'.

0. Let $A_0 = B_1 = \emptyset$, $A_1 = \{$unmatched men in $M\}$, $B_0 = \{$unmatched women in $M\}$.
1. For every edge $(y, z) \in M$ that is labeled $(+, +)$ do:
   – let $A_0 = A_0 \cup \{y\}$, $B_0 = B_0 \cup \{M(y)\}$, $B_1 = B_1 \cup \{z\}$, and $A_1 = A_1 \cup \{M(z)\}$.
2. While there exists a matched man $a \notin A_0$ adjacent in G_M to a woman in B_0 do:
   – $A_0 = A_0 \cup \{a\}$ and $B_0 = B_0 \cup \{M(a)\}$.
3. While there exists a matched woman $b \notin B_1$ adjacent in G_M to a man in A_1 do:
   – $B_1 = B_1 \cup \{b\}$ and $A_1 = A_1 \cup \{M(b)\}$.

All unmatched men are in A_1 and all unmatched women are in B_0. For every edge (y, z) that is labeled $(+, +)$, we add y and its partner to A_0 and B_0, respectively while z and its partner are added to B_1 and A_1, respectively. For any man a, if a is adjacent to a vertex in B_0 and a is not in A_0, then a and its partner get added to A_0 and B_0, respectively. Similarly, for any woman b, if b is adjacent to a vertex in A_1 and b is not in B_1, then b and its partner get added to B_1 and A_1, respectively.

The following observations are easy to see (refer to Fig. 4). Every $a \in A_1$ has an even length alternating path in G_M to either:

(1) a man unmatched in M (by Step 0 and Step 3) or
(2) a man $M(z)$ where the woman z has an edge labeled $(+, +)$ incident on it (by Step 1 and Step 3).
   Similarly, every $a \in A_0$ has an odd length alternating path in G_M to either:
(3) a woman unmatched in M (by Step 0 and Step 2) or
(4) a woman $M(y)$ where the man y has an edge labeled $(+, +)$ incident on it (by Step 1 and Step 2).

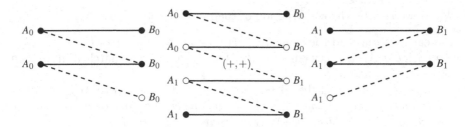

Fig. 4. Vertices get added to A_1 and A_0 by alternating paths in G_M from either unmatched vertices (first and third paths) or endpoints of edges labeled $(+,+)$ (middle path). The solid black edges are in M, and the white vertices get added to their respective sets in Steps 0 and 1.

We show the following lemma here and its proof is based on the characterization of dominant matchings in terms of conditions (i)–(iv) as given by Corollary 1. We will also use (1)–(4) observed above in our proof.

Lemma 2. $A_0 \cap A_1 = \emptyset$.

Proof. Case 1. Suppose a satisfies reasons (1) and (3) for its inclusion in A_1 and in A_0, respectively. So a is in A_1 because it is reachable via an even alternating path in G_M from an unmatched man u; also a is in A_0 because it is reachable via an odd length alternating path in G_M from an unmatched woman v. Then there is an augmenting path $\langle u, \ldots, v \rangle$ wrt M in G_M – a contradiction to the fact that M is dominant (by Lemma 1).

Case 2. Suppose a satisfies reasons (1) and (4) for its inclusion in A_1 and in A_0, respectively. So a is in A_1 because it is reachable via an even alternating path wrt M in G_M from an unmatched man u; also a is in A_0 because it is reachable via an odd length alternating path in G_M from z, where edge (y, z) is labeled $(+, +)$. Then there is an alternating path wrt M in G_M from an unmatched man u to the edge (y, z) labeled $(+, +)$ and this is a contradiction to condition (ii) of popularity.

Case 3. Suppose a satisfies reasons (2) and (3) for its inclusion in A_1 and in A_0, respectively. This case is absolutely similar to Case 2. This will cause an alternating path wrt M in G_M from an unmatched woman to an edge labeled $(+, +)$, a contradiction again to condition (ii) of popularity.

Case 4. Suppose a satisfies reasons (2) and (4) for its inclusion in A_1 and in A_0, respectively. So a is reachable via an even length alternating path in G_M from an edge labeled $(+, +)$ and $M(a)$ is also reachable via an even length alternating path in G_M from an edge labeled $(+, +)$. If it is the same edge labeled $(+, +)$ that both a and $M(a)$ are reachable from, then there is an alternating cycle in G_M with a $(+, +)$ edge – a contradiction to condition (i) of popularity. If these are 2 different edges labeled $(+, +)$, then we have an alternating path in G_M with two edges labeled $(+, +)$ – a contradiction to condition (iii) of popularity.

These four cases finish the proof that $A_0 \cap A_1 = \emptyset$. $\qquad\square$

We now describe the construction of the matching M'. Initially $M' = \emptyset$.

- For each $a \in A_0$: add the edges $(a_0, M(a))$ and $(a_1, d(a))$ to M'.
- For each $a \in A_1$: add the edge $(a_0, d(a))$ to M' and if a is matched in M then add $(a_1, M(a))$ to M'.
- For $a \notin (A_0 \cup A_1)$: add the edges $(a_0, M(a))$ and $(a_1, d(a))$ to M'.
  (Note that the men outside $A_0 \cup A_1$ are not reachable from either unmatched vertices or edges labeled $(+, +)$ via alternating paths in G_M.)

It is reasonably straightforward to show that M' is a stable matching in G'. For each $a \in A$, note that exactly one of $(a_0, d(a))$, $(a_1, d(a))$ is in M'. In order to form the set $T(M')$, the edges of M' with women in $d(A)$ are pruned and each edge $(a_i, b) \in M'$, where $b \in B$ and $i \in \{0, 1\}$, is replaced by (a, b). It is easy to see that $T(M') = M$.

This concludes the proof that every dominant matching in G can be realized as an image under T of some stable matching in G'. Thus T is surjective.

Max-Weight Popular Matchings. Our mapping T can also be used to solve the max-weight dominant matching problem in polynomial time. Here we are given a weight function $w : E \to Q$ and the problem is to find a dominant matching in G whose sum of edge weights is the highest. Using the function T, this problem can be easily reduced to computing a max-weight stable matching in G'.

If every vertex in $G = (A \cup B, E)$ has a complete preference list, then every popular matching M is dominant. This is because M is A-perfect (assuming $|A| \leq |B|$). So every vertex in A is matched in M, thus there is no augmenting path with respect to M in G (and thus in G_M). It now follows from Lemma 1 that M is dominant. Thus we can deduce Theorem 3.

Theorem 3. *Given a graph $G = (A \cup B, E)$ with strict and complete preference lists and a weight function $w : E \to \mathbb{Q}$, the problem of computing a max-weight popular matching can be solved in polynomial time.*

We now use the above result on max-weight popular matchings to efficiently solve the *popular set* problem in complete bipartite graphs. In the popular set problem, we are given a set $\{e_1, \ldots, e_k\}$ and we need to find a popular matching containing these k edges, if one exists. Else we seek a popular matching that contains as many of these edges as possible. The problem of determining if there exists a popular matching containing all these k edges can be easily posed as a max-weight popular matching problem by assigning edge weights as follows: for $1 \leq i \leq k$, set $w(e_i) = 1$ and set the weight of every other edge to be 0.

It is easy to see that under the above assignment of weights, a max-weight popular matching is exactly a popular matching that contains the largest number of edges in $\{e_1, \ldots, e_k\}$. In particular, if the weight of this popular matching is k, then there exists a popular matching that contains all these k edges. Using the max-weight stable matching algorithm of Feder [4,5] here, we can deduce the following theorem.

Theorem 4. *The popular set problem in $G = (A \cup B, E)$ with strict and complete preference lists can be solved in $O(n^3)$ time, where $|A| + |B| = n$.*

4 The Popular Edge Problem

In this section we show a decomposition for any popular matching in terms of a stable matching and a dominant matching. We use this result to design a linear time algorithm for the *popular edge* problem. Here we are given an edge $e^* = (u, v)$ in $G = (A \cup B, E)$ (with strict and possibly incomplete preference lists) and we would like to know if there exists a popular matching in G that contains e^*. We claim the following algorithm solves the above problem.

1. If there is a stable matching M_{e^*} in G that contains edge e^*, then return M_{e^*}.
2. If there is a dominant matching M'_{e^*} in G that contains edge e^*, then return M'_{e^*}.
3. Return "there is no popular matching that contains edge e^* in G".

Running time of the above algorithm. In step 1 of our algorithm, we have to determine if there exists a stable matching M_{e^*} in G that contains $e^* = (u, v)$. We modify the Gale-Shapley algorithm so that the woman v rejects all proposals from anyone worse than u. If the modified Gale-Shapley algorithm produces a matching M containing e^*, then it will be a men-optimal matching among stable matchings in G that contain e^*. Else no stable matching in G contains e^*. We refer the reader to [9, Section 2.2.2] for the correctness of the modified Gale-Shapley algorithm; it is based on the following fact:

– If G admits a stable matching that contains $e^* = (u, v)$, then exactly one of (i), (ii), (iii) occurs in any stable matching M of G: *(i)* $e^* \in M$, *(ii)* v is matched to a neighbor better than u, *(iii)* u is matched to a neighbor better than v.

In step 2 of our algorithm for the popular edge problem, we have to determine if there exists a dominant matching in G that contains $e^* = (u, v)$. This is equivalent to checking if there exists a stable matching in G' that contains either the edge (u_0, v) or the edge (u_1, v). This can be determined by using the same modified Gale-Shapley algorithm as given in the previous paragraph. Thus both steps 1 and 2 of our algorithm can be implemented in $O(m)$ time, where $m = |E|$.

We now show the correctness of our algorithm. Let M be a popular matching in G that contains edge e^*. We will use M to show that there is either a stable matching or a dominant matching that contains e^*. As before, label each edge $e = (a, b)$ outside M by the pair of votes (α_e, β_e), where α_e is a's vote for b vs. $M(a)$ and β_e is b's vote for a vs. $M(b)$.

We run the following algorithm now – this is similar to the algorithm in the previous section (where we showed T to be surjective) to build the subsets A_0, A_1 of A and B_0, B_1 of B, except that all the sets A_0, A_1, B_0, B_1 are initialized to empty sets here.

0. Initialize $A_0 = A_1 = B_0 = B_1 = \emptyset$.
1. For every edge $(a, b) \in M$ that is labeled $(+, +)$:
   - let $A_0 = A_0 \cup \{a\}$, $B_1 = B_1 \cup \{b\}$, $A_1 = A_1 \cup \{M(b)\}$, and $B_0 = B_0 \cup \{M(a)\}$.
2. While there exists a man $a' \notin A_0$ that is adjacent in G_M to a woman in B_0 do:
   - $A_0 = A_0 \cup \{a'\}$ and $B_0 = B_0 \cup \{M(a')\}$.
3. While there exists a woman $b' \notin B_1$ that is adjacent in G_M to a man in A_1 do:
   - $B_1 = B_1 \cup \{b'\}$ and $A_1 = A_1 \cup \{M(b)\}$.

All vertices added to the sets A_0 and B_1 are matched in M – otherwise there would be an alternating path from an unmatched vertex to an edge labeled $(+, +)$ and this contradicts condition (ii) of popularity of M (see Theorem 2). Note that every vertex in A_1 is reachable via an even length alternating path wrt M in G_M from some man $M(b)$ whose partner b has an edge labeled $(+, +)$ incident on it. Similarly, every vertex in A_0 is reachable via an odd length alternating path wrt M in G_M from some woman $M(a)$ whose partner a has an edge labeled $(+, +)$ incident on it. The proof of Case 4 of Lemma 2 shows that $A_0 \cap A_1 = \emptyset$.

We have $B_1 = M(A_1)$ and $B_0 = M(A_0)$ (see Fig. 5). All edges labeled $(+, +)$ are in $A_0 \times B_1$ (from our algorithm) and all edges in $A_1 \times B_0$ have to be labeled $(-, -)$ (otherwise we would contradict either condition (i) or (iii) of popularity of M).

Let $A' = A_0 \cup A_1$ and $B' = B_0 \cup B_1$. Let M_0 be the matching M restricted to $A' \cup B'$. The matching M_0 is popular on $A' \cup B'$. Suppose not and there is a matching N_0 on $A' \cup B'$ that is more popular. Then the matching $N_0 \cup (M \setminus M_0)$ is more popular than M, a contradiction to the popularity of M. Since M_0 matches all vertices in $A' \cup B'$, it follows that M_0 is dominant on $A' \cup B'$.

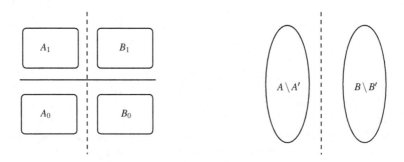

Fig. 5. M_0 is the matching M restricted to $A' \cup B'$. All unmatched vertices are in $(A \setminus A') \cup (B \setminus B')$.

Let $M_1 = M \setminus M_0$ and let $Y = A \setminus A'$ and $Z = B \setminus B'$. The matching M_1 is stable on $Y \cup Z$ as there is no edge labeled $(+, +)$ in $Y \times Z$ (all such edges are in $A_0 \times B_1$ by Step 1 of our algorithm above).

The subgraph G_M contains no edge in $A_1 \times Z$ – otherwise such a woman $z \in Z$ should have been in B_1 (by Step 3 of the algorithm above) and similarly, G_M contains no edge in $Y \times B_0$ – otherwise such a man $y \in Y$ should have been in A_0 (by Step 2 of this algorithm). We will now show Lemmas 3 and 4. These lemmas prove the correctness of our algorithm.

Lemma 3. *If $e^* \in M_0$ then there exists a dominant matching in G that contains e^*.*

Proof. Let H be the induced subgraph of G on $Y \cup Z$. We will transform the stable matching M_1 in H to a dominant matching M_1^* in H. We do this by computing a stable matching in the graph $H' = (Y' \cup Z', E')$ – the definition of H' (with respect to H) is analogous to the definition of G' (with respect to G) in Sect. 3. So for each man $y \in Y$, we have two men y_0 and y_1 in Y' and one dummy woman $d(y)$ in Z'; the set $Z' = Z \cup d(Y)$ and the preference lists of the vertices in $Y' \cup Z'$ are exactly as given in Sect. 2 for the vertices in G'.

We wish to compute a dominant matching in H, equivalently, a stable matching in H'. However we will not compute a stable matching in H' from scratch since we want to obtain a dominant matching in H using M_1. So we compute a stable matching in H' by starting with the following matching in H' (this is essentially the same as M_1):

– for each edge (y, z) in M_1, include the edges (y_0, z) and $(y_1, d(y))$ in this initial matching and for each unmatched man y in M_1, include the edge $(y_0, d(y))$ in this matching. This is a feasible starting matching as there is no blocking pair with respect to this matching.

Now run the Gale-Shapley algorithm in H' with unmatched men proposing and women disposing. Note that the starting set of unmatched men is the set of all men y_1 where y is unmatched in M_1. However as the algorithm progresses, other men could also get unmatched and propose. Let M_1' be the resulting stable matching in H'. Let M_1^* be the dominant matching in H corresponding to the stable matching M_1' in H'.

Observe that M_0 is untouched by the transformation $M_1 \rightsquigarrow M_1^*$. Let $M^* = M_0 \cup M_1^*$. Since $e^* \in M_0$, the matching M^* contains e^*. The proof that M^* is a dominant matching in G will be given in the full version of the paper. This finishes the proof of Lemma 3.   □

Lemma 4. *If the edge $e^* \in M_1$ then there exists a stable matching in G that contains e^*.*

Proof. Here will leave M_1 untouched and transform the dominant matching M_0 on $A' \cup B'$ to a stable matching M_0' on $A' \cup B'$. We do this by *demoting* all men in A_1. That is, we run the stable matching algorithm on $A' \cup B'$ with preference lists as in the original graph G, i.e., men in A_1 are not promoted over the ones in A_0. Our starting matching is M_0 restricted to edges in $A_1 \times B_1$. Since there is no blocking pair with respect to M_0 in $A_1 \times B_1$, this is a feasible starting matching.

Now unmatched men (all those in A_0) propose in decreasing order of preference to the women in B' and when a woman receives a better proposal than what she currently has, she discards her current partner and accepts the new proposal. This may make men in A_1 single and so they too propose. This is the Gale-Shapley algorithm with the only difference that our starting matching is not empty but M_0 restricted to the edges of $A_1 \times B_1$. Let M_0' be the resulting matching on $A' \cup B'$. Let $M' = M_0' \cup M_1$. This is a matching that contains the edge e^* since $e^* \in M_1$. The proof that M' is a dominant matching in G will be given in the full version of the paper. This finishes the proof of Lemma 4.   □

We have thus shown the correctness of our algorithm. Theorem 5 now follows.

Theorem 5. *Given a stable marriage instance $G = (A \cup B, E)$ with strict preference lists and an edge $e^* \in E$, we can determine in linear time if there exists a popular matching in G that contains e^*.*

We remark that identifying the whole set of popular edges in an instance also requires linear computation time. In the proof of Theorem 5 it is stated that an edge is popular if and only if it corresponds to a stable edge in G or G'. Due to [9], we can compute the set of stable edges in linear time in G and in G' as well.

Acknowledgment. Thanks to Chien-Chung Huang for useful discussions which led to the definition of dominant matchings.

References

1. Biró, P., Irving, R.W., Manlove, D.F.: Popular matchings in the marriage and roommates problems. In: Calamoneri, T., Diaz, J. (eds.) CIAC 2010. LNCS, vol. 6078, pp. 97–108. Springer, Heidelberg (2010)
2. Cseh, Á., Huang, C.-C., Kavitha, T.: Popular matchings with two-sided preferences and one-sided ties. In: Halldórsson, M.M., Iwama, K., Kobayashi, N., Speckmann, B. (eds.) ICALP 2015. LNCS, vol. 9134, pp. 367–379. Springer, Heidelberg (2015)
3. Dias, V.M.F., da Fonseca, G.D., de Figueiredo, C.M.H., Szwarcfiter, J.L.: The stable marriage problem with restricted pairs. Theor. Comput. Sci. **306**, 391–405 (2003)
4. Feder, T.: A new fixed point approach for stable networks and stable marriages. J. Comput. Syst. Sci. **45**, 233–284 (1992)
5. Feder, T.: Network flow and 2-satisfiability. Algorithmica **11**, 291–319 (1994)
6. Gale, D., Shapley, L.S.: College admissions and the stability of marriage. Am. Math. Monthly **69**, 9–15 (1962)
7. Gale, D., Sotomayor, M.: Some remarks on the stable matching problem. Discrete Appl. Math. **11**, 223–232 (1985)
8. Gärdenfors, P.: Match making: assignments based on bilateral preferences. Behav. Sci. **20**, 166–173 (1975)
9. Gusfield, D., Irving, R.W.: The Stable Marriage Problem: Structure and Algorithms. MIT Press, Cambridge (1989)
10. Huang, C.-C., Kavitha, T.: Popular matchings in the stable marriage problem. Inf. Comput. **222**, 180–194 (2013)

11. Irving, R.W., Leather, P., Gusfield, D.: An efficient algorithm for the "optimal" stable marriage. J. ACM **34**, 532–543 (1987)
12. Kavitha, T.: A size-popularity tradeoff in the stable marriage problem. SIAM J. Comput. **43**, 52–71 (2014)
13. Knuth, D.: Mariages Stables. Les Presses de L'Université de Montréal (1976). English translation in Stable Marriage and its Relation to Other Combinatorial Problems. CRM Proceedings and Lecture Notes, vol. 10. American Mathematical Society (1997)
14. McDermid, E., Irving, R.W.: Popular matchings: structure and algorithms. J. Comb. Optim. **22**(3), 339–359 (2011)
15. Rothblum, U.G.: Characterization of stable matchings as extreme points of a polytope. Math. Program. **54**, 57–67 (1992)

Semidefinite and Linear Programming Integrality Gaps for Scheduling Identical Machines

Adam Kurpisz[1], Monaldo Mastrolilli[1], Claire Mathieu[2], Tobias Mömke[3(✉)],
Victor Verdugo[2,4], and Andreas Wiese[5]

[1] Dalle Molle Institute for Artificial Intelligence Research, Manno, Switzerland
[2] Department of Computer Science,
CNRS UMR 8548, École normale supérieure, Paris, France
[3] Department of Computer Science, Saarland University, Saarbrücken, Germany
moemke@cs.uni-saarland.de
[4] Department of Industrial Engineering, Universidad de Chile, Santiago, Chile
[5] Max Planck Institute for Informatics, Saarbrücken, Germany

Abstract. Sherali-Adams [25] and Lovász-Schrijver [21] developed systematic procedures to strengthen a relaxation known as *lift-and-project* methods. They have been proven to be a strong tool for developing approximation algorithms, matching the best relaxations known for problems like Max-Cut and Sparsest-Cut. In this work we provide lower bounds for these hierarchies when applied over the configuration LP for the problem of scheduling identical machines to minimize the makespan. First we show that the configuration LP has an integrality gap of at least 1024/1023 by providing a family of instances with 15 different job sizes. Then we show that for any integer n there is an instance with n jobs in this family such that after $\Omega(n)$ rounds of the Sherali-Adams (SA) or the Lovász-Schrijver (LS$_+$) hierarchy the integrality gap remains at least 1024/1023.

1 Introduction

Scheduling

Machine scheduling is a classical family of problems in combinatorial optimization. In this paper we study the problem, known as $P||C_{\max}$, of scheduling a set J of n jobs on a set M of identical machines to minimize the *makespan*, i.e., the maximum completion time of a job, where each job $j \in J$ has a *processing time* p_j. A job cannot be preempted nor migrated to a different machine, and every job is released at time zero. This problem admits a polynomial-time approximation

Supported by the Swiss National Science Foundation project 200020-144491/1 "Approximation Algorithms for Machine Scheduling Through Theory and Experiments," by Sciex Project 12.311, by DFG grant MO 2889/1-1, and by CONICYT-PCHA/Doctorado Nacional/2014-21140930.

Q. Louveaux and M. Skutella (Eds.): IPCO 2016, LNCS 9682, pp. 152–163, 2016.
DOI: 10.1007/978-3-319-33461-5_13

scheme (PTAS) [16] and even an EPTAS [2], which is the best possible approximation result since the problem is strongly NP-hard [13]. The convex relaxations studied for the problem are weaker than those algorithmic results.

Assignment LP. A straightforward way to model $P||C_{\max}$ with a linear program (LP) is the *assignment LP* which has a variable x_{ij} for each combination of a machine $i \in M$ and a job $j \in J$, modeling whether job j is assigned to machine i. The goal is to minimize a variable T (modeling the makespan) for which we require that $\sum_{j \in J} x_{ij} \cdot p_j \leq T$ for each machine i.

$$[\text{Assign}]: \min T$$
$$\sum_{i \in M} x_{ij} \geq 1 \quad \text{for every } j \in J$$
$$\sum_{j \in J} x_{ij} p_j \leq T \quad \text{for every } i \in M$$
$$T \geq p_j \quad \text{for every } j \in J$$
$$x_{ij} \geq 0 \quad \text{for every } i \in M, j \in J.$$

Configuration LP. The assignment LP is dominated by the *configuration LP* which is, to the best of our knowledge, the strongest relaxation for the problem studied in the literature. Suppose we are given a value $T > 0$ that is an estimate on the optimal makespan, e. g., given by a binary search framework. A *configuration* corresponds to a multiset of processing times $C \subseteq \{p_j : j \in J\}$ such that $\sum_{p \in C} p \leq T$, i. e., it is a feasible assignment for a machine when the time availability is equal to T. Let, for given T, $\mathcal{C}$ denotes the set of all feasible configurations. The multiplicity function $m(p, C)$ indicates the number of times that the processing time p appears in the multiset C. For each combination of a machine i and a configuration C the configuration LP has a variable y_{iC} that models whether we want to assign exactly jobs with processing times in configuration C to machine i. Letting n_p denote the number of jobs $j \in J$ with processing time $p_j = p$, we can write:

$$[\text{clp}(T)]:$$
$$\sum_{C \in \mathcal{C}} y_{iC} = 1 \quad \text{for every } i \in M,$$
$$\sum_{i \in M} \sum_{C \in \mathcal{C}} m(p, C) y_{iC} = n_p \quad \text{for every } p \in \{p_j : j \in J\},$$
$$y_{iC} \geq 0 \quad \text{for every } i \in M, C \in \mathcal{C}.$$

We remark that in another common definition [26], a configuration is a subset, not of processing times but of jobs. We can solve that LP to a $(1 + \epsilon)$-accuracy in polynomial time [26] and similarly our LP above. The definition in terms of multisets makes sense since we are working in a setting of identical machines.

Integrality Gap. The configuration LP clp(T) does not have an objective function and instead we seek to determine the smallest value T for which it is feasible. In this context, for a convex relaxation $K(T)$ we define the *integrality gap* to be the supremum value $T_{opt}(I)/T^*(I)$ over all problem instances I, where $T_{opt}(I)$ is the optimal value and $T^*(I)$ is the minimum value T for which $K(T)$ is feasible. With the additional constraint that $T \geq \max_{j \in J} p_j$, the Assignment LP relaxation has an integrality gap of 2 (which can be shown using the analysis of the list scheduling algorithm, see e. g., [27]). Here we prove that the configuration LP has an integrality gap of at least 1024/1023 (Theorem 1(i)).

Linear Programming and Semi-definite Programming Hierarchies

Hierarchies. An interesting question is whether other convex relaxations have better integrality gaps. Convex hierarchies, parametrized by a number of *levels* or *rounds*, are systematic approaches to design improved approximation algorithms by gradually tightening the integrality gap between the integer formulation and corresponding relaxation, at the cost of increased running time. Popular among these methods are the Sherali-Adams (SA) hierarchy [25] (Definition 1), the Lovász-Schrijver (LS$_+$) semi-definite programming hierarchy [21] (Definition 4) and the Lasserre/Sum-Of-Squares hierarchy [18,22], which is the strongest of the three. For a comparison between them and their algorithmic implications we refer to [10,19,23]. In some settings, for example the Independent Set problem in sparse graphs [4], a *mixed* SA has also been considered.

Positive Results. For many problems the approximation factors of the best known algorithms match the integrality gap after performing a constant number of rounds of this hierarchies. Examples of such problems are: Max-Cut [1] and Sparsest-Cut [1,9], dense Max-Cut [11], Knapsack and Set-Cover [8]. In the scheduling context, for minimizing the makespan on two machines in the setting of unit size jobs and precedence constraints, Svensson solves the problem optimally with only one level of the linear LS hierarchy [23] (Sect. 3.1, personal communication between Svensson and the author of [23]). Furthermore, for a constant number of machines, Levey and Rothvoss give a $(1+\varepsilon)$-approximation algorithm using $(\log(n))^{\Theta(\log \log n)}$ rounds of SA hierarchy [20]. For minimizing weighted completion time on unrelated machines, one round of LS$_+$ leads to the current best algorithm [5]. Thus, hierarchies are a strong tool for approximation algorithms.

Negative Results. Nevertheless, there are known limitations on these hierarchies. Lower bounds on the integrality gap of LS$_+$ are known for Independent Set [12], Vertex Cover [3,7,14,24], Max-3-Sat and Hypergraph Vertex Cover [1], and k-Sat [6]. For the Max-Cut problem, there are lower bounds for the SA [11] and LS$_+$ [24]. For the Min-Sum scheduling problem (i. e., scheduling with job dependent cost functions on one machine) the integrality gap is unbounded even after $O(\sqrt{n})$ rounds of Lasserre [17]. In particular, that holds for the problem of minimizing the number of tardy jobs even though that problem is solvable in polynomial time, thus SDP hierarchies sometimes fail to reduce the integrality gap even on easy problems.

Our Results

Our key question in this paper is: is it possible to obtain a polynomial time $(1 + \epsilon)$-approximation algorithm based on applying the SA or the LS_+ hierarchy to one of the known LP-formulations of our problem? This would match the best known (polynomial time) approximation factor we know [2,16].

We answer this question in the negative. We prove that even after $\Omega(n)$ rounds of SA or LS_+ to the configuration LP the integrality gap of the resulting relaxation is still at least $1 + 1/1023$. Since the configuration LP dominates[1] the assignment LP, our result also holds if we apply $\Omega(n)$ rounds of SA or LS_+ to the assignment LP.

Theorem 1. *Consider the problem of scheduling identical machines to minimize the makespan, $P||C_{\max}$. For each $n \in \mathbb{N}$ there exists an instance with n jobs such that:*

(i) the configuration LP has an integrality gap of at least $1024/1023$.

(ii) after applying $r = \Omega(n)$ rounds of the SA hierarchy to the configuration LP the obtained relaxation has an integrality gap of at least $1024/1023$.

(iii) after applying $r = \Omega(n)$ rounds of the LS_+ hierarchy to the configuration LP the obtained relaxation has an integrality gap of at least $1024/1023$.

Since polynomial time approximations schemes are known [2,16] for $P||C_{\max}$, Theorem 1 implies that the SA and the LS_+ hierarchies do *not* yield the best possible approximation algorithms. We remark that for the hierarchies studied in Theorem 1, n rounds suffice to bring the integrality gap down to exactly 1, so results *(ii)* and *(iii)* are almost tight in terms of number of levels.

We prove Theorem 1 by defining a family of instances $\{I_k\}_{k \in \mathbb{N}}$ constructed from the Petersen graph (see Fig. 1). In Sect. 2 we prove that the configuration LP is feasible for $T = 1023$ while the integral optimum has a makespan of at least 1024. In Sect. 3, we show for each instance I_k that using the hypergeometric distribution we can define a fractional solution that is feasible for the polytope obtained by applying $\Omega(k)$ rounds of SA to the configuration LP parametrized by $T = 1023$. In Sect. 4 we prove the same for the semidefinite relaxations obtained with the LS_+ hierarchy, and we study the protection matrices used in the lower bound proof. In this part we work with *covariances* matrices by applying *Schur's complement* and a posterior analysis for block-symmetry matrices.

The Hard Instances

To prove the integrality gaps of $1024/1023$, for each odd $k \in \mathbb{N}$ we define an instance I_k that is inspired by the Petersen graph G (see Fig. 1) with vertex set $V = \{0, 1, \ldots, 9\}$. For each edge $e = \{u, v\}$ of G, in I_k we introduce k copies of a job $j_{\{uv\}}$ of size $2^u + 2^v$. Thus I_k has $15k$ jobs. (If n is not an odd multiple of 15, let $n = 15k + \ell$ where k is the greatest odd integer such that $15k < n$. In this case we simply add to the instance ℓ jobs that each have processing time

---

[1] The projection of the configuration LP onto the assignment space is a contained inside the polytope of the assignment LP [26].

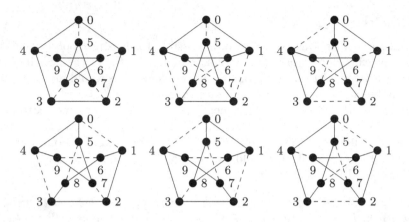

Fig. 1. The Petersen graph and its six perfect matchings (dashed lines)

equal to zero.) We define the number of machines for I_k to be $3k$. For simplicity, in the following we do not distinguish between jobs and their sizes. The graph G has exactly six perfect matchings $\bar{M}_1, \bar{M}_2, \ldots, \bar{M}_6$. Since the sum of the job sizes in a perfect matching $\bar{M}_\ell$ is

$$\sum_{e \in \bar{M}_\ell} p_{j_e} = \sum_{0 \leq u \leq 9} 2^u = 1023,$$

$\bar{M}_\ell$ corresponds to a configuration C_ℓ that contains one job corresponding to each edge in $\bar{M}_\ell$ and has makespan 1023. The configurations $C_1, \ldots, C_6$ are called *matching configurations* and we denote them by $\mathcal{D} = \{C_1, \ldots, C_6\}$.

2 Integrality Gap of the Configuration LP (Theorem 1(i))

Lemma 1. $clp[1023]$ *is feasible for I_k.*

Proof. To define the fractional solution, for every machine i and each $\ell \in \{1, 2, \ldots, 6\}$ we set $y_{iC_\ell} = 1/6$. For all other configurations C we set $y_{iC} = 0$.

The first set of constraints in $clp(T)$ (for the machines) is clearly satisfied. For the second set of constraints (for the job sizes), let p be such a job size and let e be the corresponding edge in G. The Petersen graph is such that there are exactly two perfect matchings $\bar{M}_\ell, \bar{M}_{\ell'}$ containing e, thus we get $\sum_{i=1}^{3k}(y_{iC_\ell} + y_{iC_{\ell'}}) = k$ and y is feasible. □

Lemma 2. *The optimal makespan for I_k is at least 1024.*

Proof. Assume, for a contradiction, that clp[1023] for I_k has an integer solution y. Since the total size of jobs is $k \cdot 3 \cdot 1023$ and there are $3k$ machines, only configurations C with makespan exacly equal to 1023 may have $y_{iC} \neq 0$.

Consider such a configuration C. Since $1023 = \sum_{u=0}^{9} 2^u$, considering the binary representation of 1023, by induction on u it must be that for every u, configuration C contains an odd number of jobs corresponding to edges adjacent to vertex u in G. Furthermore, since the sum does not exceed 1023, that odd number must be exactly 1. Thus C exactly corresponds to a perfect matching of G, and so the integer solution y corresponds to a 1-factorization of the multigraph G_k obtained by taking k copies of each edge in the Petersen graph.

Let $\bar{M}_1$ be the perfect matching of the Petersen graph consisting of the five edges $\{0,5\},\{1,6\},\{2,7\},\{3,8\}\{4,9\}$, called *spokes*. Let $\ell = \sum_i y_{iC_1}$. Since each spoke, which appears in exactly one other perfect matching $\bar{M}_j$ $(j > 1)$, must be contained in k matchings in total, we must have $\sum_i y_{iC_j} = k - \ell$ for each $j \in [2,6]$. Thus $\sum_{i,C} y_{iC} = 5(k-\ell)+\ell = 5k - 4\ell$. However, that sum equals $3k$, the total number of machines, and so $\ell = k/2$. Since k is odd and ℓ an integer, the contradiction follows.   □

3   Integrality Gap for SA (Theorem 1(ii))

We show that for the family of instances $\{I_k\}_{k\in\mathbb{N}}$ defined in Sect. 2, if we apply $O(k)$ rounds of SA to the configuration LP for $T = 1023$, then the resulting relaxation is feasible. Thus, after $\Omega(k)$ rounds of SA the configuration LP still has an integrality gap of at least $1024/1023$ on an instance with $O(k)$ jobs and machines. First, we define the polytope $SA^r(P)$ obtained after r rounds of SA to a polytope P that is defined via equality constraints[2].

Definition 1 (Polytope SA^r(P)). *Consider a polytope $P \subseteq [0,1]^E$ defined by equality constraints. For every constraint $\sum_{i\in E} a_{i,\ell} y_i = b_\ell$ and every $H \subseteq E$ such that $|H| \leq r$, the constraint $\sum_{i\in E} a_{i,\ell} y_{H\cup\{i\}} = b_\ell y_H$ is included in $SA^r(P)$, the level r of the Sherali-Adams hierarchy applied to P. The polytope $SA^r(P)$ lives in $\mathbb{R}^{\mathcal{P}_{r+1}(E)}$, where $\mathcal{P}_{r+1}(E) = \{A \subseteq E : |A| \leq r + 1\}$.*

For the configuration LP $clp(T)$ the variables set is $E = M \times \mathcal{C}$. Since it is defined by equality constraints, the polytope $SA^r(clp(T))$ corresponds to

$$[SA^r(clp(T))]:$$

$$\sum_{C\in\mathcal{C}} y_{H\cup\{(i,C)\}} = y_H \quad \forall\, i \in M, \forall\, H \subseteq E : |H| \leq r,$$

$$\sum_i \sum_{C\in\mathcal{C}} m(p,C)y_{H\cup\{(i,C)\}} = n_p y_H \quad \forall\, p \in \{p_j : j \in J\}, \forall\, H \subseteq E : |H| \leq r,$$

$$y_H \geq 0 \quad \forall\, H \subseteq E : |H| \leq r + 1,$$

$$y_\emptyset = 1.$$

Intuitively, the configuration LP computes a set of edges in a complete bipartite graph with vertex sets U, V where U is the set of machines and V is the set of configurations. The edges are selected such that they form a U-matching, i.e., such that each node in U is incident to at most one selected edge.

---

[2] This definition is slightly different from the one in Sherali & Adams [25]; for simplicity we give a definition that, in the case of equality constraints, is equivalent.

Definition 2. *Given two sets U, V and $F \subseteq U \times V$, the F-degree of $u \in U$ is $\delta_F(u) = |\{v : (u,v) \in F\}|$, and $\delta_F(v) = |\{u : (u,v) \in F\}|$ if $v \in V$. We say that F is an U-matching if $\delta_F(u) \leq 1$ for every $u \in U$. An element $u \in U$ is incident to F if $\delta_F(u) = 1$.*

In the following we consider the same family of instances $\{I_k : k \in \mathbb{N}, k \text{ is odd}\}$ as in Sect. 2 and $T = 1023$. For any set S we define $\mathcal{P}(S)$ to be the power set of S. We want to define a solution to $\mathrm{SA}^r(\mathrm{clp}(T))$ for $T = 1023$. To this end, we need to define a value y_A for each set $A \in \mathcal{P}_{r+1}(M \times \mathcal{C})$. In particular, for $A \in \mathcal{P}_r(M \times \mathcal{D})$, we define this value according to the *hypergeometric distribution*.

Definition 3. *Let $\phi : \mathcal{P}(M \times \mathcal{D}) \to \mathbb{R}$ be such that*

$$\phi(A) = \frac{1}{(3k)_{|A|}} \prod_{j \in [6]} (k/2)_{\delta_A(C_j)}$$

if A is an M-matching, and zero otherwise, where $(x)_a = x(x-1)\cdots(x-a+1)$, for integer $a \geq 1$, is the lower factorial function.

To get some understanding about how the distribution ϕ works, the following lemma intuitively shows the following: suppose that we know that a set A is chosen (i.e., we condition on this), then the conditional probability that also a pair (i, C_j) is chosen equals $\frac{k/2 - \delta_A(C_j)}{3k - |A|}$, assuming that $A \cup \{(i, C_j)\}$ forms an M-matching.

Lemma 3. *Let $A \subseteq M \times \mathcal{D}$ be an M-matching of size at most $3k - 1$. If $i \in M$ is not incident to A, then $\phi(A \cup \{(i, C_j)\}) = \phi(A)\frac{k/2 - \delta_A(C_j)}{3k - |A|}$.*

Proof. Given that i is not incident to A, we have $|A \cup \{(i, C_j)\}| = |A| + 1$. Furthermore, for $\ell \neq j$ we have that $\delta_{A \cup \{(i,C_j)\}}(C_\ell) = \delta_A(C_\ell)$ and $\delta_{A \cup \{(i,C_j)\}}(C_j) = \delta_A(C_j) + 1$. Therefore, $\frac{\phi(A \cup \{(i,C_j)\})}{\phi(A)} = \frac{k/2 - \delta_A(C_j)}{3k - |A|}$. $\square$

The Feasible Solution. We are ready now to define our solution to $\mathrm{SA}^r(\mathrm{clp}(T))$. It is the vector $y^\phi \in \mathbb{R}^{\mathcal{P}_{r+1}(E)}$ defined such that $y_A^\phi = \phi(A)$ if A is an M-matching in $M \times \mathcal{D}$, and zero otherwise.

Lemma 4. *For every odd k, y^ϕ is a feasible solution for $\mathrm{SA}^r(\mathrm{clp}(T))$ for the instance I_k when $r = \lfloor k/2 \rfloor$ and $T = 1023$.*

Proof. We first prove that $y^\phi \geq 0$. Consider some $H \subseteq E$. Since $y_H^\phi = \phi(H)$, using Definition 3, it is easy to check that the lower factorial stays non-negative for $r = \lfloor k/2 \rfloor$.

We next prove that y^ϕ satisfies the machine constraints in $\mathrm{SA}^r(\mathrm{clp})$. If i is a machine incident to H, then all terms in the left-hand summation are 0 except for the unique pair (i, C) that belongs to H, so the sum equals y_H^ϕ. If i is not incident to H, then by Lemma 3 we have

$$\sum_C y^\phi_{H\cup\{(i,C)\}} = \frac{\phi(H)}{3k - |H|} \sum_{j\in[6]} (k/2 - \delta_H(C_j)) = \phi(H) = y^\phi_H,$$

since $6 \cdot k/2 = 3k$ and $\sum_{j\in[6]} \delta_H(C_j) = |H|$.

Finally we prove that y^ϕ satisfies the set of constraints for every processing time. Fix p and H. Since y^ϕ is supported by six configurations, we have

$$\sum_{i\in M} \sum_{C\in\mathcal{C}} m(p,C) y^\phi_{H\cup\{(i,C)\}} = \sum_{i\in M} \sum_{j\in[6]} m(p,C_j)\phi(H \cup \{(i,C_j)\}).$$

There are exactly two configurations $C^p_1, C^p_2 \in \mathcal{D}$ such that $m(p,C^p_1) = m(p,C^p_2) = 1$, and for the others it is zero, so

$$\sum_{j\in[6]} m(p,C_j)\phi(H \cup \{(i,C_j)\}) = \phi(H \cup \{(i,C^p_1)\}) + \phi(H \cup \{(i,C^p_2)\}).$$

Let $\pi_M(H) = \{i \in M : \delta_H(i) = 1\}$ be the subset of machines incident to II. We split the sum over $i \in M$ into two parts, $i \in \pi_M(H)$ and $i \notin \pi_M(H)$. For the first part,

$$\sum_{i\in\pi_M(H)} (\phi(H \cup \{(i,C^p_1)\}) + \phi(H \cup \{(i,C^p_2)\})) = \phi(H)(\delta_H(C^p_1) + \delta_H(C^p_2))$$

since $\phi(H \cup \{(i,C^p_1)\})$ is either $\phi(H)$ or 0 depending on whether $(i,C^p_1) \in H$, and the same holds for C^p_2.

For the second part, using Lemma 3 we have that

$$\sum_{i\notin\pi_M(H)} (\phi(H \cup \{(i,C^p_1)\}) + \phi(H \cup \{(i,C^p_2)\}))$$

$$= \frac{\phi(H)}{3k - |H|} \sum_{i\notin\pi_M(H)} (k/2 - \delta_H(C^p_1) + k/2 - \delta_H(C^p_2))$$

$$= \phi(H)(k/2 - \delta_H(C^p_1) + k/2 - \delta_H(C^p_2)),$$

since $|H \setminus \pi_M(H)| = 3k - |H|$. Adding, thanks to cancellations we get precisely what we want:

$$\sum_{i\in M} \sum_{C\in\mathcal{C}} m(p,C) y^\phi_{H\cup\{(i,C)\}} = k\phi(H) = n_p y^\phi_H.$$

$\square$

Proof (of Theorem 1(ii)). Consider instance I_k as defined before, $T = 1023$ and $r = \lfloor k/2 \rfloor$. By Lemma 4 the vector $y^\phi \in SA^r(clp(T))$. $\square$

We note that in the above proof, the projection of y^ϕ onto the space of the configuration LP is the fractional solution from the proof of Lemma 1.

4   Integrality Gap for LS$_+$ (Theorem 1(iii))

Given a polytope $P \subseteq \mathbb{R}^d$, we consider the convex cone $Q = \{(a, x) \in \mathbb{R}^* \times P : x/a \in P\}$. We define an operator N_+ on convex cones $R \subseteq \mathbb{R}^{d+1}$ as follows: $y \in N_+(R)$ if and only if there exists a symmetric matrix $Y \subseteq \mathbb{R}^{(d+1) \times (d+1)}$, called the *protection matrix* of y, such that

1. $y = Y e_{\emptyset} = \mathrm{diag}(Y)$,
2. for all i, $Y e_i, Y(e_{\emptyset} - e_i) \in R$,
3. Y is positive semidefinite,

where e_i denotes the vector with a 1 in the ith coordinate and 0's elsewhere.

Definition 4. *For any $r \geq 0$ and polytope $P \subseteq \mathbb{R}^d$, level r of the LS$_+$ hierarchy, $N_+^r(Q) \subseteq \mathbb{R}^{d+1}$, is defined recursively by: $N_+^0(Q) = Q$ and $N_+^r(Q) = N_+(N_+^{r-1}(Q))$.*

To prove the integrality gap for LS$_+$ we follow an inductive argument. We start from $P = \mathrm{clp}(T)$. Along the proof, we use a special type of vectors that are integral in a subset of coordinates and fractional in the others.

The Feasible Solution. Let A be an M-matching in $M \times \mathcal{D}$. The *partial schedule* $y(A) \in \mathbb{R}^{M \times C}$ is the vector such that for every $i \in M$ and $j \in \{1, 2, \ldots, 6\}$, $y(A)_{iC_j} = \phi(A \cup \{(i, C_j)\})/\phi(A)$, and zero otherwise. Here is the key Lemma.

Lemma 5. *Let k be an odd integer and $r \leq \lfloor k/2 \rfloor$. Let Q_k be the convex cone of $\mathrm{clp}(T)$ for instance I_k and $T = 1023$. Then, for every M-matching A of cardinality $\lfloor k/2 \rfloor - r$ in $M \times \mathcal{D}$, we have $y(A) \in N_+^r(Q_k)$.*

Before proving Lemma 5, let us see how it implies the Theorem.

Proof (of Theorem 1(iii)). Consider instance I_k defined in Sect. 2, $T = 1023$ and $r = \lfloor k/2 \rfloor$. By Lemma 5 for $A = \emptyset$ we have $y(\emptyset) \in N_+^r(Q_k)$.   □

In the following two helper lemmas we describe structural properties of every partial schedule.

Lemma 6. *Let A be an M-matching in $M \times \mathcal{D}$, and let i be a machine incident to A. Then, $y(A)_{iC} \in \{0, 1\}$ for all configuration C.*

Proof. If $C \notin \mathcal{D}$ then $y(A)_{iC} = 0$ by definition. If $(i, C_j) \in A$ then $y(A)_{iC_j} = \phi(A \cup \{(i, C_j)\})/\phi(A) = \phi(A)/\phi(A) = 1$. For $\ell \neq j$, the set $A \cup \{(i, C_\ell)\}$ is not an M-matching and thus $y(A)_{iC_k} = 0$.   □

Lemma 7. *Let A be an M-matching in $M \times \mathcal{D}$ of cardinality at most $\lfloor k/2 \rfloor$. Then, $y(A) \in \mathrm{clp}(T)$.*

Proof. We note that $y(A)_{iC} = y^{\phi}_{A \cup \{(i,C)\}}/y^{\phi}_A$, and then the feasibility of $y(A)$ in $\mathrm{clp}(T)$ is implied by the feasibility of y^{ϕ} in SA$^r(\mathrm{clp}(T))$, for $r = \lfloor k/2 \rfloor$.   □

Given a partial schedule $y(A)$, let $Y(A) \in \mathbb{R}^{(|M \times \mathcal{C}|+1) \times (|M \times \mathcal{C}|+1)}$ be the matrix such that its principal submatrix indexed by $\{\emptyset\} \cup (M \times \mathcal{D})$ equals

$$\begin{pmatrix} 1 & y(A)^\top \\ y(A) & Z(A) \end{pmatrix},$$

where $Z(A)_{iC_j, \ell C_h} = \phi(A \cup \{(i, C_j), (\ell, C_h)\})/\phi(A)$. All the other entries of the matrix $Y(A)$ have value equal to zero. The matrix $Y(A)$ provides the *protection matrix* we need in the proof of the key Lemma.

Theorem 2. *For every M-matching A in $M \times \mathcal{D}$ such that $|A| \leq \lfloor k/2 \rfloor$, the matrix $Y(A)$ is positive semidefinite.*

Proof (Sketch). We prove that $Y(A)$ is positive semidefinite by performing several transformations that preserve this property. First, we remove all those zero columns and rows. Then, $Y(A)$ is positive semidefinite if and only if its principal submatrix indexed by $\{\emptyset\} \cup (M \times \mathcal{D})$ is positive semidefinite. We then construct the covariance matrix $\mathrm{Cov}(A)$ by taking the Schur's Complement of $Y(A)$ respect to the entry $(\{\emptyset\}, \{\emptyset\})$. The resulting matrix is positive semidefinite if and only if $Y(A)$ is positive semidefinite. After removing null rows and columns in $\mathrm{Cov}(A)$ we obtain a new matrix, $\mathrm{Cov}_+(A)$, which can be written using Kronecker products as $I \otimes Q + (J - I) \otimes W$, with $Q, W \in \mathbb{R}^{6 \times 6}$, $Q = \alpha W$ for some $\alpha \in (-1, 0)$ and I, J being the identity and the all-ones matrix, respectively. By applying a lemma about block matrices in [15], $Y(A)$ is positive semidefinite if and only if W is positive semidefinite. The matrix W is of the form $D_u - uu^\top$, with $u \in \mathbb{R}^6$ and D_u is a diagonal matrix such that $\mathrm{diag}(D_u) = u$. By Jensen's inequality with the function $t(y) = y^2$ it follows that W is positive semidefinite. A complete proof of the theorem can be found in the Appendix. $\square$

Lemma 8. *Let A be an M-matching in $M \times \mathcal{D}$ and i a non-incident machine to A. Then, $\sum_{j \in [6]} Y(A) e_{iC_j} = Y(A) e_\emptyset$.*

Proof. Let S be the index of a row of $Y(A)$. If $S \notin \{0\} \cup (M \times \mathcal{D})$ then that row is identically zero, so the equality is satisfied. Otherwise,

$$e_S^\top \sum_{j \in [6]} Y(A) e_{iC_j} = \sum_{j \in [6]} \frac{\phi(A \cup \{(i, C_j)\} \cup S)}{\phi(A)}.$$

If $A \cup S$ is not an M-matching then $\phi(A \cup S \cup \{i, C_j\}) = 0$ for all i and $j \in [6]$, and $e_S^\top Y(A) e_\emptyset = \phi(A \cup S) = 0$, so the equality is satisfied. If $A \cup S$ is an M-matching, then

$$\sum_{j \in [6]} \frac{\phi(A \cup \{(i, C_j)\} \cup S)}{\phi(A)} = \frac{\phi(A \cup S)}{\phi(A)} \sum_{j \in [6]} \frac{\phi(A \cup S \cup \{(i, C_j)\})}{\phi(A \cup S)}$$

$$= e_S^\top Y(A) e_\emptyset \sum_{j \in [6]} \frac{y_{A \cup S \cup \{(i, C_j)\}}^\phi}{y_{A \cup S}^\phi}$$

$$= e_S^\top Y(A) e_\emptyset,$$

since y^ϕ is a feasible solution for the SA hierarchy. $\square$

Having previous two results we are ready to prove the key Lemma.

Proof (of Lemma 5). We proceed by induction in r. The base case $r = 0$ is implied by Lemma 7, and now suppose that it is true for $r = t$. Let $y(A)$ be a partial schedule of A of cardinality $\lfloor k/2 \rfloor - t - 1$. We prove that the matrix $Y(A)$ is a protection matrix for $y(A)$. It is symmetric by definition, $y(A)e_\emptyset = \mathrm{diag}(y(A)) = y(A)$ and thanks to Theorem 2 the matrix $Y(A)$ is positive semidefinite. Let (i, C) be such that $y(A)_{iC} \in (0, 1)$. In particular, by Lemma 6 we have $(i, C) \notin A$ and $C \in \mathcal{D}$. We claim that $Y(A)e_{iC}/y(A)_{iC}$ is equal to the partial schedule $(1, y(A \cup \{(i, C)\}))$. If S indexes a row not in $M \times \mathcal{D}$ then the respective entry in both vectors is zero, so the equality is satisfied. Otherwise,

$$\frac{e_S^\top Y(A)e_{iC}}{y(A)_{iC}} = \frac{\phi(A \cup \{(i, C)\} \cup S)}{\phi(A \cup \{(i, C)\})} = y(A \cup \{(i, C)\})_S.$$

The cardinality of the M-matching $A \cup \{(i, C)\}$ is equal to $|A| + 1 = \lfloor k/2 \rfloor - t$, and therefore by induction we have that $Y(A)e_{iC}/y(A)_{iC} = (1, y(A \cup \{(i, C)\})) \in N_+^t(Q_k)$. Now we have to prove that the vectors $Y(A)(e_\emptyset - e_{iC})/(1 - y(A)_{iC})$ are feasible for $N_+^t(Q_k)$. By Lemma 8 we have that for every $\ell \in \{1, 2, \ldots, 6\}$,

$$\frac{Y(A)(e_\emptyset - e_{iC_\ell})}{1 - y(A)_{iC_\ell}} = \sum_{j \in [6] \setminus \{\ell\}} \left(\frac{y(A)_{iC_j}}{\sum_{j \in [6] \setminus \{\ell\}} y(A)_{iC_j}} \right) y(A \cup \{(i, C_j)\}),$$

and then $Y(A)(e_\emptyset - e_{iC_\ell})/(1 - y(A)_{iC_\ell})$ is a convex combination of the partial schedules $\{y(A \cup \{(i, C_j)\}) : j \in \{1, 2, \ldots, 6\} \setminus \ell\} \subset N_+^t(Q_k)$, concluding the induction. $\qquad\square$

References

1. Alekhnovich, M., Arora, S., Tourlakis, I.: Towards strong nonapproximability results in the Lovász-Schrijver hierarchy. In: STOC, pp. 294–303 (2005)
2. Alon, N., Azar, Y., Woeginger, G.J., Yadid, T.: Approximation schemes for scheduling. In: SODA, pp. 493–500 (1997)
3. Arora, S., Bollobás, B., Lovász, L., Tourlakis, I.: Proving integrality gaps without knowing the linear program. Theor. Comput. **2**, 19–51 (2006)
4. Bansal, N.: Approximating independent sets in sparse graphs. In: SODA, pp. 1–8 (2015)
5. Bansal, N., Srinivasan, A., Svensson, O.: Lift-and-round to improve weighted completion time on unrelated machines. CoRR, abs/1511.07826 (2015)
6. Buresh-Oppenheim, J., Galesi, N., Hoory, S., Magen, A., Pitassi, T.: Rank bounds and integrality gaps for cutting planes procedures. Theor. Comput. **2**, 65–90 (2006)
7. Charikar, M.: On semidefinite programming relaxations for graph coloring and vertex cover. In: SODA, pp. 616–620 (2002)
8. Chlamtáč, E., Friggstad, Z., Georgiou, K.: Lift-and-project methods for set cover and Knapsack. In: Dehne, F., Solis-Oba, R., Sack, J.-R. (eds.) WADS 2013. LNCS, vol. 8037, pp. 256–267. Springer, Heidelberg (2013)

9. Chlamtac, E., Krauthgamer, R., Raghavendra, P.: Approximating sparsest cut in graphs of bounded treewidth. In: Serna, M., Shaltiel, R., Jansen, K., Rolim, J. (eds.) APPROX 2010. LNCS, vol. 6302, pp. 124–137. Springer, Heidelberg (2010)
10. Chlamtac, E., Tulsiani, M.: Convex relaxations and integrality gaps. In: Anjos, M.F., Lasserre, J.B. (eds.) Handbook on Semidefinite, Conic and Polynomial Optimization, pp. 139–169. Springer, New York (2012)
11. Fernandez de la Vega, W., Mathieu, C.: Linear programming relaxations of maxcut. In: SODA, pp. 53–61 (2007)
12. Feige, U., Krauthgamer, R.: The probable value of the Lovász-Schrijver relaxations for maximum independent set. SIAM J. Comput. **32**, 345–370 (2003)
13. Garey, M.R., Johnson, D.S.: "Strong" NP-completeness results: motivation, examples, and implications. J. ACM **25**, 499–508 (1978)
14. Georgiou, K., Magen, A., Pitassi, T., Tourlakis, I.: Integrality gaps of 2-o(1) for vertex cover SDPs in the Lovász-Schrijver hierarchy. SIAM J. Comput. **39**, 3553–3570 (2010)
15. Gvozdenovic, N., Laurent, M.: The operator ψ for the chromatic number of a graph. SIAM J. Optim. **19**, 572–591 (2008)
16. Hochbaum, D., Shmoys, D.: Using dual approximation algorithms for scheduling problems theoretical and practical results. J. ACM **34**, 144–162 (1987)
17. Kurpisz, Adam, Leppänen, S., Mastrolilli, M.: A Lasserre lower bound for the min-sum single machine scheduling problem. In: Bansal, N., Finocchi, I. (eds.) ESA 2015. LNCS, vol. 9294, pp. 853–864. Springer, Heidelberg (2015). doi:10.1007/978-3-662-48350-3_71
18. Lasserre, J.: Global optimization with polynomials and the problem of moments. SIAM J. Optim. **11**, 796–817 (2001)
19. Laurent, M.: A comparison of the Sherali-Adams, Lovász-Schrijver, and Lasserre relaxations for 0–1 programming. Math. Oper. Res. **28**, 470–496 (2003)
20. Levey, E., Rothvoss, T.: A Lasserre-based $(1 + \varepsilon)$-approximation for $Pm|p_j = 1, \text{prec}|C_{max}$. CoRR, abs/1509.07808 (2015)
21. Lovász, L., Schrijver, A.: Cones of matrices and set-functions and 0–1 optimization. SIAM J. Optim. **1**, 166–190 (1991)
22. Parrilo, P.: Semidefinite programming relaxations for semialgebraic problems. Math. Program. **96**, 293–320 (2003)
23. Rothvoß, T.: The Lasserre hierarchy in approximation algorithms. Lecture notes for the MAPSP (2013)
24. Schoenebeck, G., Trevisan, L., Tulsiani, M.: Tight integrality gaps for Lovász-Schrijver LP relaxations of vertex cover and max cut. In: Proceedings of the Thirty-Ninth Annual ACM Symposium on Theory of Computing, pp. 302–310 (2007)
25. Sherali, H., Adams, W.: A hierarchy of relaxations between the continuous and convex hull representations for zero-one programming problems. SIAM J. Discrete Math. **3**, 411–430 (1990)
26. Verschae, J., Wiese, A.: On the configuration-LP for scheduling on unrelated machines. J. Sched. **17**, 371–383 (2014)
27. Williamson, D., Shmoys, D.: The Design of Approximation Algorithms. Cambridge University Press, New York (2011)

Stabilizing Network Bargaining Games
by Blocking Players

Sara Ahmadian[1(✉)], Hamideh Hosseinzadeh[2], and Laura Sanità[1]

[1] Department of Combinatorics and Optimization,
University of Waterloo, Waterloo, Canada
{sahmadian,lsanita}@uwaterloo.ca
[2] Faculty of Mathematical Science, Alzahra University, Tehran, Iran
hamideh.hosseinzadeh@gmail.com

Abstract. Cooperative matching games (Shapley and Shubik) and Network bargaining games (Kleinberg and Tardos) are games described by an undirected graph, where the vertices represent players. An important role in such games is played by *stable* graphs, that are graphs whose set of inessential vertices (those that are exposed by at least one maximum matching) are pairwise non adjacent. In fact, stable graphs characterize instances of such games that admit the existence of stable outcomes.

In this paper, we focus on stabilizing instances of the above games by *blocking* as few players as possible. Formally, given a graph G we want to find a minimum cardinality set of vertices such that its removal from G yields a stable graph. We give a combinatorial polynomial-time algorithm for this problem, and develop approximation algorithms for some NP-hard *weighted* variants, where each vertex has an associated non-negative weight. Our approximation algorithms are LP-based, and we show that our analysis are almost tight by giving suitable lower bounds on the integrality gap of the used LP relaxations.

1 Introduction

Game theory is an active and important area of research in the field of Theoretical Computer Science, and combinatorial optimization techniques are often crucially employed in solving game theory problems [15]. For several games defined on networks, studying the structure of the underlying graph that describes the network setting is important to identify the existence of good outcomes for the corresponding games. Prominent examples are *cooperative matching games* introduced by Shapley and Shubik [17] and *network bargaining games* studied by Kleinberg and Tardos [10]. These are games described by an undirected graph $G = (V, E)$, where the vertices represent players, and the cardinality of a maximum matching in G, denoted by $\nu(G)$, represents a total *value* that the players could gain by interacting with each other.

In an instance of a cooperative matching game [17], one seeks for an *allocation* of the value $\nu(G)$ among players, described by a vector $y \in \mathbb{R}^V_{\geq 0}$, in which no subset of players S has an incentive to form a coalition to deviate. This is formally

© Springer International Publishing Switzerland 2016
Q. Louveaux and M. Skutella (Eds.): IPCO 2016, LNCS 9682, pp. 164–177, 2016.
DOI: 10.1007/978-3-319-33461-5_14

described by the constraint $\sum_{v \in S} y_v \geq \nu(G[S])$ for all subsets S, where $G[S]$ denotes the subgraph induced by the vertices in S. Such allocation y is called *stable*. It is well-known (see e.g. [5]) that cooperative matching game instances that admit the existence of a stable allocation are precisely the set of instances described by *stable graphs*: these are graphs whose set of *inessential vertices* are pairwise non adjacent. We recall here that a vertex v of a graph G is called inessential if there exists at least one maximum matching M in G that *exposes* v, that is, v is not an endpoint of M, and it is called essential otherwise.

Network bargaining games described by Kleinberg and Tardos [10] are network extensions of the classical Nash bargaining games [14]. In an instance of a network bargaining game described by a graph G, the edges represent a set of potential *deals* of unit value that the players (vertices) could make. An outcome of the game is given by a matching M of G (representing the set of deals that the players made) together with a value allocation $y \in \mathbb{R}_{\geq 0}^V$ on each vertex (representing how the players decided to split the values of the deals they made, if any). Kleinberg and Tardos [10] introduced the notion of *stable* outcomes for such games, that are outcomes where no player has an incentive to deviate, as well as the notion of *balanced* outcomes, that are stable outcomes in which, in addition, the values are "fairly" split among the players. The authors proved that a balanced outcome exists if and only if a stable outcome exists, and this happens if and only if the graph G describing the instance is *stable*.

Since not all graphs are stable, there are instances of both network bargaining games and cooperative matching games that do not admit stable solutions. This motivated many authors in past years to address the algorithmic problem of *stabilizing* such instances by minimally modifying the underlying graph. Two very natural ways to modify a graph in order to achieve some desired properties are via *edge-removal* or *vertex-removal* operations. The authors in [4] looked at edge-removal operations, that is, stabilizing instances of the above games by blocking potential deals that the players could make. In this paper, we look at the vertex-removal counterpart, that is, stabilizing instances by *blocking players*. Formally, this translates into the following problem:

Vertex-stabilizer Problem: *Given a graph $G = (V, E)$, find a minimum cardinality* vertex-stabilizer, *that is a set $S \subseteq V$ whose removal from G yields a stable graph.*

We also generalize and study this problem in the *weighted* setting (a formal definition is in the next subsection).

In addition to the connection with game theory, the vertex-stabilizer problem is also of interest from a combinatorial optimization perspective. In fact, an alternative and equivalent characterization of stable graphs can be given using linear programming and the notion of *fractional* matchings and vertex covers, as we are now going to explain. For a graph $G = (V, E)$, a fractional matching is a feasible solution to the LP:

$$\nu_f(G) := \max \Big\{ \sum_{e \in E} x_e : \sum_{e \in \delta(v)} x_e \leq 1 \ \forall v \in V, \ x \geq 0 \Big\},$$

where $\delta(v)$ denotes the set of edges incident into v. Note that, if we add binary constraints to the above LP we obtain a formulation to find a matching of G of maximum cardinality $\nu(G)$. The dual of the above LP is:

$$\tau_f(G) := \min\left\{\sum_{v \in V} y_v : y_u + y_v \geq 1 \; \forall\{u, v\} \in E, \; y \geq 0\right\}.$$

Once again, note that if we add binary constraints to this dual LP we obtain the canonical formulation for finding a *vertex cover* of G of minimum cardinality $\tau(G)$, that is, a min-cardinality subset of vertices covering all edges of the graph. For this reason, fractional feasible solutions to the above dual LP are called *fractional* vertex covers.

By duality theory, we know that the following holds: $\nu(G) \leq \nu_f(G) = \tau_f(G) \leq \tau(G)$. In general, there are graphs for which all the above inequalities are strict (e.g. a triangle). However, for certain classes of graphs some of the above inequalities hold tight. In particular, the class of *König-Egerváry* graphs [12,18] is formed by all graphs G for which $\nu(G) = \tau(G)$, that is, all the above inequalities hold tight. Note that the class of König-Egerváry graphs is a proper superset of the class of bipartite graphs. It is known (see e.g. [10]) that stable graphs are exactly the class of graphs for which $\nu(G) = \nu_f(G) = \tau_f(G) \leq \tau(G)$, that is, graphs for which the cardinality of a maximum matching ($\nu(G)$) is equal to the minimum size of a *fractional* vertex cover ($\tau_f(G)$). We have therefore the following relation:

(Bipartite graphs) $\subsetneq$ (König-Egerváry) $\subsetneq$ (Stable graphs) $\subsetneq$ (General graphs).

The algorithmic problems of turning a general graph into a bipartite one by removing either a set of edges or a set of vertices of minimum weight/cardinality, have been studied in the literature (see e.g. [1,8]). Similarly, the algorithmic problems of turning a given graph into a König-Egerváry one by removing a min-cardinality subset of edges or of vertices have been studied (see e.g. [13]). Differently, as mentioned before, for stable graphs only the edge-removal question has been investigated so far, and this yields an additional motivation to study the vertex-removal question in this paper, both in the unweighted and in the weighted setting.

Our Results and Techniques. We study the vertex-stabilizer problem in Sect. 2. We first show a structural property of any minimal vertex-stabilizer. Namely, we prove that removing any minimal vertex-stabilizer *does not decrease* the size of a maximum matching in the resulting graph (Theorem 1). This theorem has an interesting interpretation in network bargaining and cooperative matching games: it states that it is always possible to stabilize instances by blocking a minimum number of players *without* decreasing the total value that the players could get. An analogue of Theorem 1 has been proven by Bock et al. [4] for minimal *edge*-stabilizers[1], however, their proof does not hold for the vertex-removal setting, and therefore our proof is different. Interestingly, despite this

---

[1] These are subsets of *edges* such that their removal from G yields a stable graph.

analogy, algorithmically the two problems appear to have a different complexity: while finding a min-cardinality edge-stabilizer is at least as hard as finding a minimum vertex cover [4], we here prove (Theorem 2) that finding a min-cardinality vertex-stabilizer is a polynomial-time solvable problem. In addition, we can prove (Theorem 3) that the problem of blocking as few players as possible in order to make a *given* set of deals realizable as a stable outcome is also polynomial-time solvable, once again in contrast with the edge-removal setting, where the analogous question has been studied by [4] and shown to be vertex cover-hard. These three theorems are proved using combinatorial techniques. Theorem 1 exploits the structure of maximum matchings in graphs, that follows from the seminal works in [2,7]. Using Theorem 1, one can compute a lower bound on the size of a minimum vertex-stabilizer (as is done in [4]) exploiting properties of the so-called *Edmonds-Gallai Decomposition* (EGD) of a graph (definition is in Sect. 2). By further exploiting the relation that interplays between matchings and EGD, we get algorithms that prove Theorems 2 and 3.

We study in Sect. 3 the *weighted* setting. In the vertex-stabilizer problem described before, players are all equally considered, that is, from an objective function perspective, we are assuming that blocking a player u is as costly as blocking a player v, independently on how u and v are connected to the rest of the players in the network. However, from a bargaining perspective, players might not all be equally powerful: as an example, players corresponding to essential vertices have more bargaining power than inessential ones. Moreover, blocking a player that is highly connected in the graph and therefore have the potential to enter in many deals might be more costly than blocking a less connected player. For this reason, blocking different players might have different costs. We can model this by assigning a *weight* $w_v \geq 0$ to each vertex v. In this setting, we could be interested in either *minimizing* the weight of the *blocked* players, or in *maximizing* the weight of the *remaining* players. Two optimization problems then arise:

Min-weight Vertex-stabilizer: *Given a graph $G = (V, E)$, and vertex weights $w_v \geq 0 \ \forall v \in V$, find a vertex-stabilizer S that minimizes $w(S) = \sum_{v \in S} w_v$.*

Max-weight Vertex-stabilizer: *Given a graph $G = (V, E)$, and vertex weights $w_v \geq 0 \ \forall v \in V$, find a vertex-stabilizer S that maximizes $w(V \backslash S) = \sum_{v \notin S} w_v$.*

This weighted setting poses more algorithmic challenges, and this is technically the most interesting part of the paper. We prove that both the above problems become NP-hard already if 2 different weights are involved (Theorem 4). For this reason, we focus on *approximation algorithms*. We give a 2-approximation algorithm for the max-weight vertex-stabilizer problem (Theorem 5), and a $O(\gamma)$-approximation algorithm for the min-weight vertex-stabilizer problem (Theorem 6), where γ is the size of the so-called *Tutte-set* of the graph G (a formal definition is in Sect. 2). Both our algorithms are LP-based and rely on the following strategy. As a first step, we identify a suitable LP-relaxation to use for our problems. To this extent, we show that we can reduce our problems to vertex-deletion problems in a *bipartite* graph, in which the goal is to remove a subset of vertices in order to turn some special nodes into *essential* vertices in the remaining graph. This reinterpretation of the problem allows us to write a formulation that uses a set of *flow-type* valid constraints, and exploiting the properties of this flow will be crucial to round fractional solutions into integral ones.

In addition, we show lower bounds on the integrality gap of the LP relaxations we use, that show that our analysis are almost tight[2]. We give a $\frac{3}{2}$ lower bound on the integrality gap in the max-weight case, and a $\Omega(\gamma)$ lower bound in the min-weight case, that asymptotically matches the developed approximation factor. The lower bound for the min-weight case holds even on graph with *bounded* (constant) degree, and to construct it we rely on suitable *unbalanced bipartite expander* graphs.

We conclude by showing that we can give an algorithm for the min-weight vertex-stabilizer problem whose approximation factor *is* bounded by the maximum degree of a vertex in G, *if* we have an additional information: namely, if we know which is the set of essential vertices in the final graph (Theorem 7). From a network bargaining perspective, this corresponds to stabilize instances *enforcing* that some specific players will always be able to enter in a deal in any stable outcome. Also for this latter case we show a matching lower bound on the integrality gap of the LP relaxation we use. Our lower bounds show that to improve significantly our approximation factors a different strategy or at least different formulations have to be used.

Related Works. Removing vertices or edges from a graph as to satisfy certain properties has been widely studied in the literature in many variants. The paper that is most closely related to our work is [4] that studied the edge-stabilizer problem in the unweighted setting, and in addition to the results previously mentioned, they give efficient approximation algorithms for sparse graphs and for regular graphs. Biró et al. [3] also studied the edge-stabilizer problem, but considering maximum-weight matchings instead of maximum-cardinality matchings, and showed NP-hardness for this case. Könemann et al. [11] studied a related problem of computing a minimum-cardinality *blocking set*, that is a set of edges F such that $G \backslash F$ has a fractional vertex cover of size at most $\nu(G)$ (but note that $G \backslash F$ might not be stable). They give approximation algorithms for sparse graphs. Mishra et al. [13] studied vertex-removal and edge-removal problems to turn a graph into a König-Egerváry graph. Among other results, they give an $O(\log n \log \log n)$ approximation algorithm for the vertex-removal case in the unweighted setting, and show that assuming Unique Game Conjecture, both the minimum vertex-removal and edge-removal problems do not admit a constant factor approximation algorithm. We note that their hardness results do not seem to be helpful for our setting, since the graphs used in their reductions are stable.

Finally, we note that recently Ito et al. [9] have given independently alternative proofs of Theorems 2 and 4. They also give polynomial-time algorithms to stabilize an unweighted graph by adding edges and vertices.

---

[2] The lower bound constructions are deferred to the full version of the paper.

2   Minimum Cardinality Vertex-Stabilizers

We first prove that the removal of any minimal vertex-stabilizer does not decrease the cardinality of a maximum matching in the resulting graph. Here $G\backslash S$ denotes the graph obtained by removing from $G = (V, E)$ the subset of vertices $S \subseteq V$.

Theorem 1. *For any minimal vertex-stabilizer $S \subseteq V$ of a graph $G = (V, E)$, we have $\nu(G\backslash S) = \nu(G)$.*

Before giving a proof, we need a proposition (see [10]) that follows from standard results in matching theory, and uses the notion of M-*flower* for a maximum matching M of G. An M-flower is a subgraph of G formed by a u, v-path of even length that alternates edges in $E\backslash M$ and edges in M, plus a cycle containing v of $2k + 1$ edges, for some integer $k \geq 1$, in which exactly k edges are in M.

Proposition 1 [10]. *Given graph G, the following are equivalent characterizations of a stable graph: (i) The set of inessential vertices of G are pairwise non adjacent, (ii) $\nu(G) = \tau_f(G)$, (iii) There is no M-flower in G for any maximum matching M. Moreover, if G is not stable, then for every maximum matching M there is an M-flower.*

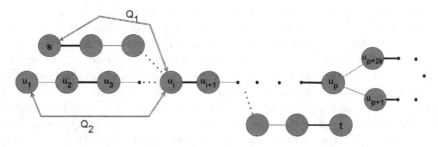

Fig. 1. M' edges are shown by bold edges. Note $M' \cap Q_2 = M \cap Q_2$ while $M \cap Q_1 = Q_1\backslash M'$.

Proof of Theorem 1. Let S be a minimal vertex-stabilizer of $G = (V, E)$, and M be a maximum matching of $G\backslash S$. Suppose by contradiction that $|M| < \nu(G)$. By classical results on matching theory [2], since M is not a maximum matching in G there exists an M-*augmenting path* P in G, that is, a path P that alternates edges from $E\backslash M$ and edges from M with endpoints s and t which are exposed by M. Clearly, we must have $|S \cap \{s, t\}| \geq 1$, otherwise P would be an augmenting path in $G\backslash S$, contradicting maximality of M. We distinguish two cases.

Case 1: $|S \cap \{s, t\}| = 1$. Without loss of generality, assume $s \in S$. In this case, we will show that $S' = S\backslash\{s\}$ is a vertex-stabilizer of G, which is a contradiction to the minimality of S. Consider the matching $M' = M \Delta P$, where Δ denotes the symmetric difference operator. M' is a matching of $G\backslash S'$ and $|M'| = |M| + 1$. Since adding one vertex to an arbitrary graph can increase the size of maximum matching by at most one, we deduce that M' is a maximum matching of $G \backslash S'$, hence $\nu(G \backslash S') = |M'|$. We now prove that $G \backslash S'$ is stable by showing that $\nu(G \backslash S') = \tau_f(G \backslash S')$. Let $y \in \mathbb{R}_{\geq 0}^{V \backslash S}$ be a minimum size fractional vertex cover of $G \backslash S$. By stability of $G \backslash S$, $\nu(G \backslash S) = \mathbf{1}^T y$. Define vector $y' \in \mathbb{R}_{\geq 0}^{V \backslash S'}$ as $y'_v = y_v$ for all $v \in V \backslash S$, and $y'_s = 1$. Obviously y' is a fractional vertex cover of $G \backslash S'$. So we have $\tau_f(G \backslash S') \leq \mathbf{1}^T y' = \mathbf{1}^T y + 1 = \nu(G \backslash S) + 1 = \nu(G \backslash S')$, i.e., $G \backslash S'$ is stable.

Case 2: $|S \cap \{s, t\}| = 2$. We first observe that $(G \backslash S) \cup \{s\}$ does not contain any M-augmenting path. Otherwise, by the same arguments as in Case 1, we can deduce that $S \setminus \{s\}$ is a vertex-stabilizer, and obtain a contradiction. Similarly, $(G \backslash S) \cup \{t\}$ does not contain any M-augmenting path. Let $S' = S \backslash \{s, t\}$, and $M' = M \Delta P$. We first show that M' is a maximum matching in $G \backslash S'$. If not, then $\nu(G \backslash S') \geq \nu(G \backslash S) + 2$. Let M'' be maximum matching in $G \backslash S'$. If we remove s from $G \backslash S'$, we delete at most one edge of M''. Therefore, $\nu((G \backslash S) \cup \{s\}) \geq \nu(G \backslash S) + 1$. However, this implies that M is not a maximum matching in $(G \backslash S) \cup \{s\}$, and therefore $(G \backslash S) \cup \{s\}$ contains an M-augmenting path contradicting our first observation. Since M' is a maximum matching in $G \backslash S'$, and $G \backslash S'$ is not stable, by Proposition 1 there exists an M'-flower F, with vertex set $u_1, \ldots, u_p$, with u_1 being the M'-exposed vertex on the even-length path. Note that F cannot be vertex disjoint from P: otherwise, F would be an M-flower as well in $G \backslash S$, contradicting stability of $G \backslash S$. It follows that $F \cup P$ is a connected subgraph of G. Let u_i be the node with the smallest index i that belongs to both F and P. Note that $i \neq 1$, since u_1 is M'-exposed and every node in P is instead M'-covered. Moreover, i is necessarily an even number: if odd, then the edge $\{u_{i-1}, u_i\}$ is in both P and F, contradicting our choice of i. Furthermore, note that the edge $\{u_i, u_{i+1}\}$ belongs to both P and F. Consider the path Q_1 that is the subpath of P connecting u_i to either s or t in $P \backslash \{u_i, u_{i+1}\}$, and the path Q_2 that is the subpath in F with vertex set $u_1, \ldots, u_i$. Their union $Q_1 \cup Q_2$ forms a path from u_1 to either s or t, say s (the other case is similar). In this case, $Q_1 \cup Q_2$ is an M-augmenting path in $(G \backslash S) \cup \{s\}$ (see Fig. 1), contradicting our first observation. $\qquad \square$

We now state our algorithm to find a minimum cardinality vertex-stabilizer, which relies on the notion of *Edmonds-Gallai Decomposition* (EGD) of a graph. The EGD of a graph $G = (V, E)$ is a partition of the set of vertices V into 3 sets (B, C, D) where B is the set of inessential vertices of G, C is the set of essential vertices of G that have at least one adjacent vertex in B, and D is the remaining essential vertices of G. The set C is called the *Tutte-set* of G.

Algorithm 1. ──
1. Compute the EGD (B, C, D) of G, and a maximum matching M^* of G that covers the maximum possible number of isolated vertices in the graph $G[B]$.
2. Let $G_1, \ldots, G_k$ be the non-singleton components of $G[B]$ with one vertex exposed by M^*. Set $S := \bigcup_{i=1}^{k} \{v_i\}$ where v_i is the M^*-exposed vertex of G_i.

Theorem 2. *Algorithm 1 is a polynomial-time algorithm to compute a minimum cardinality vertex-stabilizer S of a given graph G.*

We here sketch the main ideas of the proof. Let k be as in Algorithm 1. First, we note that k is a lower bound on the size of any minimum vertex-stabilizer. This has been proved by Bock et al. [4] for edge-stabilizers, but their proof in fact extends to vertex-stabilizers if one can assume Theorem 1. Then, we show that $G \backslash S$ is stable, by constructing a fractional vertex cover of $G \backslash S$ of size equal to $|M^*|$. This uses structural properties of maximum matchings and EGD of graphs. Since the techniques we use here are similar to [4], we defer the proof to the full version of the paper.

Finally, we consider the optimization problem of blocking as few players as possible in order to make a *given* set of deals realizable as a stable outcome. This translates into finding a minimum vertex-stabilizer S with the additional constraint that S must be element-disjoint from a given maximum matching M. We call such S an *M-vertex-stabilizer*. The proof of the next theorem is deferred to the full version of the paper.

Theorem 3. *There is a polynomial-time algorithm to compute a minimum M-vertex-stabilizer in a given graph, both in the weighted and in the unweighted setting.*

3   The Weighted Case

We here deal with the vertex-stabilizer problem in the weighted setting, that is much more challenging than the unweighted one. Due to lack of space, the proofs of the lemmas and the proofs of Theorems 4 and 7 are deferred to the full version of the paper.

Theorem 4. *The min-weight vertex-stabilizer problem and the max-weight vertex-stabilizer problem are NP-hard, even if there are only 2 distinct weights.*

Since these variants are NP-hard, we focus on approximation algorithms. To develop our approximation results, we first find a reformulation of our problems in *bipartite graphs*. The next lemma follows easily from Theorem 1.

Lemma 1. *Let (B, C, D) be the EGD of a graph G. Let $G_1, G_2, \cdots, G_p$ be the components of $G[B]$ where $G_i = (V_i, E_i)$. Let S be an optimal solution to a min-weight vertex-stabilizer (resp. max-weight vertex-stabilizer) instance defined on G. Then, (i) S is a subset of B, (ii) $|S \cap V_i| \leq 1$, (iii) if $|S \cap V_i| = 1$, then the vertex $g_i \in S$ of G_i is a minimum weight vertex in G_i.*

We can use Lemma 1 to simplify our input. If S contains a vertex from a component G_i, then it must be one of the vertices with minimum weight in G_i. Therefore, we shrink each non-singleton component G_i to a vertex g_i with minimum weight among the vertices in the component, and we call it a *pseudonode* (we remove multiple copies of the same edge created with this operation, if any).

Additionally, we know that $S \cap D = \emptyset$, so we can safely ignore these vertices and temporarily remove them from G. Lastly, we remark that it is well-known (see e.g. [16]) that every maximum matching of G matches all vertices in C to vertices in different components of $G[B]$; therefore, we ignore and remove edges between vertices in C from G. In this way we construct from G a weighted bipartite graph $G_b = (\tilde{B} \cup C, \tilde{E})$, where $\tilde{E} \subseteq E$, and $\tilde{B}$ consists of two sets of vertices: the set of pseudonodes, call this set B_1, and vertices corresponding to singletons in $G[B]$, call this set B_2. By construction and our previous remark, $\nu(G_b) = |C|$ and S naturally corresponds to a subset of $\tilde{B}$ of the same weight.

Definition 1. *Let $H = (U \cup W, F)$ be a bipartite graph and $U_1 \subseteq U$. We call $S \subseteq U$ a U_1-essentializer if all vertices in $U_1 \backslash S$ are essential in the graph $H \backslash S$.*

The next lemma basically shows that there is an approximation preserving reduction between the min-weight (resp. max-weight) vertex-stabilizer problem defined on G, and the problem of finding a suitable B_1-essentializer S that minimizes $\sum_{v \in S} w_v$ (resp. maximizes $\sum_{v \notin S} w_v$) in the weighted bipartite graph G_b.

Lemma 2. *Let $\tilde{S} \subseteq \tilde{B}$ be a B_1-essentializer of G_b that satisfies $\nu(G_b \backslash \tilde{S}) = \nu(G_b)$. Then $\tilde{S}$ corresponds to a vertex-stabilizer of G (of the same weight). Let $S \subseteq V$ be an optimal solution to a min-weight vertex-stabilizer (resp. max-weight vertex-stabilizer) instance defined on G. Then S corresponds to a B_1-essentializer in G_b (of the same weight) that satisfies $\nu(G_b \backslash S) = \nu(G_b)$.*

Next, we give an integer programming description of the set of B_1-essentializers, whose relaxation will be at the heart of our algorithms.

Integer Programming Description. Given $G_b = (\tilde{B} \cup C, \tilde{E})$, with $\tilde{B} = B_1 \cup B_2$, we introduce a binary variable z_v for $v \in \tilde{B}$ to denote if v is in a B_1-essentializer S (i.e. $z_v = 1$ if $v \in S$). We also introduce a binary variable y for $v \in \tilde{B} \cup C$ with the following meaning: for $v \in \tilde{B}$, we let $y_v = 1$ denote if v is an essential node in $G_b \backslash S$; for $v \in C$ instead, we let $y_v = 1$ denote if v is always matched to an inessential node in any maximum matching of $G_b \backslash S$. For a set of vertices T, we let $y(T) = \sum_{v \in T} y_v$, and $N(T)$ denote the set of neighbours (i.e. adjacent vertices) of T. We let

$$P_I := \Big\{ (z, y) : \qquad y_v + z_v \geq 1, \qquad \text{for } v \in B_1 \tag{1}$$

$$y_v + y_u + z_v \geq 1, \qquad \text{for } \{u, v\} \in \tilde{E}, v \in B_2, u \in C \tag{2}$$

$$y(N(A)) \geq |A| - y(A), \qquad \text{for } A \subseteq C \tag{3}$$

$$y(V) = |C|, \tag{4}$$

$$z \in \{0,1\}^{\tilde{B}}, \ y \in \{0,1\}^{\tilde{B} \cup C} \Big\}.$$

Let us give an intuition of the meaning of the linear constraints. Inequality (1) states that a vertex in B_1 is either essential in $G_b \backslash S$ or it is removed (as required by Definition 1). Inequality (2) states that if a vertex v in B_2 is not removed then either v is essential in $G_b \backslash S$ or all of its neighbours have to be matched to inessential vertices in $G_b \backslash S$. The reason is that, if v is inessential in $G_b \backslash S$ but some neighbour u of v is matched to an essential vertex v' in some maximum matching M of $G_b \backslash S$, then it is possible to construct an even length M-alternating path between some M-exposed vertex to v', contradicting the fact that v' is essential. Inequality (3) is a translation of Hall's theorem, and states that there exists a matching between vertices in C with y-value 0 and their neighbours with y-value 1, that covers all vertices in C with y-value 0. The reason is that such vertices will always be matched to essential vertices in $G_b \backslash S$ by any maximum matching. We would like to emphasize that inequalities (3) are crucial to have a meaningful formulation for our problem. Equality (4) basically ensures that there is a partition of vertices in C into those that will always be matched to inessential vertices and those that will always be matched to essential vertices of $G_b \backslash S$ by any maximum matching. The next lemma makes this intuition rigourous.

Lemma 3. P_I describes the set of B_1-essentializers of the graph G_b.

We denote by P_f the polytope obtained by relaxing the binary constraints of P_I, i.e. replacing them with $0 \leq z \leq 1$ and $0 \leq y \leq 1$. When dealing with fractional points, Inequality (3) does not correspond to Hall's theorem anymore, but it naturally ensures the existence of a *flow* of value $|C| - y(C)$ from vertices in C to vertices in $\tilde{B}$. Among other things, this also implies that although this set contains exponentially (in the size of G_b) many inequalities, we can separate over them in polynomial time.

Lemma 4. *Construct a directed network* $\mathcal{N} = (V_{\mathcal{N}}, A_{\mathcal{N}})$ *from graph* $G_b = (\tilde{B} \cup C, \tilde{E})$ *with* $V_N = \tilde{B} \cup C \cup \{s, t\}$ *and* $A_{\mathcal{N}} = \{(s, u) : u \in C\} \cup \{(v, t) : v \in \tilde{B}\} \cup \tilde{E}$ *where the edges in* $\tilde{E}$ *are oriented from* C *to* $\tilde{B}$. *Let* $(z, y) \in P_f$. *Assign* y_v *amount of capacity to each arc* (v, t), $(1 - y_u)$ *amount of capacity to each arc* (s, u), *and* ∞ *capacity to arcs in* $\tilde{E}$. *Then, there exists a maximum* $s - t$ *flow in* $\mathcal{N}$ *of value* $y(\tilde{B}) = |C| - y(C)$.

Exploiting the structure of this flow, we can derive useful properties on the extreme points of P_f. In particular, we have the following lemma:

Lemma 5. *Let* (z, y) *be an extreme point of* P_f. *There exists a maximum matching in* G_b *between the set of vertices* $\{v \in \tilde{B} : y_v > 0\}$ *and the set of vertices* $\{u \in C : y_u < 1\}$ *of cardinality* $|\{v \in \tilde{B} : y_v > 0\}|$.

Finally, we note that the problem of finding a B_1-essentializer S that maximizes $\sum_{v \notin S} w_v$, or minimizes $\sum_{v \in S} w_v$, can be formulated respectively as

$$\max \left\{ \sum_{v \in \tilde{B}} w_v (1 - z_v) : (z, y) \in P_I \right\}, \quad \text{and} \quad \min \left\{ \sum_{v \in \tilde{B}} w_v z_v : (z, y) \in P_I \right\}. \quad (5)$$

Algorithm for Max-weight Vertex-stabilizer. Given a graph $G = (V, E)$ with weights $w_v \geq 0 \; \forall v \in V$, we construct from G a weighted bipartite graph $G_b = (\tilde{B} \cup C, \tilde{E})$, with $\tilde{B} = B_1 \cup B_2$, as described in the beginning of Sect. 3. We then apply Algorithm 2 that relies on solving the LP relaxation of the maximization IP in (5).

Algorithm 2. _____

1. Let $(z^*, y^*) \leftarrow$ optimal extreme point of $\max\{\sum_{v \in \tilde{B}} (1 - z_v) : (z, y) \in P_f\}$.
2. Set $B_+ := \{v \in \tilde{B} : 0 < y_v^*\}$; $B_0^1 := \{v \in \tilde{B} : y_v^* = 0, z_v^* = 1\}$; $B_0^f := \{v \in \tilde{B} : y_v^* = 0, 0 < z_v^* < 1\}$.
3. **If** $w(B_+) \leq w(B_0^f)$ **then** set $S := (B_+ \cup B_0^1)$, **else** set $S := (B_0^f \cup B_0^1)$.
4. **While** $\nu(G_b \backslash S) < |C|$ **do**: find $v \in S$ such that $\nu(G_b \backslash (S \backslash \{v\})) > \nu(G_b \backslash S)$, and set $S := S \backslash \{v\}$.

_____

Theorem 5. *There is a polynomial-time LP-based 2-approximation algorithm for the max-weight vertex-stabilizer problem.*

Proof. We consider the set S output by Algorithm 2. Note that if S is a B_1-essentializer *and* $\nu(G_b \backslash S) = \nu(G_b)$, then it corresponds to a vertex-stabilizer in G by Lemma 2. Still, Lemma 2 implies that to prove the claimed approximation guarantee, it is enough to prove that S is a 2-approximated solution for the problem of finding a B_1-essentializer for G_b that maximizes the weight of the non selected vertices.

First, we show that (a) $\nu(G_b \backslash S) = \nu(G_b)$ and (b) every vertex in $B_1 \backslash S$ is essential in $G_b \backslash S$, i.e. S is a B_1-essentializer. Note that (a) holds by construction after step 4 (recall that $\nu(G_b) = |C|$ and $S \cap C = \emptyset$, therefore it is always possible to perform step 4 until the while condition is not satisfied anymore). Moreover, all vertices added in step 4 are essential vertices. We are left with (b). Define $C_f = \{u \in C : y_u^* < 1\}$. Furthermore, partition the set of vertices in $\tilde{B}$ in 4 sets: B_+, B_0^1, B_0^f and $B_0^0 := \{v \in \tilde{B} : z_v^* = 0 \,\&\, y_v^* = 0\}$ (the definition of the first 3 sets is given in Algorithm 2). Note that the vertices in B_1 are either in B_0^1 or B_+, so if $S = B_+ \cup B_0^1$, then $G_b \backslash S$ does not contain any B_1 vertex, and we have nothing to show. Suppose instead $S = B_0^f \cup B_0^1$. Note that there does not exist any edge between $v \in B_0^0$ and $u \in C_f$, because $y_v^* + z_v^* = 0$ holds for v and $y_u^* < 1$ holds for u, and therefore Inequality (2) will be violated for the edge $\{v, u\}$, contradicting feasibility of (z^*, y^*). Therefore, the neighbours of vertices C_f in $G_b \backslash S$ are vertices in B_+ and by Lemma 5, we know that there is a matching between C_f and B_+ covering all vertices in B_+. Since every maximum matching in $G_b \backslash S$ covers all the vertices in C, it must cover all vertices in C_f, therefore it must be the case that $|C_f| = |B_+|$ and every maximum matching in $G_b \backslash S$ covers all the vertices in B_+, i.e. all the vertices in B_+ are essential. Since $(B_1 \backslash S) \subseteq B_+$, the result follows.

To conclude the proof, we argue that the weight of the vertices in $G_b \backslash S$ is at least $\frac{1}{2}$ the optimal value of the LP. Let $w_0 = w(B_0^0)$, $w_1 = w(B_+)$, $w_2 = w(B_0^f)$. Note that the weight of the vertices in the graph $G_b \backslash S$ is at least $w_0 + \max(w_2, w_1)$ which is at least half of $w_0 + w_1 + w_2 = \sum_{v \in \tilde{B}} w_v - \sum_{v: z_v^* = 1} w_v$, which is clearly an upper bound on the optimal value of the LP.   □

Algorithm for Min-weight Vertex-stabilizer. Given a graph $G = (V, E)$ with weights $w_v \geq 0 \,\forall v \in V$, we construct a weighted bipartite graph $G_b = (\tilde{B} \cup C, \tilde{E})$, with $\tilde{B} = B_1 \cup B_2$ obtained from G as described in the beginning of Sect. 3. We then apply Algorithm 3 that relies on solving the LP relaxation of the minimization IP in (5).

Algorithm 3. _____

1. Solve the LP: min $\left\{ \sum_{v \in \tilde{B}} w_v z_v : (z, y) \in P_f \right\}$ to get an extreme point optimal solution (z, y), and set $S := \{v : z_v \geq \frac{1}{|C|+1}\}$.
2. **While** $\nu(G_b \backslash S) < |C|$ **do:** find $v \in S$ such that $\nu(G_b \backslash (S \backslash \{v\})) > \nu(G_b \backslash S)$, and set $S := S \backslash \{v\}$.

_____

Theorem 6. *There is a polynomial-time LP-based $(\gamma + 1)$-approximation algorithm for the min-weight vertex-stabilizer problem, where γ is the size of the Tutte-set of G.*

Proof. We consider the set S output by Algorithm 3. As for the max-weight case, due to Lemma 2 and step 2 of the algorithm, to prove the theorem it is enough to show that S is a $(|C| + 1)$-approximated solution for finding a B_1-essentializer for G_b that minimizes the weight of the selected vertices. Trivially, $w(S) \leq (|C|+1) \sum_{v \in \tilde{B}} w_v z_v$, therefore the approximation factor guarantee holds. It remains to show that S is in fact a B_1-essentializer for G_b.

Let $\tilde{S}$ be the set S *before* executing step 2 of the algorithm. We will prove that each $v \in B_1 \backslash \tilde{S}$ is essential in $G_b \backslash \tilde{S}$. This is enough, since every vertex added back in step 2 will be essential by construction, and this addition cannot make any vertex in B_1 inessential. Let us assume by contradiction that $v_0 \in B_1 \backslash \tilde{S}$ is inessential in $G_b \backslash \tilde{S}$. In this case, if we apply Edmonds' Blossom Algorithm [7] in $G_b \backslash \tilde{S}$, we can find a maximum matching M that exposes v_0 and a so-called *frustrated tree* $T = (V_T, E_T)$ containing v_0 with the following properties: (i) $|E_T \cap M| = |V_T \cap C|$, and all vertices in $V_T \backslash \{v_0\}$ are covered by M, and (ii) the neighbours of the set of vertices $V_T \cap \tilde{B}$ in $G_b \backslash \tilde{S}$ are all in the tree T (we refer to [6,7] for details). Note that the neighbours of $V_T \cap \tilde{B}$ in $G_b \backslash \tilde{S}$ are the same as the neighbours of $V_T \cap \tilde{B}$ in G_b, i.e. $N(V_T \cap \tilde{B}) = V_T \cap C$ as $\tilde{S} \subseteq \tilde{B}$. Feasibility of (z, y) implies that for each matching edge $\{u, v\} \in M$, we have $y_u + y_v + z_v \geq 1$. Since $\tilde{S}$ removed all vertices with z-value $\geq \frac{1}{|C|+1}$, for each edge $\{u, v\} \in M$, $y_u + y_v > 1 - \frac{1}{|C|+1}$. Let $M_T := M \cap E_T$. We have

$$y(V_T) = y_{v_0} + \sum_{\{u,v\} \in M_T} (y_u + y_v) > (1 - \frac{1}{|C|+1}) + |M_T|(1 - \frac{1}{|C|+1})$$

$$= |M_T| + 1 - \frac{|M_T| + 1}{|C|+1} \geq |M_T|,$$

where the first inequality follows from the Inequality (1) associated to v_0, and the last inequality follows from the fact that $|M_T| \leq |C|$. Furthermore, for set $A = V_T \cap C$, since $|M_T| = |A|$ by (i), we have

$$y(A) + y(N(A) \cap V_T) = y(V_T) > |M_T| = |A|. \tag{6}$$

If we consider the directed network $\mathcal{N}$ and the $s - t$ flow as in Lemma 4, (6) says that the capacity $y(N(A) \cap V_T)$ of the arcs between t and $N(A) \cap V_T$ is strictly larger than the flow sent on the arcs from s to A (that can be at most $|A| - y(A)$).

Since a maximum flow necessarily saturates *all* the edges from $N(A) \cap V_T$ to t, there is a neighbour of $(N(A) \cap V_T)$ which is not in A who sends positive flow to some vertex in $N(A) \cap V_T$, but this contradicts property (ii) of T, as $N(N(A) \cap V_T) = N(\tilde{B} \cap V_T) = A$.   □

We remark here that we can show a tight lower bound of $\Omega(\gamma)$ on the integrality gap of the minimization IP in (5) that holds even on graphs with *constant* degree. However, we can develop an algorithm whose approximation ratio is bounded by the maximum degree (δ) of a vertex in G, if we know the set of essential vertices in the final stable graph (our reduction in Theorem 4 shows that also this problem is NP-hard).

Theorem 7. *There is a δ-approximation algorithm for the min-weight vertex-stabilizer problem, if we know the set of essential vertices in the final stable graph.*

References

1. Agarwal, A., Charikar, M., Makarychev, K., Makarychev, Y.: $O(\sqrt{\log n})$ approximation algorithms for min UnCut, min 2CNF deletion, and directed cut problems. In: Proceedings of STOC 2005, pp. 573–581 (2005)
2. Berge, C.: Two theorems in graph theory. Proc. Natl. Acad. Sci. U.S.A. **43**(9), 842–844 (1957)
3. Biró, P., Bomhoff, M., Golovach, P.A., Kern, W., Paulusma, D.: Solutions for the stable roommates problem with payments. In: Golumbic, M.C., Stern, M., Levy, A., Morgenstern, G. (eds.) WG 2012. LNCS, vol. 7551, pp. 69–80. Springer, Heidelberg (2012)
4. Bock, A., Chandrasekaran, K., Könemann, J., Peis, B., Sanità, L.: Finding small stabilizers for unstable graphs. In: Lee, J., Vygen, J. (eds.) IPCO 2014. LNCS, vol. 8494, pp. 150–161. Springer, Heidelberg (2014)
5. Chalkiadakis, G., Elkind, E., Wooldridge, M.: Computational aspects of cooperative game theory. Synthesis Lectures on Artificial Intelligence and Machine Learning, 1st edn. Morgan & Claypool, San Rafael (2011)
6. Cook, W., Cunningham, W., Pulleyblank, W., Schrijver, A.: Combinatorial Optimization. Wiley, New York (1998)
7. Edmonds, J.: Paths, trees, and flowers. Can. J. Math. **17**, 449–467 (1965)
8. Garg, N., Vazirani, V.V., Yannakakis, M.: Approximate max-flow min-(multi)cut theorems and their applications. SIAM J. Comput. **25**, 698–707 (1993)
9. Ito, T., Kakimura, N., Kamiyama, N., Kobayashi, Y., Okamoto, Y.: Efficient stabilization of cooperative matching games. In: Proceedings of AAMAS (2016, to appear)
10. Kleinberg, J., Tardos, É.: Balanced outcomes in social exchange networks: In: Proceedings of STOC 2008, pp. 295–304 (2008)
11. Könemann, J., Larson, K., Steiner, D.: Network bargaining: using approximate blocking sets to stabilize unstable instances. In: Theory of Computing Systems, pp. 655–672 (2015)
12. Korach, E., Nguyen, T., Peis, B.: Subgraph characterization of Red/Blue-Split graph and könig egerváry graphs. In: Proceedings of SODA 2006, pp. 842–850 (2006)

13. Mishra, S., Raman, V., Saurabh, S., Sikdar, S., Subramanian, C.: The complexity of König subgraph problems and above-guarantee vertex cover. Algorithmica **61**(4), 857–881 (2011)
14. Nash, J.: The bargaining problem. Econometrica **18**, 155–162 (1950)
15. Nisan, N., Roughgarden, T., Tardos, É., Vazirani, V.V.: Algorithmic Game Theory. Cambridge University Press, New York (2007)
16. Schrijver, A.: Combinatorial Optimization. Springer, New York (2003)
17. Shapley, L.S., Shubik, M.: The assignment game: the core. Int. J. Game Theory **1**(1), 111–130 (1971)
18. Sterboul, F.: A characterization of the graphs in which the transversal number equals the matching number. J. Comb. Theory Ser. B **27**, 228–229 (1979)

Round-Robin Tournaments Generated by the Circle Method Have Maximum Carry-Over

Erik Lambrechts[1], Annette M.C. Ficker[2($\boxtimes$)], Dries R. Goossens[3], and Frits C.R. Spieksma[2]

[1] Department of Mechanical Engineering, KU Leuven, Leuven, Belgium
erik.lambrechts@kuleuven.be
[2] Operations Research Group, Faculty of Economics and Business, KU Leuven, Leuven, Belgium
{annette.ficker,frits.spieksma}@kuleuven.be
[3] Faculty of Economics and Business Administration, Ghent University, Gent, Belgium
dries.goossens@ugent.be

Abstract. The Circle Method is widely used in the field of sport scheduling to generate schedules for round-robin tournaments. The so-called carry-over effect value is a number that can be associated to each round-robin schedule; it represents a degree of balance of a schedule.

Here, we prove that, for an even number of teams, the Circle Method generates a schedule with maximum carry-over effect value, answering an open question.

1 Introduction

In 1847, Reverend T. Kirkman [11] published a method that can be used for constructing a schedule for round-robin competitions. This method, here called the Circle Method (aka the polygon method, or the canonical procedure; its outcome has been referred to as a Kirkman tournament, or the circle design; see Sect. 2.3 for a precise description), has been used abundantly in practice for many sports leagues around the world to construct schedules in round robin competitions (see Sect. 1.1).

In 1980, Russel [14] proposed a measure that associates to each schedule a value representing a degree of balance of a schedule; this value is called the carry-over effect value (see Sect. 2.2 for a definition).

Here, we answer the following question:

Does the Circle Method generate a schedule with maximum carry-over effect value?

Miyashiro and Matsui [13] conjecture that the answer to this question is 'yes'; we prove that to be the right answer. Even more, we show that any schedule with a maximum carry-over effect value can be generated by the Circle Method.

This work is supported by the Interuniversity Attraction Poles Programme initiated by the Belgian Science Policy Office.

1.1   Motivation

In a round-robin tournament, each pair of teams (or players) meets an equal number of times; the resulting matches are distributed over rounds such that each team plays at most a single match in each round. Organizers of round-robin tournaments face the problem of generating a schedule, i.e., to decide which match takes place in which round. Graph theory is closely connected to this problem: by having a node for each team, a match can be seen as a pair of nodes, and a round can be seen as a matching (see De Werra [4,5]). Then, the schedule boils down to a sequence of matchings, thereby partitioning the edge set of the resulting complete graph.

There are many, many different issues that can be taken into account when designing a tournament. In particular, various ways of generating a schedule exist, each resulting in a schedule with different properties; we refer to Anderson [1], Froncek [3], Januario et al. [9], and Kendall et al. [10] for introductions and (recent) overviews.

It is fair to say, however, that the so-called Circle Method is a very popular (if not the most popular) method when it comes to generating schedules for round-robin competitions. Indeed, the use of the Circle Method is well spread through different sports leagues and their organizers; for instance, Griggs and Rosa [8], followed by Goossens and Spieksma [6], documented extensively the use of the Circle Method throughout soccer leagues in Europe.

The following phenomenon is relevant in any round-robin tournament. Imagine that your team is facing some other team in an upcoming match; we will argue that the opponent of this other team in the previous match is relevant for the upcoming match. Indeed, if the team you're about to face has experienced a heavy loss in its previous match, the team may have a low morale, or be discouraged, and hence perhaps easier to beat. Then your team is receiving a so-called *carry-over effect* from the previous opponent of the team your team is about to face. Of course, the opposite is possible as well: strengthened by having beaten a weak opponent in their previous match, the team your team is about to face is full of morale, and perhaps more difficult to beat.

Thus, in each round of the competition your team receives a carry-over effect from the team that your opponent played against in its previous match (the rounds are viewed cyclically, i.e., in round 1 your team receives a carry-over effect from the team your opponent plays in the last round, see Sect. 2): we can investigate the set of teams from which your team receives a carry-over effect throughout the competition.

In one extreme case, this set of teams consists of all other teams. Then, in a single round-robin tournament, each other team gives once a carry-over effect to your team. Schedules that satisfy this property are called *balanced*, see Russel [14]. A balanced schedule need not exist; Russel [14] shows that balanced schedules exist when the number of participating teams is a power of 2; Anderson [2] exhibits balanced schedules when the number of participating teams equals 20 or 22.

In another extreme case, only very few teams give carry-over effects to your team. This gives rise to schedules that can be perceived as unbalanced or even unfair. Indeed, different cases have been reported where carry-over effects were blamed for distorting the outcome of the competition; we refer to Goossens and Spieksma [7] who describe a case in the 2007 edition of the Norwegian soccer league (Tippeligean), and a case in the 2006–2007 edition of the Belgian soccer league (ProLeague). Thus, measuring the degree of 'unbalancedness' of a schedule is relevant, and this is done by considering the square of the deviations from a balanced schedule (see Sect. 2.2). The Circle Method is known for generating unbalanced schedules; in fact, our contribution here is to show that the Circle Method actually *maximizes* the carry-over effect value.

When viewed in graph-theoretical terms, the Circle Method partitions the edge set of K_n (n even) into $n-1$ perfect matchings and arranges these matchings in a specific cyclical order (Sect. 2.3). We show here that this order maximizes a particular objective, known as the carry-over effect value.

The paper is organized as follows. In Sect. 2 we introduce our terminology and we state our result (Theorem 1), Sect. 3 formulates the building blocks of our proof, Sect. 4 finalizes the proof. Due to space limitations proofs of all facts and lemmas are omitted; they can be found in the full version of this paper [12].

2   Terminology

This section introduces terminology concerning schedules (Sect. 2.1), explains the value of the carry-over effect of a schedule (Sect. 2.2), and describes the Circle Method (Sect. 2.3).

2.1   About Schedules

Let n denote the number of teams participating in a single round-robin tournament (SRR). Throughout this paper, we assume that n is even, and that $n \geq 6$ (since the cases where $n \in \{2, 4\}$ are easy to analyze). We use N to denote the set of teams: $N = \{1, 2, \ldots, n\}$. We exclusively focus on so-called *compact* schedules, meaning that there are $n - 1$ *rounds* in an SRR; each round consists of $\frac{n}{2}$ matches (of course, a match consists of a pair of two distinct teams). A *schedule* for an SRR specifies, for each of the $n - 1$ rounds, which pairs of teams are involved in the matches.

Definition 1. *A schedule is called* feasible *if:*

(i) in each round, each team is in one match, and
(ii) after all rounds, each pair of teams has been in a match.

A schedule can be represented in the form of a table. The two tables depicted in Fig. 1 each represent a possible schedule for $n = 8$ teams. The opponent of team $i \in N$ in round r can be found on the i-th row and the r-th column ($1 \leq r \leq n - 1$).

	1	2	3	4	5	6	7
1	2	6	5	3	8	4	7
2	1	3	7	6	4	8	5
3	6	2	4	1	7	5	8
4	8	7	3	5	2	1	6
5	7	8	1	4	6	3	2
6	3	1	8	2	5	7	4
7	5	4	2	8	3	6	1
8	4	5	6	7	1	2	3

	1	2	3	4	5	6	7
1	8	3	5	7	2	4	6
2	7	8	4	6	1	3	5
3	6	1	8	5	7	2	4
4	5	7	2	8	6	1	3
5	4	6	1	3	8	7	2
6	3	5	7	2	4	8	1
7	2	4	6	1	3	5	8
8	1	2	3	4	5	6	7

Fig. 1. Two distinct schedules for $n = 8$ teams.

2.2   About the Carry-Over Effect

Consider the schedule represented in Fig. 1 on the left. In round 1, team 1 plays team 2, and in round 2, team 1 plays team 6. Thus, team 2 gives a carry-over effect (coe) to team 6 using team 1 as a carrier. Indeed, any pair of consecutive numbers on a row in a schedule indicates a coe. More generally, the opponent of one's opponent in the previous round is the originator of an effect that is passed to one's team. To capture this effect, we use the following definition.

Definition 2 [14]. *Given a feasible schedule, we say that team $i \in N$ gives a carry-over effect (coe) to team $j \in N$ in round r, if there exists a team $k \in N$ that plays team i in round $r - 1$, and plays team j in round r, $1 \leq r \leq n - 1$. We also say that team j receives a coe from team i in round r.*

It is important to realize that we view a schedule cyclically: in round 1, each team receives a coe coming from a match in round $n - 1$; and in round $n - 1$, each team gives a coe to some team playing in round 1. This is motivated by observing that, often, in practice, a double round robin schedule is found by repeating a single round robin schedule. Thus, when dealing with rounds, we compute modulo $n - 1$. Indeed, we use freely the phrase $r - 1$ or $r + 1$ with $r \in \{1, 2, \ldots, n-1\}$ (as we did in Definition 2); clearly, if $r = 1$, then $r-1 = n-1$, and if $r = n - 1$, then $r + 1 = 1$. Concluding: in each round, a team gives a coe to some team, and receives a coe from some team.

We associate to each schedule a matrix, called the *carry-over effect matrix* (the COE matrix).

Definition 3 [14]. *The COE matrix is an $n \times n$ matrix with entries $c_{i,j}$, that represent the number of times that team i gives a coe to team j in a given schedule, $i, j \in N$. The* carry-over effect value *(the COE value) of a feasible schedule is defined as* $\sum_{i \in N} \sum_{j \in N} c_{i,j}^2$.

It will be convenient to consider a team's contribution to the COE value. We define:

Definition 4. *The contribution of a team $i \in N$, denoted by $Co(i)$, to the COE value is defined as $Co(i) = \sum_{j \in N} c_{i,j}^2$.*

Observe that the COE value of a given schedule equals the sum of the contributions of the teams. The COE matrices, and their COE values, corresponding to the two schedules in Fig. 1, are given in Fig. 2, with the zero entries left blank.

$$
\begin{bmatrix}
 & 1 & 1 & 1 & 1 & 1 & 1 & 1 \\
1 & & 1 & 1 & 1 & 1 & 1 & 1 \\
1 & 1 & & 1 & 1 & 1 & 1 & 1 \\
1 & 1 & 1 & & 1 & 1 & 1 & 1 \\
1 & 1 & 1 & 1 & & 1 & 1 & 1 \\
1 & 1 & 1 & 1 & 1 & & 1 & 1 \\
1 & 1 & 1 & 1 & 1 & 1 & & 1 \\
1 & 1 & 1 & 1 & 1 & 1 & 1 &
\end{bmatrix}
\qquad
\begin{bmatrix}
 & 1 & 5 & & & & & 1 \\
 & & 1 & 5 & & & & 1 \\
 & & & 1 & 5 & & & 1 \\
 & & & & 1 & 5 & & 1 \\
 & & & & & 1 & 5 & 1 \\
5 & & & & & & 1 & 1 \\
1 & 5 & & & & & & 1 \\
1 & 1 & 1 & 1 & 1 & 1 & 1 &
\end{bmatrix}
$$

A COE matrix with COE value 56.   A COE matrix with COE value 196.

Fig. 2. Corresponding COE matrices.

As mentioned in Sect. 1.1, a schedule is called balanced if each team receives a coe from each other team exactly once. The corresponding COE matrix has all entries equal to 1, except for zero entries on the main diagonal. We see in Fig. 2 that the first schedule is balanced.

2.3   About the Circle Method

The Circle Method is a method for constructing a feasible schedule for an SRR with any (even) number of teams. An intuitive description is as follows. Select a team, say team n, and place it in the center of a circle. All other teams are placed on the circle. In round 1, the team in the center plays team 1. The neighbors of team 1 play each other, and in fact, their neighbors also play each other. This is repeated until all teams are matched up, and we have constructed the first round. To construct the next round, we "rotate" the matches, that is, team n plays team 2, the neighbors of team 2 play each other, and so on. This is illustrated in Fig. 3 for the first three rounds; the resulting schedule is represented in the right schedule in Fig. 1.

A precise description of the Circle Method is as follows. For each round $r \in \{1, \ldots, n-1\}$ we have,

- team n plays team r,
- for $i, j \in N \setminus \{r, n\}$: team i plays team j if $i + j \equiv 2r \mod (n-1)$.

Of course, permuting the teams, and next applying the Circle Method gives other schedules; we will refer to the class of schedules that can be obtained by applying the Circle Method to some permutation of the teams, as the class $\mathcal{C}$.

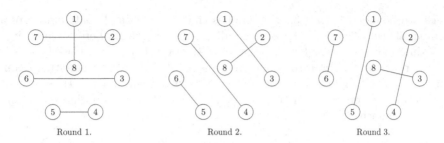

Round 1.   Round 2.   Round 3.

Fig. 3. Visual representation of a schedule constructed with the circle method.

The COE matrix corresponding to a schedule found by the Circle Method has a specific structure. Consider for example Fig. 2 on the right for the COE matrix where $n = 8$. We use this structure to find an explicit expression for the COE value of a schedule found by the Circle Method; this value is denoted by $CMCOE(n)$ (where n refers to the number of teams).

Fact 1.

$$CMCOE(n) = ((n - 3)^2 + 3)(n - 1) \text{ for each even } n \geq 6.$$

We are now well-placed to formulate our main result.

Theorem 1. *The Circle Method generates a schedule with maximum COE value.*

3   Building Blocks of the Proof

This section identifies some basic observations in Sect. 3.1, introduces two concepts in Sect. 3.2, characterizes the class of schedules that can be generated by the Circle Method in Sect. 3.3, and proves upper bounds on the contribution of teams with particular properties in Sect. 3.4.

3.1   Basic Observations

The following statements hold for any feasible schedule.

Lemma 1.

(i) *Each team gives and receives exactly $n-1$ times a coe (once in each round).*
(ii) *A team cannot give, nor receive, a coe to, or from, themselves.*
(iii) *Each team gives a coe to at least three different teams, and each team receives a coe from at least three different teams.*
(iv) *Each team gives at most $n - 3$ times a coe to a single team.*

These statements allow us to determine the maximum contribution to the COE value of a single team. Indeed, from Lemma 1, it follows that for each $i \in N$, $\sum_{j \in N} c_{i,j} = n - 1$, that at least three of these $c_{i,j}$ values are nonzero, and that all values are bounded by $n - 3$. This implies that the maximal contribution of a team equals $(n - 3)^2 + 1 + 1$, corresponding to a situation where a team gives $n - 3$ times a coe to a single team, and once to two other teams. Clearly, this is the only possible situation where a team gives $n - 3$ times a coe to another team. Thus, the second highest possible contribution of a team is $(n - 4)^2 + 2^2 + 1$, corresponding to giving $n - 4$ times a coe to a team, twice to another team, and once to yet another team.

Definition 5. *Given a feasible schedule, team $i \in N$ is called a maximally giving team (an mg team) if there exists a team j receiving $n - 3$ times a coe from team i, i.e., if there exists a j with $c_{i,j} = n - 3$. Alternatively, we call a team j a maximally receiving team.*

When considering the COE matrix on the right in Fig. 2, we see that seven out of the eight teams are mg teams. It is natural to wonder whether it is possible to have a large COE value when there are no mg teams. We can easily find the following lower bound on the number of mg teams that need to be present in a feasible schedule with a value at least $CMCOE(n)$.

Lemma 2. *Any feasible schedule with a COE value greater or equal than $CMCOE(n)$, has at least $\left\lceil \frac{n}{2} + \frac{n-6}{n-5} \right\rceil$ mg teams.*

Thus, the COE value of a schedule is bounded from above by that of the Circle Method if no more than half of the teams are maximally giving teams.

3.2   Basic Concepts: k-Chains and Bridge Teams

The existence of an mg team in a schedule has quite some impact on the structure of that schedule. Let us investigate this structure in more detail. Suppose, as an example, that team 1 gives maximally to team 2. Thus, there are $n - 3$ rounds during which team 1 gives a coe to team 2. Let us further assume team 1 plays team 2 in round 2. It is then clear that in this round, team 2 does not receive a coe from team 1. Also, in the next round, team 2 does not receive a coe from team 1. But this means that in each of the remaining $n - 3$ rounds, team 2 has to receive a coe from team 1. Thus, the opponent of team 1 in round r, and the opponent of team 2 in round $r + 1$ are the same team for each round $r \in \{3, 4, \ldots, n - 1\}$. This means that the schedule has the form that is depicted in Table 1 where we use letters as names for the other teams.

Table 1. Partial schedule in case team 1 gives maximally to team 2.

	1	2	3	4	5	6	...	$n - 2$	$n - 1$
1		2	a	b	c	d	...	y	z
2	z	1		a	b	c	...	x	y

Notice that both teams 1 and 2 still have only a single empty round; in fact, it is easily seen that the opponent of team 1 in round 1 must be the same team as the opponent of team 2 in round 3. As we have the freedom to reindex the remaining teams, let us say that the empty round is taken by team n, and let us number the other teams by considering the round in which such a team plays team 1. This gives rise to Table 2.

Table 2. The structure when a team 1 is an mg team giving maximally to team 2.

	1		2	3	4	5	6	...	$n-2$	$n-1$
1	n		2	3	4	5	6	...	$n-2$	$n-1$
2	$n-1$	1	n	3	4	5		...	$n-3$	$n-2$

We emphasize that the only crucial assumption we made for constructing Table 2 is the assumption that team 1 gives maximally to team 2.

Let us further elaborate on this example by considering what happens if team 2 is an mg team as well, giving maximally to some other team. Notice that, given the structure displayed in Table 2, there can be only one specific team to which team 2 gives maximally, namely team 3 (the team that plays team 1 in round 3). We leave verification of this claim to the reader.

The resulting implications are described in Table 3; notice the special role of team n. We will come back to the role of team n extensively.

Table 3. The structure when team 1 is an mg team giving maximally to team 2, giving maximally to team 3.

	1	2		3	4	5	6	...	$n-2$	$n-1$
1	n	2		3	4	5	6	...	$n-2$	$n-1$
2	$n-1$	1		n	3	4	5	...	$n-3$	$n-2$
3	$n-2$	$n-1$	1	2	n	4		...	$n-4$	$n-3$

We now define the concept of a k-chain, allowing us to capture the essence of the example discussed in the preceding paragraphs.

Definition 6. *Given a feasible schedule, we define a k-chain as a list consisting of $k + 1$ teams, where the first k teams are mg teams, giving maximally to the next team in the list. A k-chain in which the first and the last team are the same, is called a closed k-chain, otherwise the k-chain is called open.*

The value k of a k-chain represents the number of maximally giving teams in the chain, and is called the *length* of the k-chain. For instance, the partial schedule depicted in Table 3 exhibits a 2-chain.

Definition 7. *Suppose that, in some feasible schedule, team i is an mg team, giving maximally to team j. Suppose further that team i plays team j in round r. The opponent of team i in round $r - 1$ is called the bridge team for team i.*

For instance, in the partial schedule depicted in Table 3, team n is the bridge team for team 1, and team n is also the bridge team for team 2. Notice that when investigating a k-chain, we can freely re-index the teams. We use this freedom to choose team $n \in N$ as the bridge team for the first team in the k-chain.

Lemma 3. *Given is a feasible schedule containing a k-chain. Let team $i \in N$ be an mg team in the k-chain, giving maximally to team j, and let team $n \in N$ be the bridge team for team i.*

(i) If team i plays team j in round r, then team j plays team n in round $r + 1$.
(ii) Team j receives exactly one coe from team n.

In Table 3, team n is the bridge team for both mg-teams in the 2-chain. This is not a coincidence.

Lemma 4. *Consider a feasible schedule containing a k-chain. There is a unique team that is a bridge team for each mg team in that k-chain.*

Lemma 5. *In any feasible schedule, the only closed k-chains that can occur are those of length $n - 1$.*

3.3   Characterizing the Class $\mathcal{C}$

In this section, we characterize when a schedule is in the class $\mathcal{C}$, i.e., when a schedule can be constructed by the Circle Method. Also, we give two properties, and we show that a schedule having one of these properties is in the class $\mathcal{C}$.

Lemma 6. *A schedule can be generated with the Circle Method (i.e., is in the class $\mathcal{C}$) if and only if it contains a closed $(n - 1)$-chain.*

It is interesting to observe that if the schedule contains a k-chain that is "long enough", feasibility of the schedule allows us to argue that the schedule must, in fact, contain a closed $(n - 1)$-chain, and hence be a schedule in the class $\mathcal{C}$.

Lemma 7. *A feasible schedule that contains a k-chain of length $\frac{n}{2} - 2$ is a schedule that can be generated by the Circle Method.*

We can also show that when a team is a bridge team for "enough" teams, feasibility of the schedule allows us to argue that the schedule must contain a k-chain of length at least $\frac{n}{2} - 2$, and hence be a schedule in the class $\mathcal{C}$.

Lemma 8. *If, in a given schedule, a team is a bridge team for at least $n - 3$ other teams, then that schedule can be generated by the Circle Method.*

3.4   Upper Bounds on the Contribution of Bridge Teams

We now prove two lemmas that specify an upper bound on the contribution of a bridge team to the COE value.

Lemma 9. *Consider a feasible schedule. Let team $b \in N$ be a bridge team for ℓ distinct mg teams. Then, the contribution of team b is bounded by*

$$Co(b) \leq (n - 2 - \ell)^2 + \ell + 1.$$

In the special case where $\ell = 2$, we can improve the bound derived in the previous lemma. We prove the following statement.

Lemma 10. *Consider a feasible schedule. Let team $b \in N$ be a bridge team for two distinct mg teams. Then, the contribution of team b is bounded by*

$$Co(b) \leq (n - 5)^2 + 6.$$

4   Proving the Theorem

In this section, we 'assemble' the building blocks proven in Sect. 3, in order to prove Theorem 1. First, we deal with the case $n \in \{6, 8\}$; next, we partition in Sect. 4.1 the set of teams N into different types of subsets. In Sect. 4.2, we show how to bound the average contribution of teams in a subset, culminating in the final proof.

Lemma 11. *The Circle Method generates a schedule with maximum COE value when $n \in \{6, 8\}$.*

4.1   Identifying Subsets of Teams

In Sect. 3, we introduced two possible properties of a team: given a schedule, a team can be a maximally giving team, and a team can be a bridge team. Consider a team, say team $b \in N$, that is both maximally giving, as well as a bridge team. (One might wonder whether this is possible; however, there exist schedules containing mg bridge teams). This team has very specific properties as witnessed by the following lemma.

Lemma 12. *Let team $b \in N$ be a maximally giving bridge team, giving maximally to team $a \in N$ and a bridge team for team 1. Then: (i) team 1, being an mg team, is a bridge team for team b and (ii) team a is neither an mg team, nor a bridge team.*

The team pairs (b, a) and $(1, 2)$, where teams b and 1 are mg teams as well as bridge teams, and where teams a and 2 receive maximally from teams b and 1 respectively, are of interest to us. We now define three different types of subsets of teams.

Definition 8. *A type 1 subset is a pair of teams (b, a) where team b is an mg, bridge team, and team a receives maximally from team b.*

We use $T_1 \subseteq N$ to denote the set of teams that are in sets of Type 1.

Teams that are maximally giving bridge teams are contained in sets of type 1. Let us now discuss the remaining bridge teams. Recall that for every bridge team $b \in N$, there is a set of mg teams for which team b is a bridge team. Notice that some of these mg teams might be bridge teams as well; these teams, however, belong to a type 1 set.

Definition 9. *A* type 2 *subset is a set of teams that consists of one non-maximally giving bridge team b, and the mg teams for which team b is a bridge team.*

We use $T_2 \subseteq N$ to denote the set of teams that are in sets of Type 2. It follows from Lemma 12 that these mg teams cannot be bridge teams themselves. From this it is clear that $T_1 \cap T_2 = \emptyset$.

Each team that gives maximally or is a bridge team (or both) is now classified in a subset of type 1 or 2. And some of the non-maximally giving, non-bridge teams are classified as well. All remaining teams will form a single set, called a set of type 3.

Definition 10. *The type 3 subset contains all teams that are not part of any set of type 1 or of type 2.*

We use $T_3 \subseteq N$ to denote the set of teams that are in the type 3 set. The Definitions 8, 9 and 10 imply that the sets T_1, T_2, and T_3 form a partition of N.

4.2   Proving Theorem 1

We show how to obtain an upper bound on the average contribution of a team in a subset of a particular type. Then we compare these averages with the average contribution of a team in a schedule found by the Circle Method. Clearly, the average contribution of a team in a schedule created by the Circle Method is:

$$\frac{CMCOE(n)}{n} = \frac{((n-3)^2+3)(n-1)}{n} = n^2 - 7n + 18 - \frac{12}{n}. \tag{1}$$

We will show that, for $n \geq 10$, the average contribution of a team in any subset in a schedule not in $\mathcal{C}$ is less than the average contribution of any team in a schedule that is in $\mathcal{C}$.

Lemma 13. *In any feasible schedule not in $\mathcal{C}$, we have, for each subset T_j ($j \in \{1,2,3\}$) and for each $n \geq 10$:*

$$\frac{\sum_{i \in T_j} Co(i)}{|T_j|} < \frac{CMCOE(n)}{n}.$$

It is now easy to see that Lemma 13 implies Theorem 1: Since the sets T_1, T_2 and T_3 form a partition of N, the COE value of a schedule not in $\mathcal{C}$ is smaller than $CMCOE(n)$, the COE value found by the Circle Method. This means that the Circle Method maximizes the COE value.

Due to the fact that the average contribution of teams in every type of set is less than that of the Circle Method, the reverse holds as well. This means that a schedule that has a maximum COE value can be generated by the Circle Method.

References

1. Anderson, I.: Combinatorial Designs and Tournaments. Oxford University Press, Oxford (1997)
2. Anderson, I.: Balancing carry-over effects in tournaments. In: Holroyd, F., Quinn, K., Rowley, C., Webb, B. (eds.) Combinatorial Designs and Their Applications. Research Notes in Mathematics, vol. 403, pp. 1–16. CRC Press, Boca Raton (1997)
3. Froncek, D.: Scheduling a tournament. In: Gallian, J.A. (ed.) Mathematics and Sports. Dolciani Mathematical Expositions, vol. 43, pp. 203 216. Mathematical Association of America, Washington DC (2010)
4. De Werra, D.: Geography, games and graphs. Discrete Appl. Math. **2**, 327–337 (1980)
5. De Werra, D.: Scheduling in sports. In: Hansen, P. (ed.) Studies on Graphs and Discrete Programming. Annals of Discrete Mathematics, vol. 11, pp. 381–395. North-Holland, Amsterdam (1981)
6. Goossens, D.R., Spieksma, F.C.R.: Soccer schedules in Europe: an overview. J. Sched. **15**, 641–651 (2012)
7. Goossens, D.R., Spieksma, F.C.R.: The carry-over effect does not influence football results. J. Sports Econ. **13**, 288–305 (2012)
8. Griggs, T., Rosa, A.: A tour of european soccer schedules, or testing the popularity of GK_{2n}. Bull. Inst. Comb. Appl. **18**, 65–68 (1996)
9. Januario, T., Urrutia, S., Ribeiro, C., de Werra, D.: A Tutorial on Edge Coloring in Graphs for Sports Scheduling. http://www2.ic.uff.br/~celso/artigos/ejor_single_file.pdf. Accessed 3 Nov 2015
10. Kendall, G., Knust, S., Ribeiro, C., Urrutia, S.: Scheduling in sports: an annotated bibliography. Comput. Oper. Res. **37**, 1–19 (2010)
11. Kirkman, T.P.: On a problem in combinatorics. Camb. Dublin Math. J. **2**, 191–204 (1847)
12. Lambrechts, E., Ficker, A.M.C., Goossens, D.R., Spieksma, F.C.R.: Round-robin tournaments generated by the circle method have maximum carry-over. Research report, KBI_1603, KU Leuven
13. Miyashiro, R., Matsui, T.: Minimizing the carry-over effects value in a round-robin tournament. In: Burke, E.K., Rudová, H. (eds.) Proceedings of the 6th International Conference on the Practice and Theory of Automated Timetabling, pp. 460–463 (2006)
14. Russell, K.G.: Balancing carry-over effects in round robin tournaments. Biometrika **67**(1), 127–131 (1980)

Extreme Functions with an Arbitrary Number of Slopes

Amitabh Basu[1]($\boxtimes$), Michele Conforti[2], Marco Di Summa[2], and Joseph Paat[1]

[1] Department of Applied Mathematics and Statistics,
Johns Hopkins University, Baltimore, USA
basu.amitabh@jhu.edu
[2] Dipartimento di Matematica, Università degli Studi di Padova, Padova, Italy

Abstract. For the one dimensional infinite group relaxation, we construct a sequence of extreme valid functions that are piecewise linear and such that for every natural number $k \geq 2$, there is a function in the sequence with k slopes. This settles an open question in this area regarding a universal bound on the number of slopes for extreme functions. The function which is the pointwise limit of this sequence is an extreme valid function that is continuous and has an infinite number of slopes. This provides a new and more refined counterexample to an old conjecture of Gomory and Johnson that stated all extreme functions are piecewise linear.

1 Introduction

Let $b \in \mathbb{R} \setminus \mathbb{Z}$. The (one dimensional) *infinite group relaxation* $R_b(\mathbb{R}, \mathbb{Z})$ is the set of functions $x : \mathbb{R} \to \mathbb{Z}^+$ having finite support (that is, $\{r : x_r > 0\}$ is a finite set) satisfying:

$$\sum_{r \in \mathbb{R}} r x_r \in b + \mathbb{Z}, \ x_r \in \mathbb{Z}^+. \tag{1}$$

A function $\pi : \mathbb{R} \to \mathbb{R}^+$ is *valid* for $R_b(\mathbb{R}, \mathbb{Z})$ if

$$\sum_{r \in \mathbb{R}} \pi(r) x_r \geq 1, \ \text{for every} \ x \in R_b(\mathbb{R}, \mathbb{Z}). \tag{2}$$

Valid functions for the infinite group relaxation were first introduced by Gomory and Johnson [9,10] as means to obtain cutting planes for mixed-integer programs. This idea has recently culminated in study of *cut-generating functions* which has become one of the central aspects of modern cutting plane theory. The monographs of Basu et al. [2,4] provide a comprehensive introduction to the subject and survey the recent advances.

The most well known valid function is the Gomory mixed-integer function, defined as follows:

A. Basu and J. Paat—Supported by the NSF grant CMMI1452820.

M. Conforti and M. Di Summa—Supported by the grant "Progetto di Ateneo 2013" of the University of Padova.

Q. Louveaux and M. Skutella (Eds.): IPCO 2016, LNCS 9682, pp. 190–201, 2016.
DOI: 10.1007/978-3-319-33461-5_16

$$\pi(x) = \begin{cases} \frac{1}{b}x, & 0 \le x < b \\ \frac{1}{1-b} - \left(\frac{1}{1-b}\right)x, & b \le x < 1 \\ \pi(x-j), & x \in [j, j+1), \; j \in \mathbb{Z} \setminus \{0\}. \end{cases} \tag{3}$$

A valid function π is *minimal* if $\pi = \pi'$, for every valid function π' such that $\pi' \le \pi$. Given valid functions π and π' such that $\pi' \le \pi$ and $\pi \ne \pi'$, it holds that $\{x_r \in \mathbb{Z}^+ : \sum \pi'(r)x_r \ge 1\} \subsetneq \{x_r \in \mathbb{Z}^+ : \sum \pi(r)x_r \ge 1\}$. Therefore if a valid function is not minimal, then it is redundant.

A function $\phi \colon \mathbb{R} \to \mathbb{R}$ is *subadditive* if $\phi(r^1) + \phi(r^2) \ge \phi(r^1 + r^2)$ for all $r^1, r^2 \in \mathbb{R}$. ϕ satisfies the *symmetry condition* if $\phi(r) + \phi(b-r) = 1$ for all $r \in \mathbb{R}$. Finally, ϕ is *periodic* modulo $\mathbb{Z}$ if $\phi(r) = \phi(r+w)$ for all $r \in \mathbb{R}$ and $w \in \mathbb{Z}$.

Theorem 1 (Gomory and Johnson [9]). *Let $\pi \colon \mathbb{R} \to \mathbb{R}$ be a nonnegative function. Then π is a minimal valid function for $R_b(\mathbb{R}, \mathbb{Z})$ if and only if $\pi(w) = 0$ for all $w \in \mathbb{Z}$, π is subadditive, and π satisfies the symmetry condition. (These conditions imply that π is periodic modulo $\mathbb{Z}$ and $\pi(b+w) = 1$ for every $w \in \mathbb{Z}$.)*

It is easy to check that the Gomory mixed-integer function defined above is subadditive and satisfies the symmetry condition. Therefore, by the above theorem, it is a minimal function.

Minimal functions are the ones that are not dominated by any other function. However minimal functions may be implied by convex combinations of valid functions. Gomory and Johnson define a valid function π to be *extreme* if $\pi = \pi_1 = \pi_2$ for every pair of valid functions π_1, π_2 such that $\pi = \frac{\pi_1 + \pi_2}{2}$. If π is a valid function which is extreme, then π is easily seen to be minimal. Therefore extremality is a stronger requirement.

We say a function $\phi \colon \mathbb{R} \to \mathbb{R}$ is *piecewise linear* if there is a set of closed, non-degenerate intervals I_j, $j \in J$ such that $\mathbb{R} = \cup_{j \in J} I_j$, any bounded subset of $\mathbb{R}$ intersects only finitely many intervals, and ϕ is affine linear over each interval I_j. Note that in this definition, a piecewise linear function is continuous.

Theorem 2 (Gomory and Johnson [9]). *Let $\pi \colon \mathbb{R} \to \mathbb{R}$ be minimal valid function which is piecewise linear and has only 2 slopes. Then π is an extreme valid function.*

In particular, the above theorem implies that the Gomory mixed-integer function is extreme.

Extreme valid functions that are piecewise linear and have few slopes receive the largest number of hits is the shooting experiments of Gomory and Johnson and seem to be the most useful in practice. Indeed Gomory and Johnson [11] conjectured that every valid function that is extreme is piecewise linear. This has been disproved by Basu et al. [1].

Minimal valid functions with 3 slopes are not always extreme. However, Gomory and Johnson construct an extreme function that is piecewise linear with 3 slopes. It appears to be hard to construct extreme functions that are piecewise linear with many slopes. Indeed, until 2013, all known families of piecewise linear extreme functions had at most 4 slopes. This had led Dey and Richard to pose the question of constructing extreme functions with more than 4 slopes at a 2010 Aussois meeting [5]. In 2013, Hildebrand, in an unpublished result, constructed an extreme function that is piecewise linear with 5 slopes and very recently Köppe and Zhou [13] constructed an extreme function that is piecewise linear with 28 slopes. These functions were found through a clever computer search.

Köppe and Zhou [13] express the belief that there exist extreme functions that are piecewise linear and have an arbitrary number of slopes.[1] We prove this. More precisely, we show the following:

Theorem 3. *Let $b \in (0,1)$. For $k \geq 2$, there exists an extreme function for $R_b(\mathbb{R}, \mathbb{Z})$ that is piecewise linear with k slopes.*

Note that in Theorems 3 and 4, we may assume $b \in (0,1)$ since extreme functions are periodic with respect to $\mathbb{Z}$. The proof of Theorem 3 provided here is constructive. We define a sequence of functions $\{\pi_k\}_{k=2}^{\infty}$, where π_2 is the Gomory mixed-integer function, and π_3 is an instantiation of a construction of extreme functions that are piecewise linear and have 3 slopes provided by Gomory and Johnson. We first prove some properties about each function π_k. In Sect. 3 we use these properties to show that these functions are subadditive and satisfy the symmetry condition. Therefore each function π_k is a minimal valid function, as it satisfies the conditions of Theorem 1. Section 4 is devoted to the proof that each function π_k is extreme.

Our next result states that the function which is the pointwise limit of this sequence is an extreme function that is continuous and has an infinite number of slopes. The proof appears in Sect. 5.

Theorem 4. *Let $b \in (0,1)$. There exists a continuous extreme function π_∞ for $R_b(\mathbb{R}, \mathbb{Z})$ with an infinite number of slopes (i.e., values for the derivative of π_∞).*

In the following sections, we give constructions to establish Theorems 3 and 4 with b in the interval $(0, \frac{1}{2}]$. One may obtain extreme functions for values of $b \in [\frac{1}{2}, 1)$ by reflecting the constructions about 0. Indeed, one can check that π_b is extreme for $R_b(\mathbb{R}, \mathbb{Z})$ when $b \in (0, 1/2]$ if and only if $\pi_{1-b} : \mathbb{R} \to \mathbb{R}$ defined by $\pi_{1-b}(x) := \pi_b(-x)$ is extreme for $R_{1-b}(\mathbb{R}, \mathbb{Z})$.

2   A Construction of K-Slope Functions π_k

Let $b \in (0, \frac{1}{2}]$. Let π_2 be the mixed-integer Gomory function defined by (3).

In constructing π_k for $k \geq 3$, we use the following intervals:

$$I_1^k := [0, b(\tfrac{1}{8})^{k-2}], \qquad\qquad I_2^k := [b(\tfrac{1}{8})^{k-2}, 2b(\tfrac{1}{8})^{k-2}],$$

$$I_3^k := [2b(\tfrac{1}{8})^{k-2}, b - 2b(\tfrac{1}{8})^{k-2}], \ I_4^k := [b - 2b(\tfrac{1}{8})^{k-2}, b - b(\tfrac{1}{8})^{k-2}],$$

$$I_5^k := [b - b(\tfrac{1}{8})^{k-2}, b], \qquad\qquad I_6^k := [b, 1).$$

---

[1] This is also stated as an open question in the survey by Basu et al. [4].

Given π_{k-1}, where $k - 1 \geq 2$, define π_k to be

$$\pi_k(x) = \begin{cases} \left(\frac{2^{k-2}-b}{b-b^2}\right)x, & x \in I_1^k \\ \frac{4^{2-k}}{1-b} - \left(\frac{1}{1-b}\right)x, & x \in I_2^k \\ \frac{1-4^{2-k}}{1-b} - \left(\frac{1}{1-b}\right)x, & x \in I_4^k \\ \frac{1-2^{k-2}}{1-b} + \left(\frac{2^{k-2}-b}{b-b^2}\right)x, & x \in I_5^k \\ \pi_{k-1}(x), & x \in I_3^k \cup I_6^k \\[4pt] \pi_k(x-j), & x \in [j, j+1), \ j \in \mathbb{Z} \setminus \{0\}. \end{cases}$$

Figure 1 shows π_k for various values of k when $b = \frac{1}{2}$. The plots were generated using the help of a software package created by Hong et al. [12].

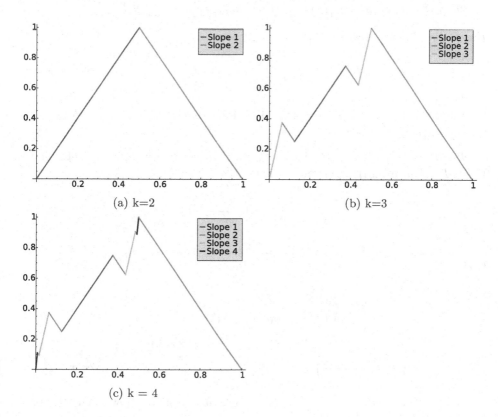

(a) k=2

(b) k=3

(c) k = 4

Fig. 1. Plots of π_k for $b = \frac{1}{2}$

Observe that π_k is built recursively with the Gomory mixed-integer function as the base case. Intuitively, π_k is created by adding to π_{k-1} a perturbation on a small interval to the right of 0 and applying a symmetric perturbation on an interval to the left of b; the interval $[b, 1)$ is kept intact. These small perturbations allow π_k to maintain much of the structure of π_{k-1}, but the number of distinct slopes is increased by one. We collect some useful properties of π_k in Propositions 1 and 2.

Proposition 1. *Let $k \geq 3$. Then*

(i) $I_1^k \cup I_2^k \subsetneq I_1^{k-1}$ and $I_4^k \cup I_5^k \subsetneq I_5^{k-1}$

(ii) *If $x \in I_3^k \cup I_6^k$, then $\pi_k(x) = \pi_{k-1}(x)$. If $x \in I_1^k \cup I_2^k$, then $\pi_k(x) \geq \pi_{k-1}(x)$. If $x \in I_4^k \cup I_5^k$, then $\pi_k(x) \leq \pi_{k-1}(x)$.*

(iii) $-\pi_k$ *is convex on $I_1^k \cup I_2^k$ and π_k is convex on $I_4^k \cup I_5^k$.*

(iv) *Let $y \in I_4^k \cup I_5^k$ such that $y \neq b$ and take $x \in [0, b - y]$. Then $\pi_k(x + y) \leq \pi_k(y) + \left(\frac{1 - \pi_k(y)}{b - y}\right) x$. Also, $\left(\frac{\pi_k(b - y)}{b - y}\right) x \leq \pi_k(x)$.*

(v) *For any $x \in (0, 1) \setminus \{b\}$, there exists some natural number N_x such that $x \in I_3^{N_x} \cup I_6^{N_x}$ and $\pi_{k_1}(x) = \pi_{k_2}(x)$ whenever $k_1, k_2 \geq N_x$.*

Proof. Proof of (i) Observe that

$$b\left(\frac{1}{8}\right)^{k-3} = 8b\left(\frac{1}{8}\right)^{k-2} > 2b\left(\frac{1}{8}\right)^{k-2}.$$

By the definitions of I_1^k, I_2^k and I_1^{k-1}, it follows that $I_1^k \cup I_2^k \subsetneq I_1^{k-1}$. A similar argument shows that $I_4^k \cup I_5^k \subsetneq I_5^{k-1}$.

Proof of (ii) Let $x \in [0, 1)$. If $x \in I_3^k \cup I_6^k$, then $\pi_k(x) = \pi_{k-1}(x)$ by definition. If $x \in I_1^k$, then from (i) it follows that $x \in I_1^{k-1}$. Note that

$$\left(\frac{2^{k-2} - b}{b - b^2}\right) x \geq \left(\frac{2^{k-3} - b}{b - b^2}\right) x,$$

and so $\pi_k(x) \geq \pi_{k-1}(x)$. If $x \in I_2^k$, then again from (i), $x \in I_1^{k-1}$ and it follows that

$$
\begin{aligned}
\frac{4^{2-k}}{1-b} - \left(\frac{1}{1-b}\right) x &= \left(\frac{1}{1-b}\right)\left(4^{2-k} - x\right) \\
&\geq \left(\frac{1}{1-b}\right)\left(4^{2-k} - 2b\left(\frac{1}{8}\right)^{k-2}\right) && \text{since } x \in I_2^k \\
&= \left(\frac{1}{b-b^2}\right)\left(2^{k-3}\left(2b\left(\frac{1}{8}\right)^{k-2}\right) - b\left(2b\left(\frac{1}{8}\right)^{k-2}\right)\right) \\
&\geq \left(\frac{2^{k-3} - b}{b - b^2}\right) x && \text{since } x \in I_2^k.
\end{aligned}
$$

Hence $\pi_k(x) \geq \pi_{k-1}(x)$ on $I_1^k \cup I_2^k$. A similar argument shows that $\pi_k(x) \leq \pi_{k-1}(x)$ on $I_4^k \cup I_5^k$.

Proof of (iii) By definition, π_k is affine linear over I_1^k with positive slope and affine linear over I_2^k with negative slope. Since π_k is continuous, it is therefore concave. So $-\pi_k$ is a convex function over $I_1^k \cup I_2^k$. The same argument shows that π_k is convex over $I_4^k \cup I_5^k$.

Proof of (iv) Fix $y \in I_4^k \cup I_5^k \setminus \{b\}$. It follows by assumption that $x+y \in [y,b]$. Therefore $\lambda = \frac{b-x-y}{b-y} \in [0,1]$. Using the facts that π_k is convex over $[y,b]$ from (iii) and $\pi_k(b) = 1$, we obtain

$$\pi_k(x+y) = \pi_k(\lambda y + (1-\lambda)b) \leq \lambda \pi_k(y) + (1-\lambda)\pi_k(b) = \pi_k(y) + \left(\frac{1-\pi_k(y)}{b-y}\right)x.$$

The other inequality follows from the fact that $-\pi_k$ is convex over $I_1^k \cup I_2^k$ by (iii).

Proof of (v) Notice that as $k \to \infty$, I_3^k converges to $(0,b)$ and thus, there exists N_x such that $x \in I_3^{N_x} \cup I_6^{N_x}$. Moreover, by definition on π_k, for any natural number N, $\pi_k(x) = \pi_N(x) \ \forall x \in I_3^N \cap I_6^N$ for every $k \geq N$.   □

Proposition 2. *For each value of k, the function π_k is piecewise linear and has k slopes taking values $-\frac{1}{1-b}$ and $\{\frac{2^{i-2}-b}{b-b^2}\}_{i=2}^k$.*

Proof. We proceed by induction. For π_2, the result follows by definition, so assume that for $k-1 \geq 2$, π_{k-1} is piecewise linear and has slopes taking values $-\frac{1}{1-b}$ and $\{\frac{2^{i-2}-b}{b-b^2}\}_{i=2}^{k-1}$.

Observe that for each value of j, π_j has a slope of $-\frac{1}{1-b}$ on the interval $(b,1)$. Therefore on the interval $[0,b)$, the function π_{k-1} must take on slope values $\{\frac{2^{i-2}-b}{b-b^2}\}_{i=2}^{k-1}$ (π_{k-1} also admits a slope of $-\frac{1}{1-b}$ on subintervals contained in $[0,b)$). By Proposition 1 (ii), $\pi_k = \pi_{k-1}$ everywhere except $I_1^k \cup I_2^k$ and $I_4^k \cup I_5^k$, on which π_k takes on slope values $\frac{2^{k-2}-b}{b-b^2}$ and $-\frac{1}{1-b}$ by definition. Since $I_1^k \cup I_2^k \subsetneq I_1^{k-1}$ and $I_4^k \cup I_5^k \subsetneq I_5^{k-1}$ by Proposition 1 (i), it follows that π_k has slopes taking values $-\frac{1}{1-b}$ and $\{\frac{2^{i-2}-b}{b-b^2}\}_{i=2}^k$.

It is left to show that π_k is piecewise linear. By Proposition 1 (ii) and the induction hypothesis, it is sufficient to show that π_k is piecewise linear on $I_1^k \cup I_2^k$ and $I_4^k \cup I_5^k$, and that $\pi_k\left(2b\left(\frac{1}{8}\right)^{k-2}\right) = \pi_{k-1}\left(2b\left(\frac{1}{8}\right)^{k-2}\right)$ and $\pi_k\left(b - 2b(\frac{1}{8})^{k-2}\right) = \pi_{k-1}\left(b - 2b\left(\frac{1}{8}\right)^{k-2}\right)$. Note that π_k is piecewise linear on $I_1^k \cup I_2^k$ and $I_4^k \cup I_5^k$ by definition. It is straightforward to check that $\pi_k\left(2b\left(\frac{1}{8}\right)^{k-2}\right) = \pi_{k-1}\left(2b\left(\frac{1}{8}\right)^{k-2}\right)$ and $\pi_k\left(b - 2b\left(\frac{1}{8}\right)^{k-2}\right) = \pi_{k-1}\left(b - 2b\left(\frac{1}{8}\right)^{k-2}\right)$. Thus π_k is piecewise linear, as desired.   □

3   Proof of Minimality of π_k

In the proof of Theorem 3, it is required to show that π_k is a minimal valid function for $R_b(\mathbb{R}, \mathbb{Z})$. Since by definition $\pi_k(0) = 0$, and π_k is periodic, by Theorem 1, it is sufficient to show that (a) $\pi_k(x) = \pi_k(b-x)$ for all $x \in [0,1)$, i.e. that π_k satisfies the symmetry condition, and (b) π_k is subadditive. We show (a) and (b) in Propositions 3 and 4, respectively.

Proposition 3. π_k *satisfies the symmetry condition for all $k \geq 2$.*

Proof. We proceed by induction on k. The Gomory mixed-integer function is known to be minimal and hence π_2 is symmetric. Assume π_{k-1} satisfies the symmetry condition for $k - 1 \geq 2$ and consider $x \in [0, 1)$. Observe that $x \in I_1^k$ if and only if $b - x \in I_5^k$. Therefore, if $x \in I_1^k$ then

$$\pi_k(x) + \pi_k(b - x) = \left(\frac{2^{k-2} - b}{b - b^2}\right) x + \frac{1 - 2^{k-2}}{1 - b} + \left(\frac{2^{k-2} - b}{b - b^2}\right)(b - x) = 1.$$

A similar argument can be used to show that π_k satisfies the symmetry condition on the intervals I_2^k and I_4^k. If $x \notin I_1^k \cup I_2^k \cup I_4^k \cup I_5^k$ then $b - x \notin I_1^k \cup I_2^k \cup I_4^k \cup I_5^k$, and so symmetry holds by induction.   □

Proposition 4. π_k *is subadditive for all* $k \geq 2$.

Proof. We proceed by induction on k. Note that π_2 is subadditive, so assume π_{k-1} is subadditive for $k - 1 \geq 2$. By periodicity of π_k, it suffices to check $\pi_k(x) + \pi_k(y) \geq \pi_k(x + y)$ for all $x, y \in [0, 1)$ and $x \leq y$.

CLAIM. *If* $y \in I_6^k = [b, 1)$, *then* $\pi_k(x + y) \leq \pi_k(x) + \pi_k(y)$.

Proof of Claim. Since π_k is piecewise linear, we may integrate it over any bounded domain. Let π_k' denote the derivative of π_k (where defined). A direct calculation shows

$$\pi_k(x + y) = \pi_k(x + (y - 1)) \qquad \text{by periodicity of } \pi_k$$

$$= \pi_k(x) + \int_x^{x-(1-y)} \pi_k'(t)dt$$

$$= \pi_k(x) + \int_{x-(1-y)}^x -\pi_k'(t)dt$$

$$\leq \pi_k(x) + \int_y^1 -\pi_k'(t)dt$$

$$= \pi_k(x) - \pi_k(1) + \pi_k(y)$$
$$= \pi_k(x) + \pi_k(y) \qquad \text{since } \pi_k(1) = 0.$$

The inequality follows from Proposition 2, as the minimum value of the slope for π_k is $-\frac{1}{1-b}$ and this is the slope over the interval $[b, 1] \supseteq [y, 1]$. This concludes the proof of the claim.   ◇

By the above claim, it suffices to consider the case $y < b$. Since $b \leq \frac{1}{2}$, this implies that $x \leq y \leq x + y < 1$.

Case 1: $x + y \in I_1^k \cup I_2^k$. By Proposition 1 (iii), the function $-\pi_k$ is convex over $I_1^k \cup I_2^k$. Therefore $\pi_k(x) + \pi_k(y) \geq \pi_k(x + y)$.

Case 2: $x + y \in I_3^k$. Since $x, y \in I_1^k \cup I_2^k \cup I_3^k$ we have that

$$\pi_k(x) + \pi_k(y) \geq \pi_{k-1}(x) + \pi_{k-1}(y) \geq \pi_{k-1}(x + y) = \pi_k(x + y),$$

where the first inequality comes from Proposition 1 (ii), the second inequality comes from the induction hypothesis, and the final inequality comes again from Proposition 1 (ii).

Case 3: $x + y \in I_4^k \cup I_5^k$. If $y \in I_1^k \cup I_2^k \cup I_3^k$ then using the induction hypothesis and Proposition 1 (ii), it follows that

$$\pi_k(x) + \pi_k(y) \geq \pi_{k-1}(x) + \pi_{k-1}(y) \geq \pi_{k-1}(x+y) \geq \pi_k(x+y).$$

If $y \in I_4^k \cup I_5^k$ then $x \in [0, b-y]$ and $b - y \in I_1^k \cup I_2^k$. Thus, $x \in I_1^k \cup I_2^k$. Note that

$$\pi_k(x+y) \leq \pi_k(y) + \left(\frac{1 - \pi_k(y)}{b-y}\right)x \quad \text{by Proposition 1 (iv)}$$

$$= \pi_k(y) + \left(\frac{\pi_k(b-y)}{b-y}\right)x \quad \text{by the symmetry property}$$

$$\leq \pi_k(y) + \pi_k(x) \qquad \text{by Proposition 1 (iv).}$$

Case 4: $x + y \in I_6^k$. π_k has a slope of $-\frac{1}{1-b}$ on the interval $[b, x+y]$. Moreover, by Proposition 2, this is the minimum slope that π_k admits. Therefore,

$$\pi_k(x+y) = \pi(b) + \int_b^{x+y} \pi'(t)dt$$
$$\leq 1 + \int_{b-x}^y \pi'(t)dt$$
$$= 1 + (\pi_k(y) - \pi_k(b-x))$$
$$= \pi_k(x) + \pi_k(y),$$

where the last equality follows by the symmetry of π_k. □

4   Proof of Extremality of π_k

We show that $\phi_1 = \phi_2 = \pi_k$ for every pair of valid functions ϕ_1, ϕ_2 such that $\pi_k = \frac{\phi_1 + \phi_2}{2}$. Our proofs are based on the following two lemmas. The first lemma is an easy consequence of the fact that a minimal valid function is subadditive. (see Lemma 2.11 (ii) in [4]).

Lemma 1. *Let ϕ be a minimal valid function and ϕ_1, ϕ_2 be valid functions such that $\phi = \frac{\phi_1 + \phi_2}{2}$. Then ϕ_1, ϕ_2 are minimal and for all $x, y \in \mathbb{R}$, $\phi(x + y) = \phi(x) + \phi(y)$ implies $\phi_i(x + y) = \phi_i(x) + \phi_i(y)$ for both $i = 1, 2$.*

The following result first appeared in [11], and was subsequently elaborated upon in [3,6–8]; see also the survey [4].

Lemma 2 (Interval Lemma). *Let U, V be non-degenerate closed intervals in $\mathbb{R}$. If $\phi : \mathbb{R} \to \mathbb{R}$ is bounded over U and V, and satisfies $\phi(x) + \phi(y) = \phi(x + y)$ for all $x \in U$, $y \in V$, then ϕ is affine over U, V and $U + V$ with the same slope.*

We will use the above lemma when ϕ is a minimal valid function. In this case ϕ is bounded, as $0 \leq \phi \leq 1$.

In the following Claims 1–4, we develop some tools towards proving extremality.

Claim 1. *Let $k \geq 3$ and let ϕ be a minimal valid function such that $\phi = \pi_k$ on I_6^k. Then for all minimal valid functions ϕ_1, ϕ_2 such that $\phi = \frac{\phi_1 + \phi_2}{2}$, we must have $\phi_1 = \phi_2 = \phi = \pi_k$ on $I_6^k \cup \{1\}$.*

Proof. Note that $I_6^k \cup \{1\} \equiv [\frac{1+b}{2}, 1] + [\frac{1+b}{2}, 1]$ (modulo 1) and $x, y \in [\frac{1+b}{2}, 1]$ implies that

$$
\begin{aligned}
\phi(x) + \phi(y) = \pi_k(x) + \pi_k(y) &= \left(\frac{1}{1-b} - \left(\frac{1}{1-b}\right)x\right) + \left(\frac{1}{1-b} - \left(\frac{1}{1-b}\right)y\right) \\
&= \frac{1}{1-b} - \left(\frac{1}{1-b}\right)(x + y - 1) \\
&= \pi_k(x + y - 1) \\
&= \pi_k(x + y) \qquad\qquad\qquad \text{by periodicity} \\
&= \phi(x + y).
\end{aligned}
$$

Therefore, Lemmas 1 and 2 together imply that each ϕ_i is affine over $I_6^k \cup \{1\}$. Since ϕ, ϕ_1, and ϕ_2 are minimal, Theorem 1 implies $\phi_1(1) = \phi_2(1) = \phi(1) = \pi_k(1) = 0$ and $\phi_1(b) = \phi_2(b) = \phi(b) = \pi_k(b) = 1$. Therefore $\phi_1 = \phi_2 = \phi = \pi_k$ on $I_6^k \cup \{1\}$. $\square$

Claim 2. *Let $k \geq 3$ and let ϕ be a minimal valid function such that $\phi = \pi_k$ on $I_3^3 = [\frac{b}{4}, \frac{3b}{4}]$. Then for all minimal valid functions ϕ_1, ϕ_2 such that $\phi = \frac{\phi_1 + \phi_2}{2}$, we must have $\phi_1 = \phi_2 = \phi = \pi_k$ on I_3^3.*

Proof. Let $A = [\frac{b}{4}, \frac{3b}{8}] \subseteq I_3^3$ and note that $A + A = [\frac{b}{2}, \frac{3b}{4}] \subseteq I_3^3$. For $x, y \in A$, we see that

$$
\phi(x) + \phi(y) = \pi_k(x) + \pi_k(y) = \frac{1}{b}x + \frac{1}{b}y = \frac{1}{b}(x + y) = \pi_k(x + y) = \phi(x + y).
$$

Using Lemmas 1 and 2, we obtain that each ϕ_i is affine over $[\frac{b}{2}, \frac{3b}{4}]$. The symmetry of ϕ_i and ϕ implies that $\phi_i(\frac{b}{2}) = \phi(\frac{b}{2}) = \pi_k(\frac{b}{2}) = \frac{1}{2}$. By subadditivity of ϕ_i, $\phi_i(\frac{b}{4}) \geq \frac{1}{2}\phi_i(\frac{b}{2}) = \frac{1}{4}$ for $i = 1, 2$. Since $\phi(\frac{b}{4}) = \pi_k(\frac{b}{4}) = \frac{1}{4}$ and $\phi(\frac{b}{4}) = \frac{\phi_1(\frac{b}{4}) + \phi_2(\frac{b}{4})}{2}$, it must be the case that $\phi_i(\frac{b}{4}) = \frac{1}{4}$ for $i = 1, 2$. By symmetry of ϕ_i, this implies $\phi_i(\frac{3b}{4}) = \frac{3}{4} = \phi(\frac{3b}{4}) = \pi_k(\frac{3b}{4})$ for $i = 1, 2$. Therefore, by the affine structure of ϕ_i over $[\frac{b}{2}, \frac{3b}{4}]$, it follows that $\phi_i = \phi$ on $[\frac{b}{2}, \frac{3b}{4}]$. The symmetric property of ϕ_i then yields that $\phi_i = \phi$ on $[\frac{b}{4}, \frac{b}{2}]$ and thus on I_3^3. $\square$

Claim 3. *Let $k \geq 3$ and let $j \in \{3, \ldots, k\}$. Let ϕ be a minimal valid function such that $\phi = \pi_k$ on $I_2^j \cup I_3^j \cup I_4^j \cup I_6^j$. Moreover, let ϕ_1, ϕ_2 be minimal valid functions such that $\phi = \phi_1 = \phi_2$ on $I_3^j \cup I_6^j$ and $\phi = \frac{\phi_1 + \phi_2}{2}$. Then $\phi_1 = \phi_2 = \phi = \pi_k$ on $I_2^j \cup I_4^j$.*

Proof. Let $A = [\frac{3}{2}b\left(\frac{1}{8}\right)^{j-2}, 2b\left(\frac{1}{8}\right)^{j-2}] \subseteq I_2^j$ and $B = [1 - \frac{1}{2}b\left(\frac{1}{8}\right)^{j-2}, 1] \subseteq I_6^j$. Observe that $A + B \equiv I_2^j$ (modulo 1). Moreover, $x \in A$ and $y \in B$ implies

$$
\begin{aligned}
\phi(x) + \phi(y) = \pi_k(x) + \pi_k(y) &= \left(\frac{4^{2-j}}{1-b} - \left(\frac{1}{1-b}\right)x\right) + \left(\frac{1}{1-b} - \left(\frac{1}{1-b}\right)y\right) \\
&= \frac{4^{2-j}}{1-b} - \left(\frac{1}{1-b}\right)(x + y - 1) \\
&= \pi_k(x + y - 1) = \pi_k(x + y) \qquad \text{by periodicity} \\
&= \phi(x + y).
\end{aligned}
$$

From Lemmas 1 and 2, it follows that each ϕ_i is affine over A, B and I_2^j with the same slope. By Claim 1, ϕ_i has slope equal to that of ϕ and π_k over $I_6^j = I_6^k$.

Since ϕ and π_k have the same slope over I_6^j and I_2^j, and ϕ_i has the same slope over $B \subseteq I_6^j$ and I_2^j, ϕ_i must have a slope over I_2^j equal to that of ϕ. Also, since $\phi_1 = \phi_2 = \phi = \pi_k$ on I_3^j by assumption, it must be that $\phi_i \left(2b \left(\frac{1}{8}\right)^{j-2}\right) = \phi \left(2b \left(\frac{1}{8}\right)^{j-2}\right)$ for $i = 1, 2$. This indicates that $\phi_i = \phi$ over I_2^j. Using symmetry, we see that $\phi_i = \phi$ over I_4^j. $\qquad\square$

Claim 4. *Let $k \geq 3$ and let $j \in \{3, \ldots, k-1\}$. Let ϕ be a minimal valid function such that $\phi = \pi_k$ on $(I_1^j \setminus \mathrm{int}(I_1^{j+1} \cup I_2^{j+1})) \cup I_2^j \cup I_3^j \cup I_4^j \cup (I_5^j \setminus \mathrm{int}(I_4^{j+1} \cup I_5^{j+1})) \cup I_6^j$. Moreover, let ϕ_1, ϕ_2 be minimal valid functions such that $\phi = \phi_1 = \phi_2$ on $I_2^j \cup I_3^j \cup I_4^j \cup I_6^j$ and $\phi = \frac{\phi_1 + \phi_2}{2}$. Then $\phi_1 = \phi_2 = \phi$ over $I_1^j \setminus \mathrm{int}(I_1^{j+1} \cup I_2^{j+1})$ and $I_5^j \setminus \mathrm{int}(I_4^{j+1} \cup I_5^{j+1})$.*

Proof. Set $I^* := I_1^j \setminus (\mathrm{int}(I_1^{j+1} \cup I_2^{j+1}) \cup \{0\}) = \left[2b \left(\frac{1}{8}\right)^{j-1}, b \left(\frac{1}{8}\right)^{j-2}\right]$. Let

$$A = \left[2b \left(\frac{1}{8}\right)^{j-1}, 4b \left(\frac{1}{8}\right)^{j-1}\right] \subseteq I^*.$$

Note that

$$A + A = \left[4b \left(\frac{1}{8}\right)^{j-1}, b \left(\frac{1}{8}\right)^{j-2}\right] \subseteq I^*$$

and $A \cup (A + A) = I^*$. A direct calculation shows that $\pi_k(x) + \pi_k(y) = \pi_k(x+y)$ and $\phi(x) + \phi(y) = \phi(x+y)$ for $x, y \in A$. By Lemmas 1 and 2, ϕ_i is affine over I^*. Let $a = 2b \left(\frac{1}{8}\right)^{j-1}$ be the left-endpoint of I^*; then $4a = b \left(\frac{1}{8}\right)^{j-2}$ is the right endpoint of I^*, which is the left endpoint of I_2^j. Since $\phi_1 = \phi_2 = \phi$ on I_2^j, we have $\phi_1(4a) = \phi_2(4a) = \phi(4a)$. By subadditivity, $\phi_i(a) \geq \frac{1}{4}\phi_i(4a)$. Since $\phi(a) = \pi_k(a) = \frac{1}{4}\pi_k(4a) = \frac{1}{4}\phi(4a)$ and $\phi = \frac{\phi_1 + \phi_2}{2}$, we obtain that $\phi_1(a) = \phi_2(a) = \phi(a)$. Therefore, $\phi_i = \phi$ over I^*. Symmetry of ϕ_i yields that $\phi_i = \phi$ over $I_5^j \setminus \mathrm{int}(I_4^{j+1} \cup I_5^{j+1})$. $\qquad\square$

Lemma 3. *Let $k \geq 3$ and $j \in \{3, \ldots, k\}$ and let ϕ be a minimal valid function such that $\phi = \pi_k$ on $I_3^j \cup I_6^j$. Then for all minimal valid functions ϕ_1, ϕ_2 such that $\phi = \frac{\phi_1 + \phi_2}{2}$, we must have $\phi_1 = \phi_2 = \phi = \pi_k$ on $I_3^j \cup I_6^j$.*

Proof. By Claim 1, we obtain $\phi_1 = \phi_2 = \phi = \pi_k$ on $I_6^j = I_6^k$. We prove $\phi_1 = \phi_2 = \phi = \pi_k$ on I_3^j by induction on j. For $j = 3$, the result follows from Claim 2. We assume the result holds for some j such that $3 \leq j \leq k - 1$ and show that it holds for $j + 1$. Note that $I_3^{j+1} \cup \{0, b\} = (I_1^j \setminus \mathrm{int}(I_1^{j+1} \cup I_2^{j+1})) \cup I_2^j \cup I_3^j \cup I_4^j \cup (I_5^j \setminus \mathrm{int}(I_4^{j+1} \cup I_5^{j+1}))$. By the induction hypothesis, the result holds for I_3^j. Since we assume $\phi = \pi_k$ on $I_3^{j+1} \cup I_6^{j+1} \supseteq I_2^j \cup I_3^j \cup I_4^j \cup I_6^j$, we can apply Claim 3, obtaining $\phi_1 = \phi_2 = \phi$ on $I_2^j \cup I_4^j$. Similarly, by applying Claim 4, we obtain that $\phi_1 = \phi_2 = \phi$ on $I_1^j \setminus \mathrm{int}(I_1^{j+1} \cup I_2^{j+1})$ and $I_5^j \setminus \mathrm{int}(I_4^{j+1} \cup I_5^{j+1})$. $\qquad\square$

Proof (Proof of Theorem 3). Consider valid functions ϕ_1, ϕ_2 such that $\frac{\phi_1 + \phi_2}{2} = \pi_k$. Since π_k is minimal, by Lemma 1, ϕ_1, ϕ_2 are also minimal. Using Lemma 3 with $\phi = \pi_k$, we obtain that $\phi_1 = \phi_2 = \pi_k$ on $I_3^k \cup I_6^k$. Using Claim 3 with $\phi = \pi_k$ and $j = k$, we obtain that $\phi_1 = \phi_2 = \pi_k$ on $I_2^k \cup I_4^k$.

It is left to show that $\phi_i = \pi_k$ on I_1^k and I_5^k. Let $U = V = \left[0, \frac{b}{2} \left(\frac{1}{8}\right)^{k-2}\right]$ and observe $U + V \equiv \left[0, b \left(\frac{1}{8}\right)^{k-2}\right]$ (modulo 1) $= I_1^k$. Since π_k is additive on I_1^k by definition, Lemma 1 implies each ϕ_i is as well. Moreover, $\pi_k = \phi_1 = \phi_2$ on $\left\{0, b \left(\frac{1}{8}\right)^{k-2}\right\}$. Using this, and applying Lemma 2 to each ϕ_i with the above choice of U and V, yields $\phi_i = \pi_k$ on I_1^k. The fact that $\phi_i = \pi_k$ on I_5^k follows by symmetry. $\qquad\square$

5   Proof of Theorem 4

Proof. Define $\pi_\infty : \mathbb{R} \to \mathbb{R}$ to be the pointwise limit of $\{\pi_i\}_{i=2}^\infty$. Since each π_k is minimal, by a standard limit argument, π_∞ is minimal (Proposition 4 in [8], Lemma 6.1 in [4]).

Using Proposition 1 (v), π_∞ is continuous over $(0, b)$ and $(b, 1)$. For $x = 0$ or $x = b$, note that, by definition of π_k, the maximum value of π_k on $I_1^k \cup I_2^k$ is $\frac{2^{4-3k}(2^k - 4b)}{1-b}$, which tends to 0 as $k \to \infty$. By symmetry, the smallest value of π_k on the interval $I_4^k \cup I_5^k$ tends to 1 as $k \to \infty$. Hence, the convergence $\pi_k \to \pi_\infty$ is actually uniform. Therefore π_∞ is continuous everywhere.

We next show that π_∞ is extreme. Suppose that $\pi_\infty = \frac{\phi_1 + \phi_2}{2}$ for valid functions ϕ_1, ϕ_2 and let $x \in [0, 1)$. Since π is minimal, by Lemma 1, ϕ_1, ϕ_2 are also minimal. If $x = 0$ or $x = b$, then $\pi_\infty(x) = \phi_1(x) = \phi_2(x)$ by the minimality of $\pi_\infty, \phi_1,$ and ϕ_2. So assume that $x \notin \{0, b\}$. By Proposition 1 (v), $x \in I_3^{N_x} \cup I_6^{N_x}$. Observe that $\pi_\infty = \pi_{N_x}$ on $I_3^{N_x} \cup I_6^{N_x}$. By applying Lemma 3 with $k = N_x$ and $\phi = \pi_\infty$, we obtain that $\phi_1(x) = \phi_2(x) = \pi_\infty(x)$.

We finally verify that π_∞ has infinitely many slopes. Note that for any $k \geq 3$, $\pi_\infty = \pi_k$ on $I_3^k \cup I_6^k$ and recall that π_k has $k - 1$ different slopes on $I_3^k \cup I_6^k$. $\qquad\square$

6   Concluding Remarks

One can ask if it is possible to create extreme functions with arbitrary number of slopes for the higher-dimensional infinite group relaxations. A trivial way to generalize to higher dimensions is to simply define $\pi_k^n : \mathbb{R}^n \to \mathbb{R}_+$ as $\pi_k^n(x_1, x_2, \ldots, x_n) = \pi_k(x_1)$ and $\pi_\infty^n : \mathbb{R}^n \to \mathbb{R}_+$ by defining $\pi_\infty^n(x_1, x_2, \ldots, x_n) = \pi_\infty(x_1)$. However, one can ask whether there are more "non-trivial" examples. In particular, one can ask whether there exist *genuinely n-dimensional* extreme functions with arbitrary number of slopes for all $n \geq 1$. A function $\theta : \mathbb{R}^n \to \mathbb{R}$ is genuinely n-dimensional if there does not exist a linear map $T : \mathbb{R}^n \to \mathbb{R}^{n-1}$ and a function $\theta' : \mathbb{R}^{n-1} \to \mathbb{R}$ such that $\theta = \theta' \circ T$. It turns out that the *sequential-merge* operation invented by Dey and Richard [7] can be used to create genuinely n-dimensional extreme functions with arbitrary number of slopes for any $n \geq 1$. Recall the Gomory mixed-integer function ξ from (3). Letting $\diamond$ denote the sequential-merge operation, it can be verified that the function

$$\pi_k \diamond \underbrace{(\xi \diamond (\xi \ldots (\xi \diamond \xi)))}_{n-1 \text{ times}}$$

is a genuinely n-dimensional extreme function with at least k different values for the gradient. The details are a little tedious, and needs a slight modification of a couple of proofs from Dey and Richard [7]. Hence, we omit these details from this extended abstract.

References

1. Basu, A., Conforti, M., Cornuéjols, G., Zambelli, G.: A counterexample to a conjecture of Gomory and Johnson. Math. Program. Ser. A **133**, 25–38 (2012)
2. Basu, A., Conforti, M., Di Summa, M.: A geometric approach to cut-generating functions. Math. Program. **151**(1), 153–189 (2015)
3. Basu, A., Hildebrand, R., Köppe, M.: Equivariant perturbation in Gomory and Johnson's infinite group problem. III. Foundations for the k-dimensional case and applications to $k = 2$. eprint: arXiv:1403.4628 [math.OC] (2014)
4. Basu, A., Hildebrand, R., Köppe, M.: Light on the infinite group problem. eprint: http://arxiv.org/abs/1410.8584 (2014)
5. Dey, S.S., Richard, J.P.P.: Gomory functions (2009). http://www2.isye.gatech.edu/sdey30/gomoryfunc2.pdf
6. Dey, S.S., Richard, J.P.P.: Facets of two-dimensional infinite group problems. Math. Oper. Res. **33**(1), 140–166 (2008). http://mor.journal.informs.org/cgi/content/abstract/33/1/140
7. Dey, S.S., Richard, J.P.P.: Relations between facets of low- and high-dimensional group problems. Math. Program. **123**(2), 285–313 (2010). http://dx.doi.org/10.1007/s10107-009-0303-8
8. Dey, S.S., Richard, J.P.P., Li, Y., Miller, L.A.: On the extreme inequalities of infinite group problems. Math. Program. **121**(1), 145–170 (2009). http://dx.doi.org/10.1007/s10107-008-0229-6
9. Gomory, R.E., Johnson, E.L.: Some continuous functions related to corner polyhedra, I. Math. Program. **3**, 23–85 (1972). http://dx.doi.org/10.1007/BF01585008
10. Gomory, R.E., Johnson, E.L.: Some continuous functions related to corner polyhedra, II. Math. Program. **3**, 359–389 (1972). http://dx.doi.org/10.1007/BF01585008
11. Gomory, R.E., Johnson, E.L.: T-space and cutting planes. Math. Program. **96**, 341–375 (2003). http://dx.doi.org/10.1007/s10107-003-0389-3
12. Hong, C., Köppe, M., Zhou, Y.: Sage program for computation and experimentation with the 1-dimensional Gomory-Johnson infinite group problem (2012). https://github.com/mkoeppe/infinite-group-relaxation-code
13. Köppe, M., Zhou, Y.: New computer-based search strategies for extreme functions of the Gomory-Johnson infinite group problem. eprint: arXiv:1506.00017 [math.OC] (2015)

Minimal Cut-Generating Functions
are Nearly Extreme

Amitabh Basu[1], Robert Hildebrand[2], and Marco Molinaro[3(✉)]

[1] Department of Applied Mathematics and Statistics,
The Johns Hopkins University, Baltimore, USA
basu.amitabh@jhu.edu
[2] IBM Research, Yorktown Heights, NY, USA
rhildeb@us.ibm.com
[3] Computer Science Department, PUC-RIO, Rio de Janeiro, Brazil
mmolinaro@inf.puc-rio.br

Abstract. We study continuous (strongly) minimal cut generating functions for the model where all variables are integer. We consider both the original Gomory-Johnson setting as well as a recent extension by Cornuéjols and Yıldız. We show that for any continuous minimal or strongly minimal cut generating function, there exists an extreme cut generating function that approximates the (strongly) minimal function as closely as desired. In other words, the extreme functions are "dense" in the set of continuous (strongly) minimal functions.

1 Introduction

Cut-generating functions are an important approach for deriving, understanding, and analyzing general-purpose cutting planes for mixed-integer programs. Given a natural number $n \in \mathbb{N}$ and a closed subset $S \subseteq \mathbb{R}^n \backslash \{0\}$, a *cut-generating function (CGF) for S* is a function $\pi \colon \mathbb{R}^n \to \mathbb{R}$ such that for every choice of natural number $k \in \mathbb{N}$ and k vectors $r_1, \ldots, r_k \in \mathbb{R}^n$, the inequality

$$\sum_{i=1}^{k} \pi(r_i) y_i \geq 1$$

is valid for the set

$$Q_0 = \left\{ y \in \mathbb{Z}_+^k : \sum_{i=1}^{k} r_i y_i \in S \right\}.$$

Note that CGFs for S only depend on n and S, and should work for all choices of $k \in \mathbb{N}$ and $r_1, \ldots, r_k \in \mathbb{R}^n$.

Amitabh Basu gratefully acknowledges partial support from NSF grant CMMI14 52820.

Most of this research was conducted while Robert Hildebrand was a postdoctoral researcher at the Institute for Operations Research, Department of Mathematics, ETH Zürich.

Q. Louveaux and M. Skutella (Eds.): IPCO 2016, LNCS 9682, pp. 202–213, 2016.
DOI: 10.1007/978-3-319-33461-5_17

Cut-generating functions were originally considered for sets S of the form $S = b + \mathbb{Z}^n$ for some $b \in \mathbb{R}^n \setminus \mathbb{Z}^n$ by Gomory and Johnson [8,9,12] under the name of the *infinite group relaxation*. Such sets S will be called *affine lattices*. In this case the connection with Integer Programming is clear: the set Q_0 is the projection of a mixed-integer set in tableaux form $Q_{\text{rel}} = \{(x,y) \in \mathbb{Z}^m \times \mathbb{Z}^k_+ : x = -b + \sum_i r_i y_i\}$ onto the non-basic variables, so CGFs give valid cuts for this set. Notice that Q_{rel} arises as a relaxation of a generic pure integer program in standard form by dropping the non-negativity of the basic variables x. So another important setting of CGFs is one that does not involve this relaxation, namely using sets S of the form $S = b + \mathbb{Z}^n_+$, where $b \in \mathbb{R}^n \setminus \mathbb{Z}^n$ and $b \le 0$ [7]; this now corresponds to the projection of the "unrelaxed" set $Q_{\text{full}} = \{(x,y) \in \mathbb{Z}^m_+ \times \mathbb{Z}^k_+ : x = -b + \sum_i r_i y_i\}$. We call such sets S *truncated affine lattices*.

Cut-generating functions have received significant attention in the literature (see the surveys [2–4] and the references therein). One important feature is that cut-generating functions capture known general purpose cuts, for example the prominent GMI cuts and more generally split cuts (when S is an affine lattice) and the *lopsided cuts* of Balas and Qualizza [1] (when S is a truncated affine lattice). Moreover, the CGF perspective gives a clean way of understanding cuts, since they abstract the finer structure of mixed integer sets and only depend on n and S (for affine and truncated affine lattices this is just the shift vector b).

Strength of Cut Generating Function and Extreme Functions. Since their introduction, there has been much interest in understanding what the "best" CGFs are – the ones that cut "most deeply". CGFs can be stratified and at the first level we have (strongly) minimal functions. An inequality $\alpha \cdot x \ge \alpha_0$ given by $\alpha \in \mathbb{R}^n$ and $\alpha_0 \in \mathbb{R}$ will be called valid for S if every $s \in S$ satisfies $\alpha \cdot s \ge \alpha_0$. A CGF π for S is called *strongly minimal* if there does not exist a *different* CGF π', a real number $\beta \ge 0$ and a valid inequality $\alpha \cdot x \ge \alpha_0$ for S, such that for all $r \in \mathbb{R}^n$, $\beta \pi'(r) + \alpha \cdot r \le \pi(r)$ and $\beta + \alpha_0 \ge 1$ [7].[1] This definition captures the standard idea of non-dominated inequalities. Gomory and Johnson characterized all continuous strongly minimal functions when S is an affine lattice.

Theorem 1 (Gomory and Johnson [8]). *Let $S = b + \mathbb{Z}^n$ for some $b \in \mathbb{R}^n \setminus \mathbb{Z}^n$. A continuous function $\pi \colon \mathbb{R}^n \to \mathbb{R}$ is a strongly minimal function if and only if all of the following hold:*

(i) π is a nonnegative function with $\pi(z) = 0$ for all $z \in \mathbb{Z}^n$,
(ii) π is subadditive, i.e., $\pi(r_1 + r_2) \le \pi(r_1) + \pi(r_2)$ for all $r_1, r_2 \in \mathbb{R}^n$, and
(iii) π satisfies $\pi(r) + \pi(b - r) = 1$ for all $r \in \mathbb{R}^n$. (This condition is known as the symmetry condition.)

Note that the first two conditions imply that π is periodic modulo $\mathbb{Z}^n$, i.e., $\pi(r) = \pi(r+z)$ for all $z \in \mathbb{Z}^n$. This characterization was also recently generalized by Cornuéjols and Yıldız to a wider class of sets S, which includes truncated affine lattices as well [7] (see Theorem 4 below).

---

[1] When S is an affine lattice this notion is equivalent to notion of *minimal inequality* used in the literature.

A stronger notion than strong minimality is that of an *extreme function*. We say that a CGF π is extreme if there do not exist distinct CGFs π_1 and π_2 such that $\pi = \frac{\pi_1 + \pi_2}{2}$. This is a subset of strongly minimal functions [7] that corresponds to a notion of "facets" in the context of CGFs (see also [3,4] for other notions of "facet" for CGFs). Because of the importance of facet-defining cuts in Integer Programming, there has been substantial interest in obtaining and understanding extreme functions (see [3,4] for a survey). For example, a celebrated result is Gomory and Johnson's 2-Slope Theorem (Theorem 5 below) that gives a *sufficient condition* for a CGF to be extreme (in the affine lattice setting with $n = 1$; see [5–7] for generalizations).

Unfortunately the structure of extreme functions seems much more complicated than that of minimal functions. For example, even verifying the extremality of a function is not completely understood (see [3,4] for preliminary steps in this direction); a simple characterization like Theorem 1 seems all the more unlikely.

Our Results. As noted above, it is easy to verify that extreme functions are always strongly minimal. We prove an approximate converse: in a strict mathematical sense, strongly minimal functions for $n = 1$ are "close" to being extreme functions. More precisely, in the affine lattice setting we prove the following.

Theorem 2. *Let $S = b + \mathbb{Z}$ for some $b \in \mathbb{Q}\backslash\mathbb{Z}$. Let $\bar{\pi}$ be a continuous strongly minimal function for S. Then for every $\epsilon > 0$ there is an extreme (2-slope) function π^* such that $\|\bar{\pi} - \pi^*\|_\infty \leq \epsilon$, where $\|\cdot\|_\infty$ is the sup norm.*

Equivalently, this states that extreme functions are dense, under the sup norm, in the set of strongly minimal functions. This surprising property of CGFs relies on their infinite-dimensional nature: for finite-dimensional polyhedra, a (non-facet) minimal inequality can never be *arbitrarily* close (under any reasonable distance) to a facet.

In the truncated affine lattice setting we prove a similar result under an additional assumption. A function $\phi \colon \mathbb{R} \to \mathbb{R}$ is *quasi-periodic (with period d)*, if there are real numbers $d > 0$ and $c \in \mathbb{R}$ such that $\phi(r + d) = \phi(r) + c$ for every $r \in \mathbb{R}$. All explicitly known CGFs from the literature are quasi-periodic. Moreover, quasi-periodic piecewise linear CGFs can be expressed using a finite number of parameters, making them attractive from a computational perspective.

Theorem 3. *Let $S = b + \mathbb{Z}_+$, where $b \in \mathbb{Q}\backslash\mathbb{Z}$ and $b \leq 0$. Let $\bar{\pi}$ be a continuous, strongly minimal function for S that is quasi-periodic with rational period. Then for any $M \in \mathbb{R}_+$ and $\epsilon > 0$ there is an extreme (2-slope) function π^* such that $|\bar{\pi}(r) - \pi^*(r)| \leq \epsilon$ for all $r \in [-M, M]$.*

Our results imply that for the purpose of cutting-planes, these strongly minimal functions perform essentially as well as extreme functions, at least for the $n = 1$ case, i.e., cutting planes from a single row. This points out the limitations of extremality as a measure for the quality of one-dimensional CGFs and suggests the need for alternative measures (see [10]). *However, we have not been able to establish such results for $n \geq 2$. The question remains open whether extremality is a more useful concept in higher dimensions, making it relevant for multi-row cuts.*

2   Preliminaries

2.1   Strongly Minimal Functions for Truncated Affine Lattices

The celebrated characterization of strongly minimal inequalities by Gomory and Johnson was recently extended by Cornuéjols and Yıldız [7] to truncated affine lattices (actually their result is more general, we only state a special case here).

Theorem 4 (Cornuéjols and Yıldız [7]). *Let $S = b + \mathbb{Z}_+^n$, where $b \in \mathbb{R}^n \setminus \mathbb{Z}^n$ and $b \leq 0$. A continuous function $\pi \colon \mathbb{R}^n \to \mathbb{R}$ is a strongly minimal function if and only if all of the following hold:*

(i) $\pi(0) = 0$, and $\pi(-e_i) = 0$ for all unit vectors e_i, $i = 1, \ldots, n$,
(ii) π is subadditive, i.e., $\pi(r_1 + r_2) \leq \pi(r_1) + \pi(r_2)$, and
(iii) π satisfies the symmetry condition, i.e., $\pi(r) + \pi(b - r) = 1$ for all $r \in \mathbb{R}^n$.

2.2   Approximations Using Piecewise Linear Functions

We say a function $\phi \colon \mathbb{R} \to \mathbb{R}$ is *piecewise linear* if there is a set of closed non-degenerate intervals I_j, $j \in J$ such that $\mathbb{R} = \bigcup_{j \in J} I_j$, any bounded subset of $\mathbb{R}$ intersects only finitely many intervals, and ϕ is affine over each interval I_j. The endpoints of the intervals I_j are called the *breakpoints* of ϕ. Note that in this definition, a piecewise linear function is continuous.

The next lemma shows that continuous strongly minimal functions can be approximated by piecewise linear functions that are also strongly minimal; this can be accomplished by restricting the function to a subgroup and performing a linear interpolation (a proof is presented in the full version of the paper). Throughout we use the following notation: Given a subset $X \subseteq \mathbb{R}$ and a function $\phi \colon \mathbb{R} \to \mathbb{R}$, we denote the restriction of ϕ to X by $\phi|_X$.

Lemma 1. *Let S be an affine lattice $b + \mathbb{Z}$ or a truncated affine lattice $b + \mathbb{Z}_+$ (for $n = 1$) with $b \in \mathbb{Q}$. Let π be uniformly continuous, strongly minimal function for S. Then for every $\epsilon > 0$ there is a continuous strongly minimal function π_{pwl} for S that is piecewise linear and satisfies $\|\pi - \pi_{\mathrm{pwl}}\|_\infty \leq \epsilon$.*

2.3   Subadditivity and Additivity

We introduce some tools for studying subadditive functions. For any function $\pi \colon \mathbb{R} \to \mathbb{R}$, define a slack function $\Delta\pi \colon \mathbb{R} \times \mathbb{R} \to \mathbb{R}$ as

$$\Delta\pi(x, y) = \pi(x) + \pi(y) - \pi(x + y). \tag{1}$$

Clearly π is subadditive if and only if $\Delta\pi \geq 0$. We will employ another concept in our analysis which we call the *additivity domain*:

$$E(\pi) = \{(x, y) : \Delta\pi(x, y) = 0\}.$$

When $\pi\colon \mathbb{R} \to \mathbb{R}$ is a piecewise linear function periodic modulo $\mathbb{Z}$ with an infinite set of breakpoints $U = \{\ldots, u_{-1}, u_0, u_1, \ldots\}$, by periodicity, we may assume that $U = \{u_0, u_1, \ldots, u_m\} + \mathbb{Z}$ where $u_0 = 0$, $u_m < 1$ and $u_i < u_{i+1}$. The function $\Delta\pi$ is affine on every set $F = \{(x, y) : u_i \leq x \leq u_{i+1}, u_j \leq y \leq u_{j+1}, u_k \leq x + y \leq u_{k+1}\}$ where (u_i, u_{i+1}), (u_j, u_{j+1}), and (u_k, u_{k+1}) are pairs of consecutive breakpoints. The set of all such F forms a polyhedral complex and will be denoted by $\Delta\mathcal{P}_U$. The vertices of any such F will be denoted by $\mathrm{vert}(F)$. If $\mathcal{F}$ is any collection of polyhedra from $\Delta\mathcal{P}_U$, then we define $\mathrm{vert}(\mathcal{F}) := \cup_{F \in \mathcal{F}} \mathrm{vert}(F)$. Note that $\mathrm{vert}(\Delta\mathcal{P}_U)$ is exactly the set of points $(x, y) \in \mathbb{R}^2$ such that either $x, y \in U$ or $x, x + y \in U$ or $y, x + y \in U$. The affine structure of $\Delta\pi$ implies the following (for example, see Fig. 1).

Lemma 2. *Let $\pi\colon \mathbb{R} \to \mathbb{R}$ be a piecewise linear function periodic modulo $\mathbb{Z}$ with breakpoints in U. Let $\mathcal{F}$ be a collection of polyhedra from $\Delta\mathcal{P}_U$. If $\Delta\pi(x, y) \geq \gamma$ for all $(x, y) \in \mathrm{vert}(\mathcal{F})$ for some $\gamma > 0$, then $\Delta\pi(x, y) \geq \gamma$ for all $(x, y) \in \cup_{F \in \mathcal{F}} F$. In particular, $\Delta\pi(x, y) \geq 0$ for all $(x, y) \in \mathrm{vert}(\Delta\mathcal{P}_U)$ if and only if π is subadditive.*

2.4   2-Slope Theorems

Piecewise linear functions where the slope takes exactly two values are referred to as 2-slope functions. We will use the following two general theorems on extreme functions to show certain 2-slope functions are extreme.

Theorem 5 (Gomory and Johnson [8]). *Let $S = b + \mathbb{Z}$ be an affine lattice and let π be a strongly minimal cut generating function for S. If π is piecewise linear and has exactly two slopes, then it is extreme.*

Recently, this theorem was extended to the case when $S = b + \mathbb{Z}_+$ using a similar proof as Gomory and Johnson used.

Theorem 6 (Cornuéjols and Yıldız [7]). *Let $S = b + \mathbb{Z}_+$ be a truncated affine lattice and let $\pi\colon \mathbb{R} \to \mathbb{R}$ be a strongly minimal cut generating function for S. If π is such that $\pi(r) \geq 0$ for all $r \geq 0$ and the restriction of π to any compact interval is piecewise linear function with exactly two slopes, then π is extreme.*

2.5   2-Slope Fill-in

Gomory and Johnson [8,9,12] described a procedure called the *2-slope fill-in* that allows us to extend subadditive functions from a subgroup of $\mathbb{R}$ to the whole of $\mathbb{R}$. Let U be a subgroup of $\mathbb{R}$. Let $g\colon \mathbb{R} \to \mathbb{R}$ be a sublinear function, *i.e.*, g is subadditive and $g(\lambda r) = \lambda g(r)$ for all $\lambda \geq 0$ and $r \in \mathbb{R}$. The two-slope fill-in of any function $\phi\colon \mathbb{R} \to \mathbb{R}$ with respect to U and g is defined as

$$\phi_{\text{fill-in}}(r) = \min_{u \in U}\{\phi(u) + g(r - u)\}.$$

Lemma 3 (Johnson (Sect. 7 in [12])). *Let U be any subgroup of $\mathbb{R}$ and let $\phi\colon \mathbb{R} \to \mathbb{R}$ be a function such that $\phi|_U$ is subadditive, i.e., $\phi(u_1 + u_2) \leq \phi(u_1) + \phi(u_2)$ for all $u_1, u_2 \in U$. Suppose g is a sublinear function such that $\phi \leq g$. Then the 2-slope fill-in $\phi_{\text{fill-in}}$ of ϕ with respect to U and g is subadditive. Moreover, $\phi_{\text{fill-in}} \geq \phi$ and $\phi_{\text{fill-in}}|_U = \phi|_U$.*

3   Proof of Theorem 2

The high-level idea is to apply the 2-slope fill-in procedure to the input function $\bar{\pi}$ and then symmetrize it to produce a 2-slope function π^* that satisfies conditions (i), (ii) and (iii) in Theorem 1, and hence is strongly minimal. Then employing Theorem 5 we have that π^* is an extreme function. Moreover, we perform the 2-slope fill-in in a way that $\|\bar{\pi} - \pi^*\|_\infty \leq \epsilon$, thus giving the desired result.

The main difficulty in pursuing this line of argument is that the symmetrization step needed after the 2-slope fill-in can easily destroy the desired subadditivity. Therefore, before applying the 2-slope fill-in plus symmetrization we perturb the original function $\bar{\pi}$ to ensure that in most places we have the strict inequality $\pi(x+y) < \pi(x) + \pi(y)$ (and with enough room).

We start describing this perturbation procedure. For the remainder of this section, we focus on the case where $S = b + \mathbb{Z}$. Also, using periodicity with respect to $\mathbb{Z}$, any function π is strongly minimal for $S = b + \mathbb{Z}$ if and only if it is strongly minimal for $S = \bar{b} + \mathbb{Z}$, where $\bar{b} = b \pmod 1$. Hence, without loss of generality, we assume $b \in (0, 1)$ throughout this section.

3.1   Equality Reducing Perturbation

The perturbation we consider produces a function with equalities (modulo $\mathbb{Z}^2$) only on the border of the unit square and on the symmetry lines $x + y = b$ and $x + y = 1 + b$. Recall that strongly minimal functions for $S = b + \mathbb{Z}$ are periodic modulo $\mathbb{Z}$ and satisfy the symmetric condition, $i.e.$, $\Delta\pi(x, y) = 0$ whenever $x + y \in b + \mathbb{Z}$. Moreover, only the lines $x + y = b + z$ for $z = 0, 1$ intersect the cube $[0, 1]^2$ since $b \in (0, 1)$. Define the sets $E_\delta = \{(x, y) : x \in [0, \delta] \cup [1 - \delta, 1]\} \cup \{(x, y) : y \in [0, \delta] \cup [1 - \delta, 1]\}$ for $\delta > 0$, $E_b = \{(x, y) \in [0, 1]^2 : b - \delta \leq x + y \leq b + \delta\}$, and $E_{1+b} = \{(x, y) \in [0, 1]^2 : (1 + b) - \delta \leq x + y \leq (1 + b) + \delta\}$. The main result of this section is the following.

Lemma 4. *Consider a piecewise linear function π that is strongly minimal for $b + \mathbb{Z}$. Then for any $\epsilon \in (0, 1)$, there is a real number $\delta > 0$ and a function π_{comb} satisfying the following:*

(1) π_{comb} is strongly minimal for $b + \mathbb{Z}$.
(2) π_{comb} is piecewise linear whose breakpoints include $\delta + \mathbb{Z}$ and $-\delta + \mathbb{Z}$. Further, π_{comb} is linear on $[0, \delta]$ and $[-\delta, 0]$.
(3) $\|\pi - \pi_{\mathrm{comb}}\|_\infty \leq \epsilon$.
(4) $E(\pi_{\mathrm{comb}}) \subseteq E_\delta \cup E_b \cup E_{1+b}$.
(5) There exists $\gamma > 0$ such that $\Delta\pi_{\mathrm{comb}}(x, y) > \gamma$ for all $(x, y) \in [0, 1]^2 \setminus (E_\delta \cup E_b \cup E_{1+b})$.

The idea behind the proof of this lemma is the observation that if we have a convex combination $\pi = \alpha\pi^1 + (1 - \alpha)\pi^2$ with $\alpha \in (0, 1)$, then $E(\pi) \subseteq E(\pi^1) \cap E(\pi^2)$. Thus, we will find a function $\hat{\pi}$ with nice equalities $E(\hat{\pi})$ and then set π_{comb} as roughly $(1 - \epsilon)\pi + \epsilon\hat{\pi}$. The nice function we use is defined for any $\delta \in (0, \min\{\frac{b}{2}, \frac{1-b}{2}\})$ as follows (see Fig. 1 for an example):

$$
\pi_\delta(r) = \begin{cases}
\frac{1}{2\delta} r & r \in [0, \delta] + \mathbb{Z}, \\
\frac{1}{2} & r \in (\delta, b - \delta] + \mathbb{Z}, \\
1 - \frac{1}{2\delta}(b - r) & r \in (b - \delta, b] + \mathbb{Z}, \\
1 + \frac{1}{2\delta}(b - r) & r \in (b, b + \delta] + \mathbb{Z}, \\
\frac{1}{2} & r \in (b + \delta, 1 - \delta] + \mathbb{Z}, \\
\frac{1}{2} + \frac{1}{2\delta}(1 - \delta - r) & r \in (1 - \delta, 1] + \mathbb{Z}.
\end{cases} \tag{2}
$$

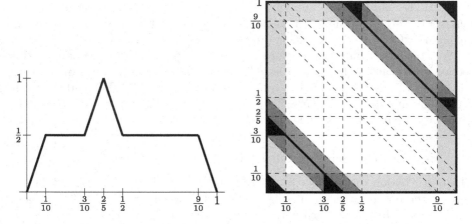

Fig. 1. On the left is a plot of $\pi_\delta \colon \mathbb{R} \to [0, 1]$ for $\delta = \frac{1}{10}$ and $b = \frac{2}{5}$. This function is periodic modulo $\mathbb{Z}$, so we only display the domain $[0, 1]$. On the right we have drawn the complex $\Delta \mathcal{P}_U$ in dashed lines on the $[0, 1]^2$ domain. The function $\Delta \pi_\delta$ is affine on each cell of $\Delta \mathcal{P}_U$. The cells of $\Delta \mathcal{P}_U$ filled in black are those contained in the set $E(\pi_\delta)$. Since π_δ is periodic, $\Delta \pi_\delta$ (and hence $E(\pi_\delta)$) is periodic modulo $\mathbb{Z}^2$. Covering the set $E(\pi_\delta)$ in a light shading are the sets E_δ, E_b, and E_{1+b}. The set E_δ around the boundary of the box is shaded in light blue, while the diagonal strips E_b and E_{1+b} are shaded in light red. Notice that $E(\pi_\delta) \subseteq E_\delta \cup E_b \cup E_{1+b}$. In fact, the remaining region $[0, 1]^2 \backslash (E_\delta \cup E_b \cup E_{1+b})$, colored white, does not intersect $E(\pi_\delta)$. Hence $\Delta \pi_\delta > \gamma > 0$ on this remaining region.

The following lemma states the key properties of π_δ; its proof, presented in the full version of the paper, uses the characterization of strong minimality from Theorem 1 and requires a case analysis based on the breakpoints of π_δ.

Lemma 5. *For all $\delta \in (0, \min\{\frac{b}{2}, \frac{1-b}{2}\})$, the function π_δ is strongly minimal for $b + \mathbb{Z}$. Furthermore, we have $E(\pi_\delta) \subseteq E_\delta \cup E_b \cup E_{1+b}$ and there exists $\gamma > 0$ such that $\Delta \pi_\delta(x, y) > \gamma$ for all $(x, y) \in [0, 1]^2 \backslash (E_\delta \cup E_b \cup E_{1+b})$.*

Proof of Lemma 4. Consider the breakpoints of π in the open interval $(0, 1)$, let $u_{\min}$ and $u_{\max}$ be respectively the smallest and the largest of these breakpoints. Choose $\delta > 0$ sufficiently small – more precisely, $\delta < \min\{u_{\min}, 1 - u_{\max}, \frac{b}{2}, \frac{1-b}{2}\}$.

By Lemma 5, π_δ is strongly minimal, and π is strongly minimal by assumption. Since the conditions (i), (ii) and (iii) in Theorem 1 are maintained under taking convex combinations, the function $\pi_{\text{comb}} = (1-\epsilon)\pi + \epsilon\pi_\delta$ is also strongly minimal. Thus, condition (1) is satisfied. By the choice of δ and the fact that $\delta + \mathbb{Z}$ and $-\delta + \mathbb{Z}$ are included in the breakpoints of π_δ, condition (2) is also satisfied. Moreover,

$$|\pi_{\text{comb}}(x) - \pi(x)| = |(1-\epsilon)\pi(x) + \epsilon\pi_\delta(x) - \pi(x)| = |-\epsilon\pi(x) + \epsilon\pi_\delta(x)| \le \epsilon,$$

where the last inequality follows from the fact that $0 \le \pi(x), \pi_\delta(x) \le 1$ for all x, since both functions are strongly minimal. Thus, condition (3) is satisfied. Finally, by Lemma 5, there exists a $\hat{\gamma} > 0$ such that $\Delta\pi_\delta(x,y) > \hat{\gamma}$ for all $(x,y) \in [0,1]^2 \setminus (E_\delta \cup E_b \cup E_{1+b})$. Since $\Delta\pi \ge 0$, it follows that $\Delta\pi_{\text{comb}}(x,y) = (1-\epsilon)\Delta\pi(x,y) + \epsilon\Delta\pi_\delta(x,y) \ge \epsilon\Delta\pi_\delta(x,y) > \hat{\gamma}\epsilon$ for all $(x,y) \in [0,1]^2 \setminus (E_\delta \cup E_b \cup E_{1+b})$. Taking $\gamma = \hat{\gamma}\epsilon$ completes the proof of conditions (4) and (5). $\square$

3.2 Symmetric 2-Slope Fill-in

We now show that we can apply the 2-slope fill-in plus a symmetrization procedure to the function π_{comb} to transform it into a strongly minimal 2-slope function (and hence extreme) while only making small changes to the function values.

Lemma 6. *Let $\epsilon > 0$ and let π_{comb} be any function that satisfies the output conditions of Lemma 4 (for some $\delta, \gamma > 0$) whose breakpoints are rational. There exists a strongly minimal 2-slope piecewise linear function π_{sym} such that $\|\pi_{\text{comb}} - \pi_{\text{sym}}\|_\infty \le \epsilon$.*

Proof. By periodicity, we focus on the [0,1] interval. Without loss of generality, we assume that $\epsilon < \frac{\gamma}{3}$ where γ is given in Lemma 4(5). Let s_+ and s_- be two slopes of the piecewise linear function π_{comb} coming from the origin, i.e., let $s_+ = \lim_{h \to 0^+} \frac{\pi_{\text{comb}}(h)}{h}$, $s_- = \lim_{h \to 0^-} \frac{\pi_{\text{comb}}(h)}{h}$. Since π_{comb} is nonnegative, it follows that $s_+ \ge 0$, $s_- \le 0$. The function $g(r) := \max(s_+ \cdot r, s_- \cdot r)$, is easily seen to be sublinear, and subadditivity of π_{comb} implies $\pi_{\text{comb}} \le g$.

Let $q \in \mathbb{Z}^+$ such that $\frac{1}{q}\mathbb{Z}$ such that the breakpoints U of π_{comb} and $\frac{b}{2}$ are contained in $\frac{1}{q}\mathbb{Z}$ and such that $\frac{1}{q}\max\{s_+, |s_-|\} < \frac{\epsilon}{2}$. Since $\pi_{\text{comb}} \le g$, by Lemma 3, the fill-in function $\pi_{\text{fill-in}}$ of π_{comb}, with respect to $\frac{1}{q}\mathbb{Z}$ and g, is subadditive. Unfortunately, $\pi_{\text{fill-in}}$ does not necessarily satisfy the symmetry condition and, therefore, is not necessarily a strongly minimal function. Hence, we further define

$$\pi_{\text{sym}}(r) = \begin{cases} \pi_{\text{fill-in}}(r) & r \in [0, \frac{b}{2}] \cup [\frac{1+b}{2}, 1], \\ 1 - \pi_{\text{fill-in}}(b - r) & r \in [\frac{b}{2}, \frac{1+b}{2}] \end{cases}$$

In the definition of π_{sym}, we have enforced the symmetry condition, possibly sacrificing the subadditivity of the function. We will show that, given the parameters used in the construction, π_{sym} is strongly minimal and actually approximates π_{comb} to the desired precision.

By Lemma 3, $\pi_{\text{fill-in}} \ge \pi_{\text{comb}}$ and $\pi_{\text{comb}}(u) = \pi_{\text{fill-in}}(u)$ for all $u \in \frac{1}{q}\mathbb{Z}$. Since π_{comb} is period modulo $\mathbb{Z}$, the function $\pi_{\text{fill-in}}$ inherits this property. Moreover, restricted to $[0, 1]$, $\pi_{\text{fill-in}}$ is the pointwise minimum of a finite collection of piecewise linear functions and therefore, $\pi_{\text{fill-in}}$ is also piecewise linear. Furthermore, the maximum slope in absolute value of $\pi_{\text{fill-in}}$ is $s = \max\{s_+, |s_-|\}$. Therefore s is also a bound on the maximum slope in absolute value for π_{comb}. Hence,

$$|\pi_{\text{fill-in}}(r) - \pi_{\text{comb}}(r)| \leq |\pi_{\text{fill-in}}(u) - \pi_{\text{comb}}(u)| + 2s|u - r| \leq \epsilon,$$

where $u \in \frac{1}{q}\mathbb{Z}$ is the closest point in $\frac{1}{q}\mathbb{Z}$ to r. Thus, we have established that $\|\pi_{\text{fill-in}} - \pi_{\text{comb}}\|_\infty \leq \epsilon$. Observe that $|\pi_{\text{sym}}(r) - \pi_{\text{comb}}(r)| = |\pi_{\text{fill-in}}(r) - \pi_{\text{comb}}(r)|$ for all $r \in [0, \frac{b}{2}] \cup [\frac{1+b}{2}, 1]$, and $|\pi_{\text{sym}}(r) - \pi_{\text{comb}}(r)| = |\pi_{\text{fill-in}}(b-r) - \pi_{\text{comb}}(b-r)|$ for all $r \in [\frac{b}{2}, \frac{1+b}{2}]$ because of the symmetry of π_{comb}. Therefore we also have that $\|\pi_{\text{sym}} - \pi_{\text{comb}}\|_\infty \leq \epsilon$.

Next, observe that π_{sym} has the same slopes as $\pi_{\text{fill-in}}$, and therefore is a 2-slope piecewise linear function.

Finally, we establish that π_{sym} is a strongly minimal function. Since it is clear that $\pi_{\text{sym}}(0) = 0$, and π_{sym} satisfies the symmetry condition, by Theorem 1, we only need to show that π_{sym} satisfies the subadditivity condition $\pi_{\text{sym}}(x + y) \leq \pi_{\text{sym}}(x) + \pi_{\text{sym}}(y)$; equivalently, $\Delta\pi_{\text{sym}}(x, y) \geq 0$. This is established by the following case analysis for each $(x, y) \in [0, 1]^2$.

Case 1. Suppose $(x, y) \in [0, 1]^2 \setminus (E_\delta \cup E_b \cup E_{1+b})$.

By Lemma 4(5), $\Delta\pi_{\text{comb}}(x, y) > \gamma > 0$. Since $\|\pi_{\text{sym}} - \pi_{\text{comb}}\|_\infty \leq \epsilon$, it follows that

$$\Delta\pi_{\text{sym}}(x, y) \geq \Delta\pi_{\text{comb}}(x, y) - 3\|\pi_{\text{sym}} - \pi_{\text{comb}}\|_\infty \geq 0.$$

Case 2. Suppose $(x, y) \in E_\delta$.

By Lemma 4(2), the slope of π_{comb} on the interval $[0, \delta]$ is s_+, while the slope on the interval $[1-\delta, 1]$ is s_-. We claim that $\pi_{\text{comb}} = \pi_{\text{fill-in}}$ on the intervals $[0, \delta]$ and $[1-\delta, \delta]$. To see this, consider any two consecutive points $u_1, u_2 \in [0, \delta] \cap \frac{1}{q}\mathbb{Z}$. For any $r \in [u_1, u_2]$, we have

$$\pi_{\text{comb}}(r) = \pi_{\text{comb}}(u_1) + s_+(r - u_1) = \pi_{\text{comb}}(u_1) + g(r - u_1)$$
$$\geq \min_{u \in U} \pi_{\text{comb}}(u) + g(r - u) = \pi_{\text{fill-in}}(r) \geq \pi_{\text{comb}}(r).$$

The first equality comes from the second part of Lemma 4(2). The last inequality comes from Lemma 3. Since this holds for any points $u_1, u_2 \in \frac{1}{q}\mathbb{Z} \cap [0, \delta]$ and $0, \delta \in \frac{1}{q}\mathbb{Z}$ by Lemma 4(2) and the assumption that all breakpoints of π_{comb} lie in $\frac{1}{q}\mathbb{Z}$, the claim holds on the interval $[0, \delta]$. A similar argument verifies the claim on the interval $[1 - \delta, 1]$ by showing that $\pi_{\text{fill-in}}$ takes slope s_- on this interval.

Therefore, for $x \in [0, \delta]$ we have

$$\pi_{\text{sym}}(x) = s_+ \cdot x \geq s_+ \cdot \alpha_1 + s_- \cdot \alpha_2 = \pi(x + y) - \pi(y),$$

where α_1 and α_2 are the lengths of the subsets of the interval $[y, x + y]$ taking slopes s_+ and s_- respectively. The inequality holds since $\alpha_1 + \alpha_2 = x$ and $s_+ \geq s_-$.

On the other hand, if $x \in [1 - \delta, 1]$, then

$$-\pi_{\text{sym}}(x) = -\pi_{\text{sym}}(x - 1) = s_- \cdot (1 - x)$$
$$\leq s_+ \cdot \alpha_1 + s_- \cdot \alpha_2 = \pi_{\text{sym}}(y) - \pi_{\text{sym}}(x + y - 1) = \pi_{\text{sym}}(y) - \pi_{\text{sym}}(x + y),$$

where α_1, α_2 are the lengths of the subsets of the interval $[x + y - 1, y]$ taking slopes s_+ and s_- respectively. The inequality holds since $\alpha_1 + \alpha_2 = 1 - x$ and $s_+ \geq s_-$. Here we used the fact that π_{sym} is periodic modulo $\mathbb{Z}$.

Case 3. Suppose $(x, y) \in E_b \cup E_{1+b}$.

Suppose first that $(x, y) \in E_b$. Then $x + y = b - \beta$ for some $\beta \in [-\delta, \delta]$. By Case 2, it follows that $\pi_{\text{sym}}(\beta) + \pi_{\text{sym}}(x) \geq \pi_{\text{sym}}(x + \beta)$. Therefore, $-(\pi_{\text{sym}}(\beta) + \pi_{\text{sym}}(x)) \leq -\pi_{\text{sym}}(x + \beta)$. Since π_{sym} satisfies the symmetry condition, we have

$$\pi_{\text{sym}}(b - \beta) = 1 - \pi_{\text{sym}}(\beta) = 1 - \pi_{\text{sym}}(\beta) - \pi_{\text{sym}}(x) + \pi_{\text{sym}}(x)$$
$$\leq 1 - \pi_{\text{sym}}(x + \beta) + \pi_{\text{sym}}(x) = \pi_{\text{sym}}(b - x - \beta) + \pi_{\text{sym}}(x)$$
$$= \pi_{\text{sym}}(x) + \pi_{\text{sym}}(y).$$

The proof is similar for $(x, y) \in E_{1+b}$.

Since Cases 1–3 cover all options for $(x, y) \in [0, 1]^2$, we see that π_{sym} is indeed subadditive. This concludes the proof.   □

3.3 Concluding the Proof of Theorem 2

Consider a strongly minimal function $\bar{\pi}$ for $S = b + \mathbb{Z}$. Since $\bar{\pi}$ is continuous and periodic with period 1, it is actually uniformly continuous. Thus, we can apply Lemma 1 to obtain a piecewise linear function π_{pwl} that is strongly minimal for S and satisfies $\|\bar{\pi} - \pi_{\text{pwl}}\|_\infty \leq \frac{\epsilon}{3}$. Then we employ the equality reduction Lemma 4 over π_{pwl} to obtain a strongly minimal function π_{comb} with $\|\pi_{\text{pwl}} - \pi_{\text{comb}}\|_\infty \leq \frac{\epsilon}{3}$. Then we can apply Lemma 6 to π_{comb} to obtain a function π_{sym} with $\|\pi_{\text{comb}} - \pi_{\text{sym}}\|_\infty \leq \frac{\epsilon}{3}$ satisfying the other properties given by the lemma. Then the 2-Slope Theorem 5 implies that π_{sym} is extreme, and triangle inequality gives $\|\bar{\pi} - \pi_{\text{sym}}\|_\infty \leq \|\bar{\pi} - \pi_{\text{pwl}}\|_\infty + \|\pi_{\text{pwl}} - \pi_{\text{comb}}\|_\infty + \|\pi_{\text{comb}} - \pi_{\text{sym}}\|_\infty \leq \epsilon$. This concludes the proof.

4 Proof of Theorem 3

The high-level idea of the proof of Theorem 3 is to take the input function $\bar{\pi}$, which is strongly minimal and quasi-periodic, and remove a linear term from it and scale the domain to obtain a function $\tilde{\pi}$ that is periodic modulo $\mathbb{Z}$ (and in fact strongly minimal). Then we can apply Theorem 2 to this transformed function $\tilde{\pi}$ to obtain an extreme function $\tilde{\pi}_{\text{sym}}$ close to it and then undo the transformation over $\tilde{\pi}_{\text{sym}}$ to obtain an extreme function π^*. The only caveat is that in this last step simply undoing the function transformation does not give us an extreme function: an extra correction step needs to take place to correct the fact that $\tilde{\pi}_{\text{sym}}$ is a (slight) modification of $\tilde{\pi}$.

One can transform a quasi-periodic function into a periodic one by removing a linear term (the proof can be readily verified).

Lemma 7. *Let ϕ be quasi-periodic with period d, and let $c \in \mathbb{R}$ be such that $\phi(x + d) = \phi(x) + c$. Then the function $\hat{\phi}(x) := \phi(x) - \frac{c}{d}x$ is periodic with period d.*

We also need the following lemma which follows from [11, Theorem 7.5.1].

Lemma 8. *Let $\phi\colon \mathbb{R} \to \mathbb{R}$ be a continuous, subadditive, periodic function and $\phi(0) = 0$. Then $\phi \geq 0$.*

We proceed with proving Theorem 3. Consider a truncated affine lattice $S = b + \mathbb{Z}_+$ with $b \in \mathbb{Q}\backslash\mathbb{Z}$ and $b \leq 0$. Consider a continuous, strongly minimal, quasi-periodic function $\bar{\pi}$ for S with rational period d and let $c \in \mathbb{R}$ be such that $\bar{\pi}(x + d) = \bar{\pi}(x) + c$. The rationality of d, combined with [7, Theorem 7], implies that $c \geq 0$. Thus, by Lemma 7, there exists $\alpha \geq 0$ such that the function $\bar{\pi}(r) - \alpha \cdot r$ is periodic with period d. Define the transformed function $\tilde{\pi}\colon \mathbb{R} \to \mathbb{R}$ by removing a linear term and scaling the domain as

$$\tilde{\pi}(r) = \frac{1}{1 - \alpha b} \cdot (\bar{\pi}(d \cdot r) - \alpha(d \cdot r)).$$

Observe that $1 - \alpha b \geq 0$ since $\alpha \geq 0$ and $b \leq 0$. Not only is $\tilde{\pi}$ periodic with period 1, it is in fact strongly minimal for an appropriately transformed set $\tilde{S}$ (a proof is presented in the full version of the paper).

Lemma 9. *The function $\tilde{\pi}$ is strongly minimal for $\tilde{S} = \frac{b}{d} + \mathbb{Z}$.*

Recall the parameters M and $\epsilon > 0$ in the statement of Theorem 3. Now set $\epsilon' > 0$ small enough so that $1 + \epsilon'b \geq \frac{1}{2}$ and

$$\max\left\{ \left(\frac{1}{1 + \epsilon'b} - 1\right), \left(1 - \frac{1}{1 - \epsilon'b}\right) \right\} \cdot \max_{y \in [-M,M]} |\bar{\pi}(y)| + \epsilon'(M + 2) \leq \epsilon.$$

Apply Theorem 2, with approximation parameter $\frac{\epsilon'}{1 - \alpha b}$, to obtain a 2-slope function $\tilde{\pi}_{\text{sym}}$ that is strongly minimal for $\tilde{S} = \frac{b}{d} + \mathbb{Z}$ and has $\|\tilde{\pi} - \tilde{\pi}_{\text{sym}}\|_\infty \leq \frac{\epsilon'}{1 - \alpha b}$.

We undo the transformation over $\tilde{\pi}_{\text{sym}}$ by rescaling the domain and function values, and adding back the linear term to define $\pi'\colon \mathbb{R} \to \mathbb{R}$ by setting

$$\pi'(r) = (1 - \alpha b) \cdot \tilde{\pi}_{\text{sym}}\left(\frac{r}{d}\right) + \alpha r.$$

Again notice that π' satisfies quasi-periodicity (with period d), subadditivity, and $\pi'(0) = 0$. Also, since $\tilde{\pi}_{\text{sym}}$ is symmetric about $\frac{b}{d}$, we obtain that π' is symmetric about b:

$$\pi'(r) + \pi'(b - r) = (1 - \alpha b) \cdot \left(\tilde{\pi}_{\text{sym}}\left(\frac{r}{d}\right) + \tilde{\pi}_{\text{sym}}\left(\frac{b - r}{d}\right)\right) + \alpha r + \alpha(b - r) = 1.$$

In addition, $\|\bar{\pi} - \pi'\|_\infty \le \epsilon'$:

$$
\begin{aligned}
|\pi'(r) - \bar{\pi}(r)| &= \left|(1 - \alpha b) \cdot \tilde{\pi}_{\mathrm{sym}}\left(\frac{r}{d}\right) + \alpha r - \bar{\pi}(r)\right| \\
&= \left|(1 - \alpha b)\left(\tilde{\pi}_{\mathrm{sym}}\left(\frac{r}{d}\right) - \tilde{\pi}\left(\frac{r}{d}\right)\right) + (1 - \alpha b)\tilde{\pi}\left(\frac{r}{d}\right) + \alpha r - \bar{\pi}(r)\right| \\
&\le (1 - \alpha b)\|\tilde{\pi}_{\mathrm{sym}} - \tilde{\pi}\|_\infty \le \epsilon'.
\end{aligned}
$$

Thus the function π' satisfies all conditions in Theorem 4, except that $\pi'(-1)$ may be different from 0, and thus it may not be strongly minimal (and hence extreme). However, we can correct this in the following way.

Let $\beta = \pi'(b) + \pi'(-1) \cdot b = 1 + \pi'(-1) \cdot b$. Since $|\pi'(-1)| \le \epsilon'$ and by choice of ϵ' we have $1 + \epsilon' b > 0$, we obtain $\beta > 0$. Define $\pi^*(r) = \frac{1}{\beta}(\pi'(r) + \pi'(-1) \cdot r)$. The proof of the following lemma is presented in the full version of the paper.

Lemma 10. *The function π^* is a piecewise linear 2-slope function that is strongly minimal for $S = b + \mathbb{Z}_+$. Furthermore, $|\bar{\pi}(r) - \pi^*(r)| \le \epsilon$ for all $r \in [-M, M]$.*

Finally, from Theorem 6 we have that π^* is extreme. This concludes the proof of Theorem 3.

References

1. Balas, E., Qualizza, A.: Monoidal cut strengthening revisited. Discrete Optim. **9**(1), 40–49 (2012)
2. Basu, A., Conforti, M., Di Summa, M.: A geometric approach to cut-generating functions. Math. Program. **151**(1), 153–189 (2015)
3. Basu, A., Hildebrand, R., Köppe, M.: Light on the infinite group relaxation I: foundations and taxonomy. 4OR **14**(1), 1–40 (2015). http://dx.org/10.1007/s10288-015-0292-9
4. Basu, A., Hildebrand, R., Köppe, M.: Light on the infinite group relaxation II: sufficient conditions for extremality, sequences, and algorithms. 4OR, 1–25 (2015). http://dx.org/10.1007/s10288-015-0293-8
5. Basu, A., Hildebrand, R., Köppe, M., Molinaro, M.: A $(k+1)$-slope theorem for the k-dimensional infinite group relaxation. SIAM J. Optim. **23**(2), 1021–1040 (2013)
6. Cornuéjols, G., Molinaro, M.: A 3-slope theorem for the infinite relaxation in the plane. Math. Program. **142**(1–2), 83–105 (2013). http://dx.org/10.1007/s10107-012-0562-7
7. Cornuéjols, G., Yıldız, S.: Cut-generating functions for integer variables (2015). http://integer.tepper.cmu.edu/webpub/draft.pdf
8. Gomory, R.E., Johnson, E.L.: Some continuous functions related to corner polyhedra, I. Math. Program. **3**, 23–85 (1972). http://dx.org/10.1007/BF01585008
9. Gomory, R.E., Johnson, E.L.: Some continuous functions related to corner polyhedra, II. Math. Program. **3**, 359–389 (1972). http://dx.org/10.1007/BF01585008
10. Gomory, R.E., Johnson, E.L.: T-space and cutting planes. Math. Program. **96**, 341–375 (2003). http://dx.org/10.1007/s10107-003-0389-3
11. Hille, E., Phillips, R.: Functional analysis and semi-groups. Am. Math. Soc. **31** (1957)
12. Johnson, E.L.: On the group problem for mixed integer programming. Math. Program. Study **2**, 137–179 (1974)

On the Mixed Binary Representability
of Ellipsoidal Regions

Alberto Del Pia[1] and Jeffrey Poskin[2(✉)]

[1] Department of Industrial and Systems Engineering & Wisconsin
Institute for Discovery, University of Wisconsin-Madison, Madison, WI, USA
delpia@wisc.edu
[2] Department of Mathematics, University of Wisconsin-Madison, Madison, WI, USA
poskin@wisc.edu

Abstract. Representability results for mixed-integer linear systems play a fundamental role in optimization since they give geometric characterizations of the feasible sets that arise from mixed-integer linear programs. We consider a natural extension of mixed-integer linear systems obtained by adding just one ellipsoidal inequality. The set of points that can be described, possibly using additional variables, by these systems are called ellipsoidal mixed binary representable. In this work, we give geometric conditions that characterize ellipsoidal mixed binary representable sets.

1 Introduction

The theory of representability starts with a paper of Dantzig [1] and studies one fundamental question: Given a specified type of algebraic constraints, which subsets of $\mathbb{R}^n$ can be represented as the feasible points of a system defined by these constraints, possibly using additional variables? Several researchers have investigated representability questions (see, e.g., [3–10]), and a systematic study for mixed-integer linear systems is mainly due to Meyer and Jeroslow.

Since the projection of a polyhedron is a polyhedron (see [11]), the sets representable by system of linear inequalities are polyhedra. More formally, a set $S \subseteq \mathbb{R}^n$ is representable as the projected solution set of a linear system

$$Dw \leq d$$
$$w \in \mathbb{R}^n \times \mathbb{R}^p$$

if and only if S is a polyhedron.

If we allow also binary extended variables, a geometric characterization has been given by Jeroslow [6]. A set $S \subseteq \mathbb{R}^n$ is representable as the projected solution set of a mixed-integer linear system

$$Dw \leq d$$
$$w \in \mathbb{R}^n \times \mathbb{R}^p \times \{0,1\}^q$$

if and only if S is the union of a finite number of polyhedra, each having the same recession cone.

© Springer International Publishing Switzerland 2016
Q. Louveaux and M. Skutella (Eds.): IPCO 2016, LNCS 9682, pp. 214–225, 2016.
DOI: 10.1007/978-3-319-33461-5_18

We are interested in giving representability results for mixed-integer sets defined not only by linear inequalities, but also by quadratic inequalities of the form $(w-c)^\top Q(w-c) \leq \gamma$, where Q is a positive semidefinite matrix. Inequalities of this type are called *ellipsoidal inequalities*, and the set of points that satisfy one of them is called an *ellipsoidal region*. Ellipsoidal inequalities arise in many practical applications. As an example, many real-life quantities are normally distributed; and for a normal distribution, a natural confidence set, containing the vast majority of the objects, is an ellipsoidal region. See, e.g., [12] for other applications of ellipsoidal inequalities.

A characterization of sets representable by an arbitrary number of ellipsoidal inequalities seems to be currently completely out of reach. In fact, it is easy to construct examples where just two ellipsoidal inequalities in $\mathbb{R}^3$ project to a semialgebraic set described by polynomials of degree four in $\mathbb{R}^2$. This can happen even without linear inequalities or binary extended variables. As a consequence, in this work we will focus on understanding the expressive power of just one ellipsoidal inequality.

Formally, we say that a set $S \subseteq \mathbb{R}^n$ is *ellipsoidal mixed binary (EMB) representable* if it can be obtained as the projection onto $\mathbb{R}^n$ of the solution set of a system of the form

$$Dw \leq d$$
$$(w - c)^\top Q(w - c) \leq \gamma \qquad (1)$$
$$w \in \mathbb{R}^{n+p} \times \{0,1\}^q,$$

where Q is positive semidefinite. There is a strong connection between EMB-representable sets and mixed-integer quadratic programming (MIQP). In a MIQP problem we aim at minimizing a quadratic function over mixed integer points in a polyhedron. Since MIQP $\in \mathcal{NP}$ [2], any MIQP with bounded objective is polynomially equivalent to a polynomial number of MIQP feasibility problems. If the objective quadratic is ellipsoidal, then each feasibility problem is a feasibility problem over a set of the form (1).

Our main result is the following geometric characterization of EMB-representable sets.

Theorem 1. *A set $S \subseteq \mathbb{R}^n$ is EMB-representable if and only if there exist ellipsoidal regions $\mathcal{E}_i \subseteq \mathbb{R}^n$, $i = 1, \ldots, k$, polytopes $\mathcal{P}_i \subseteq \mathbb{R}^n$, $i = 1, \ldots, k$, and a polyhedral cone $\mathcal{C} \subseteq \mathbb{R}^n$ such that*

$$S = \bigcup_{i=1}^{k}(\mathcal{E}_i \cap \mathcal{P}_i) + \mathcal{C}. \qquad (2)$$

An example of an EMB-representable set is given in Fig. 1.

Both directions of Theorem 1 have geometric implications. Since each set (2) can be obtained as the projection of a set described by a system (1), this means that the k ellipsoidal regions $\mathcal{E}_i$ can be expressed with just one ellipsoidal

Fig. 1. An EMB-representable set in $\mathbb{R}^3$

inequality in a higher dimension. We prove this direction of the theorem by explicitly giving an extended formulation for S.

The other direction of Theorem 1 states that the projection of each system (1) onto $\mathbb{R}^n$ is a set of the form (2). The proof of this statement essentially reduces to proving that the projection S of a set $\{x \in \mathbb{R}^{n+1} \mid Dx \leq d, \ (x-c)^\top Q(x-c) \leq \gamma\}$ onto $\mathbb{R}^n$ is a set of the form (2). In order to do so, we introduce the key concept of a *shadowing hyperplane*. This hyperplane, that will be formally introduced later, allows us to split the ellipsoidal region into two 'parts' which, in turn, allow us to decompose S as a union of subsets S_i. We will then see how each set S_i can be obtained as the projection of a set in $\mathbb{R}^{n+1}$ lying on a hyperplane. This will allow us to prove that each S_i can be described with linear inequalities and one ellipsoidal inequality.

The remainder of this paper is organized as follows. In Sect. 2, we provide a number of results relating to the intersection of an ellipsoidal region with a polyhedron and the projections of such regions. In Sect. 3, we prove Theorem 1.

2   Ellipsoidal Regions and Hyperplanes

In this section we formally define ellipsoidal regions. These regions will appear throughout our study of representability. We will prove a few results on the intersection of ellipsoidal regions with half-spaces as well as their projections. These results will be necessary for our proof of Theorem 1.

We say that a set $\mathcal{E}$ is an *ellipsoidal region* in $\mathbb{R}^n$ if there exists an $n \times n$ matrix $Q \succeq 0$ (i.e.,Q is positive semi-definite), a vector $c \in \mathbb{R}^n$, and a number $\gamma \in \mathbb{R}$, such that

$$\mathcal{E} = \{x \in \mathbb{R}^n \mid (x-c)^\top Q(x-c) \leq \gamma\}.$$

We note that if $Q \succ 0$ (i.e.,Q is positive definite) and $\gamma > 0$, then $\mathcal{E}$ is an *ellipsoid*, i.e.,the image of the unit ball $\mathcal{B} = \{x \in \mathbb{R}^n \mid ||x|| \leq 1\}$ under an invertible affine transformation.

Given a set $E \subseteq \mathbb{R}^n \times \mathbb{R}^p$ and a vector $\bar{y} \in \mathbb{R}^p$, we define the $\bar{y}$-*restriction* of E as

$$E|_{y=\bar{y}} = \{x \in \mathbb{R}^n \mid (x, \bar{y}) \in E\}.$$

Note that $E|_{y=\bar{y}}$ geometrically consists of the intersection of E with coordinate hyperplanes. Sometimes we will need to consider $E|_{y=\bar{y}}$ in the original space $\mathbb{R}^n \times \mathbb{R}^p$, thus we also define

$$\tilde{E}|_{y=\bar{y}} = \{(x, \bar{y}) \in \mathbb{R}^n \times \mathbb{R}^p \mid (x, \bar{y}) \in E\}.$$

We will also need to perform several restrictions $y_1 = \bar{y}_1, \ldots, y_k = \bar{y}_k$ at the same time. In such case we simply write $E|_{y_1=\bar{y}_1,\ldots,y_k=\bar{y}_k}$ and $\tilde{E}|_{y_1=\bar{y}_1,\ldots,y_k=\bar{y}_k}$.

In the remainder of the paper we will denote by "rec" the recession cone of a set, by "lin" the lineality space of a set, by "span" the linear space generated by a set of vectors, by "cone" the cone generated by a set of vectors, by "range" the range of a matrix, and by "ker" the kernel of a matrix.

The following observation is well-known, and we give a proof for completeness.

Observation 1. *Let* $q(x) = x^\top Q x + b^\top x$ *be a quadratic function on* $\mathbb{R}^n$ *with* $Q \succeq 0$. *Then* $q(x)$ *has a minimum on* $\mathbb{R}^n$ *if and only if* b *is in the range of* Q.

Proof. Assume $b \notin \text{range}(Q)$. Then since Q is symmetric, we can write $b = Qr + c$ with $Qc = 0$ and $c \neq 0$. Consider $x(t) = -tc$ for $t \in \mathbb{R}$. Then we have

$$q(x(t)) = b^\top x(t) = -tc^\top c.$$

Since $c \neq 0$, we see that $q(x(t)) \to -\infty$ as $t \to +\infty$. Thus, $q(x)$ has no minimum on $\mathbb{R}^n$.

Assume there exists $x_0 \in \mathbb{R}^n$ such that $\frac{1}{2}b = Qx_0$. Then

$$q(x) = (x + x_0)^\top Q(x + x_0) - x_0^\top Q x_0$$

and $q(x)$ has a minimum at any $\bar{x}$ such that $\bar{x} + x_0 \in \ker(Q)$. In particular, $-x_0$ is a minimizer and $q(-x_0) = -x_0^\top Q x_0$ is the optimal value.   □

The following lemma shows that ellipsoidal regions are closed under intersections with coordinate hyperplanes. This is equivalent to fixing a number of variables.

Lemma 1. *Let* $\mathcal{E}$ *be an ellipsoidal region in* $\mathbb{R}^n \times \mathbb{R}^p$. *Then for any* $\bar{y} \in \mathbb{R}^p$, *the set* $\mathcal{E}|_{y=\bar{y}}$ *is an ellipsoidal region in* $\mathbb{R}^n$.

Proof. Let $\mathcal{E} = \{(x, y) \in \mathbb{R}^n \times \mathbb{R}^p \mid q(x, y) \leq \gamma\}$, where $q(x, y)$ is the quadratic polynomial

$$q(x, y) = \begin{pmatrix} x - c \\ y - c' \end{pmatrix}^\top \begin{pmatrix} Q & R \\ R^\top & \bar{Q} \end{pmatrix} \begin{pmatrix} x - c \\ y - c' \end{pmatrix}.$$

For any fixed $\bar{y} \in \mathbb{R}^p$, since $Q \succeq 0$ it suffices to show there exists $c_{\bar{y}} \in \mathbb{R}^n$ and $\gamma_{\bar{y}} \in \mathbb{R}$ such that

$$\mathcal{E}|_{y=\bar{y}} = \{x \in \mathbb{R}^n \mid (x - c_{\bar{y}})^\top Q(x - c_{\bar{y}}) \leq \gamma_{\bar{y}}\}.$$

Let $\bar{y} \in \mathbb{R}^p$. Since $q(x, y)$ has a minimum on $\mathbb{R}^n \times \mathbb{R}^p$, the quadratic function

$$q(x, \bar{y}) = (x - c)^\top Q(x - c) + 2(\bar{y} - c')^\top R^\top (x - c) + (\bar{y} - c')^\top \bar{Q}(\bar{y} - c'),$$

has a minimum on $\mathbb{R}^n$. Applying Observation 1, $R(\bar{y} - c') \in \text{range}(Q)$, and there exists $\bar{x} \in \mathbb{R}^n$ such that $Q\bar{x} = R(\bar{y} - c')$. Defining $c_{\bar{y}} := c - \bar{x}$ and $\gamma_{\bar{y}} := \gamma + \bar{x}^\top Q\bar{x} - (\bar{y} - c')^\top \bar{Q}(\bar{y} - c')$ we have

$$\mathcal{E}|_{y=\bar{y}} = \{x \in \mathbb{R}^n \mid (x - c_{\bar{y}})^\top Q(x - c_{\bar{y}}) \leq \gamma_{\bar{y}}\}. \qquad \square$$

We are now ready to provide a geometric description of ellipsoidal regions. A consequence of this description is that any non-empty ellipsoidal region may be decomposed into the Minkowski sum of an ellipsoid and a linear space.

Lemma 2. *Let $\mathcal{E}$ be an ellipsoidal region in $\mathbb{R}^n$. Then*

(i) $\mathcal{E} = \emptyset$, or

(ii) $\mathcal{E}$ is an affine space, or

(iii) There exists a k-dimensional linear space $L \subseteq \mathbb{R}^n$, and k distinct indices $i_1, \ldots, i_k \in \{1, \ldots, n\}$ such that the restriction

$$\mathcal{E}|_{x_{i_1} = \bar{x}_{i_1}, \ldots, x_{i_k} = \bar{x}_{i_k}}$$

is an ellipsoid in $\mathbb{R}^{n-k}$, and

$$\mathcal{E} = \tilde{\mathcal{E}}|_{x_{i_1} = \bar{x}_{i_1}, \ldots, x_{i_k} = \bar{x}_{i_k}} + L.$$

Proof. Let $\mathcal{E} = \{x \in \mathbb{R}^n \mid (x - c)^\top Q(x - c) \leq \gamma\}$. If $\gamma < 0$, then $\mathcal{E} = \emptyset$ since Q is positive semidefinite. Thus, we may assume that $\gamma \geq 0$ and $\mathcal{E}$ is non-empty.

We now show that $\text{rec}(\mathcal{E}) = \{x \in \mathbb{R}^n \mid x^\top Q x \leq 0\} = \ker(Q)$. Since $\mathcal{E}$ is a closed convex set, $\text{rec}(\mathcal{E})$ is equal to the set of recession directions at any point $x \in \mathcal{E}$. Consider the point $c \in \mathcal{E}$. Then for any $r \in \ker(Q)$ and $\lambda > 0$ we have $c + \lambda r \in \mathcal{E}$ since $\lambda^2 r^\top Q r = 0 \leq \gamma$. Assume $r \in \mathbb{R}^n$ is a recession direction from $c \in \mathcal{E}$. Let $Q = L^\top L$ be a Cholesky decomposition of Q. Then for any $\lambda > 0$ we have $\lambda^2 r^\top Q r = \lambda^2 ||Lr||^2 \leq \gamma$, which implies $Lr = 0$ and $r \in \ker(Q)$.

Now assume $\gamma = 0$. By the above argument, $x \in \mathcal{E}$ if and only if $x \in \{c\} + \ker(Q)$. Thus $\mathcal{E} = \{c\} + \ker(Q)$ is an affine space.

Assume now $\gamma > 0$. If Q is invertible then $\mathcal{E}$ is an ellipsoid and we are done. Thus, we may assume $L := \ker(Q)$ is nontrivial. Let $\mathcal{L} = \{l_1, \ldots, l_k\}$ be a basis for L. Extend $\mathcal{L}$ to a basis $\mathcal{L}'$ of $\mathbb{R}^n$ by adding a subset of the standard basis vectors $\{e_1, \ldots, e_n\}$ of $\mathbb{R}^n$. Let $\mathcal{J} \subseteq \{1, \ldots, n\}$ be the set of indices j for which $e_j \in \mathcal{L}'$. Let $\{i_1, \ldots, i_k\} = \{1, \ldots, n\} - \mathcal{J}$. Consider $\mathcal{E}' := \mathcal{E}|_{x_{i_1} = 0, \ldots, x_{i_k} = 0}$ and $\tilde{\mathcal{E}}' := \tilde{\mathcal{E}}|_{x_{i_1} = 0, \ldots, x_{i_k} = 0}$.

We now show $\mathcal{E} = \tilde{\mathcal{E}}' + L$. Since $\tilde{\mathcal{E}}' \subseteq \mathcal{E}$ and $\mathrm{rec}(\mathcal{E}) = L$, we clearly have $\tilde{\mathcal{E}}' + L \subseteq \mathcal{E}$. Let $v \in \mathcal{E}$. Expanding v in the basis $\mathcal{L}'$, we have for some $l \in L$ and scalars $\alpha_j \in \mathbb{R}$, that $v = l + \sum_{j \in \mathcal{J}} \alpha_j e_j$. Since $L = \mathrm{rec}(\mathcal{E})$ we have $v - l = \sum_{j \in \mathcal{J}} \alpha_j e_j \in \mathcal{E}'$ and $\mathcal{E} \subseteq \tilde{\mathcal{E}}' + L$.

By Lemma 1, $\mathcal{E}'$ is an ellipsoidal region in $\mathbb{R}^{n-k}$. To show $\mathcal{E}'$ is an ellipsoid in $\mathbb{R}^{n-k}$ it remains to show that $\mathcal{E}'$ is full-dimensional and bounded. If $\mathcal{E}'$ is unbounded, then $\mathcal{E}'$ has some recession direction outside of L which contradicts the fact that $\mathrm{rec}(\mathcal{E}) = L$. We finally show that $\mathcal{E}'$ is full-dimensional. We first show that $\mathcal{E}$ is full-dimensional in $\mathbb{R}^n$. This follows since $\gamma > 0$ and there exists a vector, namely $c \in \mathbb{R}^n$, for which the continuous function $(x - c)^\top Q(x - c)$ has value 0. This implies that there exists an ϵ-ball around c, say $\mathcal{B}$, such that $\mathcal{B} \subseteq \mathcal{E}$. The fact that $\mathcal{E}'$ is full-dimensional follows by considering the intersection of $\mathcal{B} + L$ with $\mathcal{E}'$. $\qquad\square$

We make the following remark about the proof of (iii) that will be used later. If one of the standard basis vectors of $\mathbb{R}^n$, say e_n, is not contained in L, then we may assume that x_n does not occur among the fixed variables $x_{i_1}, \ldots, x_{i_k}$. To see this, note that in completing the basis $\mathcal{L}$ of L to a basis of $\mathbb{R}^n$ we may first add the standard basis vector e_n to the set $\mathcal{L}$.

It can be shown that Lemma 2 is in fact an if and only if statement. It then provides a complete geometric characterization of ellipsoidal regions. The next observation gives a description of the recession cones that will be encountered in this paper.

Observation 2. *Let $\mathcal{P}$ be a polyhedron and $\mathcal{E}$ an ellipsoidal region in $\mathbb{R}^n$. Then $\mathrm{rec}(\mathcal{E} \cap \mathcal{P})$ is a polyhedral cone.*

Proof. Clearly, $\mathrm{rec}(\mathcal{E} \cap \mathcal{P}) = \mathrm{rec}(\mathcal{E}) \cap \mathrm{rec}(\mathcal{P})$. The set $\mathrm{rec}(\mathcal{P})$ is a polyhedral cone (see, e.g., [11]), and $\mathrm{rec}(\mathcal{E})$ is a linear space by Lemma 2. As a consequence $\mathrm{rec}(\mathcal{E} \cap \mathcal{P})$ is a polyhedral cone. $\qquad\square$

The following lemma shows that to compute the projection of an ellipsoidal region $\mathcal{E}$ in $\mathbb{R}^n$, it suffices to consider the projection of $\mathcal{E} \cap H$ for a specific hyperplane $H \subseteq \mathbb{R}^n$. We will refer to such a hyperplane H as a *shadowing hyperplane*, as it contains enough information to completely describe the projection, or 'shadow', of $\mathcal{E}$.

Given a set $S \subseteq \mathbb{R}^n$, and a positive integer $k \leq n$, we will denote by $\mathrm{proj}_k(S)$ the projection of S onto its first k coordinates. Formally,

$$\mathrm{proj}_k(S) = \{x \in \mathbb{R}^k \mid \exists y \in \mathbb{R}^{n-k} \text{ with } (x, y) \in S\}.$$

Lemma 3. *Let $\mathcal{E}$ be an ellipsoidal region in $\mathbb{R}^n$. Then there exists a hyperplane $H \subseteq \mathbb{R}^n$ with $e_n \notin \mathrm{lin}(H)$ such that*

$$\mathrm{proj}_{n-1}(\mathcal{E}) = \mathrm{proj}_{n-1}(\mathcal{E} \cap H).$$

Proof. We first note that it suffices to find a hyperplane H such that for any $x \in \mathcal{E}$ there exists $\lambda \in \mathbb{R}$ such that $x + \lambda e_n \in \mathcal{E} \cap H$. The cases $\mathcal{E} = \emptyset$ and $\mathcal{E}$ an affine space are trivial. If $\mathcal{E} = \emptyset$ then any hyperplane H with $e_n \notin \text{lin}(H)$ satisfies the condition of the lemma. If $\mathcal{E} = v + L$ is an affine space, either $e_n \in \text{lin}(L)$ or $e_n \notin \text{lin}(L)$. If $e_n \in \text{lin}(L)$, we may take $H = \{x \in \mathbb{R}^n \mid x_n = 0\}$ since for any $\bar{x} \in \mathcal{E}$ there exists $\bar{\lambda} \in \mathbb{R}$, namely $\bar{\lambda} = -\bar{x}_n$, such that $\bar{x} + \bar{\lambda}e_n \in \mathcal{E} \cap H$. If $e_n \notin \text{lin}(H)$, then we may take H to be any hyperplane containing $\mathcal{E}$ with $e_n \notin \text{lin}(H)$.

We now show the lemma when $\mathcal{E}$ is an ellipsoid, say $\mathcal{E} = \{Ax + c \mid ||x|| \leq 1\}$ with A an invertible $n \times n$ matrix. Let $H' = \{x \in \mathbb{R}^n \mid x_n = 0\}$. Clearly, for any x in the standard unit ball $\mathcal{B}$ there exists $\lambda \in \mathbb{R}$ such that $x + \lambda e_n \in \mathcal{B} \cap H'$. Let U be an orthogonal transformation that maps the standard unit vector e_n to $\frac{A^{-1}e_n}{||A^{-1}e_n||}$. Let T be the invertible affine transformation defined by $T(x) = AU(x) + c$. We claim that $H := T(H')$ is an appropriate hyperplane. Let $\bar{x} \in \mathcal{E}$. Since $\mathcal{E} = A\mathcal{B} + c = T(\mathcal{B})$, we have $T^{-1}(\bar{x}) \in \mathcal{B}$. Then there exists $\bar{\lambda} \in \mathbb{R}$ such that $T^{-1}(\bar{x}) + \bar{\lambda}e_n \in \mathcal{B} \cap H'$. Applying T we have $\bar{x} + \frac{\bar{\lambda}}{||A^{-1}e_n||}e_n \in \mathcal{E} \cap H$. We have $e_n \notin \text{lin}(H)$, since otherwise $\text{proj}_{n-1}(\mathcal{E}) = \text{proj}_{n-1}(\mathcal{E} \cap H) \subseteq \text{proj}_{n-1}(H)$ would have dimension at most $n - 2$, contradicting the full-dimensionality of $\mathcal{E}$.

Assume now that $\mathcal{E}$ is a full-dimensional and unbounded ellipsoidal region, say $\mathcal{E} = \{x \in \mathbb{R}^n \mid (x - c)^\top Q(x - c) \leq \gamma\}$ for some singular positive semi-definite matrix Q, and $\gamma > 0$. Let $L = \ker(Q)$, which by Lemma 2 is the recession cone of $\mathcal{E}$. Suppose first that $e_n \in L$ and consider $H = \{x \in \mathbb{R}^n \mid x_n = 0\}$. Then for any $\bar{x} \in \mathcal{E}$, we have $\bar{x} - \bar{x}_n e_n \in \mathcal{E} \cap H$, and H has the desired property.

Thus, we may assume that $e_n \notin L$. We now apply Lemma 2, and obtain a decomposition

$$\mathcal{E} = \tilde{\mathcal{E}}|_{x_{i_1} = \bar{x}_{i_1}, \ldots, x_{i_k} = \bar{x}_{i_k}} + L.$$

Further, $\mathcal{E}' := \mathcal{E}|_{x_{i_1} = \bar{x}_{i_1}, \ldots, x_{i_k}}$ is an ellipsoid in $\mathbb{R}^{n-k}$. We note that since $e_n \notin \text{lin}(H)$, by the remark following Lemma 2, we may assume x_n is not among the variables fixed. Thus, we may assume that e'_n, the restriction of e_n obtained by dropping the fixed components, is non-zero in $\mathbb{R}^{n-k}$. We can now apply the proof of the bounded case above to the ellipsoid $\mathcal{E}'$. That is, there exists a hyperplane $H' \subseteq \mathbb{R}^{n-k}$ such that for any $x' \in \mathcal{E}'$ there exists $\lambda' \in \mathbb{R}$ such that $x' + \lambda' e'_n \in \mathcal{E}' \cap H'$.

Let $\tilde{H}'$ be obtained from H' by considering it in the original space $\mathbb{R}^n$, i.e., we have $x \in \tilde{H}'$ if and only if $x_{i_1} = \bar{x}_{i_1}, \ldots, x_{i_k} = \bar{x}_{i=k}$ and the vector consisting of components of x not among these x_{i_j} is in H'. We claim that the hyperplane $H := \tilde{H}' + L$ satisfies the condition of the lemma. By construction, $e_n \notin \text{lin}(H)$. Now for any $x \in \mathcal{E}$, there exists $l \in L$ such that $x - l \in \tilde{\mathcal{E}}'$. Then for some $\lambda \in \mathbb{R}$ we have $x - l + \lambda e_n \in \tilde{\mathcal{E}}' \cap H$ and since $L \subseteq H$ we have $x + \lambda e_n \in \mathcal{E} \cap H$.   □

With these results in hand, we are now ready to proceed to the proof of Theorem 1.

3   Proof of Theorem 1

To prove sufficiency of the condition, assume that we are given a set

$$S = \bigcup_{i=1}^{k} (\mathcal{E}_i \cap \mathcal{P}_i) + \mathcal{C},$$

where $\mathcal{E}_i = \{x \in \mathbb{R}^n \mid (x - c_i)^\top Q_i(x - c_i) \leq \gamma_i\}$ are ellipsoidal regions, $\mathcal{P}_i = \{x \in \mathbb{R}^n \mid A_i x \leq b_i\}$ are polytopes, and $\mathcal{C} = \text{cone}\{r_1, \ldots, r_t\} \subseteq \mathbb{R}^n$ is a polyhedral cone. For each ellipsoidal region $\mathcal{E}_i$, if $\gamma_i > 0$ we can normalize the right hand side of the inequality to 1. Else, $\mathcal{E}_i$ is either empty or an affine space and γ_i can be set to 1 at the cost of adding additional linear inequalities to the system $A_i x \leq b_i$. Thus, we may assume $\gamma_i = 1$ for all $i = 1, \ldots, k$.

We introduce new continuous variables $x_i \in \mathbb{R}^n$ and binary variables $\delta_i \in \{0, 1\}$, for $i = 1, \ldots, k$, that will model the individual regions $\mathcal{E}_i \cap \mathcal{P}_i + \mathcal{C}$. Then S can be described as the set of $x \in \mathbb{R}^n$ such that

$$x = \sum_{i=1}^{k} (x_i + \delta_i c_i) + \sum_{j=1}^{t} \lambda_j r_j$$

$$A_i x_i \leq \delta_i(b_i - A_i c_i) \qquad\qquad i = 1, \ldots, k$$

$$\sum_{i=1}^{k} \delta_i = 1$$

$$\begin{pmatrix} x_1 \\ x_2 \\ \vdots \\ x_k \end{pmatrix}^\top \begin{pmatrix} Q_1 & & & \\ & Q_2 & & \\ & & \ddots & \\ & & & Q_k \end{pmatrix} \begin{pmatrix} x_1 \\ x_2 \\ \vdots \\ x_k \end{pmatrix} \leq 1$$

$$x_i \in \mathbb{R}^n, \ \delta_i \in \{0, 1\} \qquad\qquad i = 1, \ldots, k$$

$$\lambda_j \in \mathbb{R}^{\geq 0} \qquad\qquad j = 1, \ldots, t.$$

Now if $\delta_1 = 1$ the remaining δ_i must be 0. Then for each x_i with $i \neq 1$, we have the constraint $A_i x_i \leq 0$ which has the single feasible point $x_i = 0$ since $\mathcal{P}_i$ is a polytope. The remaining constraints reduce to

$$x = x_1 + c_1 + \sum_{j=1}^{t} \lambda_j r_j$$

$$A_1(x_1 + c_1) \leq b_1$$

$$x_1^\top Q_1 x_1 \leq 1$$

$$x_1 \in \mathbb{R}^n$$

$$\lambda_j \in \mathbb{R}^{\geq 0} \qquad\qquad j = 1, \ldots, t.$$

By employing a change of variables $x' = x_1 + c_1$, it can be checked that the latter system describes the region $\mathcal{E}_1 \cap \mathcal{P}_1 + \mathcal{C}$. The remaining regions follow symmetrically. Therefore S is EMB-representable.

The remainder of the proof is devoted to proving necessity of the condition. We are given an ellipsoidal region $\mathcal{E}$ and a polyhedron $\mathcal{P}$ in $\mathbb{R}^{n+p+q}$, and we define

$$\bar{S} := \mathcal{E} \cap \mathcal{P} \cap (\mathbb{R}^{n+p} \times \{0, 1\}^q),$$
$$S := \mathrm{proj}_n(\bar{S}).$$

We must show the existence of ellipsoidal regions $\mathcal{E}_i \subseteq \mathbb{R}^n$, $i = 1, \ldots, k$, polytopes $\mathcal{P}_i \subseteq \mathbb{R}^n$, $i = 1, \ldots, k$, and a polyhedral cone $\mathcal{C} \subseteq \mathbb{R}^n$ such that

$$S = \bigcup_{i=1}^{k} (\mathcal{E}_i \cap \mathcal{P}_i) + \mathcal{C}.$$

Claim 1. *It suffices to find ellipsoidal regions $\mathcal{E}_i \subseteq \mathbb{R}^n$, polytopes $\mathcal{P}_i \subseteq \mathbb{R}^n$, and polyhedral cones $\mathcal{C}_i \subseteq \mathbb{R}^n$, for $i = 1, \ldots, k$, that satisfy*

$$S = \bigcup_{i=1}^{k} (\mathcal{E}_i \cap \mathcal{P}_i + \mathcal{C}_i). \tag{3}$$

Proof of Claim. Let $\tilde{S} := \mathcal{E} \cap \mathcal{P} \cap (\mathbb{R}^{n+p} \times [0, 1]^q)$. Then for every $\bar{z} \in \mathbb{R}^q$, define $\bar{S}_{\bar{z}} := \mathcal{E} \cap \mathcal{P} \cap (\mathbb{R}^{n+p} \times \{\bar{z}\})$. Clearly, for every $\bar{z} \in \{0, 1\}^q$, we have $\mathrm{rec}(\bar{S}_{\bar{z}}) = \mathrm{rec}(\tilde{S})$, and so $\mathrm{proj}_n(\mathrm{rec}(\bar{S}_{\bar{z}})) = \mathrm{proj}_n(\mathrm{rec}(\tilde{S}))$. Since projections and recession cones operators commute for closed convex sets, we obtain $\mathrm{rec}(\mathrm{proj}_n(\bar{S}_{\bar{z}})) = \mathrm{proj}_n(\mathrm{rec}(\tilde{S}))$. Let $\mathcal{C} := \mathrm{proj}_n(\mathrm{rec}(\tilde{S}))$. By Observation 2, the set $\mathrm{rec}(\tilde{S})$ is a polyhedral cone, thus so is its projection $\mathcal{C}$.

Note that $\bar{S} = \cup_{\bar{z} \in \{0,1\}^q} \bar{S}_{\bar{z}}$ implies $S = \cup_{\bar{z} \in \{0,1\}^q} \mathrm{proj}_n(\bar{S}_{\bar{z}})$, therefore $\mathrm{rec}(S) = \mathcal{C}$. This concludes the proof since $S = \bigcup_{i=1}^{k}(\mathcal{E}_i \cap \mathcal{P}_i + \mathcal{C}_i) = \bigcup_{i=1}^{k}(\mathcal{E}_i \cap \mathcal{P}_i) + \mathcal{C}$. ◇

Claim 2. *It suffices to find ellipsoidal regions $\mathcal{E}_i \subseteq \mathbb{R}^n$, polyhedra $\mathcal{P}_i \subseteq \mathbb{R}^n$, and polyhedral cones $\mathcal{C}_i \subseteq \mathbb{R}^n$, for $i = 1, \ldots, k$, that satisfy (3).*

Proof of Claim. In order to show the claim, we prove that if we have $\mathcal{E}_i \cap \mathcal{P}_i + \mathcal{C}_i$ for an ellipsoidal region $\mathcal{E}_i$, a polyhedron $\mathcal{P}_i$, and a polyhedral cone $\mathcal{C}_i$, then we may replace $\mathcal{P}_i$ with a polytope $\mathcal{R}$ without loss.

Replacing $\mathcal{C}_i$ with $\mathcal{C}_i + \mathrm{rec}(\mathcal{E}_i \cap \mathcal{P}_i)$ if necessary, we may assume that $\mathrm{rec}(\mathcal{E}_i \cap \mathcal{P}_i) \subseteq \mathcal{C}_i$. Note that the newly defined $\mathcal{C}_i$ is a polyhedral cone by Observation 2. Consider a polyhedral approximation $\mathcal{B}$ of $\mathcal{E}_i$ such that $\mathcal{B} \subseteq \mathbb{R}^n$ is a polyhedron, $\mathcal{E}_i \subseteq \mathcal{B}$, and $\mathrm{rec}(\mathcal{E}_i) = \mathrm{rec}(\mathcal{B})$. Then $\mathcal{B} \cap \mathcal{P}_i$ is a polyhedron and can be decomposed as $\mathcal{R} + \mathcal{C}_i'$ for a polytope $\mathcal{R}$ and a polyhedral cone $\mathcal{C}_i' \subseteq \mathcal{C}_i$. We claim that $\mathcal{E}_i \cap \mathcal{R} + \mathcal{C}_i = \mathcal{E}_i \cap \mathcal{P}_i + \mathcal{C}_i$.

Let $x \in \mathcal{E}_i \cap \mathcal{R} + \mathcal{C}_i$, and note that $\mathcal{R} \subseteq \mathcal{P}_i$ so that $x \in \mathcal{E}_i \cap \mathcal{P}_i + \mathcal{C}_i$. Thus, $\mathcal{E}_i \cap \mathcal{R} + \mathcal{C}_i \subseteq \mathcal{E}_i \cap \mathcal{P}_i + \mathcal{C}_i$. Let $x \in \mathcal{E}_i \cap \mathcal{P}_i + \mathcal{C}_i$. Then $x \in \mathcal{B} \cap \mathcal{P}_i + \mathcal{C}_i = \mathcal{R} + \mathcal{C}_i$ and we may write $x = r + c$ for some $r \in \mathcal{R}$, $c \in \mathcal{C}_i$. Note that $c \in \mathrm{rec}(\mathcal{E}_i)$, and since $\mathrm{rec}(\mathcal{E}_i)$ is a linear space by Lemma 2, we obtain $-c \in \mathrm{rec}(\mathcal{E}_i)$ as well. Then $x = (x - c) + c$ and $x - c = r \in \mathcal{E}_i \cap \mathcal{R}$, $c \in \mathcal{C}_i$ so $x \in \mathcal{E}_i \cap \mathcal{R} + \mathcal{C}_i$. ◇

Claim 3. *We can assume without loss of generality $q = 0$.*

Proof of Claim. Note that, using restrictions, we can write the set S in the form

$$S = \bigcup_{\bar{z} \in \{0,1\}^q} \mathrm{proj}_n(\bar{S}|_{z=\bar{z}}).$$

It suffices to show that each restriction $\bar{S}|_{z=\bar{z}} = \mathcal{E}' \cap \mathcal{P}'$ for some ellipsoidal region $\mathcal{E}' \subset \mathbb{R}^{n+p}$ and polyhedron $\mathcal{P}' \subseteq \mathbb{R}^{n+p}$. Then, assuming the result in the case $q = 0$, for each $\bar{z} \in \{0,1\}^q$ we have $\mathrm{proj}_n(\bar{S}|_{z=\bar{z}}) = \cup_{i=1}^k (\mathcal{E}_i \cap \mathcal{P}_i + \mathcal{C}_i)$. Since S is the finite union of such sets, the result follows.

Let $\bar{z} \in \{0,1\}^q$. We note $\bar{S}|_{z=\bar{z}} = \mathcal{E}|_{z=\bar{z}} \cap \mathcal{P}|_{z=\bar{z}}$. By Lemma 1, $\mathcal{E}' := \mathcal{E}|_{z=\bar{z}}$ is an ellipsoidal region in $\mathbb{R}^{n+p}$. Let $\mathcal{P} = \{(x,y,z) \in \mathbb{R}^{n+p} \times \{0,1\}^q \mid Ax + By + Cz \le d\}$. Also, $\mathcal{P}' := \mathcal{P}|_{z=\bar{z}} = \{(x,y) \in \mathbb{R}^{n+p} \mid Ax + By \le d - C\bar{z}\}$ is clearly a polyhedron.$\diamond$

Claim 4. *We can assume without loss of generality $p = 1$.*

Proof of Claim. Let $\mathcal{E} \cap \mathcal{P} \subseteq \mathbb{R}^{n+p}$. We prove that $S = \mathrm{proj}_n(\mathcal{E} \cap \mathcal{P})$ has the desired decomposition, by induction on p. For this claim, we assume the base case, $p = 1$. Now let $p = k$, and suppose the statement holds for $p < k$. Given $\mathcal{E} \cap \mathcal{P} \subseteq \mathbb{R}^{n+k}$, by the base case $p = 1$ we have

$$\mathrm{proj}_{n+k-1}(\mathcal{E} \cap \mathcal{P}) = \bigcup_{i=1}^t (\mathcal{E}_i \cap \mathcal{P}_i + \mathcal{C}_i).$$

Since the projection of a union is the union of the projections, we have

$$S = \mathrm{proj}_n(\mathcal{E} \cap \mathcal{P}) = \bigcup_{i=1}^t \mathrm{proj}_n(\mathcal{E}_i \cap \mathcal{P}_i + \mathcal{C}_i).$$

Now proj_n is a linear operator and by the induction hypothesis we have

$$S = \bigcup_{i=1}^t \left(\bigcup_{j=1}^{s_i} (\mathcal{E}_{i,j} \cap \mathcal{P}_{i,j} + \mathcal{K}_{i,j}) + \mathcal{C}_i' \right),$$

where $\mathcal{C}_i' := \mathrm{proj}_n(\mathcal{C}_i)$. Setting $\mathcal{K}_{i,j}' = \mathcal{K}_{i,j} + \mathcal{C}_i'$ for each $i = 1,\dots,t$ and $j = 1,\dots,s_i$, we have

$$S = \bigcup_{i=1}^t \left(\bigcup_{j=1}^{s_i} (\mathcal{E}_{i,j} \cap \mathcal{P}_{i,j} + \mathcal{K}_{i,j}') \right),$$

and we are done. $\diamond$

To prove Theorem 1 it remains to show the following. Assume we are given $\mathcal{E} \cap \mathcal{P} \subseteq \mathbb{R}^{n+1}$. We must show the existence of ellipsoidal regions $\mathcal{E}_i \subseteq \mathbb{R}^n$, polyhedra $\mathcal{P}_i \subseteq \mathbb{R}^n$, and polyhedral cones $\mathcal{C}_i \subseteq \mathbb{R}^n$, for $i = 1,\dots,k$, that satisfy (3).

Given a half-space $H^+ = \{x \in \mathbb{R}^n \mid a^\top x \ge b\}$, we write H for the hyperplane $\{x \in \mathbb{R}^n \mid a^\top x = b\}$ and H^- for the half-space $\{x \in \mathbb{R}^n \mid a^\top x \le b\}$. A polyhedron is the intersection of finitely many half-spaces. Thus, there exist half-spaces

$H_1^+, \ldots, H_s^+ \subseteq \mathbb{R}^{n+1}$ such that $\mathcal{P} = \cap_{i=1}^s H_i^+$. By Lemma 3, there exists a hyperplane $H_0 \subset \mathbb{R}^{n+1}$ with $e_{n+1} \notin \mathrm{lin}(H_0)$ such that $\mathrm{proj}_n(\mathcal{E}) = \mathrm{proj}_n(\mathcal{E} \cap H_0)$. Then

$$\mathcal{E} \cap \mathcal{P} = (\mathcal{E} \cap H_0^+ \cap_{i=1}^s H_i^+) \cup (\mathcal{E} \cap H_0^- \cap_{i=1}^s H_i^+),$$

and it suffices to show the statement for the region $\mathcal{E} \cap H_0^+ \cap_{i=1}^s H_i^+$.

Claim 5. *Let $\mathcal{H}$ be the collection of hyperplanes H among $H_0, \ldots, H_s$ with $e_{n+1} \notin \mathrm{lin}(H)$. Then*

$$\mathrm{proj}_n(\mathcal{E} \cap_{i=0}^s H_i^+) = \bigcup_{H \in \mathcal{H}} \mathrm{proj}_n(\mathcal{E} \cap H \cap_{i=0}^s H_i^+).$$

Proof of Claim. It suffices to show that $\mathcal{E} \cap_{i=0}^s H_i^+$ has the following property: for any $x \in \mathcal{E} \cap_{i=0}^s H_i^+$ there exists a hyperplane $H \in \mathcal{H}$ and a $\lambda \in \mathbb{R}$ such that $x + \lambda e_{n+1} \in \mathcal{E} \cap H \cap_{i=0}^s H_i^+$.

To prove the claim, we show that we can translate a point $x \in \mathcal{E} \cap_{i=0}^s H_i^+$ along the line $\{x + t e_{n+1} \mid t \in \mathbb{R}\}$, and inside the feasible region, until it meets a half-space in $\mathcal{H}$ at equality. If $e_{n+1} \in \mathrm{lin}(H_i)$ for a half-space H_i, then $x + \lambda e_{n+1} \in H_i^+$ for any $\lambda \in \mathbb{R}$.

Let $\bar{x} \in \mathcal{E} \cap_{i=0}^s H_i^+$. Then, by the existence of the shadowing hyperplane H_0, there is one direction among $\{\pm e_{n+1}\}$ along which $\bar{x}$ may be translated to intersect H_0 while staying inside $\mathcal{E}$. Thus, there exists $\bar{\lambda} \in \mathbb{R}$ such that $\bar{x} + \bar{\lambda} e_{n+1} \in \mathcal{E} \cap_{i=0}^s H_i^+$ and $\bar{x} + \bar{\lambda} e_{n+1}$ is contained in at least one hyperplane $H \in \mathcal{H}$. ◇

Then for any $H \in \mathcal{H}$ it suffices to show that there exists an ellipsoidal region $\mathcal{E}' \subseteq \mathbb{R}^n$ and a polyhedron $\mathcal{P}' \subseteq \mathbb{R}^n$ such that

$$\mathrm{proj}_n(\mathcal{E} \cap H \cap_{i=0}^s H_i^+) = \mathcal{E}' \cap \mathcal{P}'.$$

Without loss of generality, we may assume that $H_i \cap H \neq \emptyset$ for each $i = 0, \ldots, s$. If not, say $H_j \cap H = \emptyset$ for some $0 \leq j \leq s$. Then either $\mathcal{E} \cap H \cap H_j^+ = \emptyset$ and our region is empty, or $\mathcal{E} \cap H \cap H_j^+ = \mathcal{E} \cap H$ and H_j^+ is redundant and may be removed.

We now show that each half-space H_i^+ can be replaced with a different half-space M_i^+ such that $\mathcal{E} \cap H \cap H_i^+ = \mathcal{E} \cap H \cap M_i^+$ and $e_{n+1} \in \mathrm{lin}(M_i^+)$. Without loss of generality, consider H_1^+ and the region $\mathcal{E} \cap H \cap H_1^+$. Let $U = H \cap H_1$. Then U is an $(n-1)$-dimensional affine space, say $U = v + V$ for a linear space V of dimension $n-1$. Let $W = V + \mathrm{span}(e_{n+1})$. Since $e_{n+1} \notin \mathrm{lin}(U)$, $M_1 := v + W$ is a hyperplane in $\mathbb{R}^{n+1}$ that divides H into the same two regions that H_1 does. In particular, upon choice of direction, we have that M_1^+ has the desired properties. We may now replace each H_i^+ with M_i^+ in this way.

By the requirement $e_{n+1} \in \mathrm{lin}(M_i^+)$, we have that each M_i^+ is defined by a linear inequality with the coefficient of x_{n+1} equal to 0. Thus, the projection $\mathrm{proj}_n(M_i^+)$ is a half-space in $\mathbb{R}^n$ which we denote $\bar{H}_i^+$. Further, if each H_i^+ for $i = 0, \ldots, s$ is replaced in this way, we have

$$\mathrm{proj}_n(\mathcal{E} \cap H \cap_{i=0}^s H_i^+) = \mathrm{proj}_n(\mathcal{E} \cap H \cap_{i=0}^s M_i^+) = \mathrm{proj}_n(\mathcal{E} \cap H) \cap_{i=0}^s \bar{H}_i^+,$$

and we have the desired polyhedron $\mathcal{P}' := \cap_{i=0}^s \bar{H}_i^+$.

It remains to show that $\text{proj}_n(\mathcal{E} \cap H)$ is an ellipsoidal region $\mathcal{E}' \subseteq \mathbb{R}^n$. Let $H = \{(x,y) \in \mathbb{R}^{n+1} \mid a^\top(x,y) = b\}$. Then there exists a linear transformation from $\mathbb{R}^{n+1}$ to itself, defined by the matrix A whose first n rows are the first n standard unit vectors of $\mathbb{R}^{n+1}$ and whose last row is a. Moreover, A is invertible since e_{n+1} is not in $\text{lin}(H)$. Then, by construction of A, for any vector $(x,y) \in \mathbb{R}^{n+1}$ we have $A(x,y) = (x,c)$ for some $c \in \mathbb{R}$. Furthermore, $A(H)$ gets mapped to the hyperplane $\{(x,y) \in \mathbb{R}^{n+1} \mid y = b\}$. Now, since A is invertible we have

$$x \in \text{proj}_n(\mathcal{E} \cap H) \Leftrightarrow \exists y \in \mathbb{R} \text{ such that } (x,y) \in \mathcal{E} \cap H$$
$$\Leftrightarrow (x,b) \in A(\mathcal{E} \cap H)$$
$$\Leftrightarrow (x,b) \in A(\mathcal{E}).$$

This shows that $\text{proj}_n(\mathcal{E} \cap H) = A(\mathcal{E})|_{y=b}$. Ellipsoidal regions are clearly preserved under invertible linear transformations, therefore $A(\mathcal{E})$ is an ellipsoidal region. Finally, by Lemma 1, the set $A(\mathcal{E})|_{y=b}$ is an ellipsoidal region. This concludes the proof that $\text{proj}_n(\mathcal{E} \cap H)$ is an ellipsoidal region $\mathcal{E}'$. $\square$

References

1. Dantzig, G.B.: Discrete variable extremum problems. Oper. Res. **5**, 266–277 (1957)
2. Del Pia, A., Dey, S.S., Molinaro, M.: Mixed-integer quadratic programming is in NP. Manuscript (2014)
3. Glover, F.: New results on equivalent integer programming formulations. Math. Prog. **8**, 84–90 (1975)
4. Ibaraki, T.: Integer programming formulation of combinatorial optimization problems. Discrete Math. **16**, 39–52 (1976)
5. Jeroslow, R.: Representations of unbounded optimizations as integer programs. J. Optim. Theory Appl. **30**, 339–351 (1980)
6. Jeroslow, R.G.: Representability in mixed integer programming, i: characterization results. Discrete Appl. Math. **17**, 223–243 (1987)
7. Jeroslow, R.G., Lowe, J.K.: Modelling with integer variables. Math. Prog. Study **22**, 167–184 (1984)
8. Meyer, R.R.: Integer and mixed-integer programming models: general properties. J. Optim. Theory Appl. **16**(3/4), 191–206 (1975)
9. Meyer, R.R.: Mixed-integer minimization models for piecewise-linear functions of a single variable. Discrete Math. **16**, 163–171 (1976)
10. Meyer, R.R., Thakkar, M.V., Hallman, W.P.: Rational mixed integer and polyhedral union minimization models. Math. Oper. Res. **5**, 135–146 (1980)
11. Schrijver, A.: Theory of Linear and Integer Programming. Wiley, Chichester (1986)
12. Villaverde, K., Kosheleva, O., Ceberio, M.: Why ellipsoid constraints, ellipsoid clusters, and Riemannian space-time: Dvoretzky's theorem revisited. In: Ceberio, M., Kreinovich, V. (eds.) Constraint Programming and Decision Making. SCI, vol. 539, pp. 203–207. Springer, Heidelberg (2014)

Constant Factor Approximation for ATSP with Two Edge Weights

(Extended Abstract)

Ola Svensson[1], Jakub Tarnawski[1(✉)], and László A. Végh[2]

[1] École Polytechnique Fédérale de Lausanne, Lausanne, Switzerland
{ola.svensson,jakub.tarnawski}@epfl.ch
[2] London School of Economics, London, UK
L.Vegh@lse.ac.uk

Abstract. We give a constant factor approximation algorithm for the Asymmetric Traveling Salesman Problem on shortest path metrics of directed graphs with two different edge weights. For the case of unit edge weights, the first constant factor approximation was given recently in [17]. This was accomplished by introducing an easier problem called Local-Connectivity ATSP and showing that a good solution to this problem can be used to obtain a constant factor approximation for ATSP. In this paper, we solve Local-Connectivity ATSP for two different edge weights. The solution is based on a flow decomposition theorem for solutions of the Held-Karp relaxation, which may be of independent interest.

1 Introduction

The traveling salesman problem — one of finding the shortest tour of n cities — is one of the most classical optimization problems. Its definition dates back to the 19th century and since then a large body of work has been devoted to designing "good" algorithms using heuristics, mathematical programming techniques, and approximation algorithms. The focus of this work is on approximation algorithms. A natural and necessary assumption in this line of work that we also make throughout this paper is that the distances satisfy the triangle inequality: for any triple i, j, k of cities, we have $d(i, j) + d(j, k) \geq d(i, k)$ where $d(\cdot, \cdot)$ denotes the pairwise distances between cities. In other words, it is not more expensive to take the direct path compared to a path that makes a detour.

With this assumption, the approximability of TSP turns out to be a very delicate question that has attracted significant research efforts. Specifically, one of the first approximation algorithms (Christofides' heuristic [6]) was designed for the *symmetric* traveling salesman problem (STSP) where we assume symmetric distances $(d(i, j) = d(j, i))$; and, more recently, several works (see e.g. [1,3,8,9,17])

Please refer to the full version (http://arxiv.org/abs/1511.07038) for proofs and more detailed explanations (with figures).

O. Svensson—Supported by ERC Starting Grant 335288-OptApprox.

L.A. Végh—Supported by EPSRC First Grant EP/M02797X/1.

Q. Louveaux and M. Skutella (Eds.): IPCO 2016, LNCS 9682, pp. 226–237, 2016.
DOI: 10.1007/978-3-319-33461-5_19

have addressed the more general *asymmetric* traveling salesman problem (ATSP) where we make no such assumption.

However, there are still large gaps in our understanding of both STSP and ATSP. In fact, for STSP, the best approximation algorithm remains Christofides' 3/2-approximation algorithm from the 70's [6]. For the harder ATSP, the state of the art is a $\mathcal{O}(\log n / \log \log n)$-approximation algorithm by Asadpour et al. [3] and a recent $\mathcal{O}(\text{poly} \log \log n)$-estimation algorithm[1] by Anari and Oveis Gharan [1]. On the negative side, the best inapproximability results only say that STSP and ATSP are hard to approximate within factors 123/122 and 75/74, respectively [12]. Closing these gaps is a major open problem in the field of approximation algorithms (see e.g. "Problem 1" and "Problem 2" in the list of open problems in the recent book by Williamson and Shmoys [18]). What is perhaps even more intriguing about these questions is that we expect that a standard linear programming (LP) relaxation, often referred to as the Held-Karp relaxation, already gives better guarantees. Indeed, it is conjectured to give a guarantee of 4/3 for STSP and a guarantee of $\mathcal{O}(1)$ (or even 2) for ATSP.

An equivalent formulation of STSP and ATSP from a more graph-theoretic point of view is the following. For STSP, we are given a weighted undirected graph $G = (V, E, w)$ where $w : E \to \mathbb{R}_+$ and we wish to find a multisubset F of edges of minimum total weight such that (V, F) is connected and Eulerian. Recall that an undirected graph is Eulerian if every vertex has even degree. We also remark that we use the term multisubset as the solution F may use the same edge several times. An intuitive point of view on this definition is that G represents a road network, and a solution is a tour that visits each vertex at least once (and may use a single edge/road several times). The definition of ATSP is similar, with the differences that the input graph is directed and the output is Eulerian in the directed sense: the in-degree of each vertex equals its out-degree. Having defined the traveling salesman problem in this way, there are several natural special cases to consider. For example, what if G is planar? Or, what if all the edges/roads have the same length, i.e., if G is unweighted?

For planar graphs, we have much better algorithms than in general. Grigni et al. [11] first obtained a polynomial-time approximation scheme for STSP restricted to unweighted planar graphs, which was later generalized to edge-weighted planar graphs by Arora et al. [2]. More recently, ATSP on planar graphs (and more generally bounded genus graphs) was shown to admit constant factor approximation algorithms (first by Oveis Gharan and Saberi [9] and later by Erickson and Sidiropoulos [7] who improved the dependency on the genus).

In contrast to planar graphs, STSP and ATSP remain APX-hard for unweighted graphs (ones where all edges have identical weight) and, until recently, there were no better algorithms for these cases. Then, in a recent series of papers, the approximation guarantee of 3/2 was finally improved for STSP restricted to unweighted graphs. Specifically, Gharan et al. [10] first gave an

---

[1] An estimation algorithm is a polynomial-time algorithm for approximating/estimating the optimal value without necessarily finding a solution to the problem.

approximation guarantee of $1.5-\epsilon$; Mömke and Svensson [13] proposed a different approach yielding a 1.461-approximation guarantee; Mucha [14] gave a tighter analysis of this algorithm; and Sebő and Vygen [16] significantly developed the approach to give the currently best approximation guarantee of 1.4. Similarly, for ATSP, it was only very recently that the restriction to unweighted graphs could be leveraged: the first constant approximation guarantee for unweighted graphs was given by Svensson [17]. In this paper we make progress towards the general problem by addressing the simplest case left unresolved by [17]: graphs with two different edge weights.

Theorem 1.1. *There is an $\mathcal{O}(1)$-approximation algorithm for ATSP on graphs with two different edge weights.*

The paper [17] introduces an "easier" problem named Local-Connectivity ATSP, where one needs to find an Eulerian multiset of edges crossing only sets in a given partition rather than all possible sets (see next section for definitions). It is shown that an "α-light" algorithm to this problem yields a $(9 + \varepsilon)\alpha$-factor approximation for ATSP. For unweighted graphs (and slightly more generally, for node-induced weight functions[2]) it is fairly easy to obtain a 3-light algorithm for Local-Connectivity ATSP; the difficult part in [17] is the black-box reduction of ATSP to this problem. Note that [17] easily gives an $\mathcal{O}(w_{\max}/w_{\min})$-approximation algorithm in general if we take $w_{\max}$ and $w_{\min}$ to denote the largest and smallest edge weight, respectively. However, obtaining a constant factor approximation even for two different weights requires substantial further work.

In Local-Connectivity ATSP we need a lower bound function lb : $V \to \mathbb{R}_+$ on the vertices. The natural choice for node-induced weights is lb(v) = $\sum_{e \in \delta^+(v)} w(e)x_e^*$. With this weight function, every vertex is able to "pay" for the incident edges in the Eulerian subgraph we are looking for. This choice of lb does not seem to work for more general weight functions, and we need to define lb more "globally", using a new flow theorem for Eulerian graphs (Theorem 2.4). In Sect. 1.2, after the preliminaries, we give a more detailed overview of these techniques and the proof of the theorem. Our argument is somewhat technical, but it demonstrates the potential of the Local-Connectivity ATSP problem as a tool for attacking general ATSP.

Finally, let us remark that both STSP [4,15] and ATSP [5] have been studied in the case when all distances are either 1 or 2. That restriction is very different from our setting, as in those cases the input graph is complete. In particular, it is trivial to get a 2-approximation algorithm there, whereas in our setting – where the input graph is *not* complete – a constant factor approximation guarantee already requires non-trivial algorithms.

---

[2] For ATSP, we can think of a node-weighted graph as an edge-weighted graph where the weight of an edge (u, v) equals the node weight of u.

1.1   Notation and Preliminaries

We consider an edge-weighted directed graph $G = (V, E, w)$ with $w : E \to \mathbb{R}_+$. For a vertex subset $S \subseteq V$ we let $\delta^+(S) = \{(u, v) \in E : u \in S, v \in V \setminus S\}$ and $\delta^-(S) = \{(u, v) \in E : u \in V \setminus S, v \in S\}$ denote the sets of outgoing and incoming edges, respectively. For a subset of edges $E' \subseteq E$, we use $\delta^+_{E'}(S) = \delta^+(S) \cap E'$ and $\delta^-_{E'}(S) = \delta^-(S) \cap E'$. We also let $\mathcal{C}(E') = (\tilde{G}_1, \ldots, \tilde{G}_k)$ denote the set of weakly connected components of the graph (V, E'); the vertex set V will always be clear from the context. For a directed graph $\tilde{G}$ we use $V(\tilde{G})$ to denote its vertex set and $E(\tilde{G})$ the edge set. For brevity, we denote the singleton set $\{v\}$ by v (e.g. $\delta^+(v) = \delta^+(\{v\})$), and we use the notation $x(F) = \sum_{e \in F} x_e$ for a subset $F \subseteq E$ of edges. For the case of two edge weights, we use $0 \leq w_0 < w_1$ to denote the two possible values, and partition $E = E_0 \cup E_1$ so that $w(e) = w_0$ if $e \in E_0$ and $w(e) = w_1$ if $e \in E_1$. We will refer to edges in E_0 and E_1 as cheap and expensive edges, respectively.

We define ATSP as the problem of finding a connected Eulerian subgraph of minimum weight. As already mentioned in the introduction, this definition is equivalent to that of visiting each city exactly once (in the metric completion) since we assume the triangle inequality. The formal definition is as follows.

ATSP

Given: An edge-weighted (strongly connected) digraph $G = (V, E, w)$.
Find: A multisubset F of E of minimum total weight $w(F) = \sum_{e \in F} w(e)$ such that (V, F) is Eulerian and connected.

Held-Karp Relaxation. The Held-Karp relaxation has a variable $x_e \geq 0$ for every edge in G. The intended meaning is that x_e should equal the number of times e is used in the solution. The relaxation $\mathrm{LP}(G)$ is defined as follows:

$$
\begin{aligned}
\text{minimize} \quad & \sum_{e \in E} w(e) x_e \\
\text{subject to} \quad & x(\delta^+(v)) = x(\delta^-(v)) && v \in V, \\
& x(\delta^+(S)) \geq 1 && \emptyset \neq S \subsetneq V, \\
& x \geq 0.
\end{aligned}
\qquad (\mathrm{LP}(G))
$$

The first set of constraints says that the in-degree should equal the out-degree for each vertex, i.e., the solution should be Eulerian. The second set of constraints enforces that the solution is connected; they are sometimes referred to as subtour elimination constraints. Finally, we remark that although the Held-Karp relaxation has exponentially many constraints, it is well-known that we can solve it in polynomial time either by using the ellipsoid method with a separation oracle or by formulating an equivalent compact (polynomial-size) linear program. We will use x^* to denote an optimal solution to $LP(G)$ of value OPT, which is a lower bound on the value of an optimal solution to ATSP on G.

Local-Connectivity ATSP. The Local-Connectivity ATSP problem can be seen as a two-stage procedure. In the first stage, the input is an edge-weighted digraph $G = (V, E, w)$ and the output is a "lower bound" function $\text{lb} : V \to \mathbb{R}_+$ on the vertices such that $\text{lb}(V) \leq \text{OPT}$. In the second stage, the input is a partition of the vertices, and the output is an Eulerian multisubset of edges which crosses each set in the partition and where the ratio of weight to lb of every connected component is as small as possible. We now give the formal description of the second stage, assuming the lb function is already computed.

Local-Connectivity ATSP

Given: An edge-weighted digraph $G = (V, E, w)$, a function $\text{lb} : V \to \mathbb{R}_+$ with $\text{lb}(V) \leq \text{OPT}$, and a partitioning $V = V_1 \cup V_2 \cup \ldots \cup V_k$ of the vertices.

Find: A Eulerian multisubset F of E such that

$$|\delta_F^+(V_i)| \geq 1 \quad \text{for } i = 1, 2, \ldots, k \qquad \text{and} \qquad \max_{\tilde{G} \in \mathcal{C}(F)} \frac{w(\tilde{G})}{\text{lb}(\tilde{G})} \text{ is minimized.}$$

Here we used the notation that for a connected component $\tilde{G}$ of (V, F), $w(\tilde{G}) = \sum_{e \in E(\tilde{G})} w(e)$ (summation over the edges) and $\text{lb}(\tilde{G}) = \sum_{v \in V(\tilde{G})} \text{lb}(v)$ (summation over the vertices). We say that an algorithm for Local-Connectivity ATSP is α-*light* on G if it is guaranteed, for any partition, to find a solution F such that for every component $\tilde{G} \in \mathcal{C}(F)$, $w(\tilde{G})/\text{lb}(\tilde{G}) \leq \alpha$.

In [17], lb is defined as $\text{lb}(v) = \sum_{e \in \delta^+(v)} w(e)x_e^*$; note that $\text{lb}(V) = OPT$ in this case. We remark that we use the "α-light" terminology to avoid any ambiguities with the concept of approximation algorithms (an α-light algorithm does not compare its solution to an optimal solution to the given instance of Local-Connectivity ATSP).

Perhaps the main difficulty of ATSP is to satisfy the connectivity requirement, i.e., to select an Eulerian subset F of edges which connects the whole graph. Local-Connectivity ATSP relaxes this condition – we only need to find an Eulerian set F that crosses the k cuts defined by the partition. This makes it intuitively an "easier" problem than ATSP. Indeed, an α-approximation algorithm for ATSP (with respect to the Held-Karp relaxation) is trivially an α-light algorithm for Local-Connectivity ATSP for an arbitrary lb function with $\text{lb}(V) = OPT$: just return the same Eulerian subset F as the algorithm for ATSP; since the set F connects the graph, we have $\max_{\tilde{G} \in \mathcal{C}(F)} w(\tilde{G})/\text{lb}(\tilde{G}) = w(F)/\text{lb}(V) \leq \alpha$. Perhaps more surprisingly, the main technical theorem of [17] shows that the two problems are equivalent up to small constant factors.

Theorem 1.2 [17]. *Let $\mathcal{A}$ be an algorithm for Local-Connectivity ATSP. Consider an ATSP instance $G = (V, E, w)$, and let OPT denote the optimum value of the Held-Karp relaxation. If $\mathcal{A}$ is α-light on G, then there exists a tour of G with value at most 5α OPT. Moreover, for any $\varepsilon > 0$, a tour of value at most $(9+\varepsilon)\alpha$ OPT can be found in time polynomial in the number $n = |V|$ of vertices, in $1/\varepsilon$, and in the running time of $\mathcal{A}$.*

In other words, the above theorem says that in order to approximate an ATSP instance G, it is sufficient to devise a polynomial-time algorithm to calculate a lower bound lb and a polynomial time algorithm for Local-Connectivity ATSP that is $\mathcal{O}(1)$-light on G with respect to this lb function. Our main result is proved using this framework.

1.2   Technical Overview

Singleton partition. Let us start by outlining the fundamental ideas of our algorithm and comparing it to [17] for the special case of Local-Connectivity ATSP when all partition classes V_i are singletons. For unit weights, the choice $\mathrm{lb}(v) = \sum_{e \in \delta^+(v)} w(e)x_e^\star = x^\star(\delta^+(v))$ in [17] is a natural one: intuitively, every node is able to pay for its outgoing edges. We can thus immediately give an algorithm for this case: just select an arbitrary integral solution z to the circulation problem with node capacities $1 \leq z(\delta^+(v)) \leq \lceil x^\star(\delta^+(v)) \rceil$. Then for any v we have $z(\delta^+(v)) \leq x^\star(\delta^+(v)) + 1 \leq 2x^\star(\delta^+(v))$ and hence $\sum_{e \in \delta^+(v)} w(e)z_e \leq 2\,\mathrm{lb}(v)$, showing that z is a 2-light solution.

The same choice of lb does not seem to work in the presence of two different edge costs. Consider a case when every expensive edge carries only a small fractional amount of flow. Then $\sum_{e \in \delta^+(v)} w(e)x_e^\star$ can be much smaller than the expensive edge cost w_1, and thus the vertex v would not be able to "afford" even a single outgoing expensive edge. To resolve this problem, we bundle small fractional amounts of expensive flow, channelling them to reach a small set of terminals. This is achieved via Theorem 2.4, a flow result which might be of independent interest. It shows that within the fractional Held-Karp solution $x^\star$, we can send the flow from an arbitrary edge set E' to a sink set T with $|T| \leq 8x^\star(E')$; in fact, T can be any set minimal for inclusion such that it can receive the total flow from E'. We apply this theorem for $E' = E_1$, the set of expensive edges; let f be the flow from E_1 to T, and call elements of T *terminals*. Now, whenever an expensive edge is used, we will "force" it to follow f to a terminal in T, where it can be paid for. Enforcement is technically done by splitting the vertices into two copies, one carrying the f flow and the other the rest. Thus we obtain the *split graph* G_{sp} and split fractional optimal solution $x_{\mathrm{sp}}^\star$.

The design of the split graph is such that every walk in it which starts with an expensive edge must proceed through cheap edges until it reaches a terminal before visiting another expensive edge. In our terminology, expensive edges create "debt", which must be paid off at a terminal. Starting from an expensive edge, the debt must be carried until a terminal is reached, and no further debt can be taken in the meantime. The bound on the number of terminals guarantees that we can assign a lower bound function lb with $\text{lb}(V) \leq \text{OPT}$ such that (up to a constant factor) cheap edges are paid for locally, at their heads, whereas expensive edges are paid for at the terminals they are routed to. Such a splitting easily solves Local-Connectivity ATSP for the singleton partition: find an arbitrary integral circulation z_{sp} in the split graph with an upper bound $z_{\text{sp}}(\delta^+(v)) \leq \lceil 2x^\star_{\text{sp}}(\delta^+(v)) \rceil$ on every node, and a lower bound 1 on whichever copy of v transmits more flow. Note that $2x^\star_{\text{sp}}$ is a feasible fractional solution to this problem. We map z_{sp} back to an integral circulation z in the original graph by merging the split nodes, thus obtaining a constant-light solution.

Arbitrary partitions. Let us now turn to the general case of Local-Connectivity ATSP, where the input is an arbitrary partition $V = V_1 \cup \ldots \cup V_k$. For unit weights this is solved in [17] via an integer circulation problem on a modified graph. Namely, an auxiliary node A_i is added to represent each partition class V_i, and one unit of in- and outgoing flow from V_i is rerouted through A_i. In the circulation problem, we require exactly one in- and one outgoing edge incident to A_i to be selected. When we map the solution back to the original graph, there will be one incoming and one outgoing arc from every set V_i (thus satisfying the connectivity requirement) whose endpoints inside V_i violate the Eulerian condition. In [17] every V_i is assumed to be strongly connected, and therefore we can "patch up" the circulation by connecting the loose endpoints by an arbitrary path inside V_i. This argument easily gives a 3-light solution.

Let us observe that the strong connectivity assumption is in fact not needed for the result in [17]. Indeed, given a component V_i which is not strongly connected, consider its decomposition into strongly connected (sub)components, and pick a $U_i \subseteq V_i$ which is a sink (i.e. it has no edges outgoing to $V_i \setminus U_i$). We proceed by rerouting 1 unit of flow through a new auxiliary vertex just as in that algorithm, but we do this for U_i instead. This guarantees that U_i has at least one outgoing edge in our solution, and that edge must leave V_i as well.

We now turn to our result for two different edge weights. We are aiming for a similar construction as in the unit-weight case: based on the split graph G_{sp}, we construct an integer circulation problem with an auxiliary vertex A_i representing a certain subset $U_i \subseteq V_i$ for every $1 \leq i \leq k$. We then map its solution back to the original graph and patch up the loose endpoints inside every U_i by a path. However, we have to account for the following difficulties: *(i)* an edge leaving U_i should also leave V_i; *(ii)* debt should not disappear inside U_i: if the edge entering it carries debt but the edge leaving does not, we must make sure this difference can be charged to a terminal in U_i; *(iii)* the path used inside U_i must pay for all expensive edges it uses. All three issues can be appropriately tackled by defining an auxiliary graph inside V_i. Edges of the auxiliary graph represent paths containing one expensive edge and one terminal (which can pay for themselves); however, these paths may not map to paths in the split graph. We select the subset $U_i \subseteq V_i$ as a sink component in the auxiliary graph.

2   Algorithm for Local-Connectivity ATSP

We prove our main result in this section. Our claim for ATSP follows from solving Local-Connectivity ATSP:

Theorem 2.1. *There is a polynomial-time 100-light algorithm for Local-Connectivity ATSP on graphs with two edge weights.*

Together with Theorem 1.2, this implies our main result:

Theorem 2.2. *For any graph with two edge weights, the integrality gap of its Held-Karp relaxation is at most 500. Moreover, we can find an 901-approximate tour in polynomial time.*

The factor 500 comes from $5 \cdot 100$, and 901 is selected so that $(9+\varepsilon) \cdot 100 \leq 901$. Our proof of Theorem 2.1 proceeds as outlined in Sect. 1.2. In this extended abstract, we only describe the construction; the proof is given in the full version.

Recall that the edges are partitioned into the set E_0 of cheap edges and the set E_1 of expensive edges. Set $x^\star$ to be an optimal solution to the Held-Karp relaxation. We start by noting that the problem is easy if $x^\star$ assigns very small total fractional value to expensive edges. In that case, we can easily reduce the problem to the unweighted case which was solved in [17].

Lemma 2.3. *There is a polynomial-time 6-light algorithm for Local-Connectivity ATSP for graphs where $x^\star(E_1) < 1$.*

For the rest of this section, we thus assume $x^\star(E_1) \geq 1$. Our objective is to define a function lb : $V \to \mathbb{R}_+$ such that $\mathrm{lb}(V) \leq \mathrm{OPT} = w(x^\star)$ and then show how to, given a partition $V = V_1 \cup \ldots \cup V_k$, find an Eulerian set of edges F which crosses all V_i-cuts and is $\mathcal{O}(1)$-light with respect to the defined lb function.

2.1   Calculating lb and Constructing the Split Graph

Finding terminals T and flow f. For this, we use the following flow result.

Theorem 2.4. *Let $D = (V \cup \{s\}, E)$ be a directed graph, $c : E \to \mathbb{R}_+$ – a nonnegative capacity vector, and s – a source node with no incoming edges, i.e., $\delta^-(s) = \emptyset$. Assume that for all $\emptyset \neq S \subseteq V$ we have*

$$c(\delta^-(S)) \geq \max\{1, c(\delta^+(S))\}. \tag{1}$$

Consider a set $T \subseteq V$ such that there exists a flow $f \leq c$ of value $c(\delta^+(s))$ from the source s to the sink set T, and T is minimal subject to this property. Then $|T| \leq 8c(\delta^+(s))$.

Corollary 2.5. *There exist a vertex set $T \subseteq V$ and a flow $f : E \to \mathbb{R}_+$ from source set $\{\mathrm{tail}(e) : e \in E_1\}$ to sink set T of value $x^\star(E_1)$ such that: (a) $|T| \leq 8x^\star(E_1)$, (b) $f \leq x^\star$, (c) f saturates all expensive edges, i.e., $f(e) = x_e^\star$ for all $e \in E_1$, (d) for each $t \in T$, $f(E_0 \cap \delta^+(t)) = 0$ and $f(\delta^-(t)) > 0$. Moreover, T and f can be computed in polynomial time.*

Definition of lb. We set lb $: V \rightarrow \mathbb{R}_+$ to be a scaled-down variant of $\overline{\mathrm{lb}} : V \rightarrow \mathbb{R}_+$ which is defined as follows:

$$\overline{\mathrm{lb}}(v) := \begin{cases} w_0 \cdot x^\star(\delta^-(v)) & \text{if } v \notin T, \\ w_0 \cdot x^\star(\delta^-(v)) + w_1 \cdot \lceil f(\delta^-(t)) \rceil & \text{if } v \in T. \end{cases}$$

The definition of lb is now simply $\mathrm{lb}(v) = \overline{\mathrm{lb}}(v)/10$. The scaling-down is done so as to satisfy $\mathrm{lb}(V) \leq \mathrm{OPT}$ (see Lemma 2.6). Clearly we have $\overline{\mathrm{lb}}(v) \geq w_0$ for all $v \in V$ and $\overline{\mathrm{lb}}(t) \geq w_1 + w_0 \geq w_1$ for terminals $t \in T$.

The intuition behind this setting of $\overline{\mathrm{lb}}$ is that we want to pay for each expensive edge $e \in E_1$ in the terminal $t \in T$ which the flow f "assigns" to e. Indeed, in the split graph we will reroute flow (using f) so as to ensure that any path which traverses e must then visit such a terminal t to offset the cost of the expensive edge.

Lemma 2.6. $\overline{\mathrm{lb}}(V) \leq 10 \cdot \mathrm{OPT}$.

Construction of the split graph. The next step is to reroute flow so as to ensure that all expensive edges are "paid for" by the lb at terminals. To this end, we define a new *split graph* and a *split circulation* on it.

Definition 2.7. *The* split graph G_{sp} *is defined as follows. For every* $v \in V$ *we create two copies* v^0 *and* v^1 *in* $V(G_{\mathrm{sp}})$*. For every cheap edge* $(u,v) \in E_0$*:*

- *if* $x^\star(u,v) - f(u,v) > 0$*, create an edge* (u^0, v^0) *in* $E(G_{\mathrm{sp}})$ *with* $x^\star_{\mathrm{sp}}(u,v) = x^\star(u,v) - f(u,v)$,
- *if* $f(u,v) > 0$*, create an edge* (u^1, v^1) *in* $E(G_{\mathrm{sp}})$ *with* $x^\star_{\mathrm{sp}}(u,v) = f(u,v)$.

For every expensive edge $(u,v) \in E_1$ *we create one edge* (u^0, v^1) *in* $E(G_{\mathrm{sp}})$ *with* $x^\star_{\mathrm{sp}}(u,v) = f(u,v)$*. Finally, for each* $t \in T$ *we create an edge* (t^1, t^0) *in* $E(G_{\mathrm{sp}})$ *with* $x^\star_{\mathrm{sp}}(t^1, t^0) = f(\delta^-(t))$*.*

The new edges are weighted as follows: images of edges in E_0 *have weight* w_0*, the images of edges in* E_1 *have weight* w_1*, and the new edges* (t^1, t^0) *have weight* 0*. Let us denote the new weight function by* w_{sp}*.*

Vertices v^0 *will be called* free *vertices and vertices* v^1 *will be called* debt *vertices. Edges entering a free vertex will be called* free *edges, and those entering a debt vertex will be called* debt *edges.*

By construction we have that *(a)* $x^\star_{\mathrm{sp}}$ is a circulation on G_{sp}, *(b)* (the image of) every cut is still crossed by at least 1 unit of $x^\star_{\mathrm{sp}}$, and *(c)* any path in G_{sp} which begins with a debt edge and ends with a free edge must go through a terminal.

2.2   Solving Local-Connectivity ATSP

Now our algorithm is given a partition $V = V_1 \cup ... \cup V_k$. The objective is to output a set of edges F which crosses all V_i-cuts and is $\mathcal{O}(1)$-light with respect to our lb function. We do so by first defining *auxiliary graphs* that help us modify the split graph so as to force our solution to cross the cuts defined by the partition. We then use such a flow to define the set F of edges.

Construction of auxiliary graphs and modification of split graph. Our first step is to construct an *auxiliary graph* for each component V_i. The strong-connectivity structure of this graph will guide our algorithm.

Definition 2.8. *The auxiliary graph G_i^{aux} is a graph with vertex set V_i and the following edge set: for $u, v \in V_i$, $(u, v) \in E(G_i^{\mathrm{aux}})$ if any of the following three conditions is satisfied:*

- *there is a cheap edge $(u, v) \in E_0 \cap G[V_i]$ inside V_i, or*
- *there is a u-v-path in $G[V_i]$ whose first edge is expensive and all other edges are cheap, and $v \in T$ is a terminal – we then call the edge $(u, v) \in E(G_i^{\mathrm{aux}})$ a* postpaid edge *– or*
- *there is a u-v-path in $G[V_i]$ whose last edge is expensive and all other edges are cheap, and $u \in T$ is a terminal – we then call the edge $(u, v) \in E(G_i^{\mathrm{aux}})$ a* prepaid edge.

Define the preimage *of such an edge $(u, v) \in E(G_i^{\mathrm{aux}})$ to be the* shortest *path inside V_i as above (in the first case, a single edge).*

Now, for each i consider a decomposition of G_i^{aux} into strongly connected components. Let $U_i \subseteq V_i$ be the vertex set of a sink component in this decomposition. That is, there is no edge from U_i to $V_i \setminus U_i$ in the auxiliary graph G_i^{aux}. Note that G_i^{aux} is constructed based only on the original graph G and not the split graph G_{sp}. However, we will solve Local-Connectivity ATSP by solving an integral circulation problem on G'_{sp}: a modification of the split graph G_{sp}, described as follows.

For each i, define $U_i^{\mathrm{sp}} = \{v^0, v^1 : v \in U_i\} \subseteq V(G_{\mathrm{sp}})$ to be the set of vertices in the split graph corresponding to U_i. (Note that U_i^{sp} may not be strongly connected in G_{sp}.) We are going to reroute part of the $x_{\mathrm{sp}}^\star$ flow going in and out of U_i^{sp} to a new auxiliary vertex A_i. While the 3-light algorithm for unit-weight graphs rerouted flow from all boundary edges of a component U_i (see Sect. 1.2), here we will be more careful and choose only a subset of boundary edges of U_i^{sp} to be rerouted.

To this end, select a subset of edges $X_i^- \subseteq \delta^-(U_i^{\mathrm{sp}})$ with $x_{\mathrm{sp}}^\star(X_i^-) = 1/2$ such that either all edges in X_i^- are debt edges, or all are free edges.

We define the set of outgoing edges $X_i^+ \subseteq \delta^+(U_i^{\mathrm{sp}})$ to be, intuitively, the edges over which the flow that entered U_i^{sp} by X_i^- exits U_i^{sp}. That is, consider an arbitrary cycle decomposition of the circulation $x_{\mathrm{sp}}^\star$, and look at the set of cycles containing the edges in X_i^-. We define X_i^+ as the set of edges on these cycles that first leave U_i^{sp} after entering U_i^{sp} on an edge in X_i^-; clearly, $x_{\mathrm{sp}}^\star(X_i^+) = 1/2$.

Let g_i denote the flow on these cycles connecting the heads of edges in X_i^- and the tails of edges in X_i^+. We will use the following claim later in the construction.

Fact 2.9. *Assume all edges in X_i^- are debt edges but $e \in X_i^+$ is a free edge or an expensive edge. Then there exists a path in $G_{\mathrm{sp}}[U_i]$ between a vertex t^0 (for some terminal $t \in T$) and the tail of e, made up of only cheap edges.*

We now transform G_{sp} into a new graph G'_{sp} and $x^\star_{\mathrm{sp}}$ into new circulation x'_{sp} as follows. For every set V_i in the partition we introduce a new auxiliary vertex A_i and redirect all edges in X_i^- to point to A_i and those in X_i^+ to point from A_i. We further subtract the flow g_i inside U_i^{sp}; hence the resulting vector x'_{sp} will be a circulation, with $x'_{\mathrm{sp}}(\delta^-(A_i)) = 1/2$. If X_i^- is a set of free edges, then we will say that A_i is a free vertex, otherwise we say that it is a debt vertex.

Transforming x'_{sp} into an integral flow and obtaining our solution F. In the next step we round x'_{sp} to integrality while respecting degrees of vertices:

Lemma 2.10. *There exists an integral circulation y'_{sp} on G'_{sp} satisfying the following conditions: (a) $y'_{\mathrm{sp}}(\delta^-(v)) \leq \lceil 2x^\star_{\mathrm{sp}}(\delta^-(v)) \rceil$ for each $v \in V(G_{\mathrm{sp}})$, (b) $y'_{\mathrm{sp}}(\delta^-(A_i)) = 1$ for each i. Such a circulation y'_{sp} can be found in polynomial time.*

We will now transform y'_{sp} into an Eulerian set of edges F in the original graph G. We can think of this as a three-stage process.

First, we map all edges adjacent to the auxiliary vertices A_i back to their preimages in G_{sp}, obtaining from y'_{sp} an integral *pseudo-flow* y_{sp} in G_{sp}. (We use the term *pseudo-flow* as now, some vertices may not satisfy flow conservation.)

Second, we contract the two copies v^0 and v^1 of every vertex $v \in V$, thus mapping all edges back to their preimages in G. (We remove all edges (t^1, t^0) for $t \in T$.) This creates an integral pseudo-flow y in G.

Since the in- and out-degree of A_i were exactly 1 in y'_{sp}, now (in y) in each component U_i there is a pair of vertices u_i, v_i which are the head and tail, respectively, of the mapped-back edges adjacent to A_i. These are the only vertices where flow conservation in y can be violated. As the third step, to repair this, we route a walk P_i from u_i to v_i. Our Eulerian set of edges $F \subseteq E$ which we finally return is the integral pseudo-flow y plus the union (over i) of all such walks P_i, i.e., $\mathbb{1}_F = y + \sum_i \mathbb{1}_{P_i}$.

It remains to describe how we route these paths. Fix i. Recall that U_i is strongly connected in G_i^{aux}. We distinguish two cases:

- If A_i is a free vertex or the edge exiting A_i in y'_{sp} (in G'_{sp}) is a debt edge, then select a shortest u_i-v_i-path in G_i^{aux}, map each edge of this path to its preimage path (see Definition 2.8) and concatenate them to obtain a u_i-v_i-walk P_i in V_i.
- If A_i is a debt vertex but the edge exiting A_i in y'_{sp} (in G'_{sp}) is a free edge, then by Fact 2.9 there is a terminal t inside U_i, with a path from t to v_i using only cheap edges. Proceed as above to obtain a u_i-t-walk and then append this cheap t-v_i-path to it, obtaining a u_i-v_i-walk P_i in V_i.

This concludes the description of the algorithm. In the full version of the paper we prove that the returned Eulerian set of edges F has the properties we desire, i.e.,

Lemma 2.11. *For every connected component $\widetilde{G}$ of (V, F) we have $w(\widetilde{G}) \leq 10 \cdot \overline{\mathrm{lb}}(\widetilde{G})$.*

Lemma 2.12. *For every component V_i we have $|\delta_F^+(V_i)| \geq 1$.*

Lemmas 2.6 and 2.11 together prove that our algorithm is 100-light with respect to lb.

References

1. Anari, N., Gharan, S.O.: Effective-resistance-reducing flows and asymmetric TSP. CoRR, abs/1411.4613 (2014)
2. Arora, S., Grigni, M., Karger, D.R., Klein, P.N., Woloszyn, A.: A polynomial-time approximation scheme for weighted planar graph TSP. In: Proceedings of SODA, vol. 98, pp. 33–41 (1998)
3. Asadpour, A., Goemans, M.X., Madry, A., Gharan, S.O., Saberi, A.: An O(log n/ log log n)-approximation algorithm for the asymmetric traveling salesman problem. In: Proceedings of SODA, pp. 379–389 (2010)
4. Berman, P., Karpinski, M.: 8/7-approximation algorithm for (1, 2)-TSP. In: Proceedings of SODA, pp. 641–648 (2006)
5. Bläser, M.: A 3/4-approximation algorithm for maximum ATSP with weights zero and one. In: Jansen, K., Khanna, S., Rolim, J.D.P., Ron, D. (eds.) RANDOM 2004 and APPROX 2004. LNCS, vol. 3122, pp. 61–71. Springer, Heidelberg (2004)
6. Christofides, N.: Worst-case analysis of a new heuristic for the travelling salesman problem. Technical report, DTIC Document (1976)
7. Erickson, J., Sidiropoulos, A.: A near-optimal approximation algorithm for asymmetric TSP on embedded graphs. In: Proceedings of SOCG, p. 130 (2014)
8. Frieze, A.M., Galbiati, G., Maffioli, F.: On the worst-case performance of some algorithms for the asymmetric traveling salesman problem. Networks **12**(1), 23–39 (1982)
9. Gharan, S.O., Saberi, A.: The asymmetric traveling salesman problem on graphs with bounded genus. In: Proceedings of SODA, pp. 967–975. SIAM (2011)
10. Gharan, S.O., Saberi, A., Singh, M.: A randomized rounding approach to the traveling salesman problem. In: Proceedings of FOCS, pp. 550–559 (2011)
11. Grigni, M., Koutsoupias, E., Papadimitriou, C.H.: An approximation scheme for planar graph TSP. In: Proceedings of FOCS, pp. 640–645 (1995)
12. Karpinski, M., Lampis, M., Schmied, R.: New inapproximability bounds for TSP. J. Comput. Syst. Sci. **81**(8), 1665–1677 (2015)
13. Mömke, T., Svensson, O.: Approximating graphic TSP by matchings. In: 2011 Proceedings of FOCS, pp. 560–569 (2011)
14. Mucha, M.: 13/9-approximation for graphic TSP. In: Proceedings of STACS, pp. 30–41 (2012)
15. Papadimitriou, C.H., Yannakakis, M.: The traveling salesman problem with distances one and two. Math. Oper. Res. **18**(1), 1–11 (1993)
16. Sebő, A., Vygen, J.: Shorter tours by nicer ears: 7/5-approximation for the graph-TSP, 3/2 for the path version, and 4/3 for two-edge-connected subgraphs. Combinatorica **34**(5), 597–629 (2014)
17. Svensson, O.: Approximating ATSP by relaxing connectivity. In: Proceedings of FOCS (2015)
18. Williamson, D.P., Shmoys, D.B.: The Design of Approximation Algorithms. Cambridge University Press, New York (2011)

Improved Approximation Algorithms
for Hitting 3-Vertex Paths

Samuel Fiorini[1], Gwenaël Joret[2(✉)], and Oliver Schaudt[3]

[1] Département de Mathématique, Université libre de Bruxelles, Brussels, Belgium
sfiorini@ulb.ac.be
[2] Département d'Informatique, Université libre de Bruxelles, Brussels, Belgium
gjoret@ulb.ac.be
[3] Institut für Informatik, Universität zu Köln, Köln, Germany
schaudto@uni-koeln.de

Abstract. We study the problem of deleting a minimum cost set of vertices from a given vertex-weighted graph in such a way that the resulting graph has no induced path on three vertices. This problem is often called *cluster vertex deletion* in the literature and admits a straightforward 3-approximation algorithm since it is a special case of the vertex cover problem on a 3-uniform hypergraph. Very recently, You et al. [14] described an efficient 5/2-approximation algorithm for the unweighted version of the problem. Our main result is a 7/3-approximation algorithm for arbitrary weights, using the local ratio technique. We further conjecture that the problem admits a 2-approximation algorithm and give some support for the conjecture. This is in sharp constrast with the fact that the similar problem of deleting vertices to eliminate all triangles in a graph is known to be UGC-hard to approximate to within a ratio better than 3, as proved by Guruswami and Lee [7].

1 Introduction

Given a graph[1] G and cost function $c : V(G) \to \mathbb{R}_+$, the *cluster vertex deletion problem* (CLUSTER-VD) is to find a minimum cost set X of vertices such that each component of $G - X$ is a complete graph. Equivalently, $X \subseteq V(G)$ is a feasible solution if and only if $G - X$ contains no induced subgraph isomorphic to P_3, the path on three vertices.

It should be clear that the problem has a 3-approximation algorithm: Assuming unit costs for simplicity, build any inclusionwise maximal collection $\mathcal{C}$ of vertex-disjoint induced P_3's in G and include in X every vertex covered by some member of $\mathcal{C}$. If $\mathcal{C}$ contains k subgraphs then we get a lower bound of k on the optimum. On the other hand, the cost of X is $3k$.

We acknowledge support from ERC grant *FOREFRONT* (grant agreement no. 615640) funded by the European Research Council under the EU's 7th Framework Programme (FP7/2007-2013), and ARC grant AUWB-2012-12/17-ULB2 *COPHYMA* funded by the French community of Belgium.

[1] Graphs in this paper are finite, simple, and undirected.

© Springer International Publishing Switzerland 2016
Q. Louveaux and M. Skutella (Eds.): IPCO 2016, LNCS 9682, pp. 238–249, 2016.
DOI: 10.1007/978-3-319-33461-5_20

The problem admits an approximation-preserving reduction from VERTEX COVER: if H is any given graph, let G denote the graph obtained from H by adding a pendent edge to every vertex. Then solving VERTEX COVER on H is equivalent to solving CLUSTER-VD on G. Hence, known hardness and inapproximability results for VERTEX COVER apply to CLUSTER-VD as well, and in particular it is UGC-hard to approximate CLUSTER-VD within any ratio better than 2. We show that we can however come close to 2.

Theorem 1. CLUSTER-VD *admits a 7/3-approximation algorithm.*

We further conjecture that CLUSTER-VD can be 2-approximated in polynomial time, as is the case for VERTEX COVER. We give some support for this conjecture in Sect. 6, where we report on a 2-approximation algorithm for the case where the input graph does not contain a diamond (K_4 minus an edge) as an induced subgraph.

In contrast, the problem of finding a minimum cost set of vertices X such that $G - X$ has no triangle is known to be UGC-hard to approximate to within any ratio better than 3, as proved by Guruswami and Lee [7] (see also [8] for related inapproximability results).

Previous Work. CLUSTER-VD was previously mostly studied in terms of fixed parameter algorithms. Hüffner et al. [9] first gave a $O(2^k k^9 + nm)$-time fixed-parameter algorithm, parameterized by the solution size k, where n and m denote the number of vertices and edges of the graph, respectively. This was subsequently improved by Boral et al. [3], who gave a $O(1.9102^k(n + m))$-time algorithm. See also Iwata and Oka [10] for related results in the fixed parameter setting.

As for approximation algorithms, nothing better than a 3-approximation was known until the very recent work of You et al. [14], who showed that the unweighted version of CLUSTER-VD admits a 5/2-approximation algorithm. They further showed that their algorithm could be implemented efficiently, in $O(nm + n^2)$-time, using fast modular decomposition.

We note that the work of You et al. [14] and ours have been done independently. While we obtained a better approximation ratio of 7/3, let us remark that the running time of our algorithm is much larger (though still polynomial). We leave it as an open question whether it could be brought down to a small polynomial using the techniques from [14].

Incidentally, there was recent activity on another restriction of the vertex cover problem on 3-uniform hypergraph, namely, the feedback vertex set problem in tournaments. For that problem, the 5/2-approximation algorithm by Cai et al. [4] was the best known for many years, until the very recent work of Mnich et al. [12] who found a 7/3-approximation algorithm for the problem.

Our Approach. Our approximation algorithm is based on the *local ratio* technique. In order to illustrate the general approach, let us give a very simple 2-approximation algorithm for hitting all P_3-*subgraphs* (instead of induced subgraphs) in a given weighted graph (G, c), see Algorithm 1 below.

Algorithm 1. HITTING-P_3-SUBGRAPHS-APX(G, c)

Require: (G, c) a weighted graph
Ensure: X an inclusionwise minimal set of vertices hitting all the P_3 subgraphs
    if G has no P_3 subgraph **then**
        $X \leftarrow \varnothing$
    else if (G, c) has some zero-cost vertex u **then**
        $X' \leftarrow$ HITTING-P_3-SUBGRAPHS-APX$(G - u, c$ restricted to $V(G - u))$
        $X \leftarrow X'$ if $G - X'$ has no P_3 subgraph; $X \leftarrow X' \cup \{u\}$ otherwise
    else
        $u \leftarrow$ vertex of degree $d(u) \geqslant 2$, and let (H, c_H) be the weighted star centered
            at u with $V(H) := N(u) \cup \{u\}$, $c_H(u) := d(u) - 1$ and $c_H(v) := 1$ for $v \in N(u)$
        $\lambda^* \leftarrow$ maximum scalar λ s.t. $c(v) - \lambda c_H(v) \geqslant 0$ for all $v \in V(H)$
        $X \leftarrow$ HITTING-P_3-SUBGRAPHS-APX$(G, c - \lambda^* c_H)$
    end if
    return X

It can be easily verified that the set X returned by Algorithm 1 is an inclusionwise minimal feasible solution. The reason why the algorithm is a 2-approximation is that optimum cost for the weighted star (H, c_H) is $d(u) - 1$ while the solution X returned by the algorithm misses at least one of the vertices of the star, and thus has a local cost of at most $2(d(u) - 1)$.

We remark that a 2-approximation algorithm for the problem of hitting P_3-subgraphs can also be obtained via a straightforward modification of the primal/dual 2-approximation algorithm of Chudak et al. [5] for the feedback vertex set problem. (Indeed, this is exactly what was done by Tu and Zhou [13].) However, the resulting algorithm is nowhere near as simple as Algorithm 1.

It is perhaps worth pointing out that, in the case of triangle-free graphs, hitting P_3's or induced P_3's are the same problem. This was actually an important insight for the 5/2-approximation algorithm of You et al. [14]. However, for arbitrary graphs the induced version of the problem seems much more difficult. Nevertheless, we are tempted to take the simplicity of Algorithm 1 as a hint that the local ratio technique is a good approach to attack the problem.

From a high level point of view, the structure of our 7/3-approximation algorithm for CLUSTER-VD is as follows: As long as there is an induced P_3 in the graph, either we can apply a reduction operation (identifying *true twins*) that does not change the optimum, or we find some special induced subgraph H and decrease the weights of its vertices in (G, c) proportionally to a carefully chosen weighting c_H for the vertices of H, ensuring a local ratio of 7/3. (We remark that c_H depends on H only and is thus independent of the weights of vertices in G, similarly as in Algorithm 1). The crux of our proof is showing that, if no reduction can be applied, then the aforementioned special induced subgraph always exists. The list of induced subgraphs that we look for is given in Fig. 1. Every graph on the list has at most 7 vertices, and thus we can test their existence in $O(n^7)$-time.

2   Definitions and Preliminaries

Let G be a graph. Recall that the feasible solutions to CLUSTER-VD in G are the sets of vertices X that intersect every induced subgraph isomorphic to P_3. For this reason, we call such sets X *hitting sets* of G. We denote by $\mathrm{OPT}(G)$ the minimum size of a hitting set of G. The definitions extend naturally in the weighted setting: Given a weighted graph (G, c), where $c : V(G) \to \mathbb{R}_+$, we let $\mathrm{OPT}(G, c)$ denote the minimum weight of a hitting set of G. As expected, the *cost* of set $X \subseteq V(G)$ is defined as $c(X) := \sum_{v \in X} c(v)$.

For $X \subseteq V(G)$, the subgraph of G induced by X is denoted by $G[X]$. When H is an induced subgraph of G or isomorphic to an induced subgraph of G, we sometimes say that G *contains* H. If G does not contain H, we also say that G is *H-free*.

For $v \in V(G)$, the neighborhood of v is denoted by $N(v)$. From time to time, to indicate that x is a neighbor of y, we simply say that x *sees* y.

In a few occasions in the paper we resort to *trigraphs*, which are graphs with a set of special edges called the *undecided edges* (in figures, these are typically represented by wiggly edges). A trigraph is an efficient way to represent several graphs, its *instantiations*. These are the corresponding graphs in which each undecided edge may become an edge or not. Much of the terminology we use for graphs can be extended to trigraphs in a natural way. In particular, we say that graph G contains trigraph H or that H is an induced subtrigraph of G if G contains some instantiation of H.

3   Tools

3.1 α-Good Induced Subgraphs

Given a graph G, an induced subgraph H of G, and a weighting $c_H : V(G) \to \mathbb{R}_+$, we say that (H, c_H) is *α-good in G* if for every inclusionwise minimal hitting set X of G we have

$$\sum_{v \in X \cap V(H)} c_H(v) \leqslant \alpha \cdot \mathrm{OPT}(H, c_H). \tag{1}$$

Moreover, we say that an induced subgraph H of G is itself α-good in G if there exists a weighting c_H such that (H, c_H) is α-good. The first technical tool of our 7/3-approximation algorithm is the following lemma, which provides a list of α-good induced subgraphs where $\alpha \leqslant 7/3$ along with their corresponding weights, see Fig. 1. Due to length restrictions, the proof of the lemma is omitted.

Lemma 1. *Let G be a graph and H be an induced subgraph of G. Then H is 7/3-good in G whenever*

(i) H is isomorphic to C_4, W_5, $K_{1,4}$, the dart, the turtle, H_1 or H_2;

(ii) H is isomorphic to an instantiation of H_3, H_4 or H_5;

(iii) H is isomorphic to P_3, $K_{1,3}$, the gem or the bull, and there exists some vertex of H that has no neighbor in $G - V(H)$.

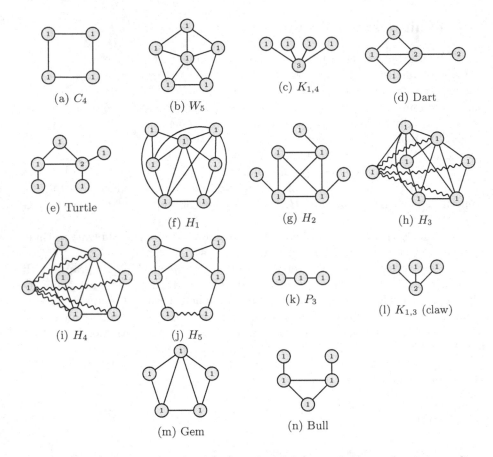

Fig. 1. The 7/3-good induced sub(tri)graphs of Lemma 1. For each corresponding induced subgraph H, a weighting c_H witnessing 7/3-goodness is shown.

Let us emphasize that in point three of the above lemma, the vertex of H having no neighbor in $G - V(H)$ is arbitrary, that is, it can be any vertex of H. For any graph G, let $\mathcal{H}(G)$ denote the collection of all weighted induced subgraphs (H, c_H) that are isomorphic to a weighted graph from Fig. 1 (the trigraph H_3 has 16 corresponding graphs). By Lemma 1, every $(H, c_H) \in \mathcal{H}(G)$ is 7/3-good in G. Notice that $\mathcal{H}(G)$ contains at most 45 isomorphism classes of weighted induced subgraphs, all of which involving graphs with at most 7 vertices. Notice also that in our collection of weighted induced subgraphs $\mathcal{H}(G)$, the induced subgraph H determines uniquely the weighting H. Thus $\mathcal{H}(G)$ contains $O(n^7)$ weighted induced graphs, where n denotes the number of vertices of G.

3.2   True Twins

Two vertices u, u' of a graph G are called *true twins* if they are adjacent and have the same neighborhood in $G - \{u, u'\}$. True twins have a particularly nice behavior regarding CLUSTER-VD, as proved in our next lemma. This is our second main technical tool.

Lemma 2. *Let (G, c) be a weighted graph and $u, u' \in V(G)$ be true twins. Let (G', c') denote the weighted graph obtained from G by transferring the whole cost of u' to u and then deleting u', that is, let $G' := G - u'$ and $c'(v) := c(v)$ if $v \in V(H'), v \neq u$ and $c'(v) := c(u) + c(u')$ if $v = u$. Then $\mathrm{OPT}(G, c) = \mathrm{OPT}(G', c')$.*

Proof. We have $\mathrm{OPT}(G, c) \leqslant \mathrm{OPT}(G', c')$ because every hitting set X' of G' yields a hitting set X of G with the same cost: we let $X := X' \cup \{u'\}$ if X contains u and $X := X'$ otherwise.

Conversely, we have $\mathrm{OPT}(G', c') \leqslant \mathrm{OPT}(G, c)$ because any inclusionwise minimal cost hitting set X of G either contains both of the true twins u and u', or none of them.   □

If G does not contain any pair of true twins, we say that it is *twin-free*.

4 Algorithm

Our 7/3-approximation algorithm is described below, see Algorithm 2. Although we could have presented as a primal-dual algorithm, we chose to present it within the local ratio framework in order to avoid some technicalities, especially those related to the elimination of true twins.

The following lemma makes explicit a simple property of CLUSTER-VD that is key when using the local ratio technique. This property is common to many minimization problems, and is often referred to as the *Local Ratio Lemma*; see e.g. the survey of Bar-Yehuda *et al.* [2].

Lemma 3 (Local Ratio Lemma). *Let (G, c) be a weighted graph with c the sum of two cost functions c' and c'', and let $\alpha \geqslant 1$. If X is a hitting set of G such that $c'(X) \leqslant \alpha \cdot \mathrm{OPT}(G, c')$ and $c''(X) \leqslant \alpha \cdot \mathrm{OPT}(G, c'')$, then $c(X) \leqslant \alpha \cdot \mathrm{OPT}(G, c)$.*

Proof. Since $c(X) = c'(X) + c''(X)$, it is enough to show that $\mathrm{OPT}(G, c') + \mathrm{OPT}(G, c'') \leqslant \mathrm{OPT}(G, c)$. To see this, let X^* be an optimal hitting set for (G, c). Then $\mathrm{OPT}(G, c) = c(X^*) = c'(X^*) + c''(X^*) \geqslant \mathrm{OPT}(G, c') + \mathrm{OPT}(G, c'')$.   □

Besides the Local Ratio Lemma, the analysis of Algorithm 2 relies on two lemmas. The first lemma guarantees that the algorithm terminates. That is, the algorithm is always able to find a 7/3-good weighted induced subgraph in Step 14. Since the number of weighted graphs in $\mathcal{H}(G)$ is polynomial, Algorithm 2 in fact runs in polynomial time. The proof of this lemma, which is the true heart of the algorithm, is given in Sect. 5.

Lemma 4 (Key Lemma). *If G is not a disjoint union of cliques and does not contain true twins, then $\mathcal{H}(G)$ is nonempty.*

Algorithm 2. CLUSTER-VD-APX(G, c)

Require: (G, c) a weighted graph
Ensure: X an inclusionwise minimal hitting set of G
1: **if** G is a disjoint union of cliques **then**
2:     $X \leftarrow \varnothing$
3: **else if** there exists $u \in V(G)$ with $c(u) = 0$ **then**
4:     $G' \leftarrow G - u$
5:     $c'(v) \leftarrow c(v)$ for $v \in V(G')$
6:     $X' \leftarrow$ CLUSTER-VD-APX(G', c')
7:     $X \leftarrow X'$ if X' is a hitting set of G; $X \leftarrow X' \cup \{u\}$ otherwise
8: **else if** there exist true twins $u, u' \in V(G)$ **then**
9:     $G' \leftarrow G - u'$
10:     $c'(v) \leftarrow c(u) + c(u')$ for $v = u$; $c'(v) \leftarrow c(v)$ for $v \in V(G') \setminus \{u\}$
11:     $X' \leftarrow$ CLUSTER-VD-APX(G', c')
12:     $X \leftarrow X'$ if X' does not contain u; $X \leftarrow X' \cup \{u'\}$ otherwise
13: **else**
14:     pick any $(H, c_H) \in \mathcal{H}(G)$
15:     $\lambda^* \leftarrow \max\{\lambda \mid \forall v \in V(H) : c(v) - \lambda c_H(v) \geqslant 0\}$
16:     $G' \leftarrow G$
17:     $c'(v) \leftarrow c(v) - \lambda^* c_H(v)$ for $v \in V(H)$; $c'(v) \leftarrow c(v)$ for $v \in V(G) \setminus V(H)$
18:     $X \leftarrow$ CLUSTER-VD-APX(G', c')
19: **end if**
20: **return** X

Combined with Lemma 4, our second lemma shows that Algorithm 2 is a 7/3-approximation algorithm.

Lemma 5. *Suppose that Algorithm 2 terminates given some weighted graph (G, c) as input, and outputs a set X. Then X is an inclusionwise minimal hitting set of G and $c(X) \leqslant \frac{7}{3} \cdot \mathrm{OPT}(G, c)$.*

Proof. The proof is by induction on the number of recursive calls. If the algorithm does not call itself, then it returns the empty set and in this case the statement trivially holds. Now assume that the algorithm calls itself at least once and that the output X' of the recursive call is an inclusionwise minimal hitting set of G' that satisfies $c'(X') \leqslant \frac{7}{3} \cdot \mathrm{OPT}(G', c')$. There are three cases to consider.

Case 1: The recursive call occurs at Step 6. Then we have $c(X) = c'(X')$ and $\mathrm{OPT}(G, c) = \mathrm{OPT}(G', c')$ because (G', c') is simply (G, c) with one zero-cost vertex removed. By construction, X is an inclusionwise minimal hitting set of G. Moreover, by what precedes, $c(X) = c'(X') \leqslant \frac{7}{3} \cdot \mathrm{OPT}(G', c') = \frac{7}{3} \cdot \mathrm{OPT}(G, c)$.

Case 2: The recursive call occurs at Step 11. Again, X is an inclusionwise minimal hitting set of G and $c(X) = c'(X') \leqslant \frac{7}{3} \cdot \mathrm{OPT}(G', c') = \frac{7}{3} \cdot \mathrm{OPT}(G, c)$, where the last equality holds by Lemma 2.

Case 3: The recursive call occurs at Step 18. In this case, $G = G'$ and $X = X'$, thus X is automatically an inclusionwise minimal hitting set of G. Let c'' denote the weighting c_H extended to $V(G)$ by letting $c''(v) := 0$ for $v \in V(G) \setminus V(H)$. We have $c'(X) \leqslant \frac{7}{3} \cdot \mathrm{OPT}(G, c')$ by induction and $\lambda^* c''(X) \leqslant \frac{7}{3} \cdot \mathrm{OPT}(G, \lambda^* c'')$ since all the weighted induced subgraphs (H, c_H) in $\mathcal{H}(G)$ are 7/3-good in G (Lemma 1). Because $c = c' + \lambda^* c''$, Lemma 3 implies $c(X) \leqslant \frac{7}{3} \cdot \mathrm{OPT}(G, c)$.   □

We are now ready to prove our main result.

Proof (of Theorem 1). By Lemmas 4 and 5, Algorithm 2 is a 7/3-approximation algorithm for CLUSTER-VD.   □

5   Finding a 7/3-Good Induced Subgraph

The aim of this section is to prove the Key Lemma, Lemma 4, which states that $\mathcal{H}(G)$ is nonempty for all twin-free graphs G that are not a disjoint union of cliques. Our approach is as follows: We consider a twin-free graph G such that our collection $\mathcal{H}(G)$ of 7/3-good induced subgraphs is empty. We first prove that, in this case, G contains no claw, then no gem, and then no cycle of length at least 4 as an induced subgraph. At that point, from a result of Kloks et al. [11], we know that G is the line graph of an acyclic multigraph. Using this, we then show that G does not contain any induced P_3, as desired.

Lemma 6. *Let G be a twin-free graph such that $\mathcal{H}(G)$ is empty. Then G is claw-free.*

Proof. Assume that G contains a claw, say on the vertex set $\{x, u, v, w\}$, where x is the central vertex. Since $G[x, u, v, w]$ is not 2-good in G, there exists a neighbor y of x that is distinct from u, v and w. If y sees none of u, v, w, then $G[\{x, y, u, v, w\}]$ is a $K_{1,4}$, a contradiction. Thus we may assume that yu is an edge.

Suppose that yv is an edge. If yw is not and edge, then $G[\{x, y, u, v, w\}]$ is a dart, a contradiction, and thus yw is an edge. Since G is twin-free, there must be a vertex z in the symmetric difference $N(x) \Delta N(y)$. By symmetry, we may assume that $z \in N(x) \setminus N(y)$.

Since G is $K_{1,4}$-free, the set $\{z, u, v, w\}$ is not stable, and hence $|N(z) \cap \{u, v, w\}| \geqslant 1$. If $|N(z) \cap \{u, v, w\}| \geqslant 2$, say both zu and zv are edges, then $G[\{y, z, u, v\}]$ is a C_4, a contradiction. So, $|N(z) \cap \{u, v, w\}| = 1$, and we may assume that zw is an edge. But now $G[\{x, y, z, u, v\}]$ is a dart, a contradiction.

Summing up, we conclude that yv is not an edge and, by symmetry, yw is not an edge.

Since u and y are not true twins in G, there is some vertex z' in the symmetric difference $N(u) \Delta N(y)$. By symmetry, we may assume that $z' \in N(y) \setminus N(u)$. If xz' is not an edge, then z' sees none of v, w, because G is C_4-free. But then the graph $G[\{x, y, z', u, v, w\}]$ is a turtle, a contradiction. Hence, xz' is an edge.

To avoid an induced dart on the vertex sets $\{x, y, z', u, v\}$ or $\{x, y, z', u, w\}$, both vz' and wz' must be edges. But then $G[\{x, z', u, v, w\}]$ is a dart, a contradiction. This completes the proof.   □

Lemma 7. *Let G be a twin-free graph such that $\mathcal{H}(G)$ is empty. Then G is gem-free.*

Proof. By Lemma 6, we know that G is claw-free. For a contradiction, assume that G contains a gem.

Let k be the maximum number of vertices of a gem contained in G that have a common neighbor outside of that gem, this maximum being taken over all gems contained in G.

Consider an induced gem in G, say with vertex set $\{v_1, v_2, v_3, v_4, v_5\}$, such that there is some vertex v outside of that gem with exactly k neighbors in the set $\{v_1, v_2, v_3, v_4, v_5\}$. Assume that the gem is made of the 5-cycle $v_1 v_2 v_3 v_4 v_5 v_1$ and the two edges $v_1 v_3$, $v_1 v_4$. Now, we will distinguish some cases depending on the value of k. Notice that $k \geqslant 1$ since the gem $G[\{v_1, v_2, v_3, v_4, v_5\}]$ is not in $\mathcal{H}(G)$.

Case 1: $k = 5$. Since v and v_1 are not true twins in G, we may assume that there is some vertex u that sees v_1 and not v.

Case 1.1: uv_2 is an edge. Then neither uv_4 nor uv_5 is an edge of G, for otherwise $G[\{u, v_2, v, v_4\}]$ or $G[\{u, v_2, v, v_5\}]$ is a C_4. Moreover, uv_3 is an edge, since otherwise $G[\{v_1, u, v_3, v_5\}]$ is a claw. But now the graph $G[\{v_1 \ldots, v_5\} \cup \{v, u\}]$ is isomorphic to the special graph H_1 (see Fig. 1), and thus belongs to $\mathcal{H}(G)$, a contradiction.

Case 1.2: uv_2 is not an edge. Then uv_4 is an edge for otherwise $G[\{v_1, u, v_2, v_4\}]$ is a claw, and similarly uv_5 is an edge for otherwise $G[\{v_1, u, v_2, v_5\}]$ is a claw. Moreover, uv_3 is not an edge, because otherwise $G[\{u, v_3, v, v_5\}]$ is a C_4. But again $G[\{v_1, \ldots, v_5\} \cup \{v, u\}]$ is isomorphic to H_1 as in Case 1.1, a contradiction.

Case 2: $k = 4$. We may assume that v sees v_1, v_2, v_3, v_4 and not v_5. Otherwise, by symmetry, we may assume that v sees either all of v_2, v_3, v_4, v_5 and $G[\{v, v_2, v_1, v_5\}]$ is a C_4, or that v sees all of v_1, v_2, v_4, v_5 and $G[\{v, v_2, v_3, v_4\}]$ is a C_4. Since v and v_3 are not true twins, we may assume that there is a vertex u that sees v but not v_3. In this case, G contains the trigraph H_3 (see Fig. 1), a contradiction.

Case 3: $k = 3$. Without loss of generality, v sees v_1, v_2, v_3, because every other (that is, non-isomorphic) possibility leads to a contradiction. Indeed, if v sees v_1, v_2, and v_5, then $G[\{v_1, \ldots, v_5, v\}]$ is a W_5. If v sees v_3, v_4, and v_5, then $G[\{v, v_3, v_1, v_5\}]$ is a C_4, and similarly we have a C_4 if v sees v_2, v_4, and v_5, or v_1, v_3, and v_5. Moreover, if v sees v_1, v_3, and v_4, the dart $G[\{v, v_1, v_2, v_3, v_5\}]$ appears.

Since v and v_2 are not true twins, we may assume that there is a vertex u seeing v but not v_2. We get a contradiction, since G contains the trigraph H_4 (see Fig. 1).

Case 4: $k \leqslant 2$. Since $G[\{v_1, v_2, v_3, v_4, v_5\}]$ is not in $\mathcal{H}(G)$, we know that there is a neighbor w of v_1 outside the gem, and w has at most one neighbor in the set $\{v_2, \ldots, v_5\}$. If w sees neither v_3 nor v_4, then w is also non-adjacent to at least one of v_2, v_5, say v_2 using symmetry, and the graph $G[\{w, v_1, v_2, v_3, v_4\}]$ is a dart. Hence, we may assume that either wv_2 or wv_3 is an edge, say wv_2 by symmetry. Then $G[\{w, v_1, v_3, v_4, v_5\}]$ is a dart, a contradiction. This completes the proof. $\square$

In the following, we need another small graph: a *diamond*, that is, K_4 minus an edge.

Lemma 8. *Let G be a twin-free graph such that such that $\mathcal{H}(G)$ is empty. Then G is diamond-free and K_4-free.*

Proof. We first prove that G is diamond-free. Suppose we have a diamond on the vertices u, v, w, and x, where ux is not an edge. By assumption, v and w are not true twins, and so we may assume that there is some $y \in N(v) \setminus N(w)$. To avoid a claw on the vertices u, v, x, and y, it must be that uy or xy is an edge. Both edges cannot be there, since otherwise $G[\{u, y, x, w\}]$ is a C_4. So we may assume that uy is an edge while xy is not. But now the graph $G[\{u, v, w, x, y\}]$ is a gem, which contradicts Lemma 7.

Next we prove that G is K_4-free. Assume not, and let u, v, w and x be four mutually adjacent vertices. Since u and x cannot be true twins, we may assume that there is some vertex u' adjacent to u but not to x. Since G is diamond-free, the only neighbor of u' in $\{u, v, w, x\}$ is u. Similarly, we obtain vertices v' and w', where v' is only adjacent to v in $\{u, v, w, x\}$ and w' is only adjacent to w in $\{u, v, w, x\}$. As G is C_4-free, the three vertices u', v', w' are pairwise non-adjacent. But now the graph $G[\{u, v, w, x, u', v', w'\}]$ is isomorphic to the special graph H_2 (see Fig. 1), a contradiction. $\square$

A *hole* in a graph is an induced cycle of length at least four.

Lemma 9. *Let G be a twin-free graph such that $\mathcal{H}(G)$ is empty. Then G is hole-free.*

Proof. Thanks to Lemmas 6 and 8, we know that G is claw-free, diamond-free, and K_4-free. By contradiction, assume that G contains a hole and let $H = v_1 v_2 \ldots v_k v_1$ be a shortest hole contained in G. Thus $k \geqslant 5$. If some vertex of H does not have a neighbor in $V(G) \setminus V(H)$, then G contains an induced P_3 whose middle vertex does not have neighbors outside the P_3, in contradiction to the assumption that $\mathcal{H}(G)$ is empty.

There cannot be a vertex outside of H having exactly one neighbor in H either, as G is claw-free. Moreover, if there is a vertex v outside of H having two or more neighbors H, they must appear consecutively. This is due to our assumption that H is a shortest hole in G, and also to the fact that G does not contain a C_4. Since G is diamond-free, this means that every vertex outside H that sees some vertex of H has exactly two neighbors in H, and they must appear consecutively on H.

Let $u \in V(G) \setminus V(H)$ be a neighbor of v_2. We may assume that u is adjacent to v_3 too. Since there is a bull on the vertices v_1, v_2, v_3, v_4, and u, there must be another neighbor v of v_2 outside of H. Note that uv is not an edge, because we cannot have a diamond or a K_4 on u, v, v_2, and v_3. Similarly, vv_3 is not an edge. Hence, v must be adjacent to v_1. But now $G[\{v_k, v_1, v_2, v_3, v_4, u, v\}]$ contains the trigraph H_5 (Fig. 1), the undecided edge being included if $k = 5$, and excluded if $k > 5$. This is a contradiction, which concludes the proof.   □

To finalize the proof, we need the following theorem.

Theorem 2 (Kloks et al. [11]). *Let G be a graph that is hole-, claw- and gem-free. Then G is the line graph of an acyclic multigraph.*

Now, we are ready to prove our key lemma.

Proof (Proof of Lemma 4). We may assume that G is connected. (Indeed, if not, simply consider a component.) So $|V(G)| \geqslant 3$. By Lemmas 6, 7 and 9, Theorem 2 gives that G is a K_4-free line graph of an acyclic multigraph, say H. Since G is twin-free, H does not have parallel edges. Also H is clearly connected, thus H is a tree.

Root the tree H at an arbitrary vertex, and let x be a leaf at maximum distance from the root. Let y be the parent of x in H. Suppose that y has a child z distinct from x. Then z is also a leaf. However, the vertices of G corresponding to edges xy, zy are true twins in G, which is not possible. Hence x is the only child of y in H. But now the edge xy is a vertex v of degree one in G and, since $|V(G)| \geqslant 3$, the vertex v is contained in an induced P_3. Since v has no neighbor outside this P_3, this is a contradiction. This completes the proof.   □

6   Conclusion

In this paper we presented a 7/3-approximation algorithm for the CLUSTER-VD problem, based on the local ratio technique. The main idea underlying the algorithm is that there exists a collection of small induced subgraphs that are on the one hand *good* in the sense that they guarantee a local ratio of at most 7/3, and on the other hand *sufficient* in the sense that one can always find and use one of them until the algorithm terminates.

As mentioned in the introduction, we conjecture that there is a 2-approximation algorithm for CLUSTER-VD. The following result gives some evidence to back up this conjecture.

Theorem 3. *There is a 2-approximation algorithm for CLUSTER-VD in the class of diamond-free graphs.*

The proof of Theorem 3 is omitted due to length constraints. The algorithm is modeled on the 7/3-approximation algorithm presented earlier, the main difference being the use of some infinite (but easy to detect) family of graphs that are 2-good. We note that Theorem 3 can be seen as a generalization of the fact that there is a 2-approximation for CLUSTER-VD in triangle-free graphs, a result that was used by You et al. [14] in their 5/2-approximation algorithm for (unweighted) CLUSTER-VD.

Finally, we point out that the analysis of our 7/3-approximation algorithm also proves that a certain $O(n^7)$-size LP relaxation for CLUSTER-VD has integrality gap at most 7/3, namely, the LP relaxation obtained by writing down at most $O(n^7)$ inequalities in the vertex variables x_v for each of the weighted graphs of Fig. 1. By considering graphs G with large girth and small stability number, we can see that the integrality gap is actually equal to 7/3, since in these graphs OPT(G) is close to n and only the graphs $H \in \{P_3, K_{1,3}, K_{1,4}\}$ are induced subgraphs of G (details are omitted due to space constraints). Thus letting $x_v := 3/7$ for all vertices v gives a feasible fractional solution, of cost $3n/7$.

References

1. Bandelt, H.-J., Mulder, H.M.: Distance-hereditary graphs. J. Comb. Theory Ser. **B 41**(2), 182–208 (1986)
2. Bar-Yehuda, R., Bendel, K., Freund, A., Rawitz, D.: Local ratio: a unified framework for approximation algorithms. ACM Comput. Surv. **36**(4), 422–463 (2004)
3. Anudhyan, B., Marek, C., Tomasz, K., Marcin, P.: A fast branching algorithm for cluster vertex deletion. In: Hirsch, Edward A., Kuznetsov, Sergei O., Pin, Jean-Éric, Vereshchagin, Nikolay K. (eds.) CSR 2014. LNCS, vol. 8476, pp. 111–124. Springer, Heidelberg (2014). arXiv:1306.3877
4. Deng, X., Zang, W.: An approximation algorithm for feedback vertex sets in tournaments. SIAM J. Comput. **30**(6), 1993–2007 (2001)
5. Chudak, F.A., Goemans, M.X., Hochbaum, D.S., Williamson, D.P.: A primal-dual interpretation of two 2-approximation algorithms for the feedback vertex set problem in undirected graphs. Oper. res. lett. **22**(4), 111–118 (1998)
6. Diestel, R.: Graph Theory. Graduate Texts in Mathematics, vol. 173. Springer, Heidelberg (2010)
7. Guruswami, V., Lee, E.: Inapproximability of feedback vertex set for bounded length cycles. ECCC:TR14-006
8. Guruswami, V., Lee, E.: Inapproximability of H-transversal/packing. arXiv:1506.06302
9. Hüffner, F., Komusiewicz, C., Moser, H., Niedermeier, R.: Fixed-parameter algorithms for cluster vertex deletion. Theor. Comput. Syst. **47**(1), 196–217 (2010)
10. Iwata, Y., Oka, K.: Fast dynamic graph algorithms for parameterized problems. In: Ravi, R., Gørtz, I.L. (eds.) SWAT 2014. LNCS, vol. 8503, pp. 241–252. Springer, Heidelberg (2014)
11. Kloks, T., Kratsch, D., Müller, H.: Dominoes. In: Mayr, E.W., Schmidt, G., Tinhofer, G. (eds.) WG 1994. LNCS, vol. 903, pp. 106–120. Springer, Heidelberg (1995)
12. Mnich, M., Williams, V.V., Végh, L.A.: A 7,3-approximation for feedback vertex sets in tournaments. arXiv:1511.01137
13. Jianhua, T., Zhou, W.: A primal-dual approximation algorithm for the vertex cover P_3 problem. Theoret. Comput. Sci. **412**(50), 7044–7048 (2011)
14. You, J., Wang, J., Cao, Y.: Approximate association via dissociation. arXiv:1510.08276

Improved Approximations for Cubic Bipartite and Cubic TSP

Anke van Zuylen[✉]

Department of Mathematics, College of William & Mary,
Williamsburg, VA, USA
anke@wm.edu

Abstract. We show improved approximation guarantees for the traveling salesman problem on cubic bipartite graphs and cubic graphs. For cubic bipartite graphs with n nodes, we improve on recent results of Karp and Ravi [10] by giving a "local improvement" algorithm that finds a tour of length at most $5/4n - 2$. For 2-connected cubic graphs, we show that the techniques of Mömke and Svensson [11] can be combined with the techniques of Correa et al. [6], to obtain a tour of length at most $(4/3 - 1/8754)n$.

Keywords: Traveling salesman problem · Approximation algorithm · Cubic bipartite graphs · Cubic graphs · Barnette's conjecture

1 Introduction

The traveling salesman problem (TSP) is one of the most famous and widely studied combinatorial optimization problems. Given a set of cities and pairwise distances, the goal is to find a tour of minimum length that visits every city exactly once. Even if we require the distances to form a metric, the problem remains NP-hard. The classic Christofides' algorithm [5] finds a tour that has length at most $\frac{3}{2}$ times the length of the optimal tour. Despite much effort in the 35 years following Christofides's result, we do not know any algorithms that improve on this guarantee.

One approach that has often been useful in designing approximation algorithms is the use of linear programming. In this context, a major open question is to determine the integrality gap of the subtour elimination linear program or Held-Karp relaxation [7,9]; the integrality gap is the worst-case ratio of the length of an optimal tour to the optimal value of the relaxation. Examples are known in which the length of the optimal tour is $\frac{4}{3}$ times the value of the Held-Karp relaxation, and a major open question is whether this is tight.

Recent years have seen some exciting progress towards answering this question on graph metrics, also called the graph-TSP. In this special case of the

A. van Zuylen—This work was supported by a grant from the Simons Foundation (#359525, Anke Van Zuylen) and by NSF Prime Award: HRD-1107147, Women in Scientific Education (WISE).

Q. Louveaux and M. Skutella (Eds.): IPCO 2016, LNCS 9682, pp. 250–261, 2016.
DOI: 10.1007/978-3-319-33461-5_21

metric TSP, we are given an unweighted graph $G = (V, E)$ in which the nodes represent the cities, and the distance between two cities is equal to the shortest path in G between the corresponding nodes. Examples are known in which the ratio between the length of the optimal tour and the Held-Karp relaxation is $\frac{4}{3}$, where the examples are in fact graph-TSP instances with an underlying graph G that is 2-connected and subcubic (every node has degree at most three).

The graph-TSP thus captures many of the obstacles that have prevented us from obtaining improved approximations for general metrics, and much recent research has focused on finding improved algorithms for the graph-TSP. The first improvement for graph-TSP metrics is due to Gamarnik et al. [8], who show an approximation guarantee strictly less than $\frac{3}{2}$ for cubic, 3-connected graphs. Aggarwal et al. [1] give a $\frac{4}{3}$-approximation algorithm for this case. Boyd et al. [3] show that there is a $\frac{4}{3}$-approximation algorithm for any cubic graph, and Mömke and Svensson [11] show this holds also for subcubic graphs.

Mömke and Svensson also show a 1.461-approximation algorithm if we make no assumptions on the underlying graph G. Mucha [12] improves their analysis to show an approximation guarantee of $\frac{13}{9}$. Sebő and Vygen [13] combine the techniques of Mömke and Svensson with a clever use of ear decompositions, which yields an approximation ratio of 1.4.

As mentioned previously, for subcubic graphs, examples exist that show that we cannot obtain better approximation guarantees than $\frac{4}{3}$ unless we use a stronger lower bound on the optimum than the Held-Karp relaxation or subtour elimination linear program. Correa et al. [6] show that this is not the case for cubic graphs. They refine the techniques of Boyd et al. [3] and show how to find a tour of length at most $\left(\frac{4}{3} - \frac{1}{61236}\right)n$ for the graph-TSP on a 2-connected cubic graph G, where n is the number of nodes. Correa et al. also consider the graph-TSP on planar cubic bipartite 3-connected graphs, and give a $\left(\frac{4}{3} - \frac{1}{18}\right)$-approximation algorithm. Planar cubic bipartite 3-connected graphs are known as Barnette graphs, and a long-standing conjecture in graph theory by Barnette [2] states that all planar cubic bipartite 3-connected graphs are Hamiltonian. Recently, Karp and Ravi [10] gave a $\frac{9}{7}$-approximation algorithm for the graph-TSP on a superset of Barnette graphs, cubic bipartite graphs.

In this paper, we give two results that improve on the results for cubic graph-TSP. For the graph-TSP on (non-bipartite) cubic graphs, we show that the techniques of Mömke and Svensson [11] can be combined with those of Correa et al. [6] to find an approximation algorithm with guarantee $\left(\frac{4}{3} - \frac{1}{8754}\right)$. We note that independent of our work, Candráková and Lukotka [4] showed very recently, using different techniques, how to obtain a 1.3-approximation algorithm for the graph-TSP on cubic graphs. For connected bipartite cubic graphs, we give an algorithm that finds a tour of length at most $\frac{5}{4}n - 2$, where n is the number of nodes. The idea behind our algorithm is the same as that of many previous papers, namely to find a cycle cover (or 2-factor) of the graph with a small number of cycles. Our algorithm is basically a simple "local improvement" algorithm. The key idea for the analysis is to assign the size of each cycle to the nodes contained in it in a clever way; this allows us to give a very simple proof that the algorithm returns a 2-factor with at most $n/8$ components. We also give an example that shows that the analysis is tight, even if we relax a certain condition in the algorithm that restricts the cases when we allow the algorithm to move to a new solution.

The remainder of this paper is organized as follows. In Sect. 2, we describe and analyze our algorithm for the graph-TSP on cubic bipartite graphs, and in Sect. 3, we give our improved result for non-bipartite cubic graphs.

2   The Graph-TSP on Cubic Bipartite Graphs

In the graph-TSP, we are given a graph $G = (V, E)$, and for any $u, v \in V$, we let the cost c_{uv} be the number of edges in the shortest path between u and v in G. The goal is to find a tour of the nodes in V that has minimum cost. A 2-factor of G is a subset of edges $F \subseteq E$, such that each node in V is incident to exactly two edges in F. Note that if F is a 2-factor, then each (connected) component of (V, F) is a simple cycle. If C is a component of (V, F), then we will use $V(C)$ to denote the nodes in C and $E(C)$ to denote the edges in C. The size of a cycle C is defined to be $|E(C)|$ (which is of course equal to $|V(C)|$). Sometimes, we consider a component of $(V, F \setminus E')$ for some $E' \subset F$. A component of such a graph is either a cycle C or a path P. We define the length of a path P to be the number of edges in P.

The main idea behind our algorithm for the graph-TSP in cubic bipartite graphs (and behind many algorithms for variants of the graph-TSP given in the literature) is to find a 2-factor F in G such that (V, F) has a a small number of cycles, say k. We can then contract each cycle of the 2-factor, find a spanning tree on the contracted graph, and add two copies of the corresponding edges to the 2-factor. This yields a spanning Eulerian (multi)graph containing $n + 2(k-1)$ edges. By finding a Eulerian walk in this graph and shortcutting, we get a tour of length at most $n + 2k - 2$. In order to get a good algorithm for the graph-TSP, we thus need to show how to find a 2-factor with few cycles, or, equivalently, for which the average size of the cycles is large.

In Sect. 2.2, we give an algorithm for which we prove in Lemma 5 that, given a cubic bipartite graph $G = (V, E)$, it returns a 2-factor with average cycle size at least 8. By the arguments given above, this implies the following result.

Theorem 1. *There exists a $\frac{5}{4}$-approximation algorithm for the graph-TSP on cubic bipartite graphs.*

Before we give the ideas behind our algorithm and its analysis in Sect. 2.1, we begin with the observation that we may assume without loss of generality that the graph has no "potential 4-cycles": a set of 4 nodes S will be called a potential 4-cycle if there exists a 2-factor in G that contains a cycle with node set exactly S. The fact that we can modify the graph so that G has no potential 4-cycles was also used by Karp and Ravi [10].

Lemma 1. *To show that every simple cubic bipartite graph $G = (V, E)$ has a 2-factor with at most $|V|/8$ components, it suffices to show that every simple cubic bipartite graph $G' = (V', E')$ with no potential 4-cycles has a 2-factor with at most $|V'|/8$ components.*

Proof. We show how to contract a potential 4-cycle S in G to get a simple cubic bipartite graph G' with fewer nodes than the original graph, and how, given a 2-factor with average component size 8 in G', we can uncontract S to get a 2-factor in G without increasing the number of components.

Let $S = \{v_1, v_2, v_3, v_4\}$ be a potential 4-cycle in G, i.e., $E[S]$ contains 4 edges, say $\{v_1, v_2\}, \{v_2, v_3\}, \{v_3, v_4\}, \{v_1, v_4\}$, and there exists no node $v_5 \notin S$ that is incident to two nodes in $\{v_1, v_2, v_3, v_4\}$ (since in that case a 2-factor containing a cycle with node set S would have v_5 as an isolated node, and this cannot be a 2-factor since v_5 must have degree 2 in a 2-factor).

We contract S, by identifying v_1, v_3 to a new node v_{odd}, and identifying v_2, v_4 to a new node v_{even}. We keep a single copy of the edge $\{v_{\text{odd}}, v_{\text{even}}\}$. The new graph G' is simple, cubic and bipartite, and $|V'| = |V| - 2$. See Fig. 1 for an illustration.

Fig. 1. The 4-cycle on the left is contracted, by identifying v_1, v_3 to a new node v_{odd}, and identifying v_2, v_4 to a new node v_{even}, and keeping a single copy of the edge $\{v_{\text{odd}}, v_{\text{even}}\}$, to obtain the simple cubic bipartite graph on the right.

Given any 2-factor in G', we can "uncontract" S and find a 2-factor in G with at most as many components as the 2-factor in G': If the 2-factor on G' does not contain $\{v_{\text{odd}}, v_{\text{even}}\}$ then it must contain the other 4 edges incident to v_{odd} and v_{even}. When uncontracting S, this gives one edge incident to each $v_i, i = 1, \ldots, 4$. Since all other node degrees are even, the graph consists of even cycles and two paths with endpoints in $\{v_1, v_2, v_3, v_4\}$. We can choose to add the edges $\{v_1, v_2\}, \{v_3, v_4\}$, or the edges $\{v_2, v_3\}, \{v_1, v_4\}$; both of these choices give a 2-factor, and at least one of the two options must give a 2-factor in which all 4 nodes in S are in the same cycle. If the 2-factor on G' does contain $\{v_{\text{odd}}, v_{\text{even}}\}$ then it must contain one other edge incident to v_{odd} and to v_{even}. When uncontracting S, this gives one edge incident to v_1 or v_3, and one edge incident to v_2 or v_4. Suppose without loss of generality the edges are incident to v_1 and v_2. Then, we add edges $\{v_2, v_3\}, \{v_3, v_4\}$, and $\{v_1, v_4\}$ to get a 2-factor. Note that it is again the case that all 4 nodes in S are in the same cycle.   □

2.1   A Local Improvement Heuristic

A cubic bipartite graph has a perfect matching (in fact, it is the case that the edge set can be decomposed into three perfect matchings), and given a cubic bipartite graph $G = (V, E)$, we can obtain a 2-factor F by simply finding a perfect matching M and letting $F = E \backslash M$. Conversely, if F is a 2-factor for G, then $E \backslash F$ is a perfect matching in G. Now, given an arbitrary 2-factor F_1, we can use these observations to build a second 2-factor F_2 such that most nodes that are in a small cycle in (V, F_1) are in a long cycle in (V, F_2): The 2-factor F_2 is constructed by taking the matching $E \backslash F_1$ and adding half of the edges from each cycle in (V, F_1). Since each cycle is even, its edges can be decomposed into two perfect matchings, and we may choose either one of them to add to F_2. We will say 2-factor F_2 is *locally optimal with respect to* F_1 if F_2 contains all edges in $E \setminus F_1$ and for each cycle C in (V, F_1), replacing F_2 by the symmetric difference of F_2 with $E(C)$ does not reduce the number of components of (V, F_2).

The essence of our algorithm is to start with an arbitrary 2-factor F_1, find a 2-factor F_2 that is locally optimal with respect to F_1, and return the 2-factor among F_1, F_2 with the smallest number of components.

If we consider a 6-cycle C in (V, F_1), and a 2-factor F_2 that is locally optimal with respect to F_1, then it is not hard to see that at least two edges of C will be part of the same cycle, say D, in (V, F_2). Moreover, the fact that the graph G has no potential 4-cycles can be shown to imply that D has size at least 10. This observation motivates the condition in Lemma 2 below that for any C in (V, F_1), there should exist D in (V, F_2) of size at least 10, such that $|V(C) \cap V(D)| \geq 4$.

In Lemma 2 we show that this condition suffices to guarantee that either (V, F_1) or (V, F_2) has at most $|V|/8$ cycles. In Lemma 3 we show the condition holds for F_2 that is locally optimal for F_1, provided that all cycles in F_1 are *chordless*: An edge $\{x, y\}$ is a *chord* for cycle C in (V, F_1), if $x, y \in C$, and $\{x, y\} \in E \setminus F_1$; a cycle will be referred to as chorded if it has at least one chord, and chordless otherwise.

A few more details are needed to deal with the general case when F_1 is not necessarily chordless; these are postponed to Sect. 2.2.

Lemma 2. *Let $G = (V, E)$ be a simple cubic bipartite graph that has no potential 4-cycles, let F_1 and F_2 be 2-factors in G, such that for any cycle C in (V, F_1), there exists a cycle D in (V, F_2) of size at least 10 such that $|V(C) \cap V(D)| \geq 4$. Then either (V, F_1) or (V, F_2) has at most $|V|/8$ components.*

Proof. Let K_i be the number of components of (V, F_i) for $i = 1, 2$. Note that it suffices to show that $\gamma K_1 + (1 - \gamma) K_2 \leq |V|/8$ for some $0 \leq \gamma \leq 1$.

In order to do this, we introduce a value $\alpha(v)$ for each node. This value is set based on the size of the cycle containing v in the second 2-factor, and they will satisfy $\sum_{v \in D} \alpha(v) = 1$ for every cycle D in the second 2-factor (V, F_2). Hence, we have that $\sum_{v \in V} \alpha(v)$ is equal to the number of cycles in (V, F_2). We will then show that the condition of the lemma guarantees that for a cycle C in (V, F_1),

$$\sum_{v \in C} \alpha(v) \leq \frac{1}{6} |V(C)| - \frac{1}{3}. \tag{$*$}$$

This suffices to prove what we want: we have $K_2 = \sum_{v \in V} \alpha(v) \leq \frac{1}{6}|V| - \frac{1}{3}K_1$, which is the same as $\frac{1}{4}K_1 + \frac{3}{4}K_2 \leq \frac{1}{8}|V|$.

The basic idea to setting the α-values is that if v is in a cycle D in (V, F_2) of size k, then we have $\alpha(v) = \frac{1}{k}$. The only exception to this rule is when D has size 10; in this case we set $\alpha(v)$ for $v \in D$ to either $\frac{1}{6}$ or $\frac{1}{12}$. There will be exactly 8 nodes with $\alpha(v) = \frac{1}{12}$ and 2 nodes with $\alpha(v) = \frac{1}{6}$. The nodes v in D with $\alpha(v) = \frac{1}{12}$ are chosen in such a way that, if there is a cycle C in (V, F_1) containing at least 4 nodes in D, then at least 4 of the nodes in $V(C) \cap V(D)$ will have $\alpha(v) = \frac{1}{12}$. It is possible to achieve this, since the fact that D has 10 nodes implies that (V, F_1) can contain at most two cycles that intersect D in 4 or more nodes.

It is easy to see that $(*)$ holds: by the condition in the lemma, any cycle C contains at least 4 nodes v such that $\alpha(v) \leq \frac{1}{12}$. Since we assumed in addition that G has no potential 4-cycles, we also know that $\alpha(v) \leq \frac{1}{6}$ for all other $v \in V(C)$. Hence $\sum_{v \in C} \alpha(v) \leq \frac{1}{6}(|V(C)| - 4) + 4 \cdot \frac{1}{12} = \frac{1}{6}|V(C)| - \frac{1}{3}$. $\qquad\square$

By Lemma 2, it is enough to find a 2-factor F_2 that satisfies that every cycle in the first 2-factor, F_1, has at least 4 nodes in some "long" cycle of F_2 (where "long" is taken to be size 10 or more). The following lemma states that a locally optimal F_2 satisfies this condition, provided that F_1 is chordless.

Lemma 3. *Let $G = (V, E)$ be a simple cubic bipartite graph that has no potential 4-cycles, let F_1 be a chordless 2-factor in G, and let F_2 be a 2-factor that is locally optimal with respect to F_1. Then for any cycle C in (V, F_1), there exists a cycle D in (V, F_2) of size at least 10 such that $|V(C) \cap V(D)| \geq 4$.*

Proof. Suppose F_2 is locally optimal with respect to F_1, and assume by contradiction that there is some cycle C in (V, F_1) such that (V, F_2) contains no cycle of size at least 10 that intersects C in at least 4 nodes. Let $F_2' = F_2 \triangle E(C)$; we will show that (V, F_2') has fewer components than (V, F_2), contradicting the fact that F_2 is locally optimal with respect to F_1.

Consider an arbitrary cycle D in (V, F_2) that intersects C. We will first show that any node v in D will be in a cycle in (V, F_2') that is at least as large as D. This shows that the number of cycles of (V, F_2') is at most the number of cycles of (V, F_2). We then show that it is not possible that for every node, its cycle in (V, F_2') is the same size as the cycle containing it in (V, F_2).

If D contains exactly one edge, say e, in C, then in (V, F_2'), the edge e in D is replaced by an odd-length path. Hence, in this case the nodes in D will be contained in a cycle in (V, F_2') that is strictly larger than D.

If D contains $k > 1$ edges in C, then D has size at most 8, since otherwise D contradicts our assumption that no cycle exists in (V, F_2) of size at least 10 that intersects C in at least 4 nodes. We now show this implies that D has size exactly 8 and $k = 2$: The size of D is either 6 or 8 since G has no potential 4-cycles. Note that D alternates edges in C and odd-length paths in $(V, F_2 \setminus E(C))$. Since C

is chordless, the paths cannot have length 1 and must thus have length at least 3. We thus have that D must consist of exactly two edges from C, say e_1, e_2, separated by two paths of length 3, say P_1, P_2.

Since P_1 and P_2 do not contain edges in C, (V, F_2') also contains the edges in P_1 and P_2. Hence, to show that all nodes in D are in cycles of size at least 8 in (V, F_2'), it now suffices to show that the cycles containing P_1 and P_2 in (V, F_2') have size at least 8. Consider the cycle in (V, F_2') containing P_1; besides P_1, the cycle contains another path, say P_1', connecting the endpoints of P_1, and this path must have odd length ≥ 3 since G is bipartite and has no potential 4-cycles. Furthermore, P_1' starts and ends with an edge in C, by definition of $F_2' = F_2 \triangle E(C)$. Note that P_1' thus cannot have length 3, as this would imply that the middle edge in P_1' is a chord for C. So P_1' has length at least 5, and the cycle in (V, F_2') containing P_1 thus has size at least 8. Similarly, the cycle containing P_2 in (V, F_2') has size at least 8.

We have thus shown that all nodes in D are in cycles of size at least $|V(D)|$ in (V, F_2'), and hence, (V, F_2') has at most as many cycles as (V, F_2). Furthermore, it follows from the argument given above that the number of cycles in (V, F_2) and (V, F_2') is the same only if all nodes in C are in cycles of size 8 in both (V, F_2) and (V, F_2') and each such cycle consists of two edges from $E(C)$ and two paths of length 3 in $(V, F_2 \setminus E(C))$.

We now show by contradiction that the latter is impossible. Suppose C is such that both in (V, F_2) and in (V, F_2') every node in $V(C)$ is contained in a cycle containing two edges from C. Then $|V(C)|$ must be a multiple of 4, say $|V(C)| = 4k$. Let the nodes of C be labeled $1, 2, \ldots, 4k \equiv 0$, such that $\{2i + 1, 2i + 2\} \in F_2$ and $\{2i, 2i + 1\} \in F_2'$ for $i = 0, \ldots, 2k - 1$. We also define a mapping $p(i)$ for every $i = 1, \ldots, 4k$, such that $(V, F_2 \setminus E(C))$ contains a path (which, by our assumption has length 3) from i to $p(i)$ for $i = 1, \ldots, 4k$. Observe that by the definition of the mapping, $p(p(i))$ must be equal to $i \pmod{4k}$ for $i = 1, \ldots, 4k$.

Let $p(1) = \ell$, then the fact that edge $\{1, 2\}$ is in a cycle with one other edge from C in (V, F_2) implies that either $\{\ell, \ell + 1\} \in F_2$ or $\{\ell, \ell - 1\} \in F_2$, and that either $p(2) = \ell + 1$ or $p(2) = \ell - 1$. In the first case, edge $\{2, 3\}$ must be in a cycle with $\{\ell + 1, \ell + 2\}$ in the second 2-factor (V, F_2'), and thus $p(3) = \ell + 2$. In the second case, $\{2, 3\}$ must be in a cycle with $\{\ell - 1, \ell - 2\}$ in (V, F_2'), and thus $p(3) = \ell - 2$. Repeating the argument shows that either $p(i) \equiv \ell + (i - 1) \pmod{4k}$ for $i = 1, \ldots, 4k$, or $p(i) \equiv \ell - (i - 1) \pmod{4k}$ for $i = 1, \ldots, 4k$.

The first case gives a contradiction to the fact that G is bipartite: note that $p(p(i)) \equiv 2\ell + i - 2 \pmod{4k}$ and this must be equal to i. Hence, $\ell \equiv 1 \pmod{2k}$; in other words, ℓ is odd, which cannot be the case since $(V, F_2 \setminus E(C))$ has a path of length 3 from 1 to ℓ (since $p(1) = \ell$) and if ℓ were odd, then C would have a path of even length from 1 to ℓ, and thus G would contain an odd cycle.

Now suppose that $p(i) \equiv \ell - (i - 1) \pmod{4k}$ for $i = 1, \ldots, 4k$. From the previous argument we know that ℓ must be even, since otherwise G is not bipartite. But then $p(\ell/2) = \ell - (\ell/2 - 1) = \ell/2 + 1$. In other words, node $\ell/2$ is connected to node $\ell/2 + 1$ in $(V, F_2 \setminus E(C))$. But since the edge $\{\ell/2, \ell/2 + 1\}$ is either in

F_2 or in F_2', $\{\ell/2, \ell/2 + 1\}$ is the only edge from C in its cycle in either (V, F_2) or (V, F_2'), contradicting the assumption on C. $\qquad\qquad\square$

It may be the case that Lemma 3 also holds for cycles in (V, F_1) that do have chords, but we have not been able to prove this. Instead, there is a simple alternative operation that ensures that a cycle in (V, F_1) with a chord intersects at least one "long" cycle of size at least 10 in (V, F_2) in 4 or more nodes. The algorithm described next will add this operation, and for technical reasons it will only modify F_2 with respect to a cycle C in (V, F_1) if the cycle C does not yet intersect a long cycle in (V, F_2) in 4 or more nodes.

2.2   2-Factor with Average Cycle Size 8

We give our algorithm in Algorithm 1. The algorithm fixes a 2-factor F_1 and initializes F_2 to contain all edges in $E \backslash F_1$. The algorithm then proceeds to modify F_2; note that F_1 is not changed. We let $\triangle$ denote the symmetric difference operator. Figure 2 illustrates the modification to F_2 in the case of a chorded cycle C_i.

Let $G = (V, E)$ be a bipartite cubic graph, with potential 4-cycles contracted using Lemma 1.
Let F_1 be an arbitrary 2-factor in G, and let $C_1, \ldots, C_k$ be the cycles in (V, F_1).
For each cycle C_i in (V, F_1), let $M(C_i) \subseteq E(C_i)$ be a perfect matching on $V(C_i)$.
Initialize $F_2 = (E \backslash F_1) \cup \bigcup_{i=1}^{k} M(C_i)$.
while *there exists a cycle C_i such that $|V(C_i) \cap V(D)| < 4$ for all cycles D in (V, F_2) of size at least 10* **do**

> **if** C_i *is a chordless cycle* **then**
> > $\vert$ $F_2 \leftarrow F_2 \triangle E(C_i)$.
>
> **else**
> > Let $\{x, y\}$ be a chord for C_i, let P_1, P_2 be the edge disjoint paths in C_i from x to y.
> > Relabel P_1 and P_2 if necessary so that P_1 starts and ends with an edge in $F_1 \setminus F_2$.
> > $F_2 \leftarrow (F_2 \triangle E(P_1)) \setminus \{x, y\}$.
>
> **end if**

end while
Uncontract the 4-cycles in (V, F_j) for $j = 1, 2$ using Lemma 1.
Return the 2-factor among F_1, F_2 with the smaller number of components.

Algorithm 1. Approximation Algorithm for Cubic Bipartite TSP

We need to prove that the set of edges F_2 remains a 2-factor throughout the course of the algorithm, that the algorithm terminates, and that upon termination, either (V, F_1) or (V, F_2) has at most $-V-/8$ components. The latter is clear: *if* the algorithm terminates, then the condition of Lemma 2 is satisfied, and therefore one of the two 2-factors has at most $-V-/8$ components.

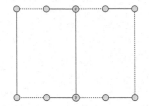 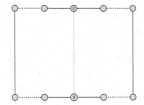

Fig. 2. The figure on the left shows a chorded cycle C in F_1 of size 10 and all edges in G that have both endpoints in C. The dashed edges are in $F_1 \setminus F_2$, and non-dashed edges are in F_2 (where not all edges in F_2 that are incident on the nodes are shown). The figure on the right shows how Algorithm 1 would update F_2.

To show that F_2 is a 2-factor and that the algorithm terminates is a little more subtle. In order to show this, it will be helpful to know that each cycle in (V, F_i) alternates edges in $F_i \cap F_{i+1}$ and edges in $F_i \setminus F_{i+1}$ for $i = 1, 2$ (where subscripts are modulo 2, so $F_3 \equiv F_1$). This is true initially, however, it is not the case that this property continues to hold for all cycles. We will show that it does hold in certain cases, which turn out to be exactly the cases "when we need it". In the following, we will say a cycle or path in (V, F_i) is *alternating* (for F_1 and F_2) if it alternates edges in $F_i \cap F_{i+1}$ and $F_i \setminus F_{i+1}$. We will say that a cycle C in (V, F_1) is *violated* if there exists no D of size at least 10 in (V, F_2) such that $|V(C) \cap V(D)| \geq 4$.

Lemma 4. *Algorithm 1 maintains that F_2 is a 2-factor that satisfies the following properties:*

(1) if C in (V, F_1) is violated, then C is alternating for F_1 and F_2;
(2) if D in (V, F_2) is not alternating for F_1 and F_2, then D has size at least 10.

Proof. We prove the lemma by induction on the algorithm. Initially, F_2 consists of $E \setminus F_1$ and $\bigcup_{i=1}^{k} M(C_i)$, which are two edge-disjoint perfect matchings on V. Hence, F_2 is a 2-factor, and the two properties hold for all cycles in (V, F_1) and (V, F_2).

Suppose the lemma holds and we modify F_2 by considering some violated cycle C. The two properties of the lemma imply the following:

Claim 1. *If C is violated, then $(V, F_2 \setminus E(C))$ consists of even cycles and odd-length paths, where paths that are not alternating for F_1 and F_2 have length at least 9.*

Proof of Claim: For each path in the graph $(V, F_2 \setminus E(C))$ there exists some cycle D in (V, F_2) such that the path results when removing $E(C) \cap E(D)$ from D. If D is alternating for F_1 and F_2, then the path must have the same property, and it must start and end with an edge in $F_2 \setminus F_1$. Hence, the path must have odd length if D is alternating. If D is not alternating then D has size at least 10 by Property (2), so C can have at most one edge in common with D, since otherwise C is not violated. Hence, the path obtained by removing the unique edge in $E(C) \cap E(D)$ has length at least 9, and its length must be odd, since D is an even cycle.

If C is chordless, then we modify F_2 to $F_2' = F_2 \triangle E(C)$. Clearly, F_2' is again a 2-factor, and Property (1) remains satisfied. Furthermore, any cycle in (V, F_2') that is not alternating for F_1 and F_2' either also existed in (V, F_2) and hence it has size at least 10, since Property (2) holds for F_2, or the cycle contains a path in $(V, F_2 \setminus E(C))$ that is not alternating for F_1 and F_2, and this path has length at least 9 by the claim. So Property (2) holds for F_2'.

Now consider the modification of F_2 when considering a chorded cycle C in (V, F_1). First consider $F_2' = F_2 \triangle E(P_1)$; every node is incident to two edges in F_2', except for x and y, which are incident to three edges in F_2', namely two edges in $E(C)$ plus the edge $\{x, y\}$. Hence, removing $\{x, y\}$ will give a new 2-factor, say F_2''. The modification from F_2 to F_2'' is exactly the modification made to F_2 by the algorithm.

We now show that the two properties are satisfied. Clearly, C is not alternating for F_1 and F_2'', so in order to maintain Property (1), we need to show that C is no longer violated. To do this, we show that (V, F_2'') contains a cycle of size at least 10 that contains x, y and their 4 neighbors in C. First, suppose by contradiction that after removing $\{x, y\}$, x and y are not in the same cycle. Consider the component of (V, F_2'') containing x: starting from x, it alternates edges in $E(C)$ and paths in $F_2 \setminus E(C)$, starting and ending with an edge in $E(C)$. By Claim 1 the paths in $(V, F_2 \setminus E(C))$ have odd length, and hence the component containing x must be an odd cycle, contradicting the fact that G is bipartite. So, x and y must be in the same cycle in (V, F_2''). This cycle must thus consist of two odd-length paths from x to y, each starting and ending with an edge in $F_2'' \cap F_1$. These paths cannot have length 3, because this would imply that the path plus the edge $\{x, y\}$ would form a potential 4-cycle. Hence, the cycle in (V, F_2'') containing x and y has size at least 10.

For Property (2), note that any cycle D in (V, F_2'') that is not alternating for F_1 and F_2'' either (i) existed in (V, F_2) and therefore has size at least 10, or (ii) contains x and y and we showed above that this cycle has size at least 10, or (iii) it contains a path in $(V, F_2 \setminus E(C))$ that is not alternating for F_1 and F_2, and by the claim this path has length at least 9. Hence, Property (2) is satisfied by F_2''. □

Lemma 5. *Given a cubic bipartite graph $G = (V, E)$, Algorithm 1 returns a 2-factor in G with at most $|V|/8$ components.*

Proof. By Lemma 1 it suffices to show that the current lemma holds if G has no potential 4-cycles. By the termination condition of Algorithm 1 and Lemma 2, the 2-factor returned by the algorithm does indeed have at most $|V|/8$ components, so it remains to show that the algorithm always returns a 2-factor.

By Lemma 4, the algorithm maintains two 2-factors F_1 and F_2. Observe that if a cycle C is not violated, then this continues to hold throughout the remainder of the algorithm: Let D be a cycle of size at least 10 in (V, F_2) such that $|V(C) \cap V(D)| \geq 4$. The only possible changes to D will be caused by a violated cycle C', which necessarily contains at most one edge in D: by Lemma 4 C' is alternating for F_1 and F_2, so if C' contains more than one edge in D, C' cannot be violated. The modification of F_2 with respect to $E(C')$ can therefore only cause the cycle D to become a larger cycle D' where $V(D') \supseteq V(D)$. So D' will have size at least 10, and $|V(C) \cap V(D')| \geq 4$.

It remains to show that if we modify F_2 with respect to some violated cycle C, then C is not violated for the new 2-factor F_2'. If C is not chordless, then this holds because C is not alternating for F_1 and the new 2-factor F_2', so by Lemma 4, C is not violated. If C is chordless and violated, then by Claim 1, $(V, F_2 \backslash E(C))$ consists of even cycles and odd-length paths. The proof of Lemma 3 then shows that taking the symmetric difference of F_2 with $E(C)$ (strictly) reduces the number of components. This implies that for $F_2' = F_2 \triangle E(C)$, cycle C is not violated: otherwise, we could apply the same arguments to show that $(V, F_2 \triangle E(C) \triangle E(C))$ has strictly fewer cycles than (V, F_2), but this is a contradiction since $F_2 \triangle E(C) \triangle E(C) = F_2$. □

It is possible to show through an example on 48 nodes that our analysis of Algorithm 1 is tight. In fact, the example is also tight for the local improvement heuristic from Sect. 2.1 and for the local improvement heuristic we obtain if we allow Algorithm 1 to modify F_2 for cycles C that are chorded and/or do have at least two edges in a cycle of size 10 or more in (V, F_2). The details of the example are omitted from this paper due to space constraints, but can be found in [14].

3   Cubic Graphs

We now consider cubic graphs, in other words, we drop the requirement that the graph is bipartite. We assume the graph is 2-connected. The best known approximation result for the graph-TSP on a 2-connected cubic graphs $G = (V, E)$ is the fact that there exists a tour of length at most $\left(\frac{4}{3} - \frac{1}{61236}\right)|V|$ by Correa et al. [6].

One obstacle for their techniques are chorded 4-cycles, i.e., nodes (v_1, v_2, v_3, v_4) such that the subgraph of G induced by $\{v_1, v_2, v_3, v_4\}$ contains edges $\{v_1, v_2\}, \{v_2, v_3\}, \{v_3, v_4\}, \{v_4, v_1\}$ and the "chord" $\{v_2, v_4\}$. The upper bound on the length of the optimal tour proved by Correa et al. [6] is $\frac{4}{3}|B| + \left(\frac{4}{3} - \frac{1}{8748}\right)(|V \backslash B|) + 2$, where $|B|$ is the number of nodes contained in a chorded 4-cycle.

On the other hand, chorded 4-cycles are "beneficial" for the analysis of the Mömke and Svensson [11] algorithm, and it is not hard to show that the upper bound on the length of the optimal tour given by their algorithm is $\frac{4}{3}|V| - \frac{1}{6}|B| - 2$.

Setting the two bounds equal to each other gives $|B| = \frac{1}{1459}|V|$ and thus shows that there exists a polynomial time algorithm for finding a tour of length at most $\left(\frac{4}{3} - \frac{1}{8754}\right)|V|$ for a graph-TSP instance on a 2-connected cubic graph $G = (V, E)$. By an observation of Mömke and Svensson [11], this also implies a

$\left(\frac{4}{3} - \frac{1}{8754}\right)$-approximation algorithm for cubic graph-TSP, i.e., the graph G does not have to be 2-connected. We thus have the following result.

Theorem 2. *There exists a $\left(\frac{4}{3} - \frac{1}{8754}\right)$-approximation algorithm for the Graph-TSP on cubic graphs.*

Acknowledgements. The author would like to thank Marcin Mucha for careful reading and pointing out an omission in a previous version, Frans Schalekamp for helpful discussions, and an anonymous reviewer for suggesting the simplified proof for the result in Sect. 3 for cubic non-bipartite graphs. Other anonymous reviewers are acknowledged for helpful feedback on the presentation of the algorithm for bipartite cubic graphs.

References

1. Aggarwal, N., Garg, N., Gupta, S.: A 4/3-approximation for TSP on cubic 3-edge-connected graphs (2011). http://arxiv.org/abs/1101.5586
2. Barnette, D.W.: Conjecture 5. In: Recent Progress in Combinatorics (1969)
3. Boyd, S., Sitters, R., van der Ster, S., Stougie, L.: The traveling salesman problem on cubic and subcubic graphs. Math. Program. **144**(1–2), 227–245 (2014)
4. Candráková, B., Lukotka, R.: Cubic TSP - a 1.3-approximation. CoRR abs/1506.06369 (2015)
5. Christofides, N.: Worst case analysis of a new heuristic for the traveling salesman problem. Report 388, Graduate School of Industrial Administration, Carnegie-Mellon University, Pittsburgh, PA (1976)
6. Correa, J.R., Larré, O., Soto, J.A.: TSP tours in cubic graphs: beyond 4/3. SIAM J. Discrete Math. **29**(2), 915–939 (2015)
7. Dantzig, G.B., Fulkerson, D.R., Johnson, S.M.: Solution of a large-scale traveling-salesman problem. Oper. Res. **2**, 393–410 (1954)
8. Gamarnik, D., Lewenstein, M., Sviridenko, M.: An improved upper bound for the TSP in cubic 3-edge-connected graphs. Oper. Res. Lett. **33**(5), 467–474 (2005)
9. Held, M., Karp, R.M.: The traveling-salesman problem and minimum spanning trees. Oper. Res. **18**, 1138–1162 (1970)
10. Karp, J., Ravi, R.: A 9/7-approximation algorithm for graphic TSP in cubic bipartite graphs. In: Approximation, Randomization, and Combinatorial Optimization (APPROX-RANDOM). LIPIcs, vol. 28, pp. 284–296. Schloss Dagstuhl - Leibniz-Zentrum fuer Informatik (2014)
11. Mömke, T., Svensson, O.: Approximating graphic TSP by matchings. In: Proceedings of the 52th Annual Symposium on Foundations of Computer Science, pp. 560–569 (2011)
12. Mucha, M.: 13/9-approximation for graphic TSP. Theory Comput. Syst. **55**(4), 640–657 (2014)
13. Sebő, A., Vygen, J.: Shorter tours by nicer ears: 7/5-approximation for the graph-TSP, 3/2 for the path version, and 4/3 for two-edge-connected subgraphs. Combinatorica **34**(5), 597–629 (2014)
14. van Zuylen, A.: Improved approximations for cubic and cubic bipartite TSP. CoRR abs/1507.07121 (2015)

An Approximation Algorithm for Uniform Capacitated k-Median Problem with $1 + \epsilon$ Capacity Violation

Jarosław Byrka[1], Bartosz Rybicki[1(✉)], and Sumedha Uniyal[2]

[1] Institute of Computer Science, University of Wrocław, Wrocław, Poland
{jby,bry}@cs.uni.wroc.pl
[2] IDSIA, University of Lugano, Lugano, Switzerland
sumedha@idsia.ch

Abstract. We study the Capacitated k-Median problem, for which all the known constant factor approximation algorithms violate either the number of facilities or the capacities. While the standard LP-relaxation can only be used for algorithms violating one of the two by a factor of at least two, Li [10,11] gave algorithms violating the number of facilities by a factor of $1 + \epsilon$ exploring properties of extended relaxations.

In this paper we develop a constant factor approximation algorithm for hard Uniform Capacitated k-Median violating only the capacities by a factor of $1 + \epsilon$. The algorithm is based on a configuration LP. Unlike in the algorithms violating the number of facilities, we cannot simply open extra few facilities at selected locations. Instead, our algorithm decides about the facility openings in a carefully designed dependent rounding process.

1 Introduction

In capacitated k-median we are given a set of potential facilities F, capacity $u_i \in \mathbb{N}^+$ for each facility $i \in F$, a set of clients C, a metric distance function d on $C \cup F$ and an integer k. The goal is to find a subset $F' \subseteq F$ of k facilities to open and an assignment $\sigma : C \to F'$ of clients to the open facilities such that $|\sigma^{-1}(i)| \leq u_i$ for every $i \in F'$, so as to minimize the connection cost $\sum_{j \in C} d(j, \sigma(j))$. In the uniform capacity case, $u_i = u, \forall i \in F$.

The standard k-median problem, where there is no restriction on the number of clients served by a facility, can be approximated up to a constant factor [3,7]. The current best is the $(2.675 + \epsilon)$-approximation algorithm of Byrka et al. [5], which is a result of optimizing a part of the algorithm of Li and Svensson [12].

Capacitated k-median is among the few remaining fundamental optimization problems for which it is not clear if there exist constant factor approximation algorithms. All the known algorithms violate either the number of facilities or the capacities. In particular, already the algorithm of Charikar et al. [7] gave 16-approximate solution for uniform capacitated k-median violating the capacities

B. Rybicki—Research supported by NCN 2012/07/N/ST6/03068 grant.
S. Uniyal—Partially supported by the ERC StG project NEWNET No. 279352.

© Springer International Publishing Switzerland 2016
Q. Louveaux and M. Skutella (Eds.): IPCO 2016, LNCS 9682, pp. 262–274, 2016.
DOI: 10.1007/978-3-319-33461-5_22

by a factor of 3. Then Chuzhoy and Rabani [8] considered general capacities and gave a 50-approximation algorithm violating capacities by a factor of 40.

Perhaps the difficulty is related to the unbounded integrality gap of the standard LP relaxation. To obtain integral solutions that are bounded w.r.t. a fractional solution to the standard LP, one has to either allow the integral solution to open twice more facilities or to violate the capacities by a factor of two. Recently, LP-rounding algorithms essentially matching these limits were obtained [1,4].

Next, Li broke this integrality gap barrier by giving a constant factor algorithm for uniform capacitated k-median by opening $(1 + \epsilon)k$ facilities [10]. The algorithm is based on rounding a fractional solution to an extended LP. More recently, he gave an algorithm working with general soft capacities and still opening $(1 + \epsilon)k$ facilities [11]. (In the soft capacitated version we can open multiple copies of the same facility, whereas in the hard version we can open at most one copy.) This new algorithm is based on an even stronger configuration LP. Notably, each of the extended linear programs is not solved exactly by the algorithms, but rather a clever "round-or-separate" technique is applied. This technique was previously used in the context of capacitated facility location in [2]. It essentially allows not to solve the strong LP upfront, but only to detect violated constraints during the rounding process. If a violated constraint is detected, it is returned back to the "feasibility-checking" ellipsoid algorithm. While it is not clear if the strong LP with all the constraints can be solved efficiently, it can be shown that the above described process terminates in polynomial time, see [2].

1.1 Our Results and Techniques

We give an algorithm for uniform capacitated k-median rounding a fractional solution to the configuration LP, from [11], via the "round-or-separate" technique. We obtain a constant factor approximate integral solution violating capacities by a factor of $1 + \epsilon$. We utilize the power of the configuration LP in effectively rounding small size facility sets, and combine it with a careful dependent rounding to coordinate the opening between these small sets. The main result of this paper is described in the following theorem.

Theorem 1. *There is a bi-factor randomized rounding algorithm for hard uniform capacitated k-median problem, with $O(1/\epsilon^2)$-approximation under $1 + \epsilon$ capacity violation.*

Our algorithm utilizes the white-grey-black tree structure from [11], but the following rounding steps are quite different. In particular, the handling of the small "black components" differs. While aiming for a solution opening $(1 + \epsilon)k$ facilities Li [11] can treat each black component independently, we are forced to precisely determine the number of open facilities. Hence we cannot allow a single black component to individually decide to open more facilities than in the fractional solution. Instead, we first use a preprocessing step which we call *massage* that reduces the variance in the number of open facilities in the fractional solution within each "black component". Then we use a form of a pipeage rounding between the "black components", that precisely preserves the total number of open facilities. The tree structure is used to route the demand to the eventually open facilities.

2 Linear Program

The following is the basic LP relaxation for the problem:

$$\min \sum_{i \in F, j \in C} d(i,j) x_{i,j} \qquad \text{s.t.} \qquad \qquad \text{(Basic LP)}$$

$$\sum_{i \in F} y_i = k; \qquad (1) \qquad \sum_{j \in C} x_{i,j} \le u_i y_i \quad \forall i \in F; \quad (3)$$

$$\sum_{i \in F} x_{i,j} = 1 \quad \forall j \in C; \quad (2) \qquad 0 \le x_{i,j} \le y_j \le 1 \forall i \in F, j \in C. \ (4)$$

In the above LP, y_i indicates whether facility i is open or not, and $x_{i,j}$ indicates whether client j is connected to facility i or not. Constraint (1) is the cardinality constraint that exactly k facilities are open, Constraint (2) requires every client j to be connected, Constraint (4) says that a client can only be connected to an open facility and Constraint (3) is the capacity constraint.

The basic LP has an unbounded integrality gap even if we allow to violate the cardinality or the capacity constraints by $2 - \epsilon$. To overcome this gap, Li introduced in [11] a stronger LP called the Configuration LP and got a constant approximation algorithm by opening $(1 + \epsilon)k$ facilities.

To formulate the configuration LP constraints, let us fix a set $B \subseteq F$ of facilities. Let $\mathcal{S} = \{S \subseteq B : |S| \le \ell_1\}$ and $\tilde{\mathcal{S}} = \mathcal{S} \cup \{\perp\}$, where ℓ_1 is some constant we will define later and $\perp$ stands for "any subset of B with size greater than ℓ_1". We treat set $\perp$ as a set that contains all the facilities $i \in B$. For every $S \in \tilde{\mathcal{S}}$, let z_S^B be an indicator variable corresponding to the event that the set of open facilities in B is exactly S and $z_\perp^B$ captures the event that the number of facilities open in B is more than ℓ_1. For every $S \in \tilde{\mathcal{S}}$ and $i \in S$, $z_{S,i}^B$ indicates the event that set S is open and i is open as well. Notice that when $i \in S \ne \perp$, we always have $z_{S,i}^B = z_S^B$. For every $S \in \tilde{\mathcal{S}}$, $i \in S$ and $j \in C$, $z_{S,i,j}^B$ indicates the event that $z_{S,i}^B = 1$ and j is connected to i. The following are valid constraints for any feasible integral solution.

$$\sum_{S \in \tilde{\mathcal{S}}} z_S^B = 1; \qquad (5) \qquad z_{S,i}^B = z_S^B, \forall S \in \mathcal{S}, i \in S; \qquad (9)$$

$$\sum_{S \in \tilde{\mathcal{S}}: i \in S} z_{S,i}^B = y_i \forall i \in B; \quad (6) \qquad \sum_{i \in S} z_{S,i,j}^B \le z_S^B \forall S \in \tilde{\mathcal{S}}, j \in C; \qquad (10)$$

$$\sum_{S \in \tilde{\mathcal{S}}: i \in S} z_{S,i,j}^B = x_{i,j} \forall i \in B, j \in C; \qquad \sum_{j \in C} z_{S,i,j}^B \le u_i z_{S,i}^B \forall S \in \tilde{\mathcal{S}}, i \in S; \qquad (11)$$

$$(7)$$

$$0 \le z_{S,i,j}^B \le z_{S,i}^B \le z_S^B \ \forall S \in \tilde{\mathcal{S}}, i \in S, j \in C; \sum_{i \in B} z_{\perp,i}^B \ge \ell_1 z_\perp^B. \qquad (12)$$

$$(8)$$

Constraint (5) says that exactly one set $S \in \tilde{S}$ is open. Constraint (6) says that if facility i is open then $z_{S,i}^B = 1$ for exactly one set $S \in \tilde{S}$. Constraint (7) says that if j is connected to i then $z_{S,i,j}^B = 1$ for exactly one set $S \in \tilde{S}$. Constraint (10) says that if $z_S^B = 1$, then j can be connected to at most 1 facility in S. Constraint (11) is the capacity constraint. Constraint (12) says that if $z_\perp^B = 1$, then at least ℓ_1 facilities in B are open.

The configuration LP is obtained by adding the above set of constraints for all subsets $B \subseteq F$. As there are exponentially many sets B, we do not know how to solve this LP. But given a fractional solution (x, y), for a fixed set B, we can construct the values of the set of variables z (see [11]) and also check the constraints in polynomial time since the total number of variables and constraints is $n^{O(\ell_1)}$. We apply method that has been used in, e.g., [10,11]. Given a fractional solution (x, y) to the basic LP relaxation, our rounding algorithm either constructs an integral solution with the desired properties or outputs a set $B \subseteq F$ for which one of the Constraints (5) to (12) is infeasible. In the latter case, we can find a violating constraint and feedback it to the ellipsoid method.

3   Rounding Algorithm

Focus on an optimal fractional solution (x, y) to the basic LP. Let $d_{av}(j) = \sum_{i \in F} d(i, j) x_{i,j}$ be the average connection cost of a client $j \in C$. Let $d(S, T) = \min_{i \in S, j \in T} d(i, j)$, for any $S, T \subseteq C \cup F$. Also let, $d(i, S) = d(\{i\}, S)$. Note that the value of the LP solution (x, y) is $\mathbf{LP} := \sum_{(i,j) \in F \times C} d(i, j) x_{i,j} = \sum_{j \in C} d_{av}(j)$. For any set $F' \subseteq F$ of facilities, let $y_{F'} := y(F') := \sum_{i \in F'} y_i$ be the volume of the set F'. For any set $F' \subseteq F$ and $C' \subseteq C$ of clients, let $x_{F',C'} := \sum_{(i,j) \in F' \times C'} x_{i,j}$. Also let, $x_{i,C'} := x_{\{i\},C'}$ and $x_{F',j} := x_{F',\{j\}}$.

Definition 1. *Let* $D_i := \sum_{j \in C} x_{i,j} d(i, j)$ *and* $D_i' := \sum_{j \in C} x_{i,j} d_{av}(j)$ *for each* $i \in F$. *Let* $D_S := D(S) := \sum_{i \in S} D_i$ *and* $D_S' := D'(S) := \sum_{i \in S} D_i'$ *for every* $S \subseteq F$, *Obviously* $D_F = D_F' = \mathbf{LP}$.

First we will partition facilities into clusters (as done in [4,10]). Each cluster will have a client v as its *representative*. We denote the set of cluster representatives by R. Each cluster will contain the set of facilities nearest to a representative $v \in R$ and the fractional number of open facilities in each cluster will be bounded below by $1 - \frac{1}{\ell}$. Let U_v be the set of facilities in the cluster corresponding to representative $v \in R$. For any set $J \subseteq R$ of representatives, we use $U_J := U(J) = \bigcup_{v \in J} U_v$. Constants $\ell := O(1/\epsilon)$ and $\ell_1 := \ell^2$ are integers, which we will define later. Since the clustering procedure is the same as in [4,10], we omit. The following Claim (see Claim 4.1 in [10]) captures the key properties of the clustering procedure.

Claim 1. The following statements hold:

1. for all $v, v' \in R$, $v \neq v'$, we have $d(v, v') > 2\ell \max\{d_{av}(v), d_{av}(v')\}$;
2. for all $j \in C$, $\exists v \in R$, such that $d_{av}(v) \leq d_{av}(j)$ and $d(v, j) \leq 2\ell d_{av}(j)$;

3. $y_{U_v} \geq 1 - 1/\ell$ for every $v \in R$;
4. for any $v \in R$, $i \in U_v$ and $j \in C$, we have $d(i, v) \leq d(i, j) + 2\ell d_{av}(j)$.

We partition the set of representatives R build a tree and color its edges in the same way as Li [11]. To partition R, we run the Kruskal's algorithm to find a minimum spanning tree of R. In the Kruskal's algorithm we maintain a partition $\mathcal{J}$ of R and the set of selected edges E_{MST}. Initially $\mathcal{J} = \{\{v\} : v \in R\}$ and E_{MST} is empty. The length of each edge $(u, v) \in \binom{R}{2}$ is the distance between u and v. We sort all edges in $\binom{R}{2}$ by length, breaking ties in an arbitrary way. For each edge (u, v) in this order if u and v are not in the same group in $\mathcal{J}$, we merge the two groups and add edge (u, v) to E_{MST}.

We now color edges of E_{MST}. For every $v \in R$, we know that $y(U_v) \geq 1 - 1/\ell$. For any subset of representatives $J \subseteq R$ we say that S is big if $y(U_J) \geq \ell$ and small otherwise. For each edge $e \in E_{MST}$ we consider the step in which edge $e = (u, v)$ was added by the Kruskal's algorithm to MST. After the iteration we merge groups J_u (containing u) and J_v (containing v) to one group $J_u \cup J_v$. If both J_u and J_v are small, then we paint edge e in black. If both are big, we paint the edge e white. Otherwise if one is small and the other is big then we direct the edge e towards the big group and paint it grey.

Consider only the black edges from E_{MST}. We define a black component of MST as a connected component in this graph. The following claim (see Claim 4.1 in [11]) is a consequence of the fact that $J \subseteq R$ appears as a group at some step of the Kruskal's algorithm.

Claim 2. Let J be a black component, then for every black edge (u, v) in $\binom{J}{2}$, we have $d(u, v) \leq d(J, R \setminus J)$.

We contract all the black components and remove all the white edges from MST. The obtained graph Υ is a forest. Each node p (vertex in the contracted graph) in Υ corresponds to a black component and each grey edge is directed. Let $J_p \subseteq R$ be the set of representatives corresponding to node p. Abusing the notation slightly, we define $U_p := U(J_p) = \bigcup_{v \in J_p} U_v$. Lets define $y_p := y(U_p)$. The following lemma follows from the way in which we create our forest. Proof can be found in [11].

Lemma 1. *For any tree $\tau \in \Upsilon$, the following statements are true:*

1. *τ has a root node r_τ such that all the grey edges in τ are directed towards r_τ;*
2. *J_{r_τ} is big and J_p is small for all other nodes in τ;*
3. *in any leaf-to-root path of τ, the lengths of grey edges form a non-increasing sequence;*
4. *for any non-root node $p \in \tau$, the length of the grey edge in τ connecting p to its parent is exactly $d(J_p, R \setminus J_p)$;*

Consider a tree $\tau \in \Upsilon$. We group the black components of τ top down into subtrees choosing grey edges in increasing order of their lengths until the volume of the group just exceeds ℓ.

Definition 2. *A black component is called a singleton component if it contains only a single node corresponding to some $v \in R$. A singleton component which is the very root of some tree $\tau \in \Upsilon$, is called a singleton root component.*

Observation 1. *Consider tree $\tau \in \Upsilon$. The root-group $\mathcal{G}$ has volume at least ℓ. If the root-group is not a singleton root component, then it has volume at most 2ℓ. The leaf-groups might have volume smaller that ℓ. All the other internal-groups have volume in the range $[\ell, 2\ell)$.*

From now on we will slightly abuse the notation and instead of z^{U_p} and x_{U_p} we will write z^p and x_p, respectively. Also, we will assume that any black component U_p corresponding to a node $p \in \Upsilon$ satisfies the Configuration LP Constraints (5) to (12). If not, then we find the violating constraint and we recompute the LP by applying the ellipsoid method.

In the next lemma we consider edges related to a group $\mathcal{G}$. The proofs for all the lemmas can be found in [6].

Lemma 2. *For any tree τ, group $\mathcal{G}$ and black component $p \in \mathcal{G}$, the following properties hold:*

1. *the total number of grey or black edges within $\mathcal{G}$ is at most $O(\ell)$;*
2. *any grey edge entering $\mathcal{G}$ is longer (or equal) to any grey or black edge in $\mathcal{G}$;*
3. *the total length of the path (including both grey and black edges), from any node $v \in J_p$ to the root r of the group $\mathcal{G}$, is at most $O(\ell)d(J_p, R \setminus J_p)$; and*
4. *the length of the path from $v \in J_p$ to the root r' of its parent group $\mathcal{G}'$ (if it exists) is $O(\ell)d(J_p, R \setminus J_p)$.*

Lemma 3. *Consider any representative $v \in R$. We can construct a new solution $\{x', y', z'\}$ such that all the facilities from set U_v are collocated with v. The cost of the new solution is at most $O(\ell)\mathbf{LP}$.*

For any black component $p \in \Upsilon$, let $\dot{y}_p = \frac{\sum_{S \in \mathcal{S}} |S| z'^p_S}{1 - z'^p_\perp} = \frac{\sum_{i \in S, S \in \mathcal{S}} z'^p_{S,i}}{1 - z'^p_\perp}$ and $y'_p = \sum_{i \in U_p} y'_i = \sum_{i \in S, S \in \tilde{\mathcal{S}}} z'^p_{S,i} = \sum_{S \in \mathcal{S}} |S| z'^p_{S,i} + \sum_{i \in U_p} z'^p_{\perp,i}$. Moreover, we define $\pi(J_p) := \sum_{j \in C} x_{p,j}(1 - x_{p,j})$ for any $p \in \Upsilon$.

Next we show that, we can pre-process each black component p by opening a set randomly from $\mathcal{S}$ and pre-assigning some clients to the open set. We send the demand of the rest of the clients that was served by p, to the root of the parent group. To do that, we first reduce the variance in the size of sets in $\mathcal{S}$.

For the $|C| = ku$ case, we perform a massage process in which we move facilities from bigger sets to smaller ones, until the size of each set is either $\lfloor \dot{y}_p \rfloor$ or $\lceil \dot{y}_p \rceil$. Using the saturation property, we can reroute the demand of clients assigned to these facilities, so that the final solution remains feasible. We scale up the opening values of these sets, so that the expected size of sets in $\mathcal{S}$ is $\dot{y}_p$.

For the general case, instead, we use a brutal massage process in which we pick a prefix of the smallest sets in set $\mathcal{S}$, such that their total opening value is at least a constant. Then we add some extra facilities to the selected sets and scale up the opening values of these sets, so that the sets have size either $\lfloor \dot{y}_p \rfloor$ or $\lceil \dot{y}_p \rceil$ and the total opening is exactly $\dot{y}_p$.

In both cases, we pick a set randomly and pre-assign some clients based on their connection values. The intuition is that the clients which are served by more than $1 - \epsilon'$ by the black component p, get assigned to the selected set with high probability and the demand of the other clients can travel to the root of the parent group by paying the total cost of $O(\ell^2)\mathbf{LP}$.

Lemma 4. *Let $p \in \Upsilon$ be a black component and U_p satisfies $y_p \leq 2\ell$ and let $Z_p \in \{0,1\}$ be a random variable, such that $E[Z_p] = \dot{y}_p - \lfloor \dot{y}_p \rfloor$. Moreover, constraints (5) to (12) are satisfied for the solution $\{x', y', z'\}$ and U_p. Then, we can pre-open a set $S \subseteq U_p$ of expected cardinality $\dot{y}_p$, where $|S| = \lfloor \dot{y}_p \rfloor + Z_p$, and pre-assign a set $C' \subseteq C$ of clients to S such that*

1. *each facility $i \in S$ is pre-assigned at most u clients*
2. *expected cost of sending not assigned demand $x_{p,C\backslash C'}$ to the root of the parent-group is at most $O(\ell^2)d(J_p, R \setminus J_p)\pi(J_p)$*
3. $\Pr[|S| = \lfloor \dot{y}_p \rfloor] = \dot{y}_p - \lfloor \dot{y}_p \rfloor$ *and* $\Pr[|S| = \lceil \dot{y}_p \rceil] = 1 - (\dot{y}_p - \lfloor \dot{y}_p \rfloor)$
4. *expected cost of pre-assignment and local moving of $x_{p,C\backslash C'}$ is at most*

$$O(\ell) \sum_{j \in C, i \in U_p} d(i,j)x'_{i,j}.$$

3.1   Dependent Rounding

We will use a dependent rounding (DR) procedure, described in [9], to decide if a particular variable should be rounded up or down. It transforms fractional vector $\{\bar{v}_i\}_{i=1}^n$ to a random integral vector $\{\hat{v}_i\}_{i=1}^n$. DR procedure has the following properties:

1. Marginal distribution: $Pr[\hat{v}_i = 1] = \bar{v}_i$
2. Sum-preservation: $\sum_{i=1}^n \hat{v}_i \in \{\lfloor \sum_{i=1}^n \bar{v}_i \rfloor, \lceil \sum_{i=1}^n \bar{v}_i \rceil\}$.

In our procedure we first fix a tree $\tau \in \Upsilon$. Then we choose a pair of fractional black components according to a predefined order. After that we increase the opening of one and decrease the opening of the other in a randomized way. After each such iteration, at least one black component has an integral opening. Based on the value of the integral opening $y'''_p \in \{\lfloor \dot{y}_p \rfloor, \lceil \dot{y}_p \rceil\}$ decided for a black component $p \in \Upsilon$, we will select a set of facilities $S \in \mathcal{S} : |S| = y'''_p$ in a random way (for details see Lemma 4). First we do dependent rounding among the black components of the children groups of each parent group. After this step each group $\mathcal{G}$ will have at most one fractional black component among all black components in its children groups, and the total opening (capacity) within these black components will be preserved. Finally, once we complete this rounding phase for all the trees in Υ, then we will do dependent rounding among all the remaining fractional black components across all the trees in Υ in an arbitrary order. The procedure will preserve the sum of facility openings, hence in the end we will open exactly k facilities.

In the "rounding among children groups" step, we will do the rounding among the black components within the children-groups of a group $\mathcal{G}$ in an order defined by non-decreasing distance of these black components to the root r of the parent group $\mathcal{G}$ (breaking ties arbitrarily). This way, we would have an extra property on the number of open facilities for every prefix in this order of the black components belonging to the children-groups of group $\mathcal{G}$.

Before we start the rounding procedure, we will send exactly $\sum_{i \in U_p} z'^p_{\perp,i} - \dot{y}_p z'^p_\perp$ opening, from each black component p, to a virtual black component $v_{\mathcal{G}}$ co-located with the root of the group. Note that since $z'^p_\perp = z^p_\perp$ and $z'^p_{\perp,i} = z^p_{\perp,i}$, we can use z instead of z'. Let us define $\dot{y}_{v_{\mathcal{G}}} = \sum_{p \in \mathcal{G}} \sum_{i \in U_p} z^p_{\perp,i} - \dot{y}_p z^p_\perp$. We will call this the *blue* opening. We will treat this blue opening, co-located with the root, as a virtual black component. Since we are in the uniform capacity case, by loosing a constant factor we can assume that $\mathcal{F} = \mathcal{C}$ [10]. Moreover we can work with soft capacitated version of the problem due to Theorem 1.2 in [10]. Hence, for the blue opening, we can simply open the decided number of co-located facilities at the virtual black component. By $BCG(\mathcal{G})$ we denote a set of all, virtual or not, black components in the children groups of group $\mathcal{G}$. Note that, from now on the group $\mathcal{G}$ also contains the virtual black component $v_{\mathcal{G}}$.

Consider the root group of the tree τ. If it is a singleton root component then we classify it as a virtual black component, otherwise we treat it as a standard black component. Note that the sum $\dot{y}_p + \sum_{i \in U_p} z'^p_{\perp,i} - \dot{y}_p z'^p_\perp = \sum_{i \in U_p} z^p_{\perp,i} + (1 - z^p_\perp)\dot{y}_p = y_p = y'_p$. Hence, the total opening across all the black components is exactly equal to k.

Lemma 5. *For any group $\mathcal{G}$, the total demand is at most $(1+O(1/\ell))\sum_{p \in \mathcal{G}} u \dot{y}_p$.*

For simplicity of exposition, we will say that *a black component p is closed*, if the procedure decides to round down the opening of that component to $\lfloor \dot{y}_p \rfloor$, otherwise we say it is *opened*.

In this dependent rounding procedure, in contrast to [4], we will also be able to *pull demand* to the black components where we decided to open an extra facility. Cost of pulling can be bounded by the LP cost for sending the demand out of a black component. This new strategy is crucial to bring down the capacity violation from $2 + \epsilon$ to $1 + \epsilon$.

3.2   Rounding Among Children Groups

Consider any tree $\tau \in \Upsilon$ and its root r. For simplicity of description, we add a fake single node parent group and attach the root r to this fake group node with a grey edge of length exactly $d(J_p, R \setminus J_p)$, where p corresponds to the only the black component in the root group. Notice that from now on even the original root group is a child-group of some other group.

In the first phase of dependent rounding, we select the deepest (w.r.t. the number of edges) leaf-group and let its parent group be $\mathcal{G}$. Let $\bar{y}_p = \dot{y}_p - \lfloor \dot{y}_p \rfloor$ for each $p \in \Upsilon$. For performing this dependent rounding procedure within children groups of $\mathcal{G}$, we use the root r of $\mathcal{G}$ as a *accumulator*, which will temporarily store all the not assigned demand from children groups of $\mathcal{G}$. Let $n_{\mathcal{G}} = |BCG(\mathcal{G})|$.

To perform the dependent rounding procedure, we would order the components in $BCG(\mathcal{G}) = \{p_1, p_2, \ldots p_{n_\mathcal{G}}\}$ by non-decreasing distance from the root r of $\mathcal{G}$, so $d(p_i, r) \leq d(p_{i+1}, r)$ for $i < n_\mathcal{G}$. We define the vectors $\dot{y}_\mathcal{G} = (\dot{y}_{p_1}, \dot{y}_{p_2}, \ldots \dot{y}_{p_{n_\mathcal{G}}})$ and $\bar{y}_\mathcal{G} = (\bar{y}_{p_1}, \bar{y}_{p_2}, \ldots \bar{y}_{p_{n_\mathcal{G}}})$. Now we apply dependent rounding between the two fractional components in the ith prefix of vector $\bar{y}_\mathcal{G}$, for each i starting from $i = 2$ until $i = n_\mathcal{G}$. Note that, after applying dependent rounding on the ith prefix of $\bar{y}_\mathcal{G}$, at most one component will remain fractional in the prefix and one will become integral. If the black component p which become integral is not virtual, we apply Lemma 4, with $r \in \mathcal{G}$ as a root and, with $Z_p = 1$ if component is open and $Z_p = 0$ if it is closed. Let the output vector be $Z_\mathcal{G} = (Z_{p_1}, Z_{p_1}, \ldots Z_{p_\mathcal{G}})$. If $\sum_i \bar{y}_\mathcal{G}(i)$ is not integral then the output vector will have one fractional variable, otherwise it will be a vector of all integral values. Notice that by the property (1) of dependent rounding $E[Z_p] = \bar{y}_p$ and by the property (2) the sum of the facility opening $\dot{y}_p$ is preserved.

From now on, we will ignore the presence of all the children of the group $\mathcal{G}$ in our procedure. We repeat this process until our tree τ has only the added fake group left. Note that the root group will contain at most one black component which is fractional. After we finish the first phase, for each group $\mathcal{G}$, at most one component of $Z_\mathcal{G}$ will be fractional.

For any vector v, let $v[i_1, i_2] = \sum_{k=i_1}^{i_2} v(k)$. Due to the ordering which we follow in the above dependent rounding procedure and the fact that we didn't move any opening out of (or into) set $BCG(\mathcal{G})$ for each group $\mathcal{G}$, the following observation holds.

Observation 2. *After the first phase of the rounding procedure, $Z_\mathcal{G}[1, i] \in [\lfloor \bar{y}_\mathcal{G}[1, i] \rfloor, \lceil \bar{y}_\mathcal{G}[1, i] \rceil]$ holds for each i and, for each non-leaf group $\mathcal{G}$. Moreover, $Z_\mathcal{G}[1, n_\mathcal{G}] = \bar{y}_\mathcal{G}[1, n_\mathcal{G}]$.*

Once we complete phase one of rounding for each tree $\tau \in \Upsilon$, the second phase of the rounding procedure starts. In the second phase of the rounding procedure we just apply dependent rounding among all the remaining fractional variables, in an arbitrary order, until everything is integral. We apply Lemma 4 to all the non-virtual components with $Z_p = 1$, if it was open, and $Z_p = 0$ otherwise. Notice that for the black component from the root group we will use the root of a fake group as an accumulator. We open $\lceil \dot{y}_{v_\mathcal{G}} \rceil$ facilities in each virtual component $v_\mathcal{G}$ if it was rounded up, and $\lfloor \dot{y}_{v_\mathcal{G}} \rfloor$ otherwise.

Since the last fractional component in $BCC(\mathcal{G})$ could be either opened or closed, the total $Z_\mathcal{G}[1, n_\mathcal{G}]$ is either $\lfloor \bar{y}_\mathcal{G}[1, n_\mathcal{G}] \rfloor$ or $\lceil \bar{y}_\mathcal{G}[1, n_\mathcal{G}] \rceil$ respectively. And since each of the components of vector Z is integral, the following observation is true.

Observation 3. *After the second phase of the rounding procedure, $Z_\mathcal{G}[1, i] \in \{\lfloor \bar{y}_\mathcal{G}[1, i] \rfloor, \lceil \bar{y}_\mathcal{G}[1, i] \rceil\}$ holds for each i and, for each non-leaf group $\mathcal{G}$.*

Lemma 6. *The cost of moving the demand from all black components to their respective accumulators can be bounded by $\sum_{p \in \Upsilon} O(\ell^2) d(J_p, R \setminus J_p) \pi(J_p)$.*

3.3   Pulling Back Demand to the Open Facilities

Now we will define a single-commodity flow corresponding to distributing the demand from the accumulator co-located with the root of some non-leaf group to the open facilities in its children groups, for each tree $\tau \in \Upsilon$. To do this, we will pull back demand to the black components in a greedy way by pulling the demand first to the component belonging to $BCG(\mathcal{G})$ which is closest to the root r. We can bound the cost of pulling demand to the open facilities by charging it to the cost of pushing the demand to the root bounded in Lemma 6. The intuition is, since we are pulling back the demand in a greedy fashion, we can argue that for every demand, the distance which it will travel in the pulling phase is at most the distance it traveled to reach the accumulator r in the pushing phase. Since the cost for pushing is bounded by $O(\ell^2)\sum_{p \in BCG(\mathcal{G})} d(J_p, R \setminus J_p)\pi(J_p)$ (see Lemma 6), hence by the above claim the cost of pulling back the demand is bounded as well.

In this procedure, we first fix a tree $\tau \in \Upsilon$. Consider a non-leaf group $\mathcal{G}$ of τ and the set $BCG(\mathcal{G})$. In the pre-assignment step (Lemma 4), let q_p be the amount of demand we assigned to the open facilities in each black component $p \in BCG(\mathcal{G})$. Notice that for any virtual black component $p \in BCG(\mathcal{G})$ we didn't assign any demand in a pre-assignment, so $q_p = 0$. Now we define vector $q_{\mathcal{G}} = (q_{p_1}, q_{p_1} \dots, q_{p_{n_g}})$, to be the vector of the pre-assigned demand, which respects the same order of the components as in vector $\dot{y}_{\mathcal{G}}$.

Now we describe the pulling back procedure which we call *the greedy pulling process*. First, we freeze $(1 + O(1/\ell))u$ units of demand at the accumulator r of group $\mathcal{G}$. Next we start pulling the rest of the demand to the black components $BCG(\mathcal{G})$. We do the pulling process in the same greedy order in which we did the dependent rounding among the black components $BCG(\mathcal{G})$, i.e. starting from the component closest to r. By definition, the vectors $\dot{y}_{\mathcal{G}}, Z_{\mathcal{G}}$ and $q_{\mathcal{G}}$ respect this ordering. We start pulling the demand equal to $(1 + O(1/\ell))(Z_{\mathcal{G}}(i) + \lfloor \dot{y}_{\mathcal{G}}(i) \rfloor)u - q_{\mathcal{G}}(i)$ from the accumulator r to the ith component starting from $i = 1$, until we have no more demand to pull. We do this process for each non-leaf group $\mathcal{G}$ in all the trees in our forest Υ.

Observation 4. *After the greedy pulling process, each black component $p \in BCG(\mathcal{G})$ has a capacity violation by a factor of at most $(1 + O(1/\ell))$.*

Lemma 7. *After the greedy pulling procedure, the left over demand at any accumulator r of some non-leaf group $\mathcal{G}$ is exactly equal to $u(1 + O(1/\ell))$; which is the demand frozen at the beginning.*

Lemma 8. *For any non-leaf group $\mathcal{G}$, the distance travelled by any demand in the greedy pulling phase is at most the distance travelled by it in the dependent rounding phase.*

Now we would distribute the demand received by any black component p to the actual open facilities (which are located at the representatives J_p), such that each facility has a capacity violation of at most $1 + O(1/\ell)$. The following lemma bounds the cost of this step.

Lemma 9. *Any demand that a black component p received in the greedy pulling back process can be distributed to the open facilities within p. The distance travelled by the demand received by p in this procedure is at most $O(\ell)d(J_p, R \setminus J_p)$.*

3.4 Distributing Frozen Demand to the Open Facilities

Now, we distribute the frozen $(1 + O(1/\ell))u$ units of demand located at the accumulators over some open facilities, such that each open facility gets at most $uO(1/\ell)$ more demand. Let us fix a tree $\tau \in \Upsilon$. To do this distribution, we first send $(1+O(1/\ell))u$ units of demand from each of the non-fake accumulator to the accumulator of his parent group. Note that, using Lemma 2, we can bound the cost for this movement by paying an additive factor of $O(\ell)d(J_p, R \setminus J_p)$ in the distance moved by this demand in Sect. 3.2. Let r be the accumulator belonging to the group $\mathcal{G}$, which received $|C_\mathcal{G}|(1 + O(1/\ell))u = O(u|C_\mathcal{G}|)$ units of demand from the accumulators of the non-leaf children groups $C_\mathcal{G}$ of the group $\mathcal{G}$ in the tree τ. Note that, $|C_\mathcal{G}| \leq n_\mathcal{G}$, since $\mathcal{G}$ may have children which are leaf-groups. We start sending $O(1/\ell)u(\lfloor \dot{y}_\mathcal{G}(i) \rfloor + Z_\mathcal{G}(i))$ units of demand to the ith black component (in the same greedy order defined by the vector $\dot{y}_\mathcal{G}$) in the $BCG(\mathcal{G})$, starting from $i = 1$, until we have no more demand left with r.

Lemma 10. *After the distribution procedure for some accumulator r belonging to the group $\mathcal{G}$, all the demand which r received from the accumulators of his non-leaf children groups will be distributed fully.*

By an argument similar to Lemma 9, we can send this demand to any open facility within p, by loosing an additive factor of $O(\ell)d(J_p, R \setminus J_p)$ in the distance traveled by the demand. Hence, this shows that we can distribute all the demand received by r from his children-accumulators, corresponding to groups $C_\mathcal{G}$, among the open facilities within black components in $BCG(\mathcal{G})$, such that each facility receives at most an extra $O(1/\ell)u$ units of demand. We keep on doing this process bottoms-up, until we reach the very root fake accumulator. Now, for the demand located in the fake accumulator, we just distribute that demand over the open facilities in the very root group of the tree, which was using this accumulator. Note that since the very root group comprises of only one black component with $O(\ell)$ opening, there will be at least $O(\ell)$ open facilities in this component and again we will send $O(1/\ell)u$ units of extra demand to all the open facilities in the very root black component. By Lemma 1, each edge in the black component p has length at most $d(J_p, R \setminus J_p)$ and by Lemma 2 the number of edges in the group is $O(\ell)$. Hence, by loosing an additive factor of $O(\ell)d(J_p, R \setminus J_p)$ in the distance traveled by the demand, we can distribute this demand over open facilities in this black component.

In the following lemma, we bound the cost of distributing the frozen demand by the upper bound which we use to bound the cost of moving this demand from black components in $BCG(\mathcal{G})$ to the accumulator.

Lemma 11. *The distance travelled by any demand from each non-fake accumulator group $\mathcal{G}$ in the above re-distribution process is bounded above by*

$O(\ell)d(J_p, R \setminus J_p)$, which is also a bound on the distance it travelled to reach the accumulator in the dependent rounding phase.

Proof. (Theorem 1). We modify the initial solution by "moving" all facilities to their respective representatives (see Lemma 3). The obtained solution has cost $O(\ell)\mathbf{LP}$. In the Lemma 4, we pre-assign some demand and all the other demand we send to the respective accumulators. The cost of this operation is $\sum_{p \in \Upsilon} \sum_{j \in C, i \in U_p} O(\ell)d(i,j)x'_{i,j} + O(\ell^2) \sum_{p \in \Upsilon} d(J_p, R \setminus J_p)\pi(J_p) \leq O(\ell^2)\mathbf{LP}$. The last inequality follows from [11]. By the Lemmas 8, 9 and 11, we can bound the distance travelled by any demand in Sects. 3.3 and 3.4 by the distance it travelled in Sect. 3.2. This implies that the cost of moving the demand in Sects. 3.3 and 3.4 is bounded by $O(\ell^2)\mathbf{LP}$. Hence, overall the connection cost of our algorithm is $O(\ell^2)\mathbf{LP}$.

From the Observation 4 we know that the capacity violation of each facility is at most $1 + O(1/\ell)$. Moreover in Sect. 3.4, we increase the capacity violation of each facility by at most $O(1/\ell)$. So the final capacity violation is $1 + O(1/\ell)$, which ends the proof of the theorem.

4   Concluding Remarks

We showed that Configuration LP helps obtaining an algorithm with $1+\epsilon$ capacity violation for uniform capacities. It remains open if a similar result is possible for general capacities. It seems that the difficulty of generalizing our algorithm to general case lies in the dependent rounding. It is hard to control the number of open facilities and the capacities at the same time.

References

1. Aardal, K., van den Berg, P.L., Gijswijt, D., Li, S.: Approximation algorithms for hard capacitated k-facility location problems. Eur. J. Oper. Res. **242**(2), 358–368 (2015)
2. An, H.-C., Singh, M., Svensson, O.: LP-based algorithms for capacitated facility location. In: IEEE 55th Annual Symposium on Foundations of Computer Science (FOCS), pp. 256–265. IEEE (2014)
3. Arya, V., Garg, N., Khandekar, R., Meyerson, A., Munagala, K., Pandit, V.: Local search heuristics for k-median and facility location problems. SIAM J. Comput. **33**(3), 544–562 (2004)
4. Byrka, J., Fleszar, K., Rybicki, B., Spoerhase, J.: Bi-factor approximation algorithms for hard capacitated k-median problems. In: Proceedings of the Twenty-Sixth Annual ACM-SIAM Symposium on Discrete Algorithms (SODA), pp. 722–736. SIAM (2015)
5. Byrka, J., Pensyl, T., Rybicki, B., Srinivasan, A., Trinh, K.: An improved approximation for k-median, and positive correlation in budgeted optimization. In: Proceedings of the Twenty-Sixth Annual ACM-SIAM Symposium on Discrete Algorithms, pp. 737–756. SIAM (2015)
6. Byrka, J., Rybicki, B., Uniyal, S.: An approximation algorithm for uniform capacitated k-median problem with $1+\epsilon$ capacity violation. CoRR abs/1511.07494 (2015)

7. Charikar, M., Guha, S., Tardos, É., Shmoys, D.B.: A constant-factor approximation algorithm for the k-median problem. In: Proceedings of the Thirty-First Annual ACM Symposium on Theory of Computing, pp. 1–10. ACM (1999)
8. Chuzhoy, J., Rabani, Y.: Approximating k-median with non-uniform capacities. In: Proceedings of the Sixteenth Annual ACM-SIAM Symposium on Discrete Algorithms, pp. 952–958. Society for Industrial and Applied Mathematics (2005)
9. Gandhi, R., Khuller, S., Parthasarathy, S., Srinivasan, A.: Dependent rounding and its applications to approximation algorithms. J. ACM **53**(3), 324–360 (2006)
10. Li, S.: On uniform capacitated k-median beyond the natural LP relaxation. In: Proceedings of the Twenty-Sixth Annual ACM-SIAM Symposium on Discrete Algorithms (SODA), pp. 696–707. SIAM (2015)
11. Li, S.: Approximating capacitated k-median with $(1 + \epsilon)k$ open facilities. In: Proceedings of the Twenty-Seventh Annual ACM-SIAM Symposium on Discrete Algorithms (SODA), pp. 786–796. SIAM (2016)
12. Li, S., Svensson, O.: Approximating k-median via pseudo-approximation. In: Proceedings of the Forty-Fifth Annual ACM Symposium on Theory of Computing, pp. 901–910. ACM (2013)

Valid Inequalities for Separable Concave Constraints with Indicator Variables

Cong Han Lim[1], Jeff Linderoth[2]([⊠]), and James Luedtke[2]([⊠])

[1] Department of Computer Sciences,
University of Wisconsin-Madison, Madison, WI, USA
conghan@cs.wisc.edu
[2] Department of Industrial and Systems Engineering,
University of Wisconsin-Madison, Madison, WI, USA
{linderoth,jim.luedtke}@wisc.edu

Abstract. We study valid inequalities for a set relevant for optimization models that have both binary indicator variables, which indicate positivity of associated continuous variables, and separable concave constraints. Such models reduce to a mixed-integer linear program (MILP) when the concave constraints are ignored, and to a nonconvex global optimization problem when the binary restrictions are ignored. In algorithms to solve such problems to global optimality, relaxations are traditionally obtained by using valid inequalities for the MILP ignoring the concave constraints, and by independently relaxing each concave constraint using the secant obtained from the bounds of the associated variable. We propose a technique to obtain valid inequalities that are based on both the MILP and the concave constraints. We begin by analyzing a low-dimensional set that contains a single binary indicator variable, a single concave constraint, and three continuous variables. Using this analysis, for the canonical Single Node Flow Set (SNFS), we demonstrate how to "tilt" a given valid inequality for the SNFS to obtain additional valid inequalities that account for separable concave functions of the arc flows. We present computational results demonstrating the utility of the new inequalities on a fixed plus concave cost transportation problem. To our knowledge, this is one of the first works that simultaneously convexifies both nonconvex functions and binary variables to strengthen the relaxations of practical mixed integer nonlinear programs.

Keywords: Mixed integer nonlinear programming · Global optimization · Valid inequalities

1 Introduction

We study a nonlinear, mixed-integer set composed of a base polyhedron, a collection of indicator variables, and the (nonconvex) epigraphs of univariate concave functions. Specifically, using notation $[n] = \{1, \ldots, n\}$, we study the intersection of a polyhedron $\mathcal{P} \subseteq \mathbb{R}^n$, the *variable bound set*,

$$\mathcal{Z} := \{(x, z) \in \mathbb{R}^n \times \mathbb{B}^n : \ell_i z_i \leq x_i \leq u_i z_i \text{ for } i \in [n]\},$$

© Springer International Publishing Switzerland 2016
Q. Louveaux and M. Skutella (Eds.): IPCO 2016, LNCS 9682, pp. 275–286, 2016.
DOI: 10.1007/978-3-319-33461-5_23

and the component-wise epigraphs of n univariate concave functions $\{f_1, \ldots, f_n\}$, $f_i : \mathbb{R} \to \mathbb{R}$,

$$\mathcal{T} = \{(x,t) \in \mathbb{R}^{2n} : t_i \geq f_i(x_i), 0 \leq x_i \leq u_i \text{ for } i \in [n]\}.$$

Finding valid inequalities for the set

$$\mathcal{X} := \{(x,t,z) \in \mathbb{R}^{2n} \times \mathbb{B}^n : x \in \mathcal{P}, (x,z) \in \mathcal{Z}, (x,t) \in \mathcal{T}\} \qquad (1)$$

is the focus of our work. For each $i \in [n]$, we assume WLOG that $f_i(0) = 0$, $\ell_i < u_i$, and for simplicity of presentation we assume that $\ell_i \geq 0$. We frequently abuse notation and perform set intersection between sets with different domains. For $\mathcal{A} \subseteq \{(a,b) \in A \times B\}$ and $\mathcal{B} \subseteq \{(b,c) \in B \times C\}$, we let $\mathcal{A} \cap \mathcal{B} = \{(a,b,c) \in A \times B \times C : (a,b) \in \mathcal{A} \text{ and } (b,c) \in \mathcal{B}\}$. With this abuse of notation, we can say that $\mathcal{X} = \mathcal{P} \cap \mathcal{Z} \cap \mathcal{T}$. Using the facts that $\text{conv}(\mathcal{P} \cap \mathcal{Z})$ is a polyhedron, and $\mathcal{T}$ is the Cartesian product of the epigraphs of univariate *concave* functions, it is easy to establish that $\text{conv}(\mathcal{X})$ is a polyhedron, see e.g., Theorem I.1 in [10].

Relaxations of the set $\mathcal{X}$ appear as substructures in many important optimization problems. For example, in the case we focus on in this paper, if $\mathcal{P}$ is a single flow constraint, $P^{\text{SNFS}} := \{x \in \mathbb{R}^n_+ : \sum_{i \in N^+} x_i - \sum_{i \in N^-} x_i \leq d\}$, then the set $P^{\text{SNFS}} \cap \mathcal{Z}$ is the well-studied *single node flow set*, which arises naturally in many important practical applications such as production planning and for which many classes of strong valid inequalities are known [7,15]. When $\mathcal{P}$ is the network-flow polytope, the set $\mathcal{P} \cap \mathcal{T}$ (i.e., without indicator variables) is the feasible region of the *minimum concave-cost network flow problem* (MCNFP) [9]. The MCNFP arises in many application areas, including production planning, communication network design, facility location, and VLSI design, where the concave functions typically model economies of scale.

The set $\mathcal{X}$ also occurs as a relaxation of formulations of engineering design problems that involve a non-linear relationship between input and output variables. For example, the set $\mathcal{X}$ arises in water and gas-network design problems [4,12,18], where the nonlinearity models the pressure loss across a pipe, and the binary variables are used to model network design decisions. Even if the nonlinear functions are *not* concave, to build a valid relaxation, we typically must understand how to convexify both the convex and concave parts of a function. Thus, the set $\mathcal{X}$ can be an important building block for building strong relaxations to many engineering design problems.

In recent years, there has been significant research studying structured mixed integer nonlinear sets, where the nonlinear functions are convex [1,5,6] There is relatively less research studying the structure of specific mixed integer nonlinear sets where the nonlinear functions are nonconvex. In [16], the authors study a set that contains a nonconvex constraint arising from product blending in combination with binary variables that are used to model fixed costs. The authors of [11] derive valid inequalities for nonlinear (nonconvex) network design problems using optimality considerations of subproblems where the integer variables are fixed. A lifting procedure is applied to include the binary variables in the inequality and to make them globally valid.

This work adds to the growing literature that studies structured, nonconvex, mixed-integer nonlinear sets. The remainder of the extended abstract is divided into five sections. In Sect. 2, we show that $\text{conv}(\mathcal{Z} \cap \mathcal{T})$ is a simple polyhedral set consisting of a strengthened version of secant inequalities from $\mathcal{T}$. Thus, we argue that in order to get stronger relaxations of $\mathcal{X}$, we must simultaneously consider each of the sets whose intersection forms $\mathcal{X} = \mathcal{P} \cap \mathcal{Z} \cap \mathcal{T}$. In Sect. 3, we build a simple, low-dimensional set derived from valid inequalities for $\mathcal{P} \cap \mathcal{Z}$, and we describe how to use this set to construct a valid inequality for $\mathcal{X}$. In Sect. 4, we demonstrate how to apply the methodology from Sect. 3 when the set $\mathcal{P} \cap \mathcal{Z}$ is the single node flow set. We derive a new class of strong valid inequalities called *Tilted Simple Generalized Flow Cover Inequalities* (TSGFCI) for $\mathcal{X}$. Section 5 contains a computational study where we demonstrate that applying the TSGFCI inequalities to a fixed plus concave cost transportation problem can result in significant speedups to state-of-the-art global optimization software. We make concluding remarks in Sect. 6. Due to space limitations, nearly all proofs are omitted from this extended abstract.

2   Motivation: The Set $\mathcal{Z} \cap \mathcal{T}$

The standard methodology used to solve optimization problems involving non-convex structures such as $\mathcal{X}$ to global optimality is to create a convex relaxation of $\mathcal{X}$ and then to refine the relaxation over the feasible region via a branch-and-bound approach. The most natural way to create a convex relaxation of $\mathcal{X}$, employed by employed by state-of-the-art software such as BARON [17], ANTIGONE [13], and SCIP [3], is to relax the integrality requirements on binary variables in $\mathcal{Z}$,

$$\mathcal{R}(\mathcal{Z}) := \{(x, z) \in \mathbb{R}^n \times [0, 1]^n : \ell_i z_i \leq x_i \leq u_i z_i \text{ for all } i \in [n]\},$$

and to underestimate the concave functions $f_i(\cdot)$ using the secant intersecting the graph of the function at the endpoints of its domain,

$$\mathcal{S}(\mathcal{T}) := \left\{(x, t) \in \mathbb{R}^{2n} : t_i \geq f(0) + \frac{f_i(u_i) - f_i(0)}{u_i - 0}(x_i - 0) \text{ for } i \in [n]\right\}.$$

The polyhedron $\mathcal{R}(\mathcal{X}) := \mathcal{P} \cap \mathcal{R}(\mathcal{Z}) \cap \mathcal{S}(\mathcal{T})$ is a relaxation that can be employed within a branch-and-bound approach to optimize over $\mathcal{X}$.

The constraints in the set $\mathcal{Z}$ enforce the logical conditions that if the binary variable $z_i = 0$, then the associated variable $x_i = 0$ as well. Using this fact, one can strengthen the relaxation $\mathcal{R}(\mathcal{X})$ using the set of *strengthened secant inequalities*

$$\mathcal{S}^+(\mathcal{T}, \mathcal{Z}) := \left\{(x, t, z) \in \mathbb{R}^{3n} : t_i \geq f(\ell_i) z_i + \frac{f_i(u_i) - f_i(\ell_i)}{u_i - \ell_i}(x_i - \ell_i z_i) \text{ for } i \in [n]\right\}.$$

The set $\mathcal{S}^+(\mathcal{T}, \mathcal{Z})$ forms the basis of the strongest possible convex relaxation of the $\mathcal{X}$ when the constraints in $\mathcal{P}$ are ignored. The proof of the following result is based on standard results on the convex hull of the union of polyhedra [2].

Proposition 1. $\operatorname{conv}(\mathcal{Z} \cap \mathcal{T}) = \mathcal{R}(\mathcal{Z}) \cap \mathcal{S}^+(\mathcal{T}, \mathcal{Z})$.

Proposition 1 implies that the strengthened secant inequalities yield the strongest relaxation we can obtain of $\mathcal{X}$ if we ignore the interaction between $\mathcal{P}$ and $\mathcal{Z} \cap \mathcal{T}$. Thus, we next investigate a methodology that can simultaneously consider portions of all components of the structure of $\mathcal{P} \cap \mathcal{Z} \cap \mathcal{T}$.

3   A Low-Dimensional Mixed-Integer Nonlinear Set

Our goal is to derive valid inequalities for the set $\mathcal{X}$, based on known valid inequalities for the set $\mathcal{P} \cap \mathcal{Z}$. To this end, in this section, we consider a set defined by two continuous variable x and s, where the variable x has associated with it a binary indicator variable z and continuous variable t used to model $f(x)$. In addition to the constraints relating x to t and z, the set contains two inequalities which we interpret as being derived from valid inequalities for the set $\mathcal{P} \cap \mathcal{Z}$. We investigate the convex hull of two closely related variants of this set. These results form the base of a general methodology for constructing valid inequalities for our the set $\mathcal{X} = \mathcal{P} \cap \mathcal{Z} \cap \mathcal{T}$, defined in (1).

Let $f : \mathbb{R} \to \mathbb{R}$ be a concave function with $f(0) = 0$, α, β be positive real numbers, and $\ell, u, \gamma \in \mathbb{R}$ with $\ell < \beta/\alpha < u$. We define the following mixed-integer linear sets

$$S_\geq := \{(s, x, z) \in \mathbb{R}^2 \times \mathbb{B} : s + \alpha x - \beta z \geq \gamma, \ s \geq \gamma, \ \ell z \leq x \leq uz\}$$

and

$$S_\leq := \{(s, x, z) \in \mathbb{R}^2 \times \mathbb{B} : s + \alpha x - \beta z \leq \gamma, \ s \leq \gamma, \ \ell z \leq x \leq uz\}.$$

We are interested in studying valid inequalities for the following mixed-integer nonlinear sets:

$$ST_\geq := \{(s, x, t, z) \in \mathbb{R}^3 \times \mathbb{B} : (s, x, z) \in S_\geq, \ t \geq f(x)\}$$

and $ST_\leq$ which is defined similarly, with $S_\geq$ replaced by $S_\leq$. We assume that

$$f(\beta/\alpha) > f(\ell) + \left(\frac{f(u) - f(\ell)}{u - \ell} \right) (\beta/\alpha - \ell) \tag{2}$$

since otherwise f is linear in the range $[\ell, u]$. The analysis of the sets $ST_\geq$ and $ST_\leq$ is nearly identical, so we focus our analysis on $ST_\geq$ and then just present the results for $ST_\leq$.

We begin by analyzing the extreme points of $\operatorname{conv}(ST_\geq)$. We first consider the set that is obtained when $z = 1$:

$$ST_\geq^1 := \{(s, x, t) \in \mathbb{R}^3 : s + \alpha x \geq \gamma + \beta, s \geq \gamma, \ell \leq x \leq u, t \geq f(x)\}$$

Figure 1 helps visualize the set $ST_\geq^1$. It follows from concavity of f that the extreme points of $\operatorname{conv}(ST_\geq^1)$ are the points $(\gamma + \beta - \alpha\ell, \ell, f(\ell))$, $(\gamma, \beta/\alpha, f(\beta/\alpha))$,

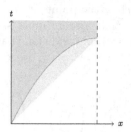
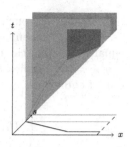
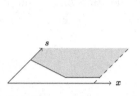

(a) The projection of $ST^1_\geq$ onto (s, x).

(b) A slice of the set $ST^1_\geq$ for a fixed value of s, along with the secant inequality.

(c) Slices of the set $ST^1_\geq$ at $s \in \{\gamma, \gamma+\beta-\alpha\ell, \omega\}$ where $\omega > \gamma + \beta - \alpha\ell$.

Fig. 1. Visualizing the set $ST^1_\geq$.

and $(\gamma, u, f(u))$. The hyperplane defined by these three points defines a valid inequality for the set $\mathrm{conv}(ST^1_\geq)$.

This discussion is formalized and extended to the set $ST_\geq$ in the following proposition.

Proposition 2. *The extreme rays of* $\mathrm{conv}(ST_\geq)$ *are given by* $(1,0,0,0)$ *and* $(0,0,1,0)$, *and the extreme points of* $\mathrm{conv}(ST_\geq)$ *are given by the points:*

$$
\begin{aligned}
v^1 &= (\quad \gamma, \quad\quad 0, \quad\quad 0, \quad 0\,), \\
v^2 &= (\,\gamma+\beta-\alpha\ell, \quad \ell, \quad f(\ell), \quad 1\,), \\
v^3 &= (\quad \gamma, \quad \beta/\alpha, f(\beta/\alpha), 1\,), \\
v^4 &= (\quad \gamma, \quad\quad u, \quad f(u), \quad 1\,).
\end{aligned}
$$

Using polarity theory, we thus obtain the following characterization of valid inequalities for $ST_\geq$.

Corollary 1. *An inequality*

$$
\lambda^s s + \lambda^x x + \lambda^t t + \lambda^z z \geq \lambda^0 \tag{3}
$$

is valid for $\mathrm{conv}(ST_\geq)$ *if and only if* $\lambda = (\lambda^s, \lambda^x, \lambda^t, \lambda^z, \lambda^0) \in C_\geq$, *where* $C_\geq$ *is the polyhedral cone*

$$
C_\geq := \{\lambda \in \mathbb{R}^5 : \lambda^s \geq 0, \lambda^t \geq 0, v^k \lambda \geq \lambda^0, \ k = 1, \ldots, 4\}.
$$

Furthermore, (3) *is a facet-defining inequality for* $\mathrm{conv}(ST_\geq)$ *if and only* λ *is an extreme ray of* $C_\geq$.

Valid inequalities for $ST_\geq$ which have $\lambda^s = 0$ are derived in Sect. 2, and valid inequalities for $ST_\geq$ which have $\lambda^t = 0$ just correspond to the inequalities defining $\mathcal{R}(S_\geq)$, the continuous relaxation of $S_\geq$. We therefore focus on valid inequalities for $ST_\geq$ which have $\lambda^t > 0$ and $\lambda^s > 0$. Under this condition, the characterization of valid inequalities for $ST_\geq$ in Corollary 1 reduces to a system of four inequalities, one for each of the points v^k, $k = 1, 2, 3, 4$, in five unknowns. Thus, the only extreme ray of that system must satisfy all four inequalities as an equality. Adding the normalization condition that $\lambda^s = 1$ and then observing that $v^1 \lambda = \lambda^0$ implies that $\lambda^0 = \gamma \lambda^s = \gamma$, we obtain the following reduced system of equations:

$$
\begin{pmatrix} \ell & 1 & f(\ell) \\ \beta/\alpha & 1 & f(\beta/\alpha) \\ u & 1 & f(u) \end{pmatrix} \begin{pmatrix} \lambda^x \\ \lambda^z \\ \lambda^t \end{pmatrix} = \begin{pmatrix} -\beta + \alpha\ell \\ 0 \\ 0 \end{pmatrix}. \tag{4}
$$

The assumption (2) together with $\ell < \beta/\alpha < u$ imply that the system (4) has a unique solution, which we denote by $(\bar{\lambda}^x, \bar{\lambda}^z, \bar{\lambda}^t)$, and can be shown to have $\bar{\lambda}^t > 0$. We thus obtain the following valid inequality.

Theorem 1. *The inequality*

$$
s + \bar{\lambda}^x x + \bar{\lambda}^z z + \bar{\lambda}^t t \geq \gamma \tag{5}
$$

is a valid and facet-defining inequality for $\mathrm{conv}(ST_\geq)$.

Combined with the previously known valid inequalities which have $\lambda^t = 0$ or $\lambda^s = 0$, we obtain a complete characterization of $\mathrm{conv}(ST_\geq)$.

Theorem 2. $\mathrm{conv}(ST_\geq)$ *is described by the set of* (s, x, t, z) *for which* $(s, x, z) \in \mathcal{R}(S_\geq)$, *and which satisfy (5) and the strengthened secant inequality*

$$
t \geq f(\ell)z + \frac{f(u) - f(\ell)}{u - \ell}(x - \ell z). \tag{6}
$$

Nearly identical arguments yield the following analogous result for $\mathrm{conv}(ST_\leq)$.

Theorem 3. *Let* $(\bar{\lambda}^x, \bar{\lambda}^z, \bar{\lambda}^t)$ *be the unique solution to the system of equations*

$$
\begin{pmatrix} \ell & 1 & f(\ell) \\ \beta/\alpha & 1 & f(\beta/\alpha) \\ u & 1 & f(u) \end{pmatrix} \begin{pmatrix} \lambda^x \\ \lambda^z \\ \lambda^t \end{pmatrix} = \begin{pmatrix} 0 \\ 0 \\ -\beta + \alpha u \end{pmatrix}. \tag{7}
$$

Then, the inequality

$$
s + \bar{\lambda}^x x + \bar{\lambda}^z z + \bar{\lambda}^t t \leq \gamma \tag{8}
$$

is valid and facet-defining for $\mathrm{conv}(ST_\leq)$, *and* $\mathrm{conv}(ST_\leq)$ *is described by the set of* (s, x, t, z) *for which* $(s, x, z) \in \mathcal{R}(S_\leq)$, *and which satisfy (6) and (8).*

4   Application to Single Node Flow Set

The *Single Node Flow Set* X^{SNFS} is

$$
X^{\mathrm{SNFS}} = \left\{ (x, z) \in \mathbb{R}^n \times \mathbb{B}^n : \sum_{i \in N^+} x_i - \sum_{i \in N^-} x_i \leq d, 0 \leq x_i \leq u_i z_i \text{ for } i \in N \right\}. \tag{9}
$$

N^+ and N^- denote the set of indices corresponding to the inflow and outflow arcs respectively, $N = N^+ \cup N^-$, and $n = |N|$.

We now define a variant of the Single Node Flow Set that incorporates concave functions of the flow variables. The *Concave Single Node Flow Set* is

$$X_f^{\text{CSNFS}} = \left\{ (x, z, t) \in \mathbb{R}^n \times \mathbb{B}^n \times \mathbb{R}^n : (x, z) \in SNFS, t_i \geq f_i(x_i) \right\}$$

where $f_i : \mathbb{R} \to \mathbb{R}$ are concave functions with $f_i(0) = 0$. Valid and facet-defining inequalities for $\text{conv}(X^{\text{SNFS}})$ are still valid and facet-defining for $\text{conv}(X_f^{\text{CSNFS}})$. We use the theory developed in Sect. 3 to derive additional valid inequalities for X_f^{CSNFS} based on valid inequalities for X^{SNFS}.

4.1   Valid Inequalities for X_f^{CSNFS}

We begin by assuming we have a valid inequality for X_f^{CSNFS} of the form

$$\sum_{i \in M^+ \setminus F} (\alpha_i x_i - \beta_i z_i) + \sum_{i \in F} (\lambda_i^x x_i + \lambda_i^z z_i + \lambda_i^t t_i) + \sum_{i \in N \setminus M^+} (\pi_i^x x_i + \pi_i^z z_i) \leq \gamma$$

(10)

where $F \subset M^+ \subseteq N^+$, and for all $i \in M^+ \setminus F$ we have $\alpha_i, \beta_i > 0$.

To apply Theorem 3, we choose $k \in M^+ \setminus F$, and write the inequality (10) as

$$s + \alpha_k x_k - \beta_k z_k \leq \gamma$$

where

$$s = \sum_{i \in M^+ \setminus (F \cup \{k\})} (\alpha_i x_i - \beta_i z_i) + \sum_{i \in F} (\lambda_i^x x_i + \lambda_i^z z_i + \lambda_i^t t_i) \sum_{i \in N \setminus M^+} (\pi_i^x x_i + \pi_i^z z_i). \quad (11)$$

Theorem 3 also requires $s \leq \gamma$ to be a valid inequality, which we now establish.

Lemma 1. *Assume* (10) *is a valid inequality for* X_f^{CSNFS}. *Then* $s \leq \gamma$ *is also a valid inequality for* X_f^{CSNFS}.

Now, we assume we know the following inequality is valid for X^{SNFS}:

$$\sum_{i \in M^+} (\alpha_i x_i - \beta_i z_i) + \sum_{i \in N \setminus M^+} (\pi_i^x x_i + \pi_i^z z_i) \leq \gamma. \quad (12)$$

By repeatedly applying Theorem 3 and Lemma 1, we derive a family of valid inequalities for X_f^{CSNFS}.

Theorem 4. *Assume* (12) *is a valid inequality for* X^{SNFS} *with* $\alpha_i, \beta_i > 0$ *for* $i \in M^+ \subseteq N^+$. *Let* $F \subseteq M^+$, *and for* $i \in F$ *let* $(\bar{\lambda}_i^x, \bar{\lambda}_i^z, \bar{\lambda}_i^t)$ *be the solution to* (7) *with* $(\alpha, \beta, \ell, u) \leftarrow (\alpha_i, \beta_i, 0, u_i)$. *Then, the following* tilted *inequality is valid for* X_f^{CSNFS}:

$$\sum_{i \in M^+ \setminus F} (\alpha_i x_i - \beta_i z_i) + \sum_{i \in F} (\bar{\lambda}_i^x x_i + \bar{\lambda}_i^z z_i + \bar{\lambda}_i^t t_i) + \sum_{i \in N \setminus M^+} (\pi_i^x x_i + \pi_i^z z_i) \leq \gamma.$$

(13)

Proof. Starting with (12) we choose $k \in M^+$ and apply Lemma 1 and Theorem 3 to obtain a valid inequality of the form (13) in which $F = \{k\}$. Proceeding inductively, given any inequality of the form (13), we can again choose $k' \in M^+ \setminus F$ and apply the same procedure, as long as $F \subset M^+$.   □

We refer to the procedure of generating an inequality (13) from a valid base inequality (12) as tilting. Given a relaxation solution $(\hat{x}, \hat{z}, \hat{t})$ and a "base" valid inequality (12), the most violated tilted inequality (13) is obtained by setting

$$F^* := \{i \in M^+ : \alpha_i \hat{x}_i - \beta_i \hat{z}_i < \bar{\lambda}_i^x \hat{x}_i + \bar{\lambda}_i^z \hat{z}_i + \bar{\lambda}_i^t \hat{t}_i\}. \tag{14}$$

We leave it is an interesting open question to determine general conditions under which the tilting procedure yields facet-defining inequalities. However, in the next section we provide such conditions when the base inequality comes from a particular class of valid inequalities for X^{SNFS}.

4.2   Tilting Flow Cover Inequalities

An important class of valid inequalities for the X^{SNFS}, are known as *flow cover inequalities* (FCI). A *generalized flow cover* is defined by sets (C^+, C^-), where $C^+ \subseteq N^+, C^- \subseteq N^-$ and $\sum_{i \in C^+} u_i - \sum_{i \in C^-} u_i = d + \mu, \mu > 0$.

There are many variants of flow cover inequalities, including FCI with inflows-only [15], simple generalized and extended generalized FCI [14,19], and lifted versions of FCI and simple generalized FCI [7,8]. As an illustration of our results, we focus on the *Simple Generalized Flow Cover Inequality* (SGFCI), which can be written as

$$\sum_{i \in C^+} (x_i - (u_i - \mu)^+ z_i) - \sum_{i \in L^-} \min(u_i, \mu) z_i - \sum_{i \in N^- \setminus (C^- \cup L^-)} x_i \leq d(C^+, C^-)$$

$$\tag{15}$$

where (C^+, C^-) is a generalized flow cover, $L^- \subseteq N^- \setminus C^-$, and $d(C^+, C^-) = d + \sum_{i \in C^-} u_i - \sum_{i \in C^+} (u_i - \mu)^+$. Van Roy and Wolsey [19] provide sufficient conditions for the SGFCI to be facet-defining. If we let $M^+ = \{i \in C^+ : u_i > \mu\}$, then the SGFCI takes the form of (12) with $\gamma = d(C^+, C^-)$. Then, applying Theorem 4 we obtain that for any $F \subseteq M^+$, the *Tilted Simple Generalized Flow Cover Inequality* (TSGFCI):

$$\sum_{i \in C^+ \setminus F} (x_i - (u_i - \mu)^+ z_i) - \sum_{i \in L^-} \min(u_i, \mu) z_i - \sum_{i \in N^- \setminus (C^- \cup L^-)} x_i$$

$$+ \sum_{i \in F} (\bar{\lambda}_i^x x_i + \bar{\lambda}_i^z z_i + \bar{\lambda}_i^t t_i) \leq d(C^+, C^-) \tag{16}$$

is valid for X_f^{CSNFS}, where for $i \in F$, $(\bar{\lambda}_i^x, \bar{\lambda}_i^z, \bar{\lambda}_i^t)$ is the solution to (7) with $(\alpha, \beta, \ell, u) \leftarrow (1, u_i - \mu, 0, u_i)$.

We next provide sufficient conditions for which these inequalities are facet-defining for X_f^{CSNFS}, which generalizes the result in [19].

Theorem 5. *Assume (i) $d > 0$, (ii) $\max_{i \in C^+ \setminus F} u_i > \mu$, (iii) $u_i > \mu$ for $i \in L^-$, (iv) $C^- = \emptyset$, and (v) $\left(\sum_{i \in C^+ \setminus F} u_i \right) - \mu > 0$. Then the TSGFCI (16) is facet-defining for X_f^{CSNFS}.*

Conditions (i)–(iv) are from [19] and are sufficient to ensure the base inequality (15) is facet-defining for X^{SNFS}. Condition (v) provides the requirement on the set F chosen for tilting.

5   Computational Results

In this section we demonstrate the effectiveness of the TSGFCI on a transportation problem in which flows incur a fixed cost plus concave cost. Given a set of facilities I with capacities b_i, $i \in I$ and a set of customers J with demands d_j, $j \in J$ the *Concave Fixed Charge Transportation Problem* (CFCTP) is the optimization problem:

$$\min_{x,z,t} \sum_{i \in I} \sum_{j \in J} (t_{ij} + p_{ij} z_{ij}) \qquad \text{(CFCTP)}$$

$$\text{s.t.} \sum_{i \in I} x_{ij} = d_j \qquad \text{for } j \in J$$

$$\sum_{j \in J} x_{ij} \le b_i \qquad \text{for } i \in I$$

$$t_{ij} \ge f_{ij}(x_{ij}) \qquad \text{for } i \in I, j \in J$$

$$0 \le x_{ij} \le u_{ij} z_{ij} \qquad \text{for } i \in I, j \in J.$$

There objective function models both a fixed charge p_{iij} associated with opening arc (i, j) and a cost that is a concave function $f_{ij}(\cdot)$ of the flow. We test our methods on randomly generated instances of the CFCTP. For a fixed problem size, we created a family of ten instances, which may be obtained at http://pages.cs.wisc.edu/~conghan/concave/.

The fixed charge network structure of (CFCTP) yields Single Node Flow Set relaxations (9) from which SGFCI may be generated. To obtain valid SGFCI that can be tilted, each instance is solved with CPLEX v12.6.0. CPLEX allows the solution of optimization problems with a nonconvex quadratic objective function, so (CFCTP) was reformulated into the equivalent formulation without the t_{ij} variables. The flow cover cuts that were applied by CPLEX at the end of the root node processing were extracted and used as the basis of our tilting procedure. CPLEX also is often able to tighten the upper bounds u_{ij} on the flow variables, and we also extract and use these improved values in computing the lifting coefficient via (4).

Using the extracted flow cover inequalities from CPLEX, TSGFCI are added in a separation loop using the optimal lifting set $F = F^*$ (defined in (14)) until no more inequalities can be separated. In addition, we also strengthen the relaxation by adding all TSGFCI for each flow cover with $|F| \le 2$. We remark that this implementation potentially underestimates the benefits from TSGFCI, since we are using flow cover inequalities that are generated without knowledge of the tilting procedure. An integrated procedure that simultaneously searches for violated flow cover inequalities along with their tilted variants could potentially identify additional violated inequalities.

We compare the performance of four different solution approaches: using CPLEX 12.6.0 with default options on the reformulation that eliminates the t variables (C), using BARON 14.4.0 on the original problem (B), using BARON 14.4.0 on the problem supplemented with implied bounds and flow covers extracted by CPLEX (BF), and finally using BARON 14.4.0 with the settings of (BF) plus the additional TSGFCI added (BT). We could not test the performance of CPLEX with the TSGFCI because CPLEX does not allow nonconvex constraints, and so the formulation used in CPLEX does not contain the necessary t variables. The experiments were performed on a heterogeneous set of servers, with each family of instances being run on the same machine. Each algorithm was limited to a single thread, a time limit of an hour, and a tolerance of 10^{-6} for relative optimality gap.

We present a summary of our computational results in Table 1. The *Number of Cuts* columns denote the average number of each class of cuts generated over the 10 instances. For a fixed problem instance, let l_M denote the best lower bound generated by method $M \in \{C, B, BF, BT\}$ and let v^* denote the objective value of the best solution amongst the four methods. The *Gap Closed* columns denote the average of the values $(l_M - l_B)/(v^* - l_B)$ over only the instances that no method was able to solve. We observe that BT solves almost twice as many instances within the time limit (32 versus 18/19 for B and BF), and for unsolved instances BT is able to close substantially more of the gap than BF. The set of instances solved for BT is a strict superset of those for BF, which is in turn a superset of those solved for B and C. In addition to these results, we note that at the root node of BARON, BT closes approximately 50 % of the initial gap over B and BF, which have the same initial lower bounds.

Table 1. Summary of computational results on the different instance families.

Instance family		Number of cuts		Instances solved				Gap closed (unsolved instances)	
Suppliers	Customers	Flow covers	Tilts	C	B	BF	BT	BF	BT
10	10	36.1	41.9	7	10	10	10	-	-
10	15	43.9	45.9	2	3	4	7	22.0 %	65.0 %
12	12	45.1	61	2	4	4	8	1.2 %	68.5 %
12	18	54.6	77.9	0	0	0	2	5.9 %	49.8 %
15	15	50.5	59.1	0	1	1	5	2.8 %	39.2 %
18	18	67.1	102.7	0	0	0	0	2.2 %	52.6 %

Figure 2 presents a plot of the cumulative distribution of solution times over the 32 instances that are solved by at least one method. Each plot indicates

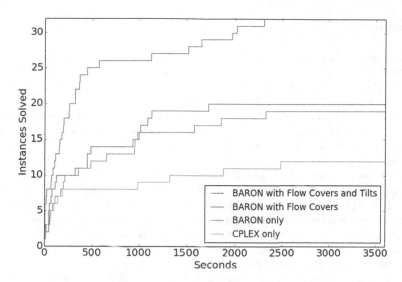

Fig. 2. Cumulative distribution plots of solution times. The plots from top to bottom correspond to the order of the labels in the legend (Color figure online).

the number of instances that have been solved by a certain time. The performance for BT significantly dominates all other methods, and BF shows a slight improvement over B, demonstrating that the flow covers make a small but significant difference, whereas the tilted flow covers yield large improvements in computation time.

6   Conclusion

We study valid inequalities for a mixed-integer nonlinear set having binary indicator variables and separable concave constraints. We derive a technique that obtains valid inequalities for this set by applying a tilting procedure to inequalities that are known for the set ignoring the concave constraints. We apply this procedure to a version of this set in which the linear constraints correspond to a network flow problem, and find that the new inequalities yield significant reductions in solution times. In future work, it will be interesting to test the proposed procedure on additional problems having this network structure, and to investigate the application of the proposed tilting procedure to other mixed-integer nonlinear structures.

Acknowledgements. The work was supported in part by the U.S. Department of Energy, Office of Science, Office of Advanced Scientific Computing Research, Applied Mathematics program under contract number DE-AC02-06CH11357.

References

1. Atamtürk, A., Narayanan, V.: Conic mixed integer rounding cuts. Math. Program. **122**, 1–20 (2010)
2. Balas, E.: Disjunctive programming. In: Annals of Discrete Mathematics 5: Discrete Optimization, pp. 3–51. North Holland (1979)
3. Berthold, T., Heinz, S., Vigerske, S.: Extending a CIP framework to solve MIQCPs. In: Lee, J., Leyffer, S. (eds.) Mixed Integer Nonlinear Programming. The IMA Volumes in Mathematics and its Applications, vol. 154, pp. 427–444. Springer, Heidelberg (2012)
4. D'Ambrosio, C., Lodi, A., Wiese, S., Bragalli, C.: Mathematical programming techniques in water network optimization. Eur. J. Oper. Res. **243**(3), 774–788 (2015)
5. Dong, H., Linderoth, J.: On valid inequalities for quadratic programming with continuous variables and binary indicators. In: Goemans, M., Correa, J. (eds.) IPCO 2013. LNCS, vol. 7801, pp. 169–180. Springer, Heidelberg (2013)
6. Frangioni, A., Gentile, C.: Perspective cuts for a class of convex 0–1 mixed integer programs. Math. Program. **106**, 225–236 (2006)
7. Gu, Z., Nemhauser, G.L., Savelsbergh, M.W.P.: Lifted flow cover inequalities for mixed 0–1 integer programs. Math. Program. **85**, 439–467 (1999)
8. Gu, Z., Nemhauser, G.L., Savelsbergh, M.W.P.: Sequence independent lifting in mixed integer programming. J. Comb. Optim. **4**(1), 109–129 (2000)
9. Guisewite, G.M., Pardalos, P.M.: Minimum concave-cost network flow problems: applications, complexity, and algorithms. Ann. Oper. Res. **25**, 75–100 (1990)
10. Horst, R., Tuy, H.: Global Optimization. Springer, New York (1993)
11. Humpola, J., Fügenschuh, A.: A new class of valid inequalities for nonlinear network design problems. Technical report 13–06, ZIB, Konrad-Zuse-Zentrum für Informationstechnik Berlin (2013)
12. Martin, A., Möller, M., Moritz, S.: Mixed integer models for the stationary case of gas network optimization. Math. Program. **105**(2), 563–582 (2006)
13. Misener, R., Floudas, C.A.: ANTIGONE: algorithms for coNTinuous/integer global optimization of nonlinear equations. J. Glob. Optim. **59**, 503–526 (2014)
14. Nemhauser, G.L., Wolsey, L.A.: Integer and Combinatorial Optimization. Wiley-Interscience, Hoboken (1988)
15. Padberg, M.W., Van Roy, T.J., Wolsey, L.A.: Valid linear inequalities for fixed charge problems. Oper. Res. **33**(4), 842–861 (1985)
16. Papageorgiou, D.J., Toriello, A., Nemhauser, G.L., Savelsbergh, M.W.P.: Fixed-charge transportation with product blending. Trans. Sci. **46**(2), 281–295 (2012)
17. Sahinidis, N.V.: BARON: a general purpose global optimization software package. J. Glob. Optim. **8**, 201–205 (1996)
18. Üster, H., Dilaveroğlu, S.: Optimization for design and operation of natural gas transmission networks. Appl. Energy **133**, 56–69 (2014)
19. Van Roy, T., Wolsey, L.A.: Valid inequalities for mixed 0–1 programs. Discrete Appl. Math. **14**(2), 199–213 (1986)

A Polyhedral Approach
to Online Bipartite Matching

Alfredo Torrico, Shabbir Ahmed, and Alejandro Toriello[✉]

H. Milton Stewart School of Industrial and Systems Engineering,
Georgia Institute of Technology, 30332 Atlanta, Georgia
atorrico3@gatech.edu, {sahmed,atoriello}@isye.gatech.edu

Abstract. We study the i.i.d. online bipartite matching problem, a
dynamic version of the classical model where one side of the bipartition
is fixed and known in advance, while nodes from the other side appear
one at a time as i.i.d. realizations of an underlying distribution, and must
immediately be matched or discarded. We consider various relaxations of
the set of *achievable* matching probabilities, introduce star inequalities
and their generalizations, and discuss when they are facet-defining. We
also show how several of these relaxations correspond to ranking policies
and their time-dependent generalizations. We finally present results of a
computational study of these relaxations and policies to determine their
empirical performance.

1 Introduction

Bipartite matching is one of the fundamental combinatorial optimization models, with a century-long history of research and applications in many areas. In
online or dynamic optimization, a widely studied variant has the right side of
the bipartition fixed and known to the decision maker ahead of time, while the
nodes in the left side appear one after another dynamically, and must immediately be matched to a remaining compatible right-hand node or discarded. This
model has application in a variety of resource allocation and revenue management areas, particularly in online search advertisement, where right-hand nodes
represent ads and left-hand nodes are search terms that the search engine wants
to show compatible ads to. Partly because of the connection to online search, the
computer science community has been interested in the model for several years,
beginning with [11], which showed that for a maximum cardinality objective, a
randomized ranking policy achieves the best possible $1 - 1/e$ competitive ratio,
assuming an adversary chooses which left-hand nodes appear; see e.g. [5] for a
corrected and simplified proof.

The adversarial model of node arrival is relatively pessimistic, and more
recent research has focused on models where arrivals are at least partly governed
by a distribution. In the simplest case, each arriving node is an i.i.d. sample from
a known distribution. Work on this model and its variants began in [8], which
showed that in the i.i.d. model it is possible to improve on the $1 - 1/e$ ratio; other

© Springer International Publishing Switzerland 2016
Q. Louveaux and M. Skutella (Eds.): IPCO 2016, LNCS 9682, pp. 287–299, 2016.
DOI: 10.1007/978-3-319-33461-5_24

subsequent results that further improve and/or generalize this approximation guarantee include [9,10,13].

While these works sometimes employ simple relaxations to design policies, to our knowledge no researchers have specifically looked at the generation of good upper bounds for the problem. Our first contribution is to study the set of matching probabilities achievable by some feasible policy, which is implicitly encoded as the projection of a doubly exponential polyhedron, and to derive relaxations by identifying various classes of valid inequalities. This focus on the *achievable region* is used in applied probability, for example to study models in queueing and multi-armed bandits (e.g. [4,6]), but to our knowledge it has not been applied in online matching. Our second contribution is then to show that optimal dual solutions of our relaxations imply specific policies; in particular, several determine simple ranking or time-dependent ranking policies. This connection is established by enforcing intuitive value function approximations on the linear programming formulation of the problem's dynamic program. This general technique dates back to [1,7,14,17], and has been used to derive relaxations and policies in many discrete dynamic optimization settings, such as economic lot scheduling [3], inventory routing [1,2] and the traveling salesman problem [15,16].

In the remainder of the paper, Sect. 2 states the problem and its formulations, and Sect. 3 discusses our proposed inequalities. Section 4 then shows how the relaxations imply policies, and Sect. 5 summarizes the results of our computational study. An Appendix has mathematical proofs not included in the body of the paper.

2   Model Description and Formulation

We are concerned with *online bipartite matching* (OBM) between two node sets, N and V, with edge set $E \subseteq N \times V$. This process occurs dynamically in the following way. The *right-hand* set V, with $|V| = m$, is known and given ahead of time. The *left-hand* set N with $|N| = n$ represents elements on the other side of the bipartition that may appear, but we do not know which ones will appear and how often. We know only that T elements in total will appear sequentially, each one drawn independently from the (stationary) uniform distribution over N; that is, at each epoch, any node in N appears with probability $1/n$ and is treated as a new copy. The assumption that the distribution is uniform is without loss of generality if the original distribution has rational probabilities, as we can duplicate nodes to put the problem in this form. Several past works on this OBM model (e.g. [8]) require $T = n$ so that each left-hand node's expected number of appearances is one, but we do not need this assumption. Each time a left-hand node appears, it must be immediately (and irrevocably) matched to an available compatible node in V or discarded. The objective is to maximize the expected number of matches. More generally, we can consider weights on each edge w_{ij}, with the objective to maximize the expected weight of matched edges; while our upper bound results extend to this case, some results in Sect. 4 depend on the cardinality objective, which we focus on for simplicity. Following convention from previous literature and the motivating application of search engine advertisement, we sometimes call $i \in N$ an *impression*, while each $j \in V$ is an *ad*.

For any impression $i \in N$, let $\Gamma(i) \subseteq V$ denote i's neighbors, and define $\Gamma(j)$ for ad $j \in V$ analogously; also, let η be the random variable with uniform distribution over N. We can now give a dynamic programming (DP) formulation for this OBM model. Let $v_t^*(i, S)$ denote the optimal expected value given that $i \in N$ appears when the set of ads $S \subseteq V$ is available and $t - 1$ draws from N still remain. Then, for all $t = 1, \ldots, T$, $i \in N$ and $S \subseteq V$,

$$v_t^*(i, S) = \max \begin{cases} \max_{j \in S \cap \Gamma(i)} \{1 + \mathbb{E}_\eta[v_{t-1}^*(\eta, S \backslash j)]\} \\ \mathbb{E}_\eta[v_{t-1}^*(\eta, S)], \end{cases} \tag{1}$$

where $v_0^*(\cdot, \cdot)$ is identically zero, and the model's optimal expected value is given by $\mathbb{E}_\eta[v_T^*(\eta, V)]$. The first term in this recursion corresponds to matching i with one of its remaining neighbors $j \in S \cap \Gamma(i)$; the second corresponds to discarding i. As with any DP, the optimal value function v^* induces an optimal policy: At any state (t, i, S), we choose an action that attains the maximum in (1). In the maximum cardinality case, it is easy to see that an optimal policy always matches a node when possible, so that the discarding action is only taken when nothing else is feasible. Furthermore, any value function approximation implies a corresponding policy, by substituting it for v^* in (1) and choosing the action that attains the maximum.

The recursion (1) can be equivalently captured with the linear program (LP)

$$\min_v \ \mathbb{E}_\eta[v_T(\eta, V)] \tag{2a}$$

$$\text{s.t. } v_t(i, S \cup j) - \mathbb{E}_\eta[v_{t-1}(\eta, S)] \geq 1, \quad t \in [T], \ i \in N, \ j \in \Gamma(i), \ S \subseteq V \backslash j \tag{2b}$$

$$v_t(i, S) - \mathbb{E}_\eta[v_{t-1}(\eta, S)] \geq 0, \quad\quad t \in [T], \ i \in N, \ S \subseteq V \tag{2c}$$

$$v \geq 0. \tag{2d}$$

The value function v^* defined in (1) is optimal for (2). Moreover, this LP is a strong dual for OBM, in the sense that any feasible v has an objective greater than or equal to $\mathbb{E}_\eta[v_T^*(\eta, V)]$.

The dual of (2) is a primal formulation; any feasible solution encodes a feasible policy and its probability of reaching any state in the DP. That formulation is the following LP:

$$\max_{x,y} \sum_{i \in N} \sum_{t \in [T]} \sum_{j \in \Gamma(i)} \sum_{S \subseteq V \setminus j} x_{i,j}^{t,S} \tag{3a}$$

$$\text{s.t.} \sum_{j \in \Gamma(j)} x_{i,j}^{T,V \setminus j} + y_i^{T,V} \le \frac{1}{n}, \quad i \in N, \tag{3b}$$

$$\sum_{j \in \Gamma(i) \cap S} x_{i,j}^{t,S \setminus j} + y_i^{t,S} - \frac{1}{n} \sum_{k \in N} y_k^{t+1,S} - \frac{1}{n} \sum_{k \in N} \sum_{j \in \Gamma(k) \cap S} x_{k,j}^{t+1,S} \le 0, \tag{3c}$$

$$t \in [1, T-1], \ i \in N, \ \emptyset \ne S \subseteq V,$$

$$\sum_{j \in \Gamma(i) \cap S} x_{i,j}^{T,S \setminus j} + y_i^{T,S} \le 0, \quad i \in N, \ S \subsetneq V, \tag{3d}$$

$$x, y \ge 0. \tag{3e}$$

Here, $x_{i,j}^{t,S}$, the variable corresponding to dual constraint (2b), represents the probability that the policy chooses to match impression i to ad j in state $(t, i, S \cup j)$, and $y_i^{t,S}$, which corresponds to (2c), similarly represents a discarding action. As with its dual, the LP (3) has exponentially many variables and constraints, and is therefore difficult to work with directly. However, we can equivalently consider optimizing over the matching probabilities achieved by a feasible policy; this corresponds to optimizing over a projection of the feasible region of (3),

$$\max \left\{ \sum_{i \in N} \sum_{j \in \Gamma(i)} z_{ij} : \exists \ (x,y) \in \ (3b)-(3e) \text{ with } z_{ij} = \sum_{t \in [T]} \sum_{S \subseteq V \setminus j} x_{i,j}^{t,S} \right\}, \tag{4}$$

where z_{ij} is the probability that impression i is *ever* matched to ad j. Any such z is a vector of matching probabilities that is *achievable* by at least one feasible policy. Let Q denote this projected polyhedron in the space of z variables, and note that Q is full-dimensional in $\mathbb{R}^E$. We consider relaxations of Q in the following section.

3   Projected Relaxations

We begin with the simplest relaxation. Recall that each ad j can be matched at most once, so this constrains all probabilities involving j to not exceed one in total. Similarly, each impression i appears in each epoch with probability $1/n$, and there are T stages, so the expected number of matches for i cannot exceed T/n. This gives us the LP

$$\max \sum_{(i,j) \in E} z_{ij} \tag{5a}$$

$$\text{s.t.} \sum_{j \in \Gamma(i)} z_{ij} \le \frac{T}{n}, \quad i \in N \tag{5b}$$

$$\sum_{i \in \Gamma(j)} z_{ij} \le 1, \quad j \in V \tag{5c}$$

$$z \ge 0. \tag{5d}$$

In particular, when $T = n$, (5) gives the deterministic bipartite matching formulation over $(N \cup V, E)$, and more generally it encodes a simple max-flow model; see e.g. [8].

We can use similar probabilistic ideas to strengthen the relaxation. An impression $i \in N$ will not appear at all with probability $(1 - 1/n)^T$, and thus

$$z_{ij} \leq 1 - (1 - 1/n)^T, \quad i \in N, \ j \in \Gamma(i) \tag{6}$$

is valid for Q; see e.g. [9]. Furthermore, these constraints in fact define facets.

Proposition 1. *Constraints (6) are facet-defining for the polyhedron of achievable probabilities Q.*

Though these inequalities were already known, the proof that they are facet-defining is new to our knowledge.

Proof (Sketch). We can construct the following $|E|$ affinely independent points corresponding to policies that satisfy (6) with equality:

- The policy that simply matches (i, j) when possible and ignores other edges.
- For any edge (i', j') that doesn't share an endpoint with (i, j), the policy that matches either edge when possible.
- For any $j' \in \Gamma(i) \backslash j$, the policy that matches (i, j) the first time i appears and then matches (i, j') the second time.
- For any $i' \in \Gamma(j) \backslash i$, the policy that matches (i, j) when possible but in the last epoch matches (i', j) if i' appears and i hasn't appeared.   □

We can generalize the previous concept to any set of impressions incident to an ad $j \in V$. Let $I \subseteq \Gamma(j)$; no nodes from this set will appear at all with probability $(1 - |I|/n)^T$, and hence the set of *right-star* inequalities

$$\sum_{i \in I} z_{ij} \leq 1 - (1 - |I|/n)^T, \qquad j \in V, \ I \subseteq \Gamma(j) \tag{7}$$

is valid. Moreover, their separation can be achieved in polynomial time by sorting the z_{ij} in non-increasing order of i, and testing each successive sum against the corresponding right-hand side. However, they are not necessarily facet-defining for Q except when $|I| = 1$. Inequality (7) also shows that (5c) can never be tight for a feasible policy unless $|\Gamma(j)| = n$.

Let us examine the analogous situation on the other side. For an impression $i \in N$, consider a set $J \subseteq \Gamma(i)$ of adjacent ads. Since i may appear more than once, the previous argument does not apply. However, we can still upper bound the corresponding probabilities by considering the expected number of matches we can hope to make with i in J. As before, i will never appear with probability $(1 - 1/n)^T$. Similarly, i will appear exactly once (and can thus only be matched once) with probability $\frac{T}{n}(1 - 1/n)^{T-1}$. This continues until we consider the event that i appears $|J|$ or more times, because we cannot match i more than these many times in J. Let $B(T, 1/n)$ denote a binomial random variable with T trials and probability of success $1/n$. The preceding argument shows that the *left-star* inequalities

$$\sum_{j \in J} z_{ij} \leq \mathbb{E}\left[\min\left\{|J|, B(T, 1/n)\right\}\right], \qquad i \in N, \ J \subseteq \Gamma(i) \tag{8}$$

are valid. In addition, the same greedy algorithm can be used to separate them, this time sorting in non-decreasing order of j.

Theorem 1. *Constraints* (8) *are facet-defining for* Q *when* $|J| < T$.

The theorem's proof is similar to Proposition 1 but more involved, and uses circulant matrices [12]. When $|J| \geq T$, we have $\mathbb{E}\left[\min\left\{|J|, B(T, 1/n)\right\}\right] = T/n$, and therefore (8) for $J \subsetneq \Gamma(i)$ is dominated by (5b). We can, however, use a similar proof to determine when this inequality is also a facet.

Corollary 1. *Constraints* (8) *for* $J = \Gamma(i)$ *are facet-defining for* Q *regardless of* T, *and thus* (5b) *is facet-defining when* $|\Gamma(i)| \geq T$.

Finally, consider two sets $I \subseteq N$ and $J \subseteq V$. In the best case, they induce a complete bipartite subgraph, and we can proceed as before. No edges within the two sets will be matched at all with probability $(1 - |I|/n)^T$, exactly one will be matched with probability $\frac{|I|}{n}(1 - |I|/n)^{T-1}$, and so forth. Generalizing, let $B(T, |I|/n)$ denote a binomial random variable with T trials and probability $|I|/n$ of success. Then

$$\sum_{(i,j) \in E \cap (I \times J)} z_{ij} \leq \mathbb{E}\left[\min\left\{|J|, B(T, |I|/n)\right\}\right], \qquad I \subseteq N, \ J \subseteq V, \tag{9}$$

are valid. This general set of inequalities contains both (7) and (8) as special cases, by respectively taking $J = \{j\}$ and $I = \{i\}$. Moreover, for any fixed I or J they can be separated using the same greedy algorithm, now applied to sums of the z variables; more generally, they can be separated with an integer program that maximizes the left-hand side for every fixed value of $|I|$ and $|J|$. These inequalities are not necessarily facet-defining, except for the cases we have already pointed out.

A natural question is whether all of Q's facet's can be written with 0–1 coefficients. However, this is not true even for very small instances. We have constructed Q in PORTA for an instance with $T = 3, N = \{1, 2, 3\}, V = \{a, b\}$ and $E = \{1a, 1b, 2b\}$. Not counting the non-negativity constraints, Q has 13 facets, but only the four identified in (8) have 0–1 coefficients.

4   Policies Derived from Bounds

We next focus on generating policies based on the relaxations in the previous section. To do so, we generate approximations of the true value function that are feasible in (2) but efficient to compute, and show that several of these approximations lead to ranking policies [11] and their generalizations. A *ranking policy* is specified by an ordering or permutation of V: Assuming we label ads in the permutation's order as $V = \{1, \ldots, m\}$, at any decision epoch (t, i, S) we match the appearing impression i to $\min\{j : j \in S \cap \Gamma(i)\}$, the lowest-indexed compatible ad that is available. Such policies are appealing from a practical perspective, as they are completely specified by a permutation and can be implemented efficiently.

As a first step, suppose we approximate the value at any state by considering the impression that has just appeared and the remaining available ads. Specifically, suppose $q_i \geq 0$ represents the value of having $i \in N$ appear and be available to match. Similarly, let $r_j \geq 0$ be the value we assign to each available ad $j \in V$. This leads to an approximation of the expected value of state (t, i, S) as

$$v_t(i, S) \approx q_i + (t - 1)\mathbb{E}_\eta[q_\eta] + \sum_{j \in S} r_j. \tag{10}$$

In the approximation, i has just appeared, and thus the state has value q_i. In addition, there are $t - 1$ more draws remaining, so we will get the expected value of q_η that many more times. Finally, each $j \in S$ still available to match contributes its value r_j.

Proposition 2. *Restricting the feasible region of (2) by forcing solutions to have the form (10) yields the dual of (5), where the q variables correspond to constraints (5b) and the r variables correspond to (5c).*

Proof (Sketch). Taking expectation in (10) yields $\mathbb{E}_\eta[v_t(\eta, S)] = t\mathbb{E}_\eta[q_\eta] + \sum_{j \in S} r_j$; this is the restricted objective (2a) when $t = T, S = V$. Similarly, plugging in the approximation into the matching constraint (2b) for $(t, i, S \cup j)$ gives

$$q_i + (t - 1)\mathbb{E}_\eta[q_\eta] + \sum_{\ell \in S \cup j} r_\ell - \mathbb{E}_\eta\left[(t - 1)q_\eta + \sum_{\ell \in S} r_\ell\right] = q_i + r_j \geq 1.$$

The discarding constraint (2c) can be shown to be redundant under the restriction, so we are left precisely with the dual of (5). $\qquad \square$

Consider the dual of (5), and let $(q^{(10)}, r^{(10)})$ be an optimal extreme point solution. Under the cardinality objective function, this dual's feasible region is the convex hull of node covers of $(N \times V, E)$; thus $(q^{(10)}, r^{(10)})$ is the incidence vector of a cover, and when $T = n$ it is a minimum cardinality cover. Suppose we are at state (t, i, S) and employ the value function approximation given by this solution in (1) to choose an action. Assuming $S \cap \Gamma(i) \neq \emptyset$, this yields the optimization problem

$$\arg \max_{j \in S \cap \Gamma(i)}\left\{1 + \mathbb{E}_\eta\left[(t - 1)q_\eta^{(10)} + \sum_{\ell \in S \setminus j} r_\ell^{(10)}\right]\right\} = \arg \min_{j \in S \cap \Gamma(i)} r_j^{(10)},$$

where the equivalence follows simply by removing terms that do not depend on j. This corresponds to a *cover ranking* policy: Given an optimal node cover, we match an arriving impression if possible to a non-cover ad, and only match it to an ad in the cover when no remaining non-cover ad is compatible. Note that any ranking that orders ads so that non-cover ads appear before ads in the cover induces a cover ranking policy.

This first approximation (10) does not capture the interaction between impressions and ads. Suppose we add a value $p_{ij} \geq 0$ to a state whenever $i \in N$ appears and $j \in S$ is one of the remaining ads. The new value function approximation is

$$v_t(i, S) \approx q_i + (t-1)\mathbb{E}_\eta[q_\eta] + \sum_{j \in S}\left(r_j + p_{ij} + \left(1 - (1 - 1/n)^{t-1}\right)\sum_{k \in N\setminus i} p_{kj}\right).$$
(11)

Since $i \in N$ is the current impression, the approximation includes a value p_{ij} for all remaining ads j. (This value will be zero if $(i, j) \notin E$, but we include it to simplify the expressions.) Furthermore, each other impression $k \in N \setminus i$ will appear at least once in the remaining epochs with probability $1 - (1 - 1/n)^{t-1}$, so we include these values as well, discounted by that probability; we only count these values once, because an ad can only be matched once.

Proposition 3. *Restricting the feasible region of* (2) *by enforcing the form* (11) *on solutions yields the dual of* (5) *with additional constraints* (6), *where the new p variables correspond to constraints* (6).

This approximation does not yield a static ranking policy, but does induce a time-dependent generalization. Let $(q^{(11)}, r^{(11)}, p^{(11)})$ be optimal for the dual of (5) with (6), and suppose we use the approximation (11) in (1). At state (t, i, S) with $S \cap \Gamma(i) \neq \emptyset$, after removing terms that do not depend on j, this results in the optimization problem

$$\underset{j \in S \cap \Gamma(i)}{\arg\max} \sum_{\ell \in S\setminus j}\left(r_\ell^{(11)} + \left(1 - (1 - 1/n)^{t-1}\right)\sum_{i \in N} p_{i\ell}^{(11)}\right) =$$
$$\underset{j \in S \cap \Gamma(i)}{\arg\min}\left\{r_j^{(11)} + \left(1 - (1 - 1/n)^{t-1}\right)\sum_{i \in N} p_{ij}^{(11)}\right\}.$$

Let $p_j^{(11)} := \sum_{i \in \Gamma(j)} p_{ij}^{(11)}$. At any epoch t, this policy's ranking is given by a linear combination of the $r_j^{(11)}$ and $p_j^{(11)}$ values; the influence of the p values in the ranking is highest in the first epoch, and decays until vanishing in the last one. As with a (static) ranking policy, these *time-dependent rankings* can be pre-computed and the policy implemented efficiently.

We can generalize this approach to include all right-star inequalities.

Theorem 2. *Consider the value function approximation*

$$v_t(i, S) \approx q_i + (t-1)\mathbb{E}_\eta[q_\eta]$$

$$+ \sum_{j \in S} \left(r_j + \sum_{\substack{I \subseteq \Gamma(j) \\ I \ni i}} p_{Ij} + \sum_{\substack{I \subseteq \Gamma(j) \\ I \not\ni i}} \left(1 - (1 - |I|/n)^{t-1}\right) p_{Ij} \right), \quad (12)$$

where $q \in \mathbb{R}_+^N$, $r \in \mathbb{R}_+^V$, *and* $p_{Ij} \in \mathbb{R}_+$ *for* $j \in V$ *and* $I \subseteq \Gamma(j)$. *Restricting the feasible region of* (2) *with this approximation yields the dual of* (5) *with the additional constraints* (7), *where the* p *variables correspond to constraints* (7).

As before, we get a time-dependent ranking policy. Let $(q^{(12)}, r^{(12)}, p^{(12)})$ be optimal for the dual of (5) with constraints (7). At state (t, i, S) with $S \cap \Gamma(i) \neq \emptyset$, using the value function approximation (12) we obtain the optimization problem

$$\arg\max_{j \in S \cap \Gamma(i)} \sum_{\ell \in S \setminus j} \left(r_\ell^{(12)} + \sum_{I \subseteq \Gamma(\ell)} (1 - (1 - |I|/n)^{t-1}) p_{I\ell}^{(12)} \right)$$

$$= \arg\min_{j \in S \cap \Gamma(i)} \left\{ r_j^{(12)} + \sum_{I \subseteq \Gamma(j)} (1 - (1 - |I|/n)^{t-1}) p_{Ij}^{(12)} \right\}.$$

As with the policy given by (11), the time-dependent rankings can be pre-computed and the policy implemented efficiently.

We can state a similar correspondence between constraints (8) and a value function approximation.

Theorem 3. *Consider the value function approximation*

$$v_t(i, S) \approx q_i + (t-1)\mathbb{E}_\eta[q_\eta] + \sum_{j \in S} r_j + \sum_{J \subseteq \Gamma(i)} p_{iJ}\mathbb{E}[\min\{|J \cap S|, B(t-1, 1/n) + 1\}]$$

$$+ \sum_{k \in N \setminus i} \sum_{J \subseteq \Gamma(k)} p_{kJ}\mathbb{E}[\min\{|J \cap S|, B(t-1, 1/n)\}], \quad (13)$$

where $q \in \mathbb{R}_+^N$, $r \in \mathbb{R}_+^V$, $p_{iJ} \in \mathbb{R}_+$ *for* $i \in N$ *and* $J \subseteq \Gamma(i)$. *Restricting the feasible region of* (2) *with this approximation yields the dual of* (5) *with the additional constraints* (8).

This approximation generalizes the intuition behind approximation (11) to subsets of ads. Suppose we model the value p_{iJ} of impression i interacting with a set of compatible ads $J \subseteq \Gamma(i)$. When i appears in epoch t, we expect no more than $\mathbb{E}[\min\{|J \cap S|, B(t-1, 1/n) + 1\}]$ matches between i and J from that point forward: The number of matches cannot exceed the number of remaining ads in the set, $|J \cap S|$, but it also cannot exceed the number of times we expect i to appear, the current appearance plus $B(t-1, 1/n)$ more. A similar argument applies to any impression that didn't appear in this epoch.

This value approximation seems not to define a dynamic ranking policy; let $(q^{(13)}, r^{(13)}, p^{(13)})$ be optimal for the dual of (5) with constraints (8). At state (t, i, S) with $S \cap \Gamma(i) \neq \emptyset$, employing the approximation (13) in (1) results in

$$\arg\max_{j \in S \cap \Gamma(i)} \left\{ \sum_{\ell \in S \setminus j} r_\ell^{(13)} + \sum_{i \in N} \sum_{J \subseteq \Gamma(i)} p_{iJ}^{(13)} \mathbb{E}[\min\{|J \cap (S \setminus j)|, B(t-1, 1/n)\}] \right\}.$$

Because the coefficients multiplying the p variables depend explicitly on the set S of remaining ads, it is impossible to compute these expressions a priori to obtain a ranking. We have nevertheless also implemented this policy in our computational experiments, outlined in the next section.

Finally, we can combine and generalize our previous approximations to derive a correspondence to (9).

Theorem 4. *Consider the value function approximation*

$$v_t(i, S) \approx q_i + (t-1)\mathbb{E}_\eta[q_\eta] + \sum_{j \in S} r_j$$

$$+ \sum_{\substack{I \subseteq N \\ I \ni i}} \sum_{J \subseteq V} p_{IJ} \mathbb{E}[\min\{|J \cap S|, B(t-1, |I|/n) + 1\}] \qquad (14)$$

$$+ \sum_{\substack{I \subseteq N \\ I \not\ni i}} \sum_{J \subseteq V} p_{IJ} \mathbb{E}[\min\{|J \cap S|, B(t-1, |I|/n)\}],$$

where $q \in \mathbb{R}_+^N, r \in \mathbb{R}_+^V, p_{IJ} \in \mathbb{R}_+$ for $I \subseteq N$ and $J \subseteq V$. Restricting the feasible region of (2) with this approximation yields the dual of (5) with additional constraints (9).

This approximation defines a policy similar to the one given by (13).

5   Computational Results

In this section we outline some of the experiments we conducted to test the bounds and policies. All of the test instances we constructed have $T = n = m$, and consist of the following:

1. A cycle of size 20 ($n = 10$).
2. A cycle of size 200 ($n = 100$).
3. 20 small instances with $n = 10$, each one randomly generated by having a possible edge in $N \times V$ be present independently with a probability of 10 %.
4. 20 large dense instances with $n = 100$, each one randomly generated by having a possible edge in $N \times V$ be present independently with a probability of 10 %.
5. 20 large sparse instances with $n = 100$, each one randomly generated by having a possible edge in $N \times V$ be present independently with probability of 2.5 %.

We tested various bounds on the instances by solving the initial relaxation (5) and then adding the other inequalities we introduced in Sect. 3. For the policies, we simulated 20,000 realizations of the small instances and 200 realizations of the large instances, and we report the sample average of each policy. To benchmark our results, for the small instances we computed the optimal expected value given by (1), and for the larger instances we calculated the sample mean of the maximum expected off-line matching, by computing the maximum cardinality matching of each simulated realization; this yields an upper bound on any policy as it affords the decision maker early access to information. As policy comparisons, we implemented two heuristics: The single-matching policy computes a maximum cardinality matching in $(N \times V, E)$, and matches only these edges, ignoring all others; this heuristic has an approximation ratio of $1 - (1 - 1/n)^T$ (approximately $1 - 1/e$ when $T = n$) [8]. The two-matching policy is a heuristic modification of the algorithm from [8] that uses a maximum cardinality 2-matching in $(N \times V, E)$.

Table 1 summarizes the results. For each instance class, in each row we present the geometric mean of each bound or policy's ratio to the best available benchmark (the DP value for small instances and the expected maximum matching for large ones). We also report the sample standard deviation of the ratios in parenthesis.

Table 1. Summary of experiment results.

Bound/policy	20-cycle	200-cycle	Small	Large dense	Large sparse
(5)	1.2681	1.2659	1.3151 (0.081)	1.0018 (0.0011)	1.2167 (0.010)
(5) + (6)	1.2681	1.2659	1.0886 (0.042)	1.0018 (0.0011)	1.1227 (0.011)
(5) + (7)	1.1319	1.0980	1.0536 (0.038)	1.0006 (0.0006)	1.0794 (0.009)
(5) + (8)	1.1606	1.1370	1.0570 (0.031)	1.0013 (0.0009)	1.0961 (0.011)
(5) + (7) + (8)	1.1319	1.0980	1.0536 (0.038)	1.0004 (0.0005)	1.0717 (0.009)
(5) + (9)	1.1319	1.0980	1.0288 (0.023)	-	-
Exp. matching	1.0514	1	1.0030 (0.005)	1	1
(1)	1	-	1	-	-
Matching	0.8259	0.8025	0.8566 (0.053)	0.6351 (0.0007)	0.7768 (0.028)
2-matching	0.9859	0.9496	0.9616 (0.037)	0.7500 (0.0023)	0.8793 (0.030)
(10)	0.9861	0.9474	0.9883 (0.020)	0.9232 (0.0048)	0.9306 (0.008)
(11)	0.9861	0.9474	0.9963 (0.005)	0.9232 (0.0048)	0.9358 (0.008)
(12)	0.9861	0.9474	0.9974 (0.005)	0.9471 (0.0091)	0.9560 (0.006)
(13)	0.9980	0.9539	0.9732 (0.032)	0.9409 (0.0036)	0.8635 (0.070)

With respect to the bounds, the right-star inequalities (7) improve the basic bound (5) more than the left-star ones (8), even though the latter are facet-defining. For small instances, the complete subgraph inequalities (9) can further cut the gap to under 3 %; however, we weren't able to compute this bound in a reasonable time for larger instances because of the significant additional computational burden. For the dense large instances, our bounds are all quite close to the expected maximum matching benchmark, which is unsurprising since in most realizations of these instances there is a perfect or near-perfect matching.

In terms of policies, the best performer overall is the time-dependent ranking policy corresponding to the right-star inequalities and approximation (12). However, the non-ranking policy corresponding to the left-star inequalities and approximation (13) does perform better on the two cycle instances. In contrast, the single-matching heuristic policy does not perform well, and even the 2-matching heuristic's performance significantly worsens for the large instances; this may indicate the benefit of having more than two choices per impression in larger graphs.

References

1. Adelman, D.: Price-directed replenishment of subsets: methodology and its application to inventory routing. Manuf. Serv. Oper. Manage. **5**, 348–371 (2003)
2. Adelman, D.: A price-directed approach to stochastic inventory/routing. Oper. Res. **52**, 499–514 (2004)
3. Adelman, D., Barz, C.: A unifying approximate dynamic programming model for the economic lot scheduling problem. Math. Oper. Res. **39**, 374–402 (2014)
4. Bertsimas, D., Niño-Mora, J.: Conservation laws, extended polymatroids and multi-armed bandit problems; a polyhedral approach to indexable systems. Math. Oper. Res. **21**, 257–306 (1996)
5. Birnbaum, B., Mathieu, C.: On-line bipartite matching made simple. ACM SIGACT News **39**, 80–87 (2008)
6. Coffman Jr., E., Mitrani, I.: A characterization of waiting time performance realizable by single-server queues. Oper. Res. **28**, 810–821 (1980)
7. de Farias, D., van Roy, B.: The linear programming approach to approximate dynamic programming. Oper. Res. **51**, 850–865 (2003)
8. Feldman, J., Mehta, A., Mirrokni, V., Muthukrishnan, S.: Online stochastic matching: beating $1 - 1/e$. In: Proceedings of the 50th Annual IEEE Symposium on Foundations of Computer Science (FOCS), pp. 117–126. IEEE (2009)
9. Haeupler, B., Mirrokni, V.S., Zadimoghaddam, M.: Online stochastic weighted matching: improved approximation algorithms. In: Chen, N., Elkind, E., Koutsoupias, E. (eds.) WINE 2011. LNCS, vol. 7090, pp. 170–181. Springer, Heidelberg (2011)
10. Jaillet, P., Lu, X.: Online stochastic matching: new algorithms with better bounds. Math. Oper. Res. **39**, 624–646 (2014)
11. Karp, R., Vazirani, U., Vazirani, V.: An optimal algorithm for on-line bipartite matching. In: Proceedings of the 22nd Annual ACM Symposium on the Theory of Computing (STOC), pp. 352–358. ACM, New York (1990)
12. Kra, I., Simanca, S.: On circulant matrices. Not. AMS **59**, 368–377 (2012)
13. Manshadi, V., Oveis Gharan, S., Saberi, A.: Online stochastic matching: online actions based on offline statistics. Math. Oper. Res. **37**, 559–573 (2012)
14. Schweitzer, P., Seidmann, A.: Generalized polynomial approximations in markovian decision processes. J. Math. Anal. Appl. **110**, 568–582 (1985)

15. Toriello, A.: Optimal toll design: a lower bound framework for the asymmetric traveling salesman problem. Math. Program. **144**, 247–264 (2014)
16. Toriello, A., Haskell, W., Poremba, M.: A dynamic traveling salesman problem with stochastic arc costs. Oper. Res. **62**, 1107–1125 (2014)
17. Trick, M., Zin, S.: Spline approximations to value functions: a linear programming approach. Macroecon. Dyn. **1**, 255–277 (1997)

On Some Polytopes Contained in the 0,1 Hypercube that Have a Small Chvátal Rank

Gérard Cornuéjols[(✉)] and Dabeen Lee

Tepper School of Business, Carnegie Mellon University, Pittsburgh, USA
{gc0v,dabeenl}@andrew.cmu.edu

Abstract. In this paper, we consider polytopes P that are contained in the unit hypercube. We provide conditions on the set of infeasible 0,1 vectors that guarantee that P has a small Chvátal rank. Our conditions are in terms of the subgraph induced by these infeasible 0,1 vertices in the skeleton graph of the unit hypercube. In particular, we show that when this subgraph contains no 4-cycle, the Chvátal rank is at most 3; and when it has tree width 2, the Chvátal rank is at most 4. We also give polyhedral decomposition theorems when this graph has a vertex cutset of size one or two.

1   Introduction

Let $H_n := [0,1]^n$ denote the 0,1 hypercube in $\mathbb{R}^n$. Let $P \subseteq H_n$ be a polytope. Let $S := P \cap \{0,1\}^n$ denote the set of 0,1 vectors in P. If an inequality $cx \geq d$ is valid for P for some $c \in \mathbb{Z}^n$, then $cx \geq \lceil d \rceil$ is valid for conv(S) since it holds for any $x \in P \cap \mathbb{Z}^n$. Chvátal [4] introduced an elegant notion of closure as follows.

$$P' = \bigcap_{c \in \mathbb{Z}^n} \{x \in \mathbb{R}^n : cx \geq \lceil \max\{cx : x \in P\} \rceil\}$$

is the *Chvátal closure* of P. Chvátal [4] proved that the closure of a rational polyhedron is, again, a rational polyhedron. Recently, Dadush et al. [7] showed that the Chvátal closure of any convex compact set is a rational polytope. Let $P^{(0)}$ denote P and $P^{(t)}$ denote $(P^{(t-1)})'$ for $t \geq 1$. Then $P^{(t)}$ is the tth Chvátal closure of P, and the smallest k such that $P^{(k)} = \text{conv}(S)$ is called the *Chvátal rank* of P. Chvátal [4] proved that the Chvátal rank of every rational polytope is finite, and Schrijver [11] later proved that the Chvátal rank of every rational polyhedron is also finite.

Eisenbrand and Schulz [8] proved that the Chvátal rank of any polytope $P \subseteq H_n$ is $O(n^2 \log n)$. Rothvoss and Sanitá [10] constructed a polytope $P \subseteq H_n$ whose Chvátal rank is $\Omega(n^2)$. However, some special polytopes arising in combinatorial optimization problems have small Chvátal rank; for example, the fractional matching polytope has Chvátal rank 1. Hartmann et al. [9] gave a

This work was supported in part by NSF grant CMMI1263239 and ONR grant N00014-12-10032.

© Springer International Publishing Switzerland 2016
Q. Louveaux and M. Skutella (Eds.): IPCO 2016, LNCS 9682, pp. 300–311, 2016.
DOI: 10.1007/978-3-319-33461-5_25

necessary and sufficient condition for a facet-defining inequality of conv(S) to have rank 1. In this paper, we investigate 0,1 polytopes with a Chvátal rank that is a small constant or grows slowly with n.

The *skeleton* of H_n is the graph $G := (V, E)$ whose vertices correspond to the 2^n extreme points of H_n and whose edges correspond to the 1-dimensional faces of H_n, namely the $n2^{n-1}$ line segments joining two extreme points of H_n that differ in exactly one coordinate. Let $\bar{S} := \{0,1\}^n \setminus S$ denote the set of 0,1 vectors that are not in P. Consider the subgraph $G(\bar{S})$ of G induced by the vertices in $\bar{S}$. In this paper, we give conditions on $G(\bar{S})$ that guarantee a small Chvátal rank. For example, we show that when $\bar{S}$ is a stable set in G, the Chvátal rank of P is at most 1; when each connected component of $G(\bar{S})$ is a cycle of length greater than 4 or a path, the Chvátal rank is at most 2; when $G(\bar{S})$ contains no 4-cycle, the Chvátal rank is at most 3; in particular when $G(\bar{S})$ is a forest, the Chvátal rank is at most 3; when the tree width of $G(\bar{S})$ is 2, the Chvátal rank is at most 4. In.Sect. 4, we give polyhedral decomposition theorems for conv(S) when $G(\bar{S})$ contains a vertex cutset of cardinality 1 or 2. These decomposition theorems are used to prove the results on forests and on graphs of tree width two mentioned above. In Sect. 5, we give an upper bound on the Chvátal rank of P that depends on the cardinality of $\bar{S}$. In particular, we show that if only a constant number of 0,1 vectors are infeasible, then the Chvátal rank of P is also a constant. We also give a superpolynomial range on the number of infeasible 0,1 vectors where the upper bound of $O(n^2 \log n)$ on the Chvátal rank obtained by Eisenbrand and Schulz can be slightly improved to $O(n^2 \log \log n)$. Finally, in Sect. 6, we show that optimizing a linear function over S is polynomially solvable when the Chvátal rank of Q_S is constant.

Although our results are mostly of theoretical interest, we mention two applications. The first is to the theory of clutters with the packing property. Abdi et al. [1] constructed a class of minimal nonpacking clutters from 0,1 polytopes with Chvátal rank at most 2. In particular, a 0,1 polytope in $[0,1]^5$ where the infeasible 0,1 vectors induce 2 cycles of length 8 and the remaining 16 points are feasible lead to the discovery of a new minimally nonpacking clutter on 10 elements. Another application occurs when S is the set of 0,1 vectors whose sum of entries is congruent to i modulo k. The cases $k = 2$ and $k = 3$ are discussed in Sects. 2.1 and 3.

2   Some Polytopes with Small Chvátal Rank

To prove results on a polytope $P \subset [0,1]^n$, we will work with a canonical polytope Q_S that has exactly the same set S of feasible $0,1$ vectors. The description of Q_S is as follows.

$$Q_S := \{x \in [0,1]^n : \sum_{j=1}^n (\bar{x}_j(1 - x_j) + (1 - \bar{x}_j)x_j) \geq 1/2 \text{ for } \bar{x} \in \bar{S}\}.$$

The reason for working with Q_S is that the Chvátal rank of P is always less than or equal to the Chvátal rank of Q_S. Furthermore, we have a good handle on the kth Chvátal closure $Q_S^{(k)}$ because of the following lemma.

Lemma 1 (CCH [5]). *The middle points of all $k + 1$ dimensional faces of H_n belong to $Q_S^{(k)}$ for $0 \leq k \leq n - 1$.*

Chvátal, Cook and Hartmann proved this result when $S = \emptyset$. The result clearly follows for general $S \subseteq \{0,1\}^n$ since $Q_\emptyset \subseteq Q_S$ implies $Q_\emptyset^{(k)} \subseteq Q_S^{(k)}$. We also make repeated use of the two following results in our proofs.

Lemma 2. *Consider a half-space $D := \{x \in \mathbb{R}^n : dx \geq d_0\}$. Let $T := D \cap \{0,1\}^n$ and $\bar{T} := \{0,1\}^n \setminus T$. For every face F of H_n, the graph $G(F \cap \bar{T})$ is connected. In particular $G(\bar{T})$ is a connected graph.*

Theorem 1 (AADK [2]). *Let P be a polytope and let $G = (V, E)$ be its skeleton. Let $S \subset V$, $\bar{S} = V \setminus S$, and $\bar{S}_1, \ldots, \bar{S}_m$ be a partition of $\bar{S}$ such that there are no edges of G connecting $\bar{S}_i, \bar{S}_j$ for all $1 \leq i < j \leq m$. Then $conv(S) = \bigcap_{i=1}^m conv(V \setminus \bar{S}_i)$.*

Theorem 1, due to Angulo et al. [2], shows that we can consider each connected component of $G(\bar{S})$ separately when studying $conv(S)$. In Sect. 4, we give similar theorems in the case where $P \subset [0,1]^n$ and $G(\bar{S})$ contains a vertex cutset of cardinality 1 or 2. In this section, we provide the descriptions for $Q_S^{(1)}, Q_S^{(2)}, Q_S^{(3)}$.

2.1   Chvátal Rank 1

Theorem 2. *The polytope P has Chvátal rank at most 1 when $\bar{S}$ is a stable set in G.*

In particular, if S contains all the 0,1 vertices of H_n with an even (odd resp.) number of 1s, then P has Chvátal rank at most 1. Theorem 2 is obtained by characterizing $Q_S^{(1)}$. For each $\bar{x} \in \bar{S}$, we call

$$\sum_{j=1}^n (\bar{x}_j(1 - x_j) + (1 - \bar{x}_j)x_j) \geq 1 \tag{1}$$

the *vertex inequality* corresponding to $\bar{x}$. For example, when $\bar{x} = 0$, the corresponding vertex inequality is $x_1 + x_2 + \ldots + x_n \geq 1$. Note that each vertex inequality cuts off exactly the vertex $\bar{x}$ and it goes through all the neighbors of $\bar{x}$ on H_n.

Theorem 3. *$Q_S^{(1)}$ is the intersection of $[0,1]^n$ with the half-spaces defined by the vertex inequalities (1) for $\bar{x} \in \bar{S}$.*

2.2   Chvátal Rank 2

The next theorem characterizes $Q_S^{(2)}$. Let $N := \{1, \ldots, n\}$.

Throughout the paper, we will use the following notation. For a 0,1 vector $\bar{x}$, we denote by $\bar{x}^i$ the 0,1 vector that differs from $\bar{x}$ only in coordinate $i \in N$, and more generally, for $J \subseteq N$, we denote by $\bar{x}^J$ the 0,1 vector that differs from $\bar{x}$ exactly in the coordinates J. We denote by e^i the ith unit vector for $i \in N$.

Let $\bar{x}, \bar{y} \in \bar{S}$ be two vertices of $G(\bar{S})$ such that they differ in exactly one coordinate, say $\bar{y} = \bar{x}^i$. The inequality

$$\sum_{j \in N \setminus \{i\}} (\bar{x}_j(1 - x_j) + (1 - \bar{x}_j)x_j) \geq 1 \tag{2}$$

is called the *edge inequality* corresponding to edge $\bar{x}\bar{y}$. For example, when $\bar{x} = 0$ and $\bar{y} = e^1$, the corresponding edge inequality is $x_2 + x_3 + \ldots + x_n \geq 1$. The inequality (2) is the strongest inequality that cuts off $\bar{x}$ and $\bar{y}$ but no other vertex of H_n. Indeed, its boundary contains all $2(n-1)$ neighbors of $\bar{x}$ or $\bar{y}$ on H_n (other than $\bar{x}$ and $\bar{y}$ themselves). The next theorem states that vertex and edge inequalities are sufficient to describe the second Chvátal closure of Q_S.

Theorem 4. $Q_S^{(2)}$ is the intersection of $Q_S^{(1)}$ with the half-spaces defined by the edge inequalities (2) for $\bar{x}, \bar{y} \in \bar{S}$ such that $\bar{x}\bar{y}$ is an edge of H_n.

Note that the edge inequality (2) dominates the vertex inequalities for $\bar{x} \in \bar{S}$ and for $\bar{y} \in \bar{S}$. Thus vertex inequalities are only needed for the isolated vertices of $G(\bar{S})$.

Theorem 5. For $n \geq 3$, the Chvátal rank of Q_S is 2 if and only if $G(\bar{S})$ contains a connected component of cardinality at least 2, and each connected component of $G(\bar{S})$ is either a cycle of length greater than 4 or a path.

2.3   Chvátal Rank 3

Theorem 6 below is the main result of this section. It characterizes $Q_S^{(3)}$.

4-cycles of $G(\bar{S})$ correspond to 2-dimensional faces of H_n that are squares. Using our notation, if $\bar{x}, \bar{x}^i, \bar{x}^\ell, \bar{x}^{i\ell} \in \bar{S}$, we say that $(\bar{x}, \bar{x}^i, \bar{x}^\ell, \bar{x}^{i\ell})$ is a square. Note that

$$\sum_{j \in N \setminus \{i,\ell\}} (\bar{x}_j(1 - x_j) + (1 - \bar{x}_j)x_j) \geq 1 \tag{3}$$

is the strongest inequality cutting off exactly the four points of the square $(\bar{x}, \bar{x}^i, \bar{x}^\ell, \bar{x}^{i\ell})$. Indeed, the $4(n-2)$ neighbors of $\bar{x}, \bar{x}^i, \bar{x}^\ell, \bar{x}^{i\ell}$ in H_n (other than $\bar{x}, \bar{x}^i, \bar{x}^\ell, \bar{x}^{i\ell}$ themselves) all satisfy (3) at equality. We call (3) a *square inequality*. As an example, if $(0, e^1, e^2, e^1 + e^2)$ is a square contained in $G(\bar{S})$, the corresponding square inequality is $x_3 + x_4 + \ldots + x_n \geq 1$.

If $\bar{x}$ and $t \geq 3$ of its neighbors $\bar{x}^{i_1} := \bar{x} + (1 - 2\bar{x}_{i_1})e^{i_1}, \ldots, \bar{x}^{i_t} := \bar{x} + (1 - 2\bar{x}_{i_t})e^{i_t}$ all belong to $\bar{S}$, then we say that $(\bar{x}, \bar{x}^{i_1}, \ldots, \bar{x}^{i_t})$ is a *star*. The following *star inequality* is valid for conv(S)

$$\sum_{r=1}^{t} (\bar{x}_{i_t}(1 - x_{i_t}) + (1 - \bar{x}_{i_t})x_{i_t}) + 2 \sum_{j \neq i_1, \ldots, i_t} (\bar{x}_j(1 - x_j) + (1 - \bar{x}_j)x_j) \geq 2. \tag{4}$$

It cuts off the vertices of the star, and goes through the other $n - t$ neighbors of $\bar{x}$ on H_n and the $t(t-1)/2$ neighbors of two vertices among $\bar{x}^{i_1}, \ldots, \bar{x}^{i_t}$. For example, if $(0, e^1, \ldots, e^t)$ is a star, then (4) is $x_1 + \ldots + x_t + 2(x_{t+1} + \ldots + x_n) \geq 2$.

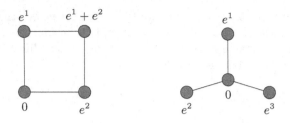

Fig. 1. Square and star with $\bar{x} = 0$

Theorem 6. $Q_S^{(3)}$ *is the intersection of* $Q_S^{(2)}$ *with the half-spaces defined by the square inequalities* (3) *and the star inequalities* (4).

To illustrate our proof techniques, we will prove Theorem 6 in this extended abstract. The proof uses the following lemma, which gives the linear description of conv(S) when $\bar{S}$ is a star.

Lemma 3. *Let* $n \geq 3$. *If* $\bar{S}$ *is a star, then* conv(S) *is completely defined by the corresponding star inequality together with the edge inequalities and the bounds* $0 \leq x \leq 1$.

Proof. We may assume that $\bar{x} = 0$, $\bar{S} = \{0, e^1, \ldots, e^t\}$ and $I := \{1, \ldots, t\}$.

If $t = n$, then S is the set of 0,1 vectors satisfying the system $\sum_{j=1}^{n} x_j \geq 2$ with $0 \leq x \leq 1$. This constraint matrix is totally unimodular. Therefore it defines an integral polytope, which must be conv(S).

If $t = 2$, we observe similarly that $\{x \in [0,1]^n : \sum_{j \in N \setminus \{r\}} (\bar{x}_j(1 - x_j) + (1 - \bar{x}_j)x_j) \geq 1$ for $r = 1, 2\}$ is an integral polytope. Indeed, the corresponding constraint matrix is also totally unimodular.

If $3 \leq t < n$, it is sufficient to show that $A := \{x \in [0,1]^n : \sum_{i \in I} x_i + 2 \sum_{j \in N \setminus I} x_j \geq 2, \sum_{j \in N \setminus \{r\}} x_j \geq 1$ for $1 \leq r \leq t\}$ is an integral polytope. Let v be an extreme point of A. We will show that v is an integral vector. Since we assumed $n \geq 3$, A has dimension n and there exist n linearly independent inequalities active at v.

First, consider the case when the star inequality is active at v. If no edge inequality is active at v, then $n - 1$ inequalities among $0 \leq x \leq 1$ are active at v. Since $\sum_{i \in I} v_i + 2 \sum_{j \in N \setminus I} v_j = 2$, it follows that all coordinates of v are integral. Thus we may assume that an edge inequality $\sum_{j \in N \setminus \{1\}} x_j \geq 1$ is active at v. Consider the face F of A defined by setting this edge inequality and the star inequality as equalities. Clearly v is a vertex of F. Observe that the two equations defining F can be written equivalently as $\sum_{j \in N \setminus \{1\}} x_j = 1$ and $x_1 + \sum_{j \in N \setminus I} x_j = 1$. Furthermore, any other edge inequality $\sum_{j \in N \setminus \{r\}} x_j \geq 1$ is implied by $x \geq 0$ since it can be rewritten as $\sum_{j \in I \setminus \{1,r\}} x_j \geq 0$ using $x_1 + \sum_{j \in N \setminus I} x_j = 1$. This means that F is entirely defined by $0 \leq x \leq 1$ and the two equations $x_1 + \sum_{j \in N \setminus I} x_j = 1$ and $\sum_{j \in N \setminus \{1\}} x_j = 1$. This constraint matrix is totally unimodular, showing that v is an integral vertex.

Assume now that the star inequality is not active at v, namely $\sum_{i \in I} v_i + 2\sum_{j \in N \setminus I} v_j > 2$. If at most one edge inequality is tight at v, then v is obviously integral. Thus, we may assume that $k \geq 2$ edge inequalities are tight at v, say $\sum_{j \in N \setminus \{r\}} x_j \geq 1$ for $1 \leq r \leq k$. Then $v_1 = \ldots = v_k$. If v_1 is fractional, v has at least k fractional coordinates. We assumed that only k inequalities other than $0 \leq x \leq 1$ are active at v, so the other coordinates are integral. Hence, $v_j = 0$ for $j \notin \{1, \ldots, k\}$ and $v_1 = \ldots = v_k = \frac{1}{k-1}$. Then $\sum_{r=1}^{t} v_r + 2\sum_{j \in N \setminus I} v_j = \frac{k}{k-1} \leq 2$. However, this contradicts the assumption that $\sum_{i \in I} v_i + 2\sum_{j \in N \setminus I} v_j > 2$. $\square$

Proof of Theorem 6: Applying the Chvátal procedure to inequalities defining $Q_S^{(2)}$, it is straightforward to show the validity of the inequalities (3) and (4) for $Q_S^{(3)}$.

To complete the proof of the theorem, we need to show that all other valid inequalities for $Q_S^{(3)}$ are implied by those defining $Q_S^{(2)}$, (3) and (4).

Consider a valid inequality for $Q_S^{(3)}$ and let $\bar{T}$ denote the set of 0,1 vectors cut off by this inequality. If $\bar{T} = \emptyset$, then the inequality is implied by $0 \leq x \leq 1$. Thus, we assume that $\bar{T} \neq \emptyset$. Let $T := \{0,1\}^n \setminus \bar{T}$. By the definition of a Chvátal inequality, there exists an inequality $ax \geq b$ valid for $Q_S^{(2)}$ that cuts off exactly the vertices in $\bar{T}$. By Lemma 1, the center points of the cubes of H_n all belong to $Q_S^{(2)}$. This means $ax \geq b$ does not cut off any of them. By Lemma 2, $G(\bar{T})$ is a connected graph. We claim that the distance between any two vertices in $G(\bar{T})$ is at most 2. Indeed, otherwise $G(\bar{T})$ contains two opposite vertices of a cube, and therefore its center satisfies $ax < b$, a contradiction.

We consider 3 cases: $|\bar{T}| \leq 3$, $G(\bar{T})$ contains a square, and $G(\bar{T})$ contains no square.

If $|\bar{T}| \leq 3$, then $G(\bar{T})$ is either an isolated vertex, an edge, or a path of length two. Then vertex and edge inequalities with the bounds $0 \leq x \leq 1$ are sufficient to describe $\text{conv}(T)$ by Lemma 3.

If $G(\bar{T})$ contains a square $(\bar{x}, \bar{x}^i, \bar{x}^\ell, \bar{x}^{i\ell})$, it cannot cut off any other vertex of H_n (otherwise, by Lemma 2 there would be another vertex of $\bar{T}$ adjacent to the square, and thus in a cube, and cut off by the inequality, a contradiction). Thus, $\bar{T} = \{\bar{x}, \bar{x}^i, \bar{x}^\ell, \bar{x}^{i\ell}\}$. Since

$$\text{conv}(T) = \{x \in [0,1]^n : \sum_{j \in N \setminus \{i, \ell\}} (\bar{x}_j(1 - x_j) + (1 - \bar{x}_j)x_j) \geq 1\},$$

a Chvátal inequality derived from $ax \geq b$ will therefore be implied by the square inequality that corresponds to $(\bar{x}, \bar{x}^i, \bar{x}^\ell, \bar{x}^{i\ell})$ and the bounds $0 \leq x \leq 1$.

Assume that $G(\bar{T})$ contains no square and $|\bar{T}| \geq 4$. Note that a cycle of H_n that is not a square has length at least six. Since the distance between any two vertices in $G(\bar{T})$ is at most two, $G(\bar{T})$ contains no cycle of H_n. Thus, $G(\bar{T})$ is a tree. In fact, $G(\bar{T})$ is a star since the distance between any two of its vertices is at most two. Thus $\bar{T} = \{\bar{x}, \bar{x}^{i_1}, \ldots, \bar{x}^{i_t}\}$ for some $t \geq 3$. By Lemma 3, $\text{conv}(T)$ is described by edge and star inequalities with the bounds $0 \leq x \leq 1$. Any Chvátal inequality that one can obtain from $ax \geq b$ is therefore implied by the edge inequalities corresponding to the edges $(\bar{x}, \bar{x}^{i_1}), \ldots, (\bar{x}, \bar{x}^{i_t})$ and the star inequality that corresponds to the star $(\bar{x}, \bar{x}^{i_1}, \ldots, \bar{x}^{i_t})$. $\square$

Note that, if an edge $\bar{x}\bar{y}$ of $G(\bar{S})$ belongs to a square of $G(\bar{S})$, the corresponding inequality is not needed in the description of $Q_S^{(3)}$ since it is dominated by the square inequality. On the other hand, if an edge belongs to a star $(\bar{x}, \bar{x}^{i_1}, \ldots, \bar{x}^{i_t})$ of $G(\bar{S})$ with $t < n$, there is no domination relationship between the corresponding edge inequality and star inequality by Lemma 3.

3   Chvátal Rank 4

In this section, we give the characterization of $Q_S^{(4)}$. It is somewhat more involved than the results for $Q_S^{(1)}$, $Q_S^{(2)}$ and $Q_S^{(3)}$, but it is in the same spirit.

Consider any cube with vertices in $G(\bar{S})$. Specifically, for $\bar{x} \in \{0,1\}^n$, recall that we use the notation $\bar{x}^i$ to denote the 0,1 vertex that differs from $\bar{x}$ only in coordinate i, and more generally, for $J \subseteq N$, let $\bar{x}^J$ denote the 0,1 vector that differs from $\bar{x}$ exactly in the coordinates J. If the 8 points $\bar{x}, \bar{x}^i, \bar{x}^k, \bar{x}^\ell, \bar{x}^{ik}, \bar{x}^{i\ell}, \bar{x}^{k\ell}, \bar{x}^{ik\ell}$ all belong to $\bar{S}$, then we say that these points form a *cube*. Note that

$$\sum_{j \in N \setminus \{i,k,\ell\}} (\bar{x}_j(1 - x_j) + (1 - \bar{x}_j)x_j) \geq 1 \tag{5}$$

is a valid inequality for $\text{conv}(S)$ and that it cuts off exactly 8 vertices of H_n, namely the 8 corners of the cube. In fact, it is the strongest such inequality since it is satisfied at equality by all $8(n-3)$ of their neighbors in H_n. We call (5) a *cube inequality*.

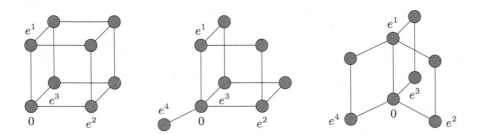

Fig. 2. Cube, tulip, and propeller with $\bar{x} = 0$

If $\bar{x}, \bar{x}^{i_1}, \bar{x}^{i_2}, \bar{x}^{i_3}, \bar{x}^{i_1 i_2}, \bar{x}^{i_2 i_3}, \bar{x}^{i_3 i_1}, \bar{x}^{i_4}, \ldots, \bar{x}^{i_t}$ all belong to $\bar{S}$ for some $t \geq 4$, then we say that these points form a *tulip*. Let $I_T := \{i_1, \ldots, i_t\}$. Note that

$$\sum_{k=1}^{3} (\bar{x}_{i_k}(1 - x_{i_k}) + (1 - \bar{x}_{i_k})x_{i_k}) + 2 \sum_{r=4}^{t} (\bar{x}_{i_r}(1 - x_{i_r}) + (1 - \bar{x}_{i_r})x_{i_r})$$

$$+ 3 \sum_{j \notin I_T} (\bar{x}_j(1 - x_j) + (1 - \bar{x}_j)x_j) \geq 3 \tag{6}$$

is a valid inequality for conv(S) that cuts off exactly these points. We call it a *tulip inequality*. For example, if $\bar{x} = 0$, and $\bar{x}^{i_k} = e^k$ for $k = 1, \ldots, t$, (6) is
$$x_1 + x_2 + x_3 + 2(x_4 + \ldots + x_t) + 3(x_{t+1} + \ldots + x_n) \geq 3.$$
If $\bar{x}, \bar{x}^{i_1}, \bar{x}^{i_2}, \ldots, \bar{x}^{i_t}, \bar{x}^{i_{t+1}}, \bar{x}^{i_1 i_{t+1}}, \bar{x}^{i_2 i_{t+1}}, \ldots, \bar{x}^{i_t i_{t+1}}$ all belong to $\bar{S}$ for some $t \geq 3$, then we say that these points form a *propeller*. Let $I_P := \{i_1, \ldots, i_{t+1}\}$. Note that

$$\sum_{r=1}^{t} (\bar{x}_{i_r}(1 - x_{i_r}) + (1 - \bar{x}_{i_r})x_{i_r}) + 2 \sum_{j \notin I_P} (\bar{x}_j(1 - x_j) + (1 - \bar{x}_j)x_j) \geq 2 \quad (7)$$

is a valid inequality that cuts off exactly the above points. We call it a *propeller inequality*. For example, if $\bar{x} = 0$, $\bar{x}^{i_{t+1}} = e^1$ and $\bar{x}^{i_k} = e^{k+1}$ for $k = 1, \ldots, t$, the propeller inequality is $x_2 + \ldots + x_{t+1} + 2(x_{t+2} + \ldots + x_n) \geq 2$.

Theorem 7. $Q_S^{(4)}$ *is the intersection of* $Q_S^{(3)}$ *and the half spaces defined by all cube, tulip, and propeller inequalities.*

Corollary 1. *Let* $P \subseteq [0,1]^n$ *be a polytope,* $S = P \cap \{0,1\}^n$ *and* $\bar{S} = \{0,1\}^n \setminus S$. *If* $G(\bar{S})$ *contains no 4-cycle, then* P *has Chvátal rank at most 3.*

The set of vertices $\bar{T}$ cut off by a linear inequality induces a connected graph by Lemma 2. One can show that if $G(\bar{T})$ contains vertices at distance greater than 2, then it contains a 4-cycle. Therefore, if $G(\bar{T})$ contains no 4-cycle, it is a star in the bipartite graph $G(H_n)$ with one vertex on one side and at most n on the other.

Remark 1. Let $P \subseteq [0,1]^n$ be given by a system of k inequalities. If $G(\bar{S})$ contains no 4-cycle, then $|\bar{S}| \leq k(n+1)$. It follows that optimizing a linear function over S can be solved in polynomial time in this case.

Corollary 2. *Let* $n \geq 3$ *and* $i = 0, 1$ *or* 2. *For* $S \supseteq \{x \in \{0,1\}^n : \sum_{j=1}^n x_j = i \pmod 3\}$, *the set conv($S$) is entirely described by vertex, edge, star inequalities and bounds* $0 \leq x \leq 1$.

We note that, for $n \geq 5$, $i = 0, 1, 2, 3$ and $S \supseteq \{x \in \{0,1\}^n : \sum_{j=1}^n x_j = i \pmod 4\}$, conv($S$) might contain an inequality with Chvátal rank 5 in its linear description.

4   Vertex Cutsets

Corollary 1 implies that if $G(\bar{S})$ induces a forest, the Chvátal rank of P is at most 3. This can also be proved directly using a vertex cutset decomposition theorem in the spirit of Theorem 1. We present it below in Sect. 4.1.

Trees can be generalized using the notion of tree width. A connected graph has tree width one if and only if it is a tree. Next, we focus our attention on the case when $G(\bar{S})$ has tree width two. Instead of working directly with the definition of tree width, we will use the following characterization: A graph has tree width at most two if and only if it contains no K_4-minor; furthermore a graph with no K_4-minor and at least four vertices always has a vertex cut of size two.

The main result of this section is that P has Chvátal rank at most 4 when $G(\bar{S})$ has tree width two.

Theorem 8. *Let $P \subseteq [0,1]^n$, $S = P \cap \{0,1\}^n$ and $\bar{S} = \{0,1\}^n \setminus S$. If $G(\bar{S})$ has tree width 2, the Chvátal rank of P is at most 4.*

The proof follows from a 2-vertex cutset decomposition theorem, which we state below in Sect. 4.2.

4.1   1-Vertex Cutset

The next theorem shows that $conv(S)$ can be decomposed when $G(\bar{S})$ contains a vertex cut. This result is in the spirit of the theorem of Angulo, Ahmed, Dey and Kaibel (Theorem 1) but it is specific to polytopes contained in the unit hypercube.

Let $G = (V, E)$ be a graph and let $X \subseteq V$. For $v \in X$, let $N_X[v]$ denote the closed neighborhood of v in the graph $G(X)$. That is $N_X[v] := \{v\} \cup \{u \in X : uv \in E\}$.

Theorem 9. *Let $S \subseteq \{0,1\}^n$ and $\bar{S} = \{0,1\}^n \setminus S$. Let v be a cut vertex in $G(\bar{S})$ and let $\bar{S}_1, \ldots, \bar{S}_\ell$ denote the connected components of $G(\bar{S} \setminus \{v\})$. Then $conv(S) = \bigcap_{i=1}^{\ell} conv(\{0,1\}^n \setminus (N_{\bar{S}}[v] \cup \bar{S}_i))$.*

Furthermore, if v does not belong to any 4-cycle in $G(\bar{S})$, then $conv(S) = conv(\{0,1\}^n \setminus N_{\bar{S}}[v]) \cap \bigcap_{i=1}^{\ell} conv(\{0,1\}^n \setminus (\{v\} \cup \bar{S}_i))$.

Theorem 9 cannot be extended to general polytopes, as shown in the following example.

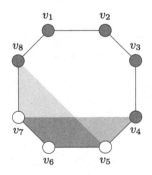

Fig. 3. An example in $\mathbb{R}^2$

Example 1. Let P be the polytope in $\mathbb{R}^2$ shown in Fig. 3. Let $V := \{v_1, \ldots, v_8\}$ denote its vertex set and let $G = (V, E)$ be its skeleton graph. Let $S := \{v_5, v_6, v_7\}$ and $\bar{S} := V \setminus S$. In the figure the set of white vertices is S, while the set of black vertices is $\bar{S}$. Note that v_2 is a cut vertex of $G(\bar{S})$, and $N_{\bar{S}}[v_2] = \{v_1, v_2, v_3\}$. Therefore, $\bar{S}_1 := \{v_1, v_8\}$ and $\bar{S}_2 := \{v_3, v_4\}$ induce two distinct connected components of $G(\bar{S} \setminus \{v_2\})$.

Note that $\text{conv}(S)$ is a triangle, but the intersection of $\text{conv}(V \setminus \{v_1, v_2, v_3, v_4\})$ and $\text{conv}(V \setminus \{v_1, v_2, v_3, v_8\})$ is a parallelogram. Therefore, we get that

$$\text{conv}(S) \neq \text{conv}(V \setminus (N_{\bar{S}}[v_2] \cup \bar{S}_1)) \cap \text{conv}(V \setminus (N_{\bar{S}}[v_2] \cup \bar{S}_2)). \qquad \square$$

4.2 2-Vertex Cut

A key step in proving Theorem 8 is the next theorem.

Theorem 10. *Let $S \subseteq \{0,1\}^n$ and $\bar{S} = \{0,1\}^n \setminus S$. Let $\{v_1, v_2\}$ be a vertex cut of size two in $G(\bar{S})$. Let $\bar{S}_1, \ldots, \bar{S}_k$ denote the connected components of $G(\bar{S} \setminus \{v_1, v_2\})$. Then $\text{conv}(S) = \bigcap_{i=1}^k \text{conv}(\{0,1\}^n \setminus (N_{\bar{S}}[v_1] \cup N_{\bar{S}}[v_2] \cup \bar{S}_i))$.*

It is natural to ask whether this theorem can be extended to vertex cuts of larger sizes. The 3-vertex cut case is open, but it turns out that Theorem 10 cannot be generalized to 4-vertex cutsets as shown by the following example.

Example 2. Consider $\bar{S} = ((\{0,1\}^4 \times \{0\}) \setminus \{e^1 + e^2 + e^3 + e^4\}) \cup \{e^5\}$. Then $x_1 + x_2 + x_3 + x_4 + 3x_5 \geq 4$ is a facet-defining inequality for $\text{conv}(S)$. Note that it cuts off all points in $\bar{S}$. In addition, $\bar{C} := \{e^1, e^2, e^3, e^4\}$ is a vertex cut of cardinality four in $\bar{S}$. Then $\bar{S}_1 := \{0, e^5\}$ and $\bar{S}_2 := \{e^1 + e^2 + e^3, e^1 + e^2 + e^4, e^1 + e^3 + e^4, e^2 + e^3 + e^4\}$ induce two connected components of $G(\bar{S} \setminus \bar{C})$. However,

$$\text{conv}(S) \neq \bigcap_{i=1}^2 \text{conv}(\{0,1\}^5 \setminus (N_{\bar{S}}[e^1] \cup \ldots \cup N_{\bar{S}}[e^4] \cup \bar{S}_i))$$

since $x_1 + x_2 + x_3 + x_4 + 3x_5 \geq 4$ is not valid for $\text{conv}(\{0,1\}^5 \setminus (N_{\bar{S}}[e^1] \cup \ldots \cup N_{\bar{S}}[e^4] \cup \bar{S}_i))$ for $i = 1, 2$. $\qquad \square$

4.3 Implication for the Chvátal Rank

Theorems 9 and 10 imply bounds on the Chvátal rank of P when $G(\bar{S})$ has a vertex cutset of size one or two.

Corollary 3. *Let $P = \bigcap_{i=1}^t P_i$, where $P_i \subseteq [0,1]^n$ are polytopes. Let $V_i = P_i \cap \{0,1\}^n$, $S = P \cap \{0,1\}^n$ and $\bar{S} = \{0,1\}^n \setminus S$.*

(i) Let v be a cut vertex in $G(\bar{S})$, let $\bar{S}_1, \ldots, \bar{S}_t$ induce the connected components of $G(\bar{S} \setminus \{v\})$. Assume $V_i = \{0,1\}^n \setminus (N_{\bar{S}}[v] \cup \bar{S}_i)$. Then the Chvátal rank of P is no greater than the maximum Chvátal rank of P_i, $i = 1, \ldots, t$.

(ii) Let $\{v_1, v_2\}$ be a vertex cut of size two in $G(\bar{S})$. Let $\bar{S}_1, \ldots, \bar{S}_t$ induce the connected components of $G(\bar{S} \setminus \{v_1, v_2\})$. Assume $V_i = \{0, 1\}^n \setminus (N_{\bar{S}}[v_1] \cup N_{\bar{S}}[v_2] \cup \bar{S}_i)$. Then the Chvátal rank of P is no greater than the maximum Chvátal rank of P_i, $i = 1, \ldots, t$.

5 Dependency on the Cardinality of the Infeasible Set

One can derive an upper bound on the Chvátal rank as a function of $|\bar{S}|$ using the result of Eisenbrand and Schulz [8] showing that the Chvátal rank of a 0,1 polytope is at most $n^2(1 + \log_2 n)$.

Theorem 11. *If $|\bar{S}| = k$ for some $k \leq n$, then the Chvátal rank of P is at most $k^2(1 + \log_2 k)$.*

This theorem implies that if the number of infeasible 0,1 vectors is a constant, then P is of constant Chvátal rank.

The next theorem shows that the Chvátal rank of P can be guaranteed to be smaller than the upper bound of $O(n^2 \log n)$ when the cardinality of $\bar{S}$ is bounded above by a subexponential but superpolynomial function of n. The proof uses a result of Eisenbrand and Schulz [8] stating that, if $cx \geq c_0$ is a valid inequality for $\text{conv}(S)$, where the c_js are relatively prime integers, then the Chvátal rank of P is at most $n^2 + 2n \log_2 \|c\|_\infty$.

Theorem 12. *If $|\bar{S}| < n^{f_k(n)}$ where $f_k(n) \leq (\log_2 n)^k$ for some positive constant k, then the Chvátal rank of P is $O(n^2 \log \log n)$.*

6 Optimization Problem Under Small Chvátal Rank

Let $P \subseteq [0, 1]^n$ and $S = P \cap \{0, 1\}^n$. Even when the Chvátal rank of P is just 1, it is still an open question whether optimizing a linear function over S is polynomially solvable or not [6]. In this section, we prove a weaker result.

Theorem 13. *Let $P \subseteq [0, 1]^n$ and $S = P \cap \{0, 1\}^n$. If the Chvátal rank of Q_S is constant, then there is a polynomial algorithm to optimize a linear function over S.*

Proof. The optimization problem is of the form $\min\{cx : x \in S\}$ where $c \in \mathbb{R}^n$. By complementing variables, we may assume $c \geq 0$. By hypothesis, $\text{conv}(S) = Q_S^{(k)}$ for some constant k. We claim that an optimal solution can be found among the 0,1 vectors with at most $k + 1$ nonzero components. This will prove the theorem since there are only polynomially many such vectors. Indeed, if an optimal solution $\bar{x}$ has more than $k+1$ nonzero components, any 0,1 vector $\bar{z} \leq \bar{x}$ with exactly $k+1$ nonzero components satisfies $c\bar{z} \leq c\bar{x}$. Because $\text{conv}(S) = Q_S^{(k)}$ Lemma 1 implies that the face of H_n of dimension $k + 1$ that contains 0 and $\bar{z}$ contains a feasible point $\bar{y} \in S$. Since $c\bar{y} \leq c\bar{z} \leq c\bar{x}$, the solution $\bar{y}$ is an optimal solution. $\square$

References

1. Abdi, A., Cornuéjols, G., Pashkovich, K.: Delta minors in clutters (work in progress)
2. Angulo, G., Ahmed, S., Dey, S.S., Kaibel, V.: Forbidden vertices. Math. Oper. Res. **40**, 350–360 (2015)
3. Bockmayr, A., Eisenbrand, F., Hartmann, M., Schulz, A.S.: On the Chvátal rank of polytopes in the 0/1 cube. Discrete Appl. Math. **98**, 21–27 (1999)
4. Chvátal, V.: Edmonds polytopes and a hierarchy of combinatorial problems. Discrete Math. **4**, 305–337 (1973)
5. Chvátal, V., Cook, W., Hartmann, M.: On cutting-plane proofs in combinatorial optimization. Linear Algebra Appl. **114/115**, 455–499 (1989)
6. Cornuéjols, G., Li, Y.: Deciding emptiness of the Gomory-Chvátal closure is NP-complete, even for a rational polyhedron containing no integer point. In: Louveaux, Q., Skutella, M. (eds.) IPCO 2016, vol. 9682, pp. 387–397. Springer, Switzerland (2016)
7. Dadush, D., Dey, S.S., Vielma, J.P.: On the Chvátal-Gomory closure of a compact convex set. Math. Program. Ser. A **145**, 327–348 (2014)
8. Eisenbrand, F., Schulz, A.S.: Bounds on the Chvátal rank of polytopes in the 0/1 cube. Combinatorica **23**, 245–261 (2003)
9. Hartmann, M.E., Queyranne, M., Wang, Y.: On the Chvátal rank of certain inequalities. In: Cornuéjols, G., Burkard, R.E., Woeginger, G.J. (eds.) IPCO 1999. LNCS, vol. 1610, pp. 218–233. Springer, Heidelberg (1999)
10. Rothvoß, T., Sanitá, L.: 0/1 polytopes with quadratic Chvátal rank. In: Goemans, M., Correa, J. (eds.) IPCO 2013. LNCS, vol. 7801, pp. 349–361. Springer, Heidelberg (2013)
11. Schrijver, A.: On cutting planes. Ann. Discrete Math. **9**, 291–296 (1980)

Robust Monotone Submodular Function Maximization

James B. Orlin, Andreas S. Schulz, and Rajan Udwani[✉]

MIT, Cambridge, USA
{jorlin,schulz,rudwani}@mit.edu

Abstract. We consider a robust formulation, introduced by Krause et al. (2008), of the classic cardinality constrained monotone submodular function maximization problem, and give the first constant factor approximation results. The robustness considered is w.r.t. adversarial removal of a given number of elements from the chosen set. In particular, for the fundamental case of single element removal, we show that one can approximate the problem up to a factor $(1 - 1/e) - \epsilon$ by making $O(n^{\frac{1}{\epsilon}})$ queries, for arbitrary $\epsilon > 0$. The ideas are also extended to more general settings.

1 Introduction

A set function $f : 2^N \to \mathbb{R}$ on the ground set N is called submodular if,

$$f(A + a) - f(A) \leq f(B + a) - f(B) \text{ for all } B \subseteq A \subseteq N \text{ and } a \in N \backslash A.$$

The function is monotone if $f(B) \leq f(A)$ for all $B \subseteq A$. We also impose $f(\emptyset) = 0$, which combined with monotonicity implies non-negativity. Optimization problems with submodular objective functions have received a lot of interest due to several applications where instances of these problems arise naturally. However, unlike the (unconstrained) minimization of submodular functions, for which polytime algorithms exist [18,25], even the simplest maximization versions are NP-hard [9–11,29]. In fact, they encompass some very fundamental hard problems, such as max-cut, max-k-coverage, max-dicut and variations of max-SAT and max-welfare.

A long line of beautiful work has culminated in fast and tight approximation algorithms for many settings of the problem. As an example, for unconstrained maximization of non-monotone submodular functions, Feige et al. in [11], provided an algorithm with approximation ratio of 0.4 and showed an inapproximability threshold of 1/2 in the value-oracle model. Extensions by Gharan and Vondrák [14] and subsequently by Feldman et al. [13] led to further improvement of the guarantee (roughly 0.41 and 0.42, respectively). Finally, Buchbinder et al. in [7] gave a tight randomized 1/2 approximation algorithm, and this was recently derandomized [6].

Here we are interested in the problem of maximizing a monotone submodular function subject to a cardinality constraint, written as: $P1 := \max\limits_{A \subseteq N, |A| \leq k} f(A)$.

© Springer International Publishing Switzerland 2016
Q. Louveaux and M. Skutella (Eds.): IPCO 2016, LNCS 9682, pp. 312–324, 2016.
DOI: 10.1007/978-3-319-33461-5_26

The problem has been well studied and instances of P1 arise in several important applications, two of them being:

Sensor Placement [17,19–21]: Given a large number of locations N, we would like to place up to k sensors at certain locations so as to maximize the *coverage*. Many commonly used coverage functions measure the cumulative information gathered in some form, and are thus monotone (more sensors is better) and submodular (decreasing marginal benefit of adding a new sensor).

However, as highlighted in [20], it is important to ask what happens if some sensors were to fail. Will the remaining sensors have good coverage regardless of which sensors failed, or is a small crucial subset responsible for most of the coverage?

Feature Selection [15,20,22,27]: In many machine learning models, adding a new feature to an existing set of features always improves the modeling power (monotonicity) and the marginal benefit of adding a new feature decreases as we consider larger sets (submodularity). Given a large set of features, we would like to select a small subset such that, we reduce the problem dimensionality while retaining most of the information.

However, as discussed in [15,20], in situations where the nature of underlying data is uncertain, leading to non-stationary feature distributions, it is important to not have too much dependence on a few features. Taking a concrete example from [15], in document classification, features may take not standard values due to small sample effects or in fact, the test and training data may come from different distributions. In other cases, a feature may even be deleted at some point, due to input sensors failures for example. Thus, similar questions arise here too and we would like to have an 'evenly spread' dependence on the set of chosen features. With such issues in mind, consider the following *robust* variant of the problem, introduced in [20],

$$P2 := \max_{A \subseteq N, |A| \leq k} \min_{Z \subseteq A, |Z| \leq \tau} f(A - Z).$$

Note that the parameter τ controls the degree of robustness of the chosen set since the larger τ is, the larger the size of subset Z that can be adversarially removed from the chosen set A. For $\tau = 0$, $P2$ reduces to the $P1$. Since this formulation optimizes the worst case scenario, a natural variation is to optimize the average case failure scenario [16]. However, this is not suitable for some applications. For instance, we may have no prior on the failure/deletion mechanism and furthermore, in critical applications, such as sensor placement for outbreak detection [20,21], we want protection against the worst case. This form of worst case analysis has been of great interest in operations research and beyond, under the umbrella of *robust optimization* (e.g., [2–5]). The idea is to formulate the uncertainty in model parameters through a deterministic *uncertainty set*. However, much work in this area assumes that the uncertainty set is a connected, if not convex, set and in contrast, the uncertainty set in $P2$, when $\tau = 1$ for instance, is the disconnected set of canonical unit vectors $e_i \in \mathbb{R}^N$ (1 at entry i, 0 otherwise).

Previous Work on $P1$ *and* $P2$. The first rigorous analysis of $P1$ was by Nemhauser et al. [23,24] in the late 70's, where they showed that the greedy algorithm gives a guarantee of $(1-1/e)$ and that this is best possible in the value-oracle model. Later, Feige [10] showed that this is also the best possible under standard complexity assumptions (through the special case of Max-k-cover). On the algorithmic side, Badanidiyuru and Vondrák [1] recently gave a faster algorithm for $P1$ that improved the quadratic query complexity of the classical greedy algorithm to nearly linear complexity, by trading off on the approximation guarantee. However, the optimality (w.r.t. approximation guarantee) of the greedy algorithm is rather coincidental, and for many complex settings of the problem (monotone or not), the greedy algorithm tends to be sub-optimal (there are exceptions, like [26]). An approach first explored by Chekuri et al. [8], that has been very effective, is to perform optimization on the *multilinear relaxation* of the submodular function, followed by clever *rounding* to get a solution to the original problem. Based on this framework, tremendous progress has been made over the last decade for both monotone and non-monotone versions with various kinds of constraints [8,12,28–30]. In fact, a general framework for establishing hardness of many of these variants [9,29], also relies intricately on properties of this relaxation.

Moving on to $P2$, as we will see, the well known greedy algorithm and also the above mentioned *continuous* greedy approach for $\tau = 0$, can be arbitrarily bad even for $\tau = 1$. In fact, many natural approaches do not have a constant factor guarantee for $\tau \geq 1$. The paper by Krause et al. [20], which formally introduced the problem, actually gives a bi-criterion approximation to the much more general and inapproximable problem: $\max_{A \subseteq N, |A| \leq k} \min_{i \in \{1,2,...,m\}} f_i(A)$, where $f_i(.)$ is monotone submodular for every i. Their algorithm, which is based on the greedy algorithm for $P1$, when specialized to $P2$, guarantees optimality by allowing sets up to size $k(1+\Theta(\log(\tau k \log n)))$ and has runtime exponential in τ. To the best of our knowledge, no stronger/constant factor approximation results were known for $P2$ prior to our work.

Our Contributions: We work in the value oracle model and give constant factor guarantees for $P2$ with combinatorial, 'greedy like' algorithms. Our major focus is on the case $\tau = 1$, since it is a fundamental step towards understanding this form of robustness for the maximization problem. For this case, we propose: (i) an algorithm with parameter m, that guarantees a ratio of $(1 - 1/e) - \Omega(1/m)$ using $O(n^{m+1})$ queries and (ii) a fast and practical 0.5547 approximation. Intuitively, the central idea is that we would like to be protected against removal of important elements in our set; leveraging properties of the greedy algorithm, we find that only a few elements are actually critical in this sense. Thus, we ensure that we have a set that is robust to removal of any one of those few elements. For $\tau = o(\sqrt{k})$, we give a fast and asymptotically 0.285 approximate algorithm. In the more general case, where we wish to find a robust set A in an independence system, we extend some of the ideas from the cardinality case into an enumerative procedure that yields an $\alpha/(\tau + 1)$ approximation using an α

approximation algorithm for $\tau = 0$ as a subroutine. However, the runtime scales as $n^{\tau+1}$.

The outline for the rest of the paper is as follows: In Sect. 2, we introduce some notation and see how several natural ideas fail to give any approximation guarantees. In Sect. 3, we focus on $\tau = 1$ and starting with a special case slowly build up to an algorithm with asymptotic guarantee $(1 - 1/e) - \epsilon$. Finally, in Sect. 4, we extend some of the ideas to $\tau = o(\sqrt{k})$ and to more general constraints. Section 5 concludes with some open questions. Also, due to space constraints we defer most proofs to the full version of this paper.

2 Preliminaries

2.1 Definitions

We denote an instance of $P2$ on ground set N with cardinality constraint parameter k and robustness parameter τ by (k, N, τ). Subsequently, we use $OPT(k, N, \tau)$ to denote an optimal set for the instance (k, N, τ). For any given set A, we call a subset Z a *minimizer* if $f(A - Z) = \min_{B \subseteq A; |B| \leq \tau} f(A - B)$. Also, let $Z(A)$ be the set of minimizers of A. When $\tau = 1$, we often use the letter z for minimizers. In fact, we generally refer to singleton sets without braces $\{\}$ and use $+$ and $\cup$ interchangeably. Also, we call the subset $A - Z$ of A, an *active* subset of A. Next, consider the set function $g_\tau(A) = \min_{B \subseteq A; |B| \leq \tau} f(A - B)$, i.e., $g_\tau(A) = f(A - Z)$, the value of an active subset of A. We simply use $g(.)$, when τ is clear from context. Also, define the marginal increase in value due to a set X, when added to the set A as $f(X|A) = f(A \cup X) - f(A)$. Similarly, $g(X|A) = g(A \cup X) - g(A)$.

Let $\beta(\eta, \alpha) = \frac{e^\alpha - 1}{e^\alpha - \eta}$ for $\eta \in [0, 1]$, $\alpha \geq 0$. Note that $\beta(0, 1) = (1 - 1/e)$. This function appears naturally in our analysis and will be very useful for expressing approximation guarantees of the algorithms. Finally, consider the basic greedy algorithm defined below:

Algorithm 1. Greedy Algorithm

1: Initialize $A = \emptyset$
2: **while** $|A| < k$ **do** $A = A + \operatorname*{argmax}_{x \in N - A} f(x|A)$
3: Output: A

If we run the above algorithm for $k = n$, we end up with the set N. Using this we index the elements in the order they were added. So $N = \{a_1, a_2, \ldots, a_n\}$, where a_1 is the first element picked by the greedy algorithm and thus also a highest value element.

It can be shown that we cannot approximate $P2$ better than $P1$ (approximable up to a factor of $\beta(0, 1)$, details in full version). Before going into positive results, we will see how some natural ideas fail to give any guarantees for the robust problem, starting with the greedy algorithm itself.

2.2   Negative Results

The example below demonstrates why the greedy algorithm that does well for instances of $P1$, fails for $P2$. However, the weakness will also indicate a property which will be useful later.

Example: Consider a ground set N of size $2k$ such that $f(a_1) = 1$, $f(a_i) = 0$, $\forall 2 \leq i \leq k$ and $f(a_j) = \frac{1}{k}$, $\forall j \geq k + 1$. Also, for all $j \geq k + 1$, let $f(a_j|X) = \frac{1}{k}$ if $X \cap \{a_1, a_j\} = \emptyset$ and 0 otherwise. Consider the set $S = \{a_{k+1}, \cdots, a_{2k}\}$ and let the set picked by the greedy algorithm (with arbitrary tie-breaking) be $A = \{a_1, \cdots, a_k\}$. Then we have that $f(A - a_1) = 0$ and $f(S - a_j) = 1 - \frac{1}{k}$ for every $a_j \in S$. The insight here is that greedy may select a set such that only the first few elements contribute most of the value in the set, which makes it non-robust. However, as we discuss more formally later, such a concentration of value implies that only the first two elements $\{a_1, a_2\}$ are critical and protecting against removal of either of those two suffices for best possible guarantees.

In fact, many natural variations fail to give an approximation ratio better than $1/(k - 1)$. Indeed, a guarantee of this order, i.e. $1/(k - \tau)$, is achievable for any τ by the following naïve algorithm: *Pick the k largest value elements.* It is also important to examine if the function g is super/sub-modular, since that would make existing techniques useful. It turns out not However, it is monotonic. Despite this, it is interesting to examine a natural variant of the greedy algorithm, where we greedily pick w.r.t g, but that variant can also be arbitrarily bad. We defer details/proofs omitted above to the full version.

3   Main Result: Algorithms for $\tau = 1$

3.1   Special Case of "Copies"

We first consider a special case, which will serve two purposes. First, it will simplify things and the insights gained for this case can be generalized to get stronger results. Secondly, since this case may arise in practical scenarios, it is worthwhile to discuss the special algorithms as they are much simpler than the general Algorithms 3 and 4 discussed later on.

Given an element $x \in N$, we call another element x' a *copy* of element x if,

$$f(x') = f(x) \text{ and } f(x'|x) = 0.$$

This implies $f(x|x') = f(\{x, x'\}) - f(x') = f(x) + f(x'|x) - f(x') = 0$. In fact, it can be shown that $f(x'|A) = f(x|A)$ for every $A \subset N$ not containing x or x'. This is a useful case for robust sensor placement, if we were allowed to place multiple/duplicate sensors at certain locations that are critical for coverage.

For the rest of this sub-section, assume that each element in N has a copy and denote the copy of element a_i by a_i'. As indicated previously, we would like to make our set robust to removal of critical elements. In the presence of copies, adding a copy of these elements achieves that. So as a first step, let's construct a set that includes a copy of each element, and so is unaffected by single element removal. One way to do this is to run the greedy algorithm for $k/2$ iterations and then add a copy of each of the $k/2$ elements. Then, one can show that $g(S) = f(S) \geq \beta(0, 0.5) f(OPT(k, N, 0)) \geq (1 - \frac{1}{\sqrt{e}}) g(OPT(k, N, 1))$, where the last inequality follows from substitution and because $f(.) \geq g(.)$. Hence, we have ≈ 0.393-approximation and the bound is tight. Can we improve this? One way to do better is to think about whether we really need to copy all $k/2$ elements. Turns out, just copying $\{a_1, a_2\}$ is enough. The intuition here is that if the greedy set has value nicely spread out, we could get away without copying anything. So in such a case, copying the first two elements does not hurt much. Otherwise, as in the example from Sect. 2, if greedy concentrates its value on the first few elements, then copying just them suffices.

Now as a step towards showing that the first two elements are enough, consider the algorithm where we add a copy of just the first element a_1 and then continue adding greedily to get the set $\{a_1, a_1', a_2, \cdots, a_{k-1}\}$. We call this the 1-Copy algorithm and it guarantees a ratio 0.5 for $k \geq 7$ (proof in full version). In fact, we cannot do better if we copy just one element. Now, consider the algorithm that copies the first two elements and thus outputs: $\{a_1, a_1', a_2, a_2', a_3, \ldots, a_{k-2}\}$. Call this the 2-Copy algorithm. We sketch the proof here since it captures some of the essence of the much more technical analysis for later algorithms.

Theorem 1. *For the case with copies, 2-Copy is $\beta(0, \frac{k-5}{k-1})$ approximate.*

Proof (sketch). Let z be a minimizer of the output $A = \{a_1, a_1', a_2, a_2', a_3, \ldots, a_{k-2}\}$. So if $z \in \{a_1, a_2\}$ (or their copies), then using properties of the greedy algorithm and by definition of g, it can be shown that, $f(A - z) = f(\{a_1, a_2, \ldots, a_{k-2}\}) \geq \beta(0, \frac{k-2}{k-1}) f(OPT(k-1, N, 0)) \geq \beta(0, \frac{k-2}{k-1}) g(OPT(k, N, 1))$,

If $z \notin \{a_1, a_2\}$ (or their copies) then let $f(z|A - z) = \eta f(A)$. We have due to greedy additions and submodularity, $f(a_3|\{a_1, a_2\}) \geq f(z|A - z)$ and $f(\{a_1, a_2\}) \geq 2f(z|A - z)$, which in turn implies that $f(\{a_1.a_2, a_3\}) \geq 3\eta f(A)$. This relates the value removed by a minimizer z to the value concentrated on the first 3 elements $\{a_1, a_2, a_3\}$. Higher the value removed, higher the concentration and closer the value of $f(A)$ to $f(OPT(k-1, N, 0))$. More formally, we can show that,

$$f(A) \geq \beta(3\eta, \frac{k-5}{k-1}) f(OPT(k-1, N, 0)) \geq \beta(3\eta, \frac{k-5}{k-1}) g(OPT(k, N, 1)).$$

Which implies that $g(A) = (1 - \eta) f(A) \geq (1 - \eta)\beta(3\eta, \frac{k-5}{k-1}) g(OPT(k, N, 1))$ Now, we show that the factor $(1 - \eta)\beta(3\eta, \frac{k-5}{k-1})$ on the RHS is minimized for $\eta = 0$, which finishes the proof. For ease of notation, let e' denote $e^{\frac{k-5}{k-1}}$, then we have,

$$(1 - \eta)\beta(3\eta, \frac{k-5}{k-1}) = \frac{1}{3}(3 - 3\eta)\frac{e' - 1}{e' - 3\eta} = \frac{1}{3}(e' - 1)(1 + \frac{3 - e'}{e' - 3\eta})$$

$$\geq \frac{1}{3}(e' - 1)(1 + \frac{3 - e'}{e'}) = (1 - 1/e') = \beta(0, \frac{k-5}{k-1})$$

As an example, for $k \geq 55$ the value of the ratio is ≥ 0.6 $\square$

The result also implies that if for large k, the output of the greedy algorithm has a minimizer a_i with $i \geq 3$, then the greedy set is $(1 - 1/e)$ approximate for $(k, N, 1)$. This is because, for such an instance we can assume that the set contains copies of a_1, a_2 and then apply Theorem 1. Moreover, if we copy the first i elements for $i \geq 3$, we get the same guarantee but with worse asymptotics, so copying more than first two does not result in a gain.

Naturally, the problem we must deal with now is that in general we won't have copies. However, as we argued above, the main takeaway from the special case is that if we can make the greedy set robust to removal of either one of the first two elements, we get the best possible guarantee (except for small k). Indeed, we will now essentially focus on how to make a greedy set that is robust in such a manner without relying on copies.

3.2   Guarantees in Absence of "Copies"

We start by discussing how one could construct a greedy set that is robust to the removal of just a_1. One approach would be to pick a_1 and then pick the rest of the elements greedily while ignoring a_1, since that would allow a copy of a_1 to be chosen and if no copy exists, it will still allow useful elements which have small marginal on a_1, but large value in absence of it (and so would not have been considered otherwise). Formally:

Algorithm 2. 0.387 Algorithm

1: Initialize $A = \{a_1\}$
2: **while** $|A| < k$ **do** $A = A + \underset{x \in N - A}{\operatorname{argmax}} f(x|A - \{a_1\})$
3: Output: A

This simple algorithm is in fact, asymptotically 0.387 approximate (0.357-approximate for all $k \geq 4$, we omit the proof here). However, a clear issue with the algorithm is that it is oblivious to the minimizers of the set at any iteration. It ignores a_1 throughout, even if a_1 stops being the minimizer after a few iterations of ignoring it. Thus, if we track the minimizer and stop ignoring a_1 when it is not the minimizer, we do better and can in fact, achieve an approximation ratio of 0.5 for $k \geq 7$ (details in full version). Next, just as in the presence of copies, in order to get even better guarantees, we need to look at the set $\{a_1, a_2\}$. A direct generalization to a rule that ignores both a_1 and a_2, i.e. $\underset{x \in N - A}{\operatorname{argmax}} f(x|A - \{a_1, a_2\})$, can be shown to have an upper bound less than 0.5. In fact, many natural addition rules result in upper bounds ≤ 0.5. Algorithm 3 avoids looking at both elements simultaneously and instead ignores a_1 until its marginal becomes sufficiently small and then does the same for a_2, if required.

Algorithm 3. $0.5547-\Omega(1/k)$ Algorithm

1: Initialize $A = \{a_1, a_2\}$
    Phase 1:
2: **while** $|A| < k$ **and** $f(a_1|A - a_1) > \frac{f(A)}{3}$ **do** $A = A + \underset{x \in N-A}{\operatorname{argmax}} f(x|A - a_1)$
    Phase 2:
3: **while** $|A| < k$ **and** $f(a_2|A - a_2) > \frac{f(A)}{3}$ **do** $A = A + \underset{x \in N-A}{\operatorname{argmax}} f(x|A - a_2)$
    Phase 3:
4: **while** $|A| < k$ **do** $A = A + \underset{x \in N-A}{\operatorname{argmax}} f(x|A)$
5: Output: A

The algorithm is asymptotically 0.5547-approximate (as an example, guarantee >0.5 for $k \geq 50$, details deferred to full version) and note that it's minimizer oblivious and only uses greedy as subroutine, which makes it fast and easy to implement and use. Additionally, in each of the phases of Algorithm 3, we can replace the step-wise rule $A = A + \underset{x \in N-A}{\operatorname{argmax}} f(x|A-y)$, where y is either in $\{a_1, a_2\}$ or $\emptyset$, by the more efficient thresholding rule in [1] where, given a threshold w, we add a new element x if $f(x|A - y) \geq w$, with $y \in \{a_1, a_2, \emptyset\}$ depending on the phase, as before. This improves the query/run time to $O(\frac{n}{\epsilon} \log \frac{n}{\epsilon})$, with the loss of a factor of $(1 - \epsilon)$ in the guarantee.

In order to improve upon the guarantee of Algorithm 3, we would like to devise a way to add new elements while paying attention to both a_1 and a_2 simultaneously. To this end, observe that while a_1 is a minimizer, the addition rule of Algorithm 2 is equivalent to $\underset{x \in N-A}{\operatorname{argmax}} g(x|A)$. By extension, we now propose an algorithm which asymptotically guarantees $(1-1/e)-\Omega(1/m)$ approximation, by using the addition rule $\underset{|S| \leq m; S \subseteq N-A}{\operatorname{argmax}} g(S|A)$, for $m \geq 1$ (while $z(A) \subseteq \{a_1, a_2\}$), and thus making $O(n^{m+1})$ queries. Formally:

Algorithm 4. $(1 - 1/e) - \epsilon$ Algorithm

---

input: m

1: Initialize $A = \{a_1, a_2\}$
    Phase 1:
2: **while** $|A| < k$ **and** $z(A) \subseteq \{a_1, a_2\}$ **do**
3:     $l = \min\{m, k - |A|\}$
4:     $A = A \cup \underset{|S|=l; S \subseteq N-A}{\operatorname{argmax}} g(S|A)$
    Phase 2:
5: **while** $|A| < k$ **do** $A = A + \underset{x \in N-A}{\operatorname{argmax}} f(x|A)$
6: Output: A

Theorem 2. *Algorithm 4 is $\beta\left(0, \frac{(m-1)(k-2m-2)}{mk}\right)$ approximate.*

Note that, for $k \gg m$ we have, $\frac{(m-1)(k-2m-2)}{mk} \approx \frac{m-1}{m}$, which translates the above factor to $1 - \frac{1}{e^{(1-\frac{1}{m})}} \geq (1 - 1/e) - \frac{1}{e(m-1)}$. For instance, when $k \geq 1000$, choosing $m = 20$ guarantees a ratio >0.6. The outline of the analysis is similar to that of the 2-Copy algorithm (proof of Theorem 1) however, with a lot more technical details. So instead, here we discuss the rationale behind such an addition rule.

Recall, through Phase 1 in case of Algorithm 4 and Phases 1 and 2 in case of Algorithm 3, we try to make the final set robust to removal of either one of the two elements $\{a_1, a_2\}$. The reason we resort to m-tuples instead of singletons in Algorithm 4 is because, for $m = 1$ we cannot guarantee improvements at each iteration, as there need not be any element that has marginal value on both a_1 and a_2. However, for larger m we can show improving guarantees. More concretely, consider an instance where $f(a_1) = f(a_2) = 1$, a_1 has a copy a_1' and additionally both a_1 and a_2 have 'partial' copies, $f(a_i^j) = \frac{1}{k}$ and $f(a_i^j | a_i) = 0$ for $j \in \{1, \ldots, k\}$, $i \in \{1, 2\}$. Also, let there be a set G of $k - 2$ garbage elements with $f(G) = 0$. Finally, let $f(\{a_1, a_2\}) = 2$ and $f(a_i^j | X) = \frac{1}{k}$ if $\{a_i, a_i'\} \cap X = \emptyset$. Running the algorithm with: (i) $m = 1$ outputs $\{a_1, a_2\} \cup G$ in the worst case, with (ii) $m = 2$ outputs a_1^j, a_2^j on step j of Phase 1 and thus, 'partially' copies both a_1 and a_2. Instead, if we run the algorithm with (iii) $m = 3$, the algorithm picks up a_1' and then copies a_2 almost completely with $\{a_2^1, \ldots, a_2^{k-3}\}$. Thus, while $\{a_1, a_2\}$ are minimizers, we find that adding m-tuples allows us to guarantee that at each step we increase $g(A)$ by $\frac{m-1}{m} \frac{1}{k}$ times the difference from optimal. When m is large enough that $\frac{m-1}{m} \approx 1$, this in turn allows us to (almost) replicate the guarantee that we see in presence of copies (Theorem 1).

4   Extensions

4.1   Constant Factor Guarantee for $\tau = o(\sqrt{k})$

Here, we present an algorithm for the general case, which is an extension of the 0.387 algorithm for $\tau = 1$ and effectively uses the greedy algorithm as a subroutine $\tau + 1$ times. It achieves an asymptotic approximation ratio of 0.285 for $\tau = o(\sqrt{k})$ and the guarantee degrades proportionally to $1 - \frac{\tau^2}{2k}$, as τ approaches $\sqrt{2k}$.

Algorithm 5. Algorithm for $\tau \leq \sqrt{2k}$

1: Initialize $A_0 = A_1 = \emptyset$, $i = 1$.
2: **while** $|A_0| < \frac{\tau(\tau+1)}{2}$ **do**
3:     $X = \{ i$ **steps of Greedy Algorithm** on ground set $N - A_0$, starting with $\emptyset\}$
4:     $A_0 = A_0 \cup X; i = i + 1$
5: **while** $|A_1| < k - \frac{\tau(\tau+1)}{2}$ **do** $A_1 = A_1 + \underset{x \in N - (A_0 \cup A_1)}{\mathrm{argmax}} f(x | A_1)$
6: Output: $A_0 \cup A_1$

Note that for the special case where we have at least τ copies for each element available, we can get an asymptotic guarantee of $(1 - 1/e)$ for $\tau = o(\sqrt{k})$. While we defer the details to the full version, to get some intuition, consider the case $\tau = 2$ with at least two copies available for each of a_1, a_2. As before, to get close to $(1-1/e)$, we need to make the set unaffected by removal of a_1 and a_2. We thus include two copies each of a_1, a_2, filling up the rest of the set greedily. This will give us the desired ratio, but notice that one may also build such a set by running the greedy algorithm for two steps to get $\{a_1, a_2\}$ and then ignoring these and running the greedy algorithm again for two steps on the reduced ground set to get a set of copies for a_1, a_2 and finally, ignoring the previous four elements while adding $k - 4$ elements greedily. This scheme easily generalizes for larger τ.

The approximation ratio converges to a constant only asymptotically because we must compare the value of $OPT(k, N, \tau)$, which has size $k - \tau$, to a set of size $k - \Theta(\tau^2)$ (since the $\Theta(\tau^2)$ elements added as copies do not contribute any real value). Now note that $\frac{k-\Theta(\tau^2)}{k}$ converges to 1 only for $\tau = o(\sqrt{k})$ and it is this degradation that creates the threshold of $o(\sqrt{k})$. Finally, as we mentioned for Algorithm 3, the greedy steps can be replaced by thresholding steps in [1], to get a runtime of $O((\tau + 1)\frac{n}{\epsilon} \log \frac{n}{\epsilon})$ at the cost of losing a factor of $(1 - \epsilon)$ in the guarantee.

4.2   General Constraints

So far, we have looked at a robust formulation of $P1$, where we have a cardinality constraint. However, there are more sophisticated applications where we find instances of budget or even matroid constraints. In particular, consider the generalization $\max\limits_{A \in \mathcal{I}} \min\limits_{|B| \leq \tau} f(A \backslash B)$, for some independence system $\mathcal{I}$. By definition, for any feasible set $A \in \mathcal{I}$, all subsets of the form $A \backslash B$ are feasible as well, so the formulation is sensible. Let's briefly discuss the case of $\tau = 1$ and suppose that we are given an α approximation algorithm $\mathcal{A}$, with query/run time $O(R)$ for the $\tau = 0$ case. Let G_0 denote its output and z_0 be a minimizer of G_0. Consider the restricted system $\mathcal{I}_{z_0} = \{A : z_0 \in A, A \in \mathcal{I}\}$. Now, in order to be able to pick elements that have small marginal on z_0 but large value otherwise, we can generalize the notion of ignoring z_0 by maximizing the monotone submodular function $f(.\backslash z_0)$ subject to the independence system $\mathcal{I}_{z_0}$. However, unlike the cardinality constraint case, where this algorithm gives a guarantee of 0.387, the algorithm can be arbitrarily bad in general (because of severely restricted $\mathcal{I}_{z_0}$, for instance). We tackle this issue by adopting an enumerative procedure.

Let $\mathcal{A}_j$ denote the algorithm for $\tau = j$ and let $\mathcal{A}_j(N, Z)$ denote the output of $\mathcal{A}_j$ on ground set N and subject to restricted system $\mathcal{I}_Z$. Finally, let $\hat{z}(A) = \arg\max\limits_{x \in A} f(x)$. With this, we have for general constraints:

Algorithm 6. $\mathcal{A}_\tau : \frac{\alpha}{\tau+1}$ for General Constraints

---
1: Initialize $i = 0, Z = \emptyset$
2: **while** $N - Z \neq \emptyset$ **do**
3:    $G_i = \mathcal{A}_0(N - Z, \emptyset)$
4:    $z_i \in \hat{z}(G_i);\quad Z = Z \cup z_i$
5:    $M_i = z_i \cup \mathcal{A}_{\tau-1}(N - Z, z_i);\quad i = i + 1$
6: Output: $\mathrm{argmax}\{g_\tau(S)|S \in \{G_j\}_{j=0}^i \cup \{M_j\}_{j=0}^i\}$

The above scheme has an approximation guarantee of $\alpha/(\tau + 1)$ with query time $O(n^\tau R + n^{\tau+1})$. To understand the basic idea behind the algorithm, assume that z_0 is in an optimal solution for the given τ. Then, given the algorithm $\mathcal{A}_{\tau-1}$, if a minimizer of the set $M_0 = z_0 \cup \mathcal{A}_{\tau-1}(N - z_0, z_0)$ includes z_0, it only removes $\tau - 1$ elements from $\mathcal{A}_{\tau-1}(N - z_0, z_0)$. On the other hand, if a minimizer doesn't include z_0, $g_\tau(M_0) \geq f(z_0) \geq \frac{f(M_0) - g_\tau(M_0)}{\tau}$. These two cases yield the desired ratio, however, since z_0 need not be in an optimal solution, we systematically enumerate. Finally, for the cardinality constraint case, the algorithm can be simplified to get runtime polynomial in (n, τ) and guarantee that scales as $\frac{1}{\tau}$, which for $\sqrt{2k} \leq \tau = o(k)$, is a better guarantee than the naive one of $\frac{1}{k-\tau}$ from Sect. 2.2.

5   Conclusion and Open Problems

We looked at a robust version of the classical monotone submodular function maximization problem, where we want sets that are robust to the removal of any τ elements. In particular, we focused on the pivotal case of single element removal and gave several approximation algorithms with the best asymptotic performance lower bound approaching $(1 - 1/e)$. We then also gave an asymptotically 0.285-approximation algorithm for the case, where up to $o(\sqrt{k})$ elements could be removed. It is not known if we can get exactly $(1 - 1/e)$ in polynomial time or if we can do much better than 0.285 for larger τ. Another interesting open question is, whether a constant factor approximation algorithm exists for $\tau = \Omega(\sqrt{k})$ (as opposed to the $\frac{1}{\tau}$ guarantee discussed at the end of the previous section). Also, similar robustness versions can be considered for maximization subject to independence system constraints and we gave an enumerative black box approach that leads to an $\frac{\alpha}{\tau+1}$ approximation algorithm with query time scaling as $n^{\tau+1}$, given an α approximation algorithm for the non-robust case.

References

1. Badanidiyuru, A., Vondrák, J.: Fast algorithms for maximizing submodular functions. In: SODA 2014, pp. 1497–1514. SIAM (2014)
2. Ben-Tal, A., El Ghaoui, L., Nemirovski, A.: Robust Optimization. Princeton University Press, Princeton (2009)

3. Bertsimas, D., Brown, D., Caramanis, C.: Theory and applications of robust optimization. SIAM Rev. **53**(3), 464–501 (2011)
4. Bertsimas, D., Sim, M.: Robust discrete optimization and network flows. Math. Program. **98**(1–3), 49–71 (2003)
5. Bertsimas, D., Sim, M.: The price of robustness. Oper. Res. **52**(1), 35–53 (2004)
6. Buchbinder, N., Feldman, M.: Deterministic algorithms for submodular maximization problems. CoRR, abs/1508.02157 (2015)
7. Buchbinder, N., Feldman, M., Naor, J.S., Schwartz, R.: A tight linear time (1/2)-approximation for unconstrained submodular maximization. In: FOCS 2012, pp. 649–658 (2012)
8. Calinescu, G., Chekuri, C., Pál, M., Vondrák, J.: Maximizing a monotone submodular function subject to a matroid constraint. SIAM J. Comput. **40**(6), 1740–1766 (2011)
9. Dobzinski, S., Vondrák, J.: From query complexity to computational complexity. In: STOC 2012, pp. 1107–1116. ACM (2012)
10. Feige, U.: A threshold of ln n for approximating set cover. J. ACM (JACM) **45**(4), 634–652 (1998)
11. Feige, U., Mirrokni, V.S., Vondrak, J.: Maximizing non-monotone submodular functions. SIAM J. Comput. **40**(4), 1133–1153 (2011)
12. Feldman, M., Naor, J.S., Schwartz, R.: A unified continuous greedy algorithmfor submodular maximization. In: FOCS 2011, pp. 570–579. IEEE
13. Feldman, M., Naor, J.S., Schwartz, R.: Nonmonotone submodular maximization via a structural continuous greedy algorithm. In: Aceto, L., Henzinger, M., Sgall, J. (eds.) ICALP 2011, Part I. LNCS, vol. 6755, pp. 342–353. Springer, Heidelberg (2011)
14. Gharan, S.O., Vondrák, J.: Submodular maximization by simulatedannealing. In: SODA 2011, pp. 1098–1116. SIAM
15. Globerson, A., Roweis, S.: Nightmare at test time: robust learning by feature deletion. In Proceedings of the 23rd International Conference on Machine Learning, pp. 353–360. ACM (2006)
16. Golovin, D., Krause, A.: Adaptive submodularity: theory and applications in active learning and stochastic optimization. J. Artif. Intell. Res. **42**, 427–486 (2011)
17. Guestrin, C., Krause, A., Singh, A.P.: Near-optimal sensor placements in Gaussian processes. In: Proceedings of the 22nd International Conference on Machine Learning, pp. 265–272. ACM (2005)
18. Iwata, S., Fleischer, L., Fujishige, S.: A combinatorial strongly polynomial algorithm for minimizing submodular functions. J. ACM (JACM) **48**(4), 761–777 (2001)
19. Krause, A., Guestrin, C., Gupta, A., Kleinberg, J.: Near-optimal sensor placements: Maximizing information while minimizing communication cost. In: Proceedings of the 5th International Conference on Information Processing in Sensor Networks, pp. 2–10. ACM (2006)
20. Krause, A., McMahan, H.B., Guestrin, C., Gupta, A.: Robust submodular observation selection. J. Mach. Learn. Res. **9**, 2761–2801 (2008)
21. Leskovec, J., Krause, A. Guestrin, C., Faloutsos, C., VanBriesen, J., Glance, N.: Cost-effective outbreak detection in networks. In: Proceedings of the 13th ACM SIGKDD International Conference on Knowledge Discovery and Data Mining, pp. 420–429. ACM (2007)
22. Liu, Y., Wei, K., Kirchhoff, K., Song, Y., Bilmes, J.: Submodular feature selection for high-dimensional acoustic score spaces. In: 2013 IEEE International Conference on Acoustics, Speech and Signal Processing (ICASSP), pp. 7184–7188. IEEE (2013)

23. Nemhauser, G.L., Wolsey, L.A.: Best algorithms for approximating the maximum of a submodular set function. Math. Oper. Res. **3**(3), 177–188 (1978)
24. Nemhauser, G.L., Wolsey, L.A., Fisher, M.L.: An analysis of approximations for maximizing submodular set functions–I. Math. Program. **14**(1), 265–294 (1978)
25. Schrijver, A.: A combinatorial algorithm minimizing submodular functions in strongly polynomial time. J. Comb. Theor. Ser. B **80**(2), 346–355 (2000)
26. Sviridenko, M.: A note on maximizing a submodular set function subject to a knapsack constraint. Oper. Res. Lett. **32**(1), 41–43 (2004)
27. Thoma, M., Cheng, H., Gretton, A., Han, J., Kriegel, H.P., Smola, A.J., Song, L., Philip, S.Y., Yan, X., Borgwardt, K.M.: Near-optimal supervised feature selection among frequent subgraphs. In: SDM, pp. 1076–1087. SIAM (2009)
28. Vondrák, J.: Optimal approximation for the submodular welfare problem in the value oracle model. In: STOC 2008, pp. 67–74. ACM
29. Vondrák, J.: Symmetry and approximability of submodular maximization problems. SIAM J. Comput. **42**(1), 265–304 (2013)
30. Vondrák, J., Chekuri, C., Zenklusen, R.: Submodular function maximization via the multilinear relaxation and contention resolution schemes. In: STOC 2011, pp. 783–792. ACM (2011)

Maximizing Monotone Submodular Functions over the Integer Lattice

Tasuku Soma[1(✉)] and Yuichi Yoshida[2]

[1] Graduate School of Information Science and Technology,
The University of Tokyo, Tokyo, Japan
`tasuku_soma@mist.i.u-tokyo.ac.jp`
[2] National Institute of Informatics and Preferred Infrastructure, Inc., Tokyo, Japan
`yyoshida@nii.ac.jp`

Abstract. The problem of maximizing non-negative monotone submodular functions under a certain constraint has been intensively studied in the last decade. In this paper, we address the problem for functions defined over the integer lattice.

Suppose that a non-negative monotone submodular function $f : \mathbb{Z}_+^n \to \mathbb{R}_+$ is given via an evaluation oracle. Assume further that f satisfies the diminishing return property, which is not an immediate consequence of the submodularity when the domain is the integer lattice. Then, we show polynomial-time $(1 - 1/e - \epsilon)$-approximation algorithm for cardinality constraints, polymatroid constraints, and knapsack constraints. For a cardinality constraint, we also show a $(1 - 1/e - \epsilon)$-approximation algorithm with slightly worse time complexity that does not rely on the diminishing return property.

Our algorithms for a polymatroid constraint and a knapsack constraint first extend the domain of the objective function to the Euclidean space and then run the continuous greedy algorithm. We give two different kinds of continuous extensions, one is for polymatroid constraints and the other is for knapsack constraints, which might be of independent interest.

1 Introduction

Submodular functions have been studied intensively in various areas of operations research and computer science since submodularity naturally arises in many problems [12,14,18]. In the last decade, *maximization* of submodular functions has attracted particular interest. For example, one can find novel applications of submodular function maximization in the spread of influence through social networks [17], text summarization [19,20], and optimal budget allocation for advertisements [1].

In most previous works, submodular functions defined over a set are considered, that is, they take a subset of a ground set as the input and return a real value. However, in many practical scenarios, it is more natural to consider submodular functions over a multiset, or equivalently, submodular functions over

© Springer International Publishing Switzerland 2016
Q. Louveaux and M. Skutella (Eds.): IPCO 2016, LNCS 9682, pp. 325–336, 2016.
DOI: 10.1007/978-3-319-33461-5_27

the integer lattice $\mathbb{Z}^E$ for some finite set E. We say that a function $f : \mathbb{Z}^E \to \mathbb{R}$ is *(lattice) submodular* if $f(\boldsymbol{x}) + f(\boldsymbol{y}) \geq f(\boldsymbol{x} \vee \boldsymbol{y}) + f(\boldsymbol{x} \wedge \boldsymbol{y})$ for all $\boldsymbol{x}, \boldsymbol{y} \in \mathbb{Z}^E$, where $\boldsymbol{x} \vee \boldsymbol{y}$ and $\boldsymbol{x} \wedge \boldsymbol{y}$ denote the coordinate-wise maximum and minimum, respectively. Such a generalized form of submodularity arises in maximizing the spread of influence with partial incentives [9], optimal budget allocation, sensor placement, and text summarization [25].

When designing algorithms for maximizing submodular functions, the *diminishing return property* often plays a crucial role. A set function $f : 2^E \to \mathbb{R}$ is said to satisfy the diminishing return property if $f(X + i) - f(X) \geq f(Y + i) - f(Y)$ for all $X \subseteq Y \subseteq E$ and $i \notin Y$. For example, the simple greedy algorithm for cardinality constraints by Nemhauser *et al.* [22] works because of this property. For set functions, it is well known that the submodularity is equivalent to the diminishing return property. For functions over the integer lattice, however, the lattice submodularity only implies a weaker variant of the inequality. This nature causes difficulty in designing approximation algorithms; even for a single cardinality constraint, we need a more complicated approach such as partial enumeration [1,25].

Fortunately, objective functions appearing in practical applications admit the diminishing return property in the following sense. We say that a function $f : \mathbb{Z}^E \to \mathbb{R}$ is *diminishing return submodular (DR-submodular)* if $f(\boldsymbol{x} + \chi_i) - f(\boldsymbol{x}) \geq f(\boldsymbol{y} + \chi_i) - f(\boldsymbol{y})$ for arbitrary $\boldsymbol{x} \leq \boldsymbol{y}$ and $i \in E$, where χ_i is the ith unit vector. Any DR-submodular function is lattice submodular; i.e., DR-submodularity is stronger than the lattice submodularity.[1] The problem of maximizing DR-submodular functions over $\mathbb{Z}^E$ naturally appears in the submodular welfare problem [16,23] and the budget allocation problem with the decreasing influence probabilities [25]. Nevertheless, there are only a few works on this problem. In fact, it was not known whether we can compute $(1 - 1/e)$-approximation in polynomial time under a single cardinality constraint.

1.1   Main Results

In this paper, we give polynomial-time approximation algorithms for maximizing monotone DR-submodular functions under cardinality constraints, polymatroid constraints, and knapsack constraints. Let $f : \mathbb{Z}^E \to \mathbb{R}$ be a non-negative monotone DR-submodular function unless explicitly stated otherwise. Then given any small constant $\epsilon > 0$, our algorithms find $(1 - 1/e - \epsilon)$-approximate solutions under these constraints. The details are described below.

Cardinality Constraint: The objective is to maximize $f(\boldsymbol{x})$ subject to $\boldsymbol{0} \leq \boldsymbol{x} \leq \boldsymbol{c}$ and $\boldsymbol{x}(E) \leq r$, where $\boldsymbol{c} \in \mathbb{Z}_+^E$, $r \in \mathbb{Z}_+$, and $\boldsymbol{x}(E) = \sum_{e \in E} \boldsymbol{x}(e)$. We design a deterministic approximation algorithm whose running time is $O(\frac{n}{\epsilon} \log \|\boldsymbol{c}\|_\infty \log \frac{r}{\epsilon})$, which is the first polynomial time algorithm for this problem.

---

[1] Note that f is DR-submodular if and only if it is lattice submodular and satisfies the *coordinate-wise concave condition*: $f(\boldsymbol{x} + \chi_i) - f(\boldsymbol{x}) \geq f(\boldsymbol{x} + 2\chi_i) - f(\boldsymbol{x} + \chi_i)$ for any $\boldsymbol{x}$ and $i \in E$ (see [26, Lemma 2.3]).

Cardinality Constraint (lattice submodular case): For cardinality constraints, we also show a $(1 - 1/e - \epsilon)$-approximation algorithm for a monotone *lattice submodular* function f. This algorithm runs in $O(\frac{n}{\epsilon^2} \log \|c\|_\infty \log \frac{r}{\epsilon} \log \tau)$ time, where τ is the ratio of the maximum value of f to the minimum positive value of f.

Polymatroid Constraint: The objective is to maximize $f(x)$ subject to $x \in P \cap \mathbb{Z}_+^E$, where P is a polymatroid given via an *independence oracle*. Our algorithm runs in $\widetilde{O}(\frac{n^3}{\epsilon^5} \log^2 r + n^8)$ time, where r is the maximum value of $x(E)$ for $x \in P$. This is the first polynomial time $(1 - 1/e - \epsilon)$-approximation algorithm for this problem.

Knapsack Constraint: The objective is to maximize $f(x)$ subject to a single knapsack constraint $w^\top x \leq 1$, where $w \in \mathbb{R}_+^E$. We devise an approximation algorithm with $\widetilde{O}(\frac{n^2}{\epsilon^{18}} \log \frac{1}{w})(\frac{1}{\epsilon})^{O(1/\epsilon^8)}$ running time, which is the first polynomial time algorithm for this problem.

In order to achieve polynomial-time algorithms instead of pseudo-polynomial time algorithms, we need to combine several techniques carefully. Our algorithms adapt the "decreasing threshold greedy" framework, recently introduced by Badanidiyuru and Vondrák [2], and work in the following way. We maintain a feasible solution $x \in \mathbb{R}^E$ and a threshold $\theta \in \mathbb{R}$ during the algorithm. Starting from $x = 0$, we greedily increase each component of x if the average gain of the increase is above the threshold θ, with consideration of constraints. Slightly decreasing the threshold θ, we repeat this greedy process until θ becomes sufficiently small. Except for the cardinality constraint, our algorithms follow the "continuous greedy" approach [7]; instead of the discrete problem, we consider the problem of maximizing a continuous extension of the original objective function. After the greedy phase, we then round the current fractional solution to an integral solution if needed.

1.2 Technical Contribution

Although our algorithms share some ideas with the algorithms of [2,7], we achieve several improvements as well as new ideas, mainly due to the essential difference between set functions and functions over the integer lattice.

Binary Search in the Greedy Phase: In most previous algorithms, the greedy step works as follows; find the direction of maximum marginal gain and move the current solution along the direction with a *unit* step size. However, it turns out that a naive adaptation of this greedy strategy only gives a pseudo-polynomial time algorithm. To circumvent this issue, we perform a binary search to determine the step size in the greedy phase. Combined with the decreasing threshold framework, this technique reduces the time complexity significantly.

New Continuous Extensions: To carry out the continuous greedy algorithm, we need a continuous extension of functions over the integer lattice. Note that the *multilinear extension* cannot be directly used since the domain of the multilinear extension is only the hypercube $[0,1]^E$. In this paper, we propose two different kinds of new continuous extensions of a function over the integer lattice, one of which is for polymatroid constraints and the other is for knapsack constraints. These continuous extensions have similar properties to the multilinear extension when f is DR-submodular, and are carefully designed so that we can round fractional solutions without violating polymatroid or knapsack constraints. To the best of our knowledge, these continuous extensions in $\mathbb{R}_+^E$ have not been known up to this paper.

Rounding without violating polymatroid and knapsack constraints: It is non-trivial how to round fractional solutions in $\mathbb{R}_+^E$ without violating polymatroid or knapsack constraints. For polymatroid constraints, we show that the rounding can be reduced to rounding in a matroid polytope, and therefore we can use existing rounding methods for a matroid polytope. For knapsack constraints, we design a new simple rounding method based on our continuous extension.

1.3 Related Work

Studies on maximizing monotone submodular functions were pioneered by the paper of Neumhauser, Wolsey, and Fisher [22]. They showed that a *greedy* algorithm achieves a $(1 - 1/e)$-approximation for maximizing a monotone and submodular set function under a cardinality constraint, and a $1/2$-approximation under a matroid constraint. Their algorithm provides a prototype for following works. For knapsack constraints, Sviridenko [27] devised the first $(1 - 1/e)$-approximation algorithm whose running time is $O(n^5)$ time. Whereas these algorithms are combinatorial and deterministic, the best known algorithms for matroid constraints are based on a continuous and randomized method. The first $(1 - 1/e)$-approximation algorithm for a matroid constraint was provided by [6], which employed the *continuous greedy* approach; first solve a continuous relaxation problem and obtain a fractional approximate solution, then round it to an integral feasible solution. In their framework, the *multilinear extension* of a submodular set function is used as the objective function in the relaxation problem. They also provided the *pipage rounding* to obtain an integral feasible solution. Chekuri, Vondrák, and Zenklusen [7] designed a simple rounding method, called *swap rounding*, based on the exchange property of matroid base families. Recently, Badanidiyuru and Vondrák [2] devised $(1 - 1/e - \epsilon)$-approximation algorithms for any fixed constraint $\epsilon > 0$, with significantly lower time complexity for various constraints. For the inapproximability side, Nemhauser et al. [22] proved that no algorithm making polynomially many queries to a value oracle of f cannot achieve an approximation ratio better than $1 - 1/e$ under any of constraints mentioned so far. Also, Feige [10] showed that, even if f is given explicitly, $(1 - 1/e)$-approximation is the best possible unless $P = NP$.

Generalized forms of submodularity have been studied in various contexts. Fujishige [12] discusses submodular functions over a distributive lattice and its related polyhedra. In the theory of discrete convex analysis by Murota [21], a subclass of submodular functions over the integer lattice is considered.

The maximization problem also has been studied for variants of submodular functions. Shioura [23] investigates the maximization of discrete convex functions. Soma et al. [25] provides a $(1-1/e)$-approximation algorithm for maximizing a monotone lattice submodular function under a knapsack constraint. However, its running time is pseudo-polynomial. Although the present work focuses on monotone submodular functions, there are a large body of work on maximization of *non-monotone* submodular functions [3–5,11]. Gottschalk and Peis [13] provided a 1/3-approximation algorithm for maximizing a lattice submodular function over (bounded) integer lattice. Recently, *bisubmodular* functions and *k-submodular* functions, other generalizations of submodular functions, have been studied, and approximation algorithms for maximizing these functions can be found in [15,24,28].

1.4 Organization of This Paper

The rest of this paper is organized as follows. In Sect. 2, we provide our notations and basic facts on submodular functions and polymatroid. Section 3 describes our algorithm for cardinality constraints. In Sect. 4, we provide the continuous extension for polymatroid constraints and our approximate algorithm. We present our algorithm for knapsack constraints as well as another continuous extension in Sect. 5. Omitted details and proofs are deferred to the full version.

2 Preliminaries

Notation. We denote the sets of non-negative integers and non-negative reals by $\mathbb{Z}_+$ and $\mathbb{R}_+$, respectively. For a positive integer k, $[k]$ denotes the set $\{1, \ldots, k\}$. Throughout this paper, E denotes a ground set of size n. For $f : \mathbb{R}^E \to \mathbb{R}$ and $x, y \in \mathbb{R}^E$, we define $f(x \mid y) := f(x + y) - f(y)$. For $x \in \mathbb{R}^E$ and $X \subseteq E$, we denote $x(X) := \sum_{i \in X} x(i)$. For a vector $x \in \mathbb{R}^E$, $\mathrm{supp}^+(x)$ denotes the set $\{e \in E \mid x(e) > 0\}$. For $x \in \mathbb{Z}_+^E$, $\{x\}$ denotes the multiset where the element e appears $x(e)$ times. For arbitrary two multisets $\{x\}$ and $\{y\}$, we define $\{x\} \backslash \{y\} := \{(x - y) \vee 0\}$. For a multiset $\{x\}$, we define $|\{x\}| := x(E)$.

For the error parameter $\epsilon > 0$, we always assume that $\frac{1}{\epsilon}$ is an integer (otherwise we can slightly decrease it without changing the asymptotic time complexity and the approximation ratio).

Lattice and DR-Submodularity. We say that a function $f : \mathbb{Z}_+^E \to \mathbb{R}$ is *lattice submodular* if it satisfies $f(x) + f(y) \geq f(x \vee y) + f(x \wedge y)$ for all $x, y \in \mathbb{Z}^E$, where $x \vee y$ and $x \wedge y$ denote the coordinate-wise maximum and minimum, respectively, i.e., $(x \vee y)(e) = \max\{x(e), y(e)\}$ and $(x \wedge y)(e) = \min\{x(e), y(e)\}$ for each $e \in E$. A function $f : \mathbb{Z}_+^E \to \mathbb{R}$ is *monotone* if $f(x) \leq f(y)$ for all x and y with $x \leq y$. We say that $f : \mathbb{Z}^n \to \mathbb{R}$ is *diminishing return submodular (DR-submodular)* if $f(x + \chi_i) - f(x) \geq f(y + \chi_i) - f(y)$ for every $x \leq y$ and $i \in E$, where χ_i denotes the ith unit vector. We note that the lattice submodularity of

---

Algorithm 1. Cardinality Constraint/DR-Submodular

Input: $f : \mathbb{Z}_+^E \to \mathbb{R}_+$, $c \in \mathbb{Z}_+^E$, $r \in \mathbb{Z}_+$, and $\epsilon > 0$. **Output:** $y \in \mathbb{Z}_+^E$.
1: $y \leftarrow 0$ and $d \leftarrow \max_{e \in E} f(\chi_e)$.
2: **for** $(\theta = d; \theta \geq \frac{\epsilon}{r} d; \theta \leftarrow \theta(1 - \epsilon))$ **do**
3:     **for** all $e \in E$ **do**
4:         Find maximum $k \leq \min\{c(e) - y(e), r - y(E)\}$ with $f(k\chi_e \mid y) \geq k\theta$ with
        binary search.
5:         **If** such k exists **then** $y \leftarrow y + k\chi_e$
6: **return** y.

---

f does not imply the DR-submodularity when the domain is the integer lattice. Throughout this paper, we assume that $f(0) = 0$ without loss of generality.

If a function $f : \mathbb{Z}^E \to \mathbb{R}$ satisfies $f(x \vee k\chi_i) - f(x) \geq f(y \vee k\chi_i) - f(y)$ for any $i \in E$, $k \in \mathbb{Z}_+$, x and y with $x \leq y$, then we say that f satisfies the *weak diminishing return property*. Any monotone lattice submodular function satisfies the weak diminishing return property [25].

Polymatroid. Let $\rho : 2^E \to \mathbb{Z}_+$ be a monotone submodular set function with $\rho(\emptyset) = 0$. The (integral) *polymatroid* associated with ρ is the polytope $P = \{x \in \mathbb{R}_+^E : x(X) \leq \rho(X) \quad \forall X \subseteq E\}$, and ρ is called the *rank function* of P.

3 Cardinality Constraint

We give approximation algorithms for maximizing monotone DR-submodular and lattice submodular functions in Sects. 3.1 and 3.2, respectively.

3.1 Maximization of Monotone DR-Submodular Function

We start with the case of a DR-submodular function. Let $f : \mathbb{Z}_+^E \to \mathbb{R}_+$ be a monotone DR-submodular function. Let $c \in \mathbb{Z}_+^E$ and $r \in \mathbb{Z}_+$. We want to maximize $f(x)$ under the constraints $0 \leq x \leq c$ and $x(E) \leq r$. The pseudocode description of our algorithm is shown in Algorithm 1.

The following lemma can be obtained by adopting analysis for the greedy algorithm (a proof is deferred to the full version).

Lemma 1. *Let x^* be an optimal solution. When adding $k\chi_e$ to the current solution y in Line 5, the average gain satisfies the following.*

$$\frac{f(k\chi_e \mid y)}{k} \geq \frac{(1 - \epsilon)}{r} \sum_{s \in \{x^*\} \setminus \{y\}} f(\chi_s \mid y)$$

Theorem 1. *Algorithm 1 achieves the approximation ratio of $1 - \frac{1}{e} - \epsilon$ in $O(\frac{n}{\epsilon} \log \|c\|_\infty \log \frac{r}{\epsilon})$ time.*

Algorithm 2. Binary Search for Cardinality Constraint/Lattice Submodular

---

Input: $f : \mathbb{Z}_+^E \to \mathbb{R}_+$, $e \in E$, $\theta > 0$, $k_{\max} \in \mathbb{Z}_+$, $\epsilon > 0$. **Output:** $0 \le k \le k_{\max}$ or **fail**.
1: Find $k_{\min}$ with $0 \le k_{\min} \le k_{\max}$ such that $f(k_{\min}\chi_e) > 0$ by binary search.
2: **if** no such $k_{\min}$ exists **then fail**.
3: **for** $(h = f(k_{\max}\chi_e); h \ge (1 - \epsilon)f(k_{\min}\chi_e); h = (1 - \epsilon)h)$ **do**
4:     Find the smallest $k_{\min} \le k \le k_{\max}$ such that $f(k\chi_e) \ge h$ by binary search.
5:         **If** $f(k\chi_e) \ge (1 - \epsilon)k\theta$ **then return** k.
6: **fail**.

Proof. Let $\boldsymbol{y}$ be the output of Algorithm 1. Without loss of generality, we can assume that $\boldsymbol{y}(E) = r$. To see this, consider a modified version of the algorithm in which the threshold is updated until $\boldsymbol{y}(E) = r$. Let $\boldsymbol{y}'$ be the output of this modified algorithm. Since the marginal gain of increasing any coordinate of $\boldsymbol{y}$ by one is at most $\epsilon\frac{d}{r}$, we have $f(\boldsymbol{y}') - f(\boldsymbol{y}) \le \epsilon d \le \epsilon\text{OPT}$. Therefore, it suffices to show that $\boldsymbol{y}'$ is a $(1 - 1/e - \epsilon)$-approximate solution.

Let $\boldsymbol{y}_i$ be the vector after i steps. Let $k_i\chi_{e_i}$ be the vector added in the i-th step. That is, $\boldsymbol{y}_i = \sum_{j=1}^i k_j\chi_{e_j}$. By Lemma 1, we have

$$\frac{f(k_{i+1}\chi_{e_{i+1}} \mid \boldsymbol{y}_i)}{k_{i+1}} \ge \frac{1 - \epsilon}{r} \sum_{s \in \{\boldsymbol{x}^*\} \backslash \{\boldsymbol{y}_i\}} f(\chi_s \mid \boldsymbol{y}_i).$$

By DR-submodularity, $\sum_{s \in \{\boldsymbol{x}^*\} - \{\boldsymbol{y}_i\}} f(\chi_s \mid \boldsymbol{y}_i) \ge f(\boldsymbol{x}^* \vee \boldsymbol{y}_i) - f(\boldsymbol{y}_i)$ holds. Therefore by monotonicity, we have

$$f(\boldsymbol{y}_{i+1}) - f(\boldsymbol{y}_i) = f(k_{i+1}\chi_{e_{i+1}} \mid \boldsymbol{y}_i) \ge \frac{(1 - \epsilon)k_{i+1}}{r}(f(\boldsymbol{x}^* \vee \boldsymbol{y}_i) - f(\boldsymbol{y}_i))$$

$$\ge \frac{(1 - \epsilon)k_{i+1}}{r}(\text{OPT} - f(\boldsymbol{y}_i)).$$

Hence, we can show by induction that $f(\boldsymbol{y}) \ge \left(1 - \prod_i \left(1 - \frac{(1-\epsilon)k_i}{r}\right)\right)\text{OPT}$. Since $\prod_i \left(1 - \frac{(1-\epsilon)k_i}{r}\right) \le \prod_i \exp\left(-\frac{(1-\epsilon)k_i}{r}\right) = \exp\left(-\frac{(1-\epsilon)\sum_i k_i}{r}\right) = e^{-(1-\epsilon)} \le \frac{1}{e} + \epsilon$, we obtain $(1 - \frac{1}{e} - \epsilon)$-approximation. $\square$

3.2   Maximization of Monotone Lattice Submodular Function

We now consider the case that $f : \mathbb{Z}_+^E \to \mathbb{R}_+$ is a monotone lattice submodular function. The pseudocode description of our algorithm is shown in Algorithm 3.

The main issue is that we cannot find k such that $f(k\chi_e \mid \boldsymbol{x}) \ge k\theta$ by naive binary search. However, we can find k such that $f(k\chi_e \mid \boldsymbol{x}) \ge (1 - \epsilon)k\theta$ in polynomial time (if exists). The idea is guessing the value of $f(k\chi_e)$ by iteratively decreasing the threshold and checking whether the desired k exists by binary search. See Algorithm 2 for the details.

Basically replacing the binary search in Algorithm 1 with Algorithm 2 yields the same approximation guarantee. However analysis gets more involved, and therefore is deferred to the full version.

Algorithm 3. Cardinality Constraint/Lattice Submodular

Input: $f : \mathbb{Z}_+^E \to \mathbb{R}_+$, $\boldsymbol{c} \in \mathbb{Z}_+^E$, $r \in \mathbb{Z}_+$, $\epsilon > 0$.
1: $\boldsymbol{y} \leftarrow \boldsymbol{0}$ and $d_{\max} \leftarrow \max_{e \in E} f(c(e)\chi_e)$.
2: **for** $(\theta = d_{\max}; \theta \geq \frac{\epsilon}{r} d_{\max}; \theta \leftarrow \theta(1 - \epsilon))$ **do**
3:     **for** all $e \in E$ **do**
4:         Invoke Algorithm 2 with $f(\cdot \mid \boldsymbol{y})$, e, θ, $\min\{c(e) - y(e), r - y(E)\}$, and ϵ.
5:         **If** Algorithm 2 outputs $k \in \mathbb{Z}_+$ **then** $\boldsymbol{y} \leftarrow \boldsymbol{y} + k\chi_e$
6: **return** $\boldsymbol{y}$.

Theorem 2. *Algorithm 3 achieves an approximation ratio of $1 - \frac{1}{e} - O(\epsilon)$ in $O(\frac{n}{\epsilon^2} \log \|\boldsymbol{c}\|_\infty \log \frac{r}{\epsilon} \log \tau)$ time, where $\tau = \frac{\max_{e \in E} f(c(e)\chi_e)}{\min\{f(\boldsymbol{x}) : 0 \leq \boldsymbol{x} \leq \boldsymbol{c}, f(\boldsymbol{x}) > 0\}}$.*

4 Polymatroid Constraint

Let P be a polymatroid with a ground set E and the rank function $\rho : 2^E \to \mathbb{Z}_+$. The objective is to maximize $f(\boldsymbol{x})$ subject to $\boldsymbol{x} \in P \cap \mathbb{Z}^E$, where f is a DR-submodular function. In what follows, we denote $\rho(E)$ by r.

4.1 Continuous Extension for Polymatroid Constraints

For $\boldsymbol{x} \in \mathbb{R}^E$, let $\lfloor \boldsymbol{x} \rfloor$ denote the vector obtained by rounding down each entry of $\boldsymbol{x}$. For $a \in \mathbb{R}$, let $\langle a \rangle$ denote the fractional part of a, that is, $\langle a \rangle := a - \lfloor a \rfloor$. For $\boldsymbol{x} \in \mathbb{R}^E$, we define $C(\boldsymbol{x}) := \{\boldsymbol{y} \in \mathbb{R}^E \mid \lfloor \boldsymbol{x} \rfloor \leq \boldsymbol{y} \leq \lfloor \boldsymbol{x} \rfloor + \boldsymbol{1}\}$ as the hypercube $\boldsymbol{x}$ belongs to.

For $\boldsymbol{x} \in \mathbb{R}^E$, we define $\mathcal{D}(\boldsymbol{x})$ as the distribution from which we sample $\bar{\boldsymbol{x}}$ such that $\bar{\boldsymbol{x}}(i) = \lfloor \boldsymbol{x}(i) \rfloor$ with probability $1 - \langle \boldsymbol{x}(i) \rangle$ and $\bar{\boldsymbol{x}}(i) = \lceil \boldsymbol{x}(i) \rceil$ with probability $\langle \boldsymbol{x}(i) \rangle$, for each $i \in E$. We define the continuous extension $F : \mathbb{R}_+^E \to \mathbb{R}_+$ of $f : \mathbb{Z}_+^E \to \mathbb{R}_+$ as follows. For each $\boldsymbol{x} \in \mathbb{R}_+^E$, we define

$$F(\boldsymbol{x}) := \mathop{\mathbf{E}}_{\bar{\boldsymbol{x}} \sim \mathcal{D}(\boldsymbol{x})} [f(\bar{\boldsymbol{x}})] = \sum_{S \subseteq E} f(\lfloor \boldsymbol{x} \rfloor + \chi_S) \prod_{i \in S} \langle \boldsymbol{x}(i) \rangle \prod_{i \notin S} (1 - \langle \boldsymbol{x}(i) \rangle). \quad (1)$$

We call this type of continuous extensions *the continuous extension for polymatroid constraints*. Note that F is obtained by gluing the multilinear extension of f restricted to each hypercube. If $f : \{0, 1\}^E \to \mathbb{R}_+$ is a monotone submodular function, then it is known that its multilinear extension is monotone and concave along nonnegative directions. We can show similar properties for the continuous extension of a function $f : \mathbb{Z}_+^E \to \mathbb{R}_+$ if f is monotone and DR-submodular. A proof is deferred to the full version.

Lemma 2. *For a monotone DR-submodular function f, the continuous extension F for the polymatroid constraint is a nondecreasing concave function along any line of direction $\boldsymbol{d} \geq 0$.*

Algorithm 4.

Input: $f : \mathbb{Z}_+^E \to \mathbb{R}_+$, $x \in \mathbb{R}_+^E$, $e \in E$, $\theta > 0$, $\alpha, \beta, \delta \in (0, 1)$, $k_{\max} \in \mathbb{Z}_+$ **Output:** $k \in \mathbb{Z}_+$.

1: $\ell \leftarrow 1$, $u \leftarrow k_{\max}$.
2: **while** $\ell < u$ **do**
3:    $m = \lfloor \frac{\ell+u}{2} \rfloor$.
4:    $\widetilde{F}(m\chi_e \mid x) \leftarrow$ Estimates of $F(m\chi_e \mid x)$ by averaging $O\left(\frac{\log(k_{\max}/\delta)}{\alpha\beta}\right)$ random samples, respectively.
5:    **If** $\widetilde{F}(m\chi_e \mid x) \geq m\theta$ **then** $\ell \leftarrow m + 1$.
6:    **Else** $u \leftarrow m$.
7: $k \leftarrow \ell - 1$.
8: **return** k.

Algorithm 5. Decreasing-Threshold

Input: $f : \mathbb{Z}_+^E \to \mathbb{R}_+$, $x \in \mathbb{R}_+^E$, $\epsilon \in [0, 1]$, $P \subseteq \mathbb{R}_+^E$. **Output:** A vector $y \in P \cap \mathbb{Z}_+^E$.

1: $y \leftarrow \mathbf{0}$ and $d \leftarrow \max_{e \in E} f(\chi_e)$.
2: $N \leftarrow$ the solution to $N = n\lceil \log_{1/1-\epsilon} \frac{N}{\epsilon} \rceil$. Note that $N = O(\frac{n}{\epsilon} \log \frac{n}{\epsilon})$.
3: **for** $(\theta = d;\ \theta \geq \frac{\epsilon d}{N};\ \theta \leftarrow \theta(1 - \epsilon))$ **do**
4:    **for** all $e \in E$ **do**
5:       Invoke Algorithm 4 to find maximum $0 \leq k \leq k_{\max}$ such that $\widetilde{F}(k\chi_e \mid x + \epsilon y) \geq k\theta$, where $k_{\max} = \max\{y + k\chi_e \in P\}$, with additive error $\alpha_5 = \epsilon$, multiplicative error $\beta_5 = \frac{\epsilon}{N(n+1)}$, and failure probability $\delta_5 = \frac{\epsilon}{N}$.
6:       **If** such k exists **then** $y \leftarrow y + k\chi_e$.
7: **return** y.

4.2 Continuous Greedy Algorithm for Polymatroid Constraint

In this section, we describe our algorithm for the polymatroid constraint, whose pseudocode description is presented in Algorithm 6. For a continuous extension F for polymatroid constraints and $x, y \in \mathbb{R}^E$, we define $F(x \mid y) = F(x+y) - F(y)$.

At a high level, our algorithm produces a sequence $x^0 = \mathbf{0}, \dots, x^{1/\epsilon}$ in P. Given x^t, Decreasing-Threshold determines an update direction y^t (here our continuous extension F comes into play), and we update as $x^{t+1} := x^t + \epsilon y^t$. Finally, we perform a rounding algorithm to the fractional solution $x^{1/\epsilon}$ and obtain an integral solution $\bar{x}$.

Lemma 3. *In the end of Algorithm 6, we have $F(x) \geq (1 - 1/e - O(\epsilon))\mathrm{OPT}$ with probability at least 2/3.*

The analysis basically follows [2], but we need to replace the multilinear extension with the continuous extension for polymatroid constraints. Fortunately, Lemma 2 ensures that a similar argument still works. As the analysis is fairly technical, we defer it to the full version.

Algorithm 6. Polymatroid Constraint/DR-Submodular

Input: $f : \mathbb{Z}_+^E \to \mathbb{R}_+, P \subseteq \mathbb{R}_+^E$. **Output:** A vector $\bar{x} \in P \cap \mathbb{Z}_+^E$.
1: $x \leftarrow 0$.
2: **for** $(t \leftarrow 1; t \leq \lfloor \frac{1}{\epsilon} \rfloor; t \leftarrow t+1)$ **do**
3:     $y \leftarrow$ Decreasing-Threshold(f, x, ϵ, P), $x \leftarrow x + \epsilon y$.
4: $\bar{x} \leftarrow$ Rounding(x, P).
5: **return** $\bar{x}$.

4.3 Rounding

We need a rounding procedure that takes a real vector x as the input and returns an integral vector $\bar{x}$ such that $\mathbf{E}[f(\bar{x})] \geq F(x)$. There are several rounding algorithm in the $\{0,1\}^E$ case [6,7]. However, generalizing these rounding algorithm over integer lattice is a nontrivial task. Here, we show that rounding in the integer lattice can be reduced to rounding in the $\{0,1\}^E$ case.

Suppose that we have a fractional solution x. Then $P \cap C(x)$ is a translation of a matroid polytope P_M. The independence oracle of the corresponding matroid is just the independence oracle of P restricted to $C(x)$. Thus, the pipage rounding algorithm for P' yields an integral solution $\bar{x}$ with $\mathbf{E}[f(\bar{x})] \geq F(x)$ in strongly polynomial time. Slightly faster rounding can be achieved by swap rounding. Swap rounding requires that the given fractional solution x is represented by a convex combination of extreme points of the matroid polytope. In our setting, we can represent x as a convex combination of extreme points of $P \cap C(x)$, using the algorithm of Cunningham [8]. Then we run the swap rounding algorithm for the convex combination and $P \cap C(x)$. The running time of this rounding algorithm is dominated by the complexity of finding a convex combination for x, which is $O(n^8)$ time.

Theorem 3. *Algorithm 6 finds an $(1 - 1/e - \epsilon)$-approximate solution (in expectation) with high probability in $O(\frac{nr}{\epsilon^3} \log \frac{r}{\epsilon} \log \|c\|_\infty \log \frac{rn \log(r/\epsilon)}{\epsilon^2} + n^8)$ time.*

5 Knapsack Constraint

In this section, we give an efficient algorithm for maximizing DR-submodular functions under knapsack constraints. The main difficulty of this case is that we cannot obtain a feasible solution by applying the swap rounding or the pipage rounding to a fractional solution of the multilinear extension considered in Sect. 4. To overcome this issue, we introduce another multilinear extension in Sect. 5.1. Then, we describe our continuous greedy algorithm using this extension in Sect. 5.2.

5.1 Multilinear Extension for Knapsack Constraints

Let $X : I \to \mathbb{R}_+^E$ be a multiset of vectors (indexed by a set I). We say that X is a *multi-vector* if, for every $i \in I$, $X(i)$ is of the form $k\chi_e$, where $k \in \mathbb{R}_+$ and $e \in E$. For $J \subseteq I$, let $X(J) := \sum_{j \in J} X(j) \in \mathbb{R}_+^E$. Let $p \in [0,1]^I$. Then we define

$$F_{\boldsymbol{p}}(\boldsymbol{X}) = \sum_{J \subseteq I} \prod_{i \in J} \boldsymbol{p}(i) \prod_{i \notin J} (1 - \boldsymbol{p}(i)) f(\boldsymbol{X}(J))$$

In this section, $\oplus$ denotes the concatenation. In our setting, we often consider $F_{\boldsymbol{p}}$ for $\boldsymbol{p} = \epsilon\mathbf{1}$, i.e., each element of $\boldsymbol{X}$ is independently sampled in the same probability ϵ. In this case, we will use the shorthand notation $F_\epsilon(\boldsymbol{X}) := F_{\epsilon\mathbf{1}}(\boldsymbol{X})$. In addition, we define $F_\epsilon(k\chi_e \mid \boldsymbol{X}) := F_\epsilon(\boldsymbol{X} \oplus k\chi_e) - F_\epsilon(\boldsymbol{X})$. Also, we define $F_\epsilon^{\oplus 1}(k\chi_e \mid \boldsymbol{X}) := F_{\epsilon\mathbf{1}\oplus 1}(\boldsymbol{X} \oplus k\chi_e) - F_\epsilon(\boldsymbol{X})$, where $\epsilon\mathbf{1}$ corresponds to $\boldsymbol{X}$ and 1 corresponds to $k\chi_e$. For a multi-vector $\boldsymbol{X}$ and a vector $\boldsymbol{w} : E \to \mathbb{R}$, we define $\boldsymbol{w}(\boldsymbol{X}) = \sum_{k\chi_e \in \boldsymbol{X}} k\boldsymbol{w}(e)$ as the sum of weights of vectors in $\boldsymbol{X}$.

Lemma 4. *Let $f : \mathbb{Z}_+ \to \mathbb{R}_+$ be a monotone DR-submodular function. Fix a multi-vector $\boldsymbol{X} : I \to \mathbb{Z}_+^E$. Then, the function $F_{\boldsymbol{p}}(\boldsymbol{X})$, as a function of $\boldsymbol{p}$, is a monotone concave function along any line of direction $\boldsymbol{d} \geq 0$.*

Lemma 5. *Let $f : \mathbb{Z}_+ \to \mathbb{R}_+$ be a monotone DR-submodular function. Then, $F_\epsilon(\boldsymbol{Y} \mid \boldsymbol{X}) \geq F_\epsilon(\boldsymbol{Y} \mid \boldsymbol{X} \oplus \boldsymbol{X}')$ for any multi-vectors $\boldsymbol{X}$, $\boldsymbol{X}'$, and $\boldsymbol{Y}$.*

5.2 Algorithm (Sketch)

We now sketch our algorithm for knapsack constraints. Suppose that $\boldsymbol{x}^*$ is of the form $\sum_{e \in E} k_e^* \chi_e$. We call an item e *small* if it satisfies $k_e^* f(\chi_e) \leq \epsilon^6 f(\boldsymbol{x}^*)$ and $\boldsymbol{w}(e) \leq \epsilon^4$. (In order to obtain the upper bound on the number of copies to be included, we only need an estimate for $f(\boldsymbol{x}^*)$. So we can discretize the values in between $\max_{e \in E} \frac{f(e)}{\boldsymbol{w}(e)}$ and $n \max_{e \in E} \frac{f(e)}{\boldsymbol{w}(e)}$ into $\frac{\log n}{\epsilon}$ values and check each one of them). Other items are called *large*. Let $\ell \leq \frac{1}{\epsilon^6} + \frac{1}{\epsilon^4} = O(\frac{1}{\epsilon^6})$ be the number of large items. Let the optimal solution be $\boldsymbol{x}^* = \boldsymbol{x}_L^* + \boldsymbol{x}_S^*$ where $\boldsymbol{x}_L^* = \sum_{i=1}^\ell k_i^* \chi_{e_i^*}$ is the vector corresponding to large items and $\boldsymbol{x}_S^* = \boldsymbol{x}^* - \boldsymbol{x}_L^*$ is the vector corresponding to small items.

Our overall approach is a variation of the continuous greedy method. However, we deal with large items and small items separately. This is because, when there are large items, we cannot round a fractional solution to a feasible solution without losing significantly in the approximation ratio. To get around this issue, we guess value sets taken by these large sets. Since the detail of our algorithm is very involved, we defer it to the full version.

References

1. Alon, N., Gamzu, I., Tennenholtz, M.: Optimizing budget allocation among channels and influencers. In: Proceedings of WWW, pp. 381–388 (2012)
2. Badanidiyuru, A., Vondrák, J.: Fast algorithms for maximizing submodular functions. In: Proceedings of SODA, pp. 1497–1514 (2013)
3. Buchbinder, N., Feldman, M., Naor, J., Schwartz, R.: A tight linear time (1/2)-approximation for unconstrained submodular maximization. In: Proceedings of FOCS, pp. 649–658 (2012)
4. Buchbinder, N., Feldman, M.: Deterministic algorithms for submodular maximization problems. In: Proceedings of SODA, pp. 392–403 (2016)

5. Buchbinder, N., Feldman, M., Naor, J.S., Schwartz, R.: Submodular maximization with cardinality constraints. In: Proceedings of SODA, pp. 1433–1452 (2014)
6. Calinescu, G., Chekuri, C., Pál, M., Vondrák, J.: Maximizing a monotone submodular function subject to a matroid constraint. SIAM J. Comput. **40**, 1740–1766 (2011)
7. Chekuri, C., Vondrák, J., Zenklusen, R.: Dependent randomized rounding via exchange properties of combinatorial structures. In: Proceedings of FOCS, pp. 575–584 (2010)
8. Cunningham, W.H.: Testing membership in matroid polyhedra. J. Comb. Theor. Ser. B **188**, 161–188 (1984)
9. Demaine, E.D., Hajiaghayi, M., Mahini, H., Malec, D.L., Raghavan, S., Sawant, A., Zadimoghadam, M.: How to influence people with partial incentives. In: Proceedings of WWW, pp. 937–948 (2014)
10. Feige, U.: A threshold of $\ln n$ for approximating set cover. J. ACM **45**, 634–652 (1998)
11. Feige, U., Mirrokni, V.S., Vondrak, J.: Maximizing non-monotone submodular functions. SIAM J. Comput. **40**(4), 1133–1153 (2011)
12. Fujishige, S.: Submodular Functions and Optimization, 2nd edn. Elsevier, New York (2005)
13. Gottschalk, C., Peis, B.: Submodular function maximization on the bounded integer lattice. ArXiv preprint (2015)
14. Iwata, S.: Submodular function minimization. Math. Program. **112**(1), 45–64 (2007)
15. Iwata, S., Tanigawa, S., Yoshida, Y.: Bisubmodular function maximization and extensions. Mathematical Engineering Technical Reports (2013)
16. Kapralov, M., Post, I., Vondrak, J.: Online submodular welfare maximization: Greedy is optimal. In: Proceedings of SODA, pp. 1216–1225 (2012)
17. Kempe, D., Kleinberg, J., Tardos, E.: Maximizing the spread of influence through a social network. In: Proceedings of KDD, pp. 137–146 (2003)
18. Krause, A., Golovin, D.: Submodular function maximization. In: Tractability: Practical Approaches to Hard Problems, pp. 71–104. Cambridge University Press (2014)
19. Lin, H., Bilmes, J.: Multi-document summarization via budgeted maximization of submodular functions. In: Proceedings of NAACL, pp. 912–920 (2010)
20. Lin, H., Bilmes, J.: A class of submodular functions for document summarization. In: Proceedings of NAACL, pp. 510–520 (2011)
21. Murota, K.: Discrete Convex Analysis. SIAM, Philadelphia (2003)
22. Nemhauser, G.L., Wolsey, L.A., Fisher, M.L.: An analysis of approximations for maximizing submodular set functions - II. Math. Program. Studies **8**, 73–87 (1978)
23. Shioura, A.: On the pipage rounding algorithm for submodular function maximization – a view from discrete convex analysis–. Discrete Math. Algorithms Appl. **1**(1), 1–23 (2009)
24. Singh, A., Guillory, A., Bilmes, J.: On bisubmodular maximization. In: Proceedings of AISTATS, pp. 1055–1063 (2012)
25. Soma, T., Kakimura, N., Inaba, K., Kawarabayashi, K.: Optimal budget allocation: theoretical guarantee and efficient algorithm. In: Proceedings of ICML (2014)
26. Soma, T., Yoshida, Y.: A generalization of submodular cover via the diminishing return property on the integer lattice. In: Proceedings of NIPS (2015)
27. Sviridenko, M.: A note on maximizing a submodular set function subject to a knapsack constraint. Oper. Res. Lett. **32**(1), 41–43 (2004)
28. Ward, J., Živný, S.: Maximizing bisubmodular and k-submodular functions. In: Proceedings of SODA, pp. 1468–1481 (2014)

Submodular Unsplittable Flow on Trees

Anna Adamaszek[1], Parinya Chalermsook[2(✉)], Alina Ene[3], and Andreas Wiese[2]

[1] University of Copenhagen, Copenhagen, Denmark
anad@di.ku.dk
[2] Max-Planck-Institut Für Informatik, Saarbrücken, Germany
{parinya,awiese}@mpi-inf.mpg.de
[3] University of Warwick, Coventry, UK
a.ene@dcs.warwick.ac.uk

Abstract. We study the Unsplittable Flow problem (UFP) on trees with a submodular objective function. The input to this problem is a tree with edge capacities and a collection of tasks, each characterized by a source node, a sink node, and a demand. A subset of the tasks is feasible if the tasks can simultaneously send their demands from the source to the sink without violating the edge capacities. The goal is to select a feasible subset of the tasks that maximizes a submodular objective function.

Our main result is an $O(k \log n)$-approximation algorithm for Submodular UFP on trees where k denotes the pathwidth of the given tree. Since every tree has pathwidth $O(\log n)$, we obtain an $O(\log^2 n)$ approximation for arbitrary trees. This is the first non-trivial approximation guarantee for the problem and it matches the best approximation known for UFP on trees with a linear objective function.

Our main technical contribution is a new geometric relaxation for UFP on trees that builds on the recent work of [Bonsma et al., FOCS 2011; Anagnostopoulos et al., SODA 2014] for UFP on paths with a linear objective. Our relaxation is very structured and we can combine it with the contention resolution framework of [Chekuri et al., STOC 2011]. Our approach is robust and extends to several related problems, such as UFP with bag constraints and the Storage Allocation Problem.

Additionally, we study the special case of UFP on trees with a linear objective and upward instances where, for each task, the source node is a descendant of the sink node. Such instances generalize UFP on paths. We build on the work of [Bansal et al., STOC 2006] for UFP on paths and obtain a QPTAS for upward instances when the input data is quasi-polynomially bounded. We complement this result by showing that, unlike the path setting, upward instances are APX-hard if the input data is arbitrary.

1 Introduction

Submodular functions are a rich class of functions with many applications both in theory and in practice. On the theoretical side, submodularity is a key concept

Partially supported by the Danish Council for Independent Research DFF-MOBILEX mobility grant.

© Springer International Publishing Switzerland 2016
Q. Louveaux and M. Skutella (Eds.): IPCO 2016, LNCS 9682, pp. 337–349, 2016.
DOI: 10.1007/978-3-319-33461-5_28

in combinatorial optimization and economics with deep mathematical consequences. On the practical side, submodular functions arise naturally in a variety of settings such as data summarization, sensor placement, inference in graphical models, image segmentation, social networks, auctions, and exemplar clustering [6,14,16,17,19–22].

One of the main reasons for the success of submodularity is that it combines a significant modeling power with a certain degree of tractability. This delicate balance between generality and tractability has made submodular functions very appealing, and there has been a significant interest in optimizing submodular functions subject to a variety of constraints.

The traditional approach to submodular maximization makes extensive use of the classical Greedy algorithm of Nemhauser, Wolsey, and Fisher [23]. The Greedy algorithm and its continuous counterparts are well-suited for constraints such as cardinality, matroids, and knapsack, but they fail to handle other types of natural constraints. Thus there is an increasing need to develop algorithms for general constraints.

A major contribution in this direction comes from the work of Chekuri et al. [13] which has developed a powerful framework for submodular function maximization with general constraints. Their framework leverages the power of mathematical programming relaxations coupled with structured rounding schemes called *contention resolution (CR)* schemes. In particular, it unifies several previous results for special cases (e.g., matroids or knapsack constraints) and thus captures the types of constraints for which we know how to optimize submodular functions. This has led to the following very interesting meta-question: *For which type of constraints can we provide structured relaxations that admit good CR schemes?* In this paper, we address this question in the specific case of the unsplittable flow problem (UFP). In this setting, we are given an edge-capacitated, undirected graph and a collection of tasks; each task is specified by a source vertex, a sink vertex, and a demand. The goal is to select a maximum profit subset of the tasks that can be routed unsplittably, i.e., the task's demand is routed along a single path from the source to the sink subject to the edge capacities.

The problem is well-studied, and most of the results focus on linear objectives. Despite its apparent simplicity, already UFP on paths captures several well-studied problems, including the knapsack problem (when the graph is a single edge) and resource allocation problems. UFP is quite challenging even on paths and trees, and one of the main reasons for the difficulty is the lack of LP relaxations with small integrality gaps. The natural LP relaxation for the problem has an $\Omega(n)$ integrality gap even on paths [9], and standard approaches for strengthening the LP by adding valid inequalities fail to improve the integrality gap significantly [12]. Chekuri et al. [12] gave a novel LP relaxation for UFP on paths that strengthens the standard LP using clique type of constraints, and they showed that it has an $O(\log n)$ integrality gap. The relaxation of [12] can also be extended to trees, and understanding this relaxation has been an interesting and challenging open question[1].

The design of good relaxations for UFP is motivated not only by the goal of obtaining better approximations for linear objectives, but also by the need of handling more general constraints and objective functions. In particular, the current approaches for submodular objectives rely on structured relaxations with good CR schemes. As a result, there is a discrepancy between the approximation guarantees for linear and submodular objectives. There has been a long line of work for UFP on paths with a linear objective that led to a constant factor approximation [4,5]; these approaches combine the standard LP relaxation with dynamic programming techniques. Chekuri *et al.* [12] give a combinatorial greedy algorithm for UFP on trees with a linear objective that achieves an $O(\log^2 n)$ approximation. In contrast, for UFP with a submodular objective, only an $O(\log n)$ approximation is known for paths and *no non-trivial approximation was known for trees prior to our work*. Chekuri *et al.* [13] consider instances of submodular UFP on trees that satisfy a certain assumption, called the no-bottleneck assumption (NBA)[2], and they give a constant factor approximation for such instances. However, the no-bottleneck assumption is very restrictive and removing this restriction poses several technical challenges, particularly for the design of mathematical programming relaxations.

Thus, there has been an extensive work on UFP on paths but relatively fewer results on trees. Since UFP models the allocation of communication bandwidths in networks, we believe that it is worthwhile to develop a better understanding for more complex network topologies, such as trees. Also, submodular objective functions are much richer than linear objectives, and can model for instance linear objective functions with additional constraints.

Our Contributions. We give the first approximation algorithm for submodular UFP on trees and the first relaxation with a matching integrality gap. Our algorithm achieves an approximation ratio of $O(k \log n)$ on trees with pathwidth k. As each tree has pathwidth $O(\log n)$, this gives an $O(\log^2 n)$-approximation for arbitrary trees, matching the best known result for linear objective functions [12]. For several special cases of the problem, such as paths, spiders, and caterpillars, our approximation ratio improves to $O(\log n)$ (since in those cases $k = O(1)$), and such a ratio was not even known for linear objectives. Thus our result generalizes and improves the best approximations known for UFP on paths with a submodular objective and UFP on trees with a linear objective.

Theorem 1. *There is a $O(k \cdot \log n)$ approximation for Submodular UFP on trees, where k is the pathwidth of the tree and n is the number of nodes in the tree. Additionally, there is a polynomial-sized relaxation for the problem with a matching integrality gap.*

---

[1] Friggstad and Zao [15] showed an $O(\log^2 n)$ upper bound on the integrality gap of the relaxation of [12] for UFP on trees with a linear objective. This upper bound is shown via a primal-dual analysis which is not suitable for designing a CR scheme.

[2] The no-bottleneck assumption states that the maximum demand of any task is at most the minimum capacity of any edge.

We obtain our result via a new geometric LP relaxation for UFP on trees that is very different from the clique-based approach of [12]. Our relaxation builds on a powerful two-dimensional geometric viewpoint developed in the context of the UFP problem on paths with a linear objective [5]. This viewpoint connects UFP to structured instances of the Maximum Independent Set of Rectangles (MISR) problem [1,10,11], which in turn allows one to handle instances of UFP on paths for which the standard LP relaxation fails. The geometry was exploited to obtain a combinatorial algorithm for such instances that is based on dynamic programming. A related two-dimensional visualization was used in [2], again as the basis of a dynamic program. These approaches, however, break down for submodular UFP on trees; in the two-dimensional viewpoints, an input path corresponds to a subinterval of the x-axis and this is no longer meaningful for trees. Also, dynamic programming approaches are not suitable for submodular objective functions. In contrast to previous work, the focus in this paper is to translate these geometric insights to an LP relaxation for UFP on trees. We give a CR scheme for our relaxation that can be combined with the framework of [13] to obtain approximation guarantees for submodular objectives. The core of our reasoning is that our LP-formulation not only decides which tasks to select, but also computes a drawing of them as non-overlapping rectangles on suitable subpaths of the tree. We remark that our LP is a polynomial-sized extended formulation and, to the best of our knowledge, this is the first time that an extended formulation is used in the context of CR schemes.

A very important feature of the CR scheme framework is that it allows one to combine several constraints, thus extending the applicability of our approach to two generalized settings. First, in the Submodular Bag-UFP on trees problem, the input tasks are partitioned into bags and a feasible solution is allowed to select at most one task per bag [8][3]. We obtain an $O(k \log n)$ approximation for Submodular Bag-UFP on trees of pathwidth k. Second, we obtain an $O(\log n)$ approximation for the Submodular Storage Allocation Problem on trees. This problem has the same input as UFP, with additional requirements that each selected task gets a private subinterval of width equal to the demand, contained in $[0, u_e)$ for each edge e used by the task. We require that these subintervals are disjoint for any two tasks sharing an edge of the tree. Intuitively, this models that each task gets a contiguous portion of the resource spectrum.

Finally, we round up our contributions with the following results for a special case of UFP on trees with a linear objective function. An instance of UFP on tree is an *upward instance* if the input tree is rooted and, for every task, the source node of the task is an ancestor of the sink node (or vice-versa).

Theorem 2. *There is a $(1 + \epsilon)$ approximation algorithm for upward instances of UFP on trees with running time $n^{\text{poly}(\log(n/\epsilon)) \log(d_{\max}/d_{\min})}$. In particular, if the demands are quasi-polynomially bounded, this gives a QPTAS.*

---

[3] For linear objective functions, the bag constraints can be "glued" with the objective function, yielding an instance of Submodular UFP. It is not clear though whether this holds in general for any initial submodular objective function.

Unlike for UFP on paths [4], we show that the dependency of the running time on the term $\log(d_{\max}/d_{\min})$ can *not* be removed for upward instances of UFP on trees. In fact, assuming the *Exponential Time Hypothesis (ETH)*, the running time of our approximation scheme is essentially tight. This illustrates an inherent distinction between paths and upward instances on trees. Also, it shows that this is one of the very rare problems that allows a QPTAS on quasi-polynomially bounded input data but becomes APX-hard on general instances.

Theorem 3. *There is a universal constant ϵ_0 such that for all $\delta > 0$ any $(1+\epsilon_0)$-approximation algorithm for upward instances of UFP on trees runs in time of at least $n^{\mathrm{poly}(\log n)\log^{1-\delta}(d_{\max}/d_{\min})}$, unless ETH fails. Also, the problem is APX-hard.*

Other Related Work. The problem of maximizing submodular functions subject to various constraints is very well-studied and several results are known; we refer the reader to [13] for an overview. UFP with a linear objective is also extensively studied. Due to space limitation, we omit a detailed discussion of these results. The best approximation is a $(2+\varepsilon)$ approximation [2] and a QPTAS [3] for UFP on paths, and an $O(\log^2 n)$ approximation for UFP on trees [12].

Formal problem definitions. We consider the Unsplittable Flow problem on trees (UFP-tree). The input consists of an undirected tree $T = (V, E)$ with edge capacities $u_e \in \mathbb{Z}_+$, and a set of tasks $\mathcal{T}$. Each task $i \in \mathcal{T}$ is characterized by a start vertex $s_i \in V$, an end vertex $t_i \in V$, a demand $d_i \in \mathbb{Z}_+$, and a profit $w_i \in \mathbb{Z}_+$. For each task $i \in \mathcal{T}$ denote by p_i the unique path between s_i and t_i in T. A feasible solution is a subset of the tasks $\mathcal{T}' \subseteq \mathcal{T}$ satisfying the capacity constraints $\sum_{i \in \mathcal{T}' : p_i \ni e} d_i \le u_e$ for each edge $e \in E$. The goal is to find a feasible solution maximizing $w(\mathcal{T}') := \sum_{i \in \mathcal{T}'} w_i$.

The Submodular UFP-tree problem is a generalization of UFP-tree, where instead of a linear weight function w we are given a submodular objective function $f : 2^{\mathcal{T}} \to \mathbb{R}_+$ and the goal is to select a feasible subset $\mathcal{T}' \subseteq \mathcal{T}$ maximizing $f(\mathcal{T}')$. A function $f : 2^{\mathcal{T}} \to \mathbb{R}_+$ is *submodular* if $f(A) + f(B) \ge f(A \cap B) + f(A \cup B)$ for any two subsets $A, B \subseteq \mathcal{T}$. We assume that f is given as a value oracle, i.e., we are given access to an oracle that takes as input any set S and outputs $f(S)$.

In the Bag-UFP-tree problem, in addition to the input of UFP-tree, the input tasks are partitioned into sets called *bags* and we are allowed to select at most one task from each bag. We also consider the Storage Allocation Problem (SAP-tree). The input to SAP-tree is the same as for UFP-tree, with additional requirement that for each selected task i in $\mathcal{T}' \subseteq \mathcal{T}$ we have to compute a value $h(i) \ge 0$ such that $h(i)+d_i \le u_e$ for each edge $e \in p_i$, and $[h(i), h(i)+d_i) \cap [h(i'), h(i')+d_{i'}) = \emptyset$ for any two tasks $i, i' \in \mathcal{T}'$ with $p_i \cap p_{i'} \ne \emptyset$. This corresponds to giving each task $i \in \mathcal{T}'$ the portion $[h(i), h(i) + d_i)$ of the resource spectrum.

2 Geometric Relaxation for Submodular UFP on Trees

In this section, we present our $O(k \cdot \log n)$ approximation algorithm for Sub-modular UFP-tree. We first describe a pseudo-polynomial sized LP relaxation for UFP-tree with a linear objective function. In Sect. 2.2, we show how to reduce the size of the LP to polynomial. In Sect. 2.3, we extend our algorithm to a submodular objective function. We defer the description of our results for the Submodular Bag-UFP and SAP problems to the full version of this paper.

2.1 A Pseudo-Polynomial Sized Relaxation

In the following, we give a geometric LP-relaxation for UFP-tree with a linear objective function. The relaxation has pseudo-polynomial size.

Reduction to Intersecting Instances. First, we reduce the general case to the case in which the path of each task contains the root of the tree. We call such instances *intersecting instances*. Chekuri *et al.* [12] showed that, via a standard centroid decomposition, we can reduce an arbitrary instance to a collection of intersecting instances at a loss of $O(\log n)$ in the approximation ratio.

Lemma 1 (Chekuri *et al.* [12]). *Suppose that there is a polynomial time algorithm for* UFP-tree *that achieves an α-approximation on intersecting instances. Then there is a polynomial time $O(\alpha \cdot \log n)$ approximation algorithm for the problem on arbitrary trees. Moreover, this holds for the generalization of the problem in which the objective function is sub-additive[4].*

Partitioning into Paths. In the remainder of this section, we assume that we are given an intersecting instance on a tree T of pathwidth k. Intuitively, a graph has pathwidth k if it has a tree-decomposition of width k in which the tree describing the decomposition is a path (see e.g. [18] for a formal definition). Note that every tree has pathwidth at most $O(\log n)$ [18]. Our goal is to compute a $O(k)$-approximation for such instances, so that using Lemma 1 we obtain a $O(k \log n)$-approximation for the general problem. First, we split a given tree into a collection $\mathcal{P}$ of paths such that each input task shares an edge with at most $O(k)$ paths in $\mathcal{P}$. Then, we define a new LP relaxation for the problem with a randomized rounding with alteration strategy. The relaxation will be based on a two-dimensional geometric viewpoint for each path in $\mathcal{P}$.

For our path partition $\mathcal{P}$ we require that each path $P \in \mathcal{P}$ is an *upward path*, i.e., one endpoint of the path is an ancestor in T of the other endpoint. The following observation follows from the property of an intersecting instance.

Observation 1. *For each task i and each upward path P, if i uses an edge of P then it uses the top edge of P.*

---

[4] A set function $f : 2^V \to \mathbb{R}$ is sub-additive if $f(A \cup B) \le f(A) + f(B)$ for any two disjoint sets A and B. Note that a non-negative submodular function is sub-additive.

Definition 1. *Consider an intersecting instance of* UFP-tree *on a rooted tree T. Let* $\mathcal{P} = \{P_1, \ldots, P_\ell\}$ *be a collection of paths in T. We say that* $\mathcal{P}$ *is a K-nice splitting if it has the following properties:*

- *The paths in* $\mathcal{P}$ *are edge-disjoint, upward paths, partitioning the edges of T.*
- *Each task uses an edge of at most K paths in* $\mathcal{P}$.

The next lemma shows the existence of a $O(k)$-nice splitting where k is the pathwidth of T.

Lemma 2. *Consider an intersecting instance I of* UFP-tree *on a rooted tree T of pathwidth k. There is a polynomial time algorithm that constructs an* $O(k)$- *nice splitting for I.*

Geometric Viewpoint. Let $\mathcal{P} = \{P_1, \ldots, P_\ell\}$ be an $O(k)$-nice splitting of the instance that is guaranteed by Lemma 2. We use $\mathcal{P}$ to write an LP relaxation for the problem, based on the following geometric viewpoint. For a path $P \in \mathcal{P}$, let $\mathcal{T}_P$ be the set of tasks from T using an edge of P.

If we restrict the tasks in our instance to a path P in T, we get an instance of the Unsplittable Flow problem on paths (UFP-path) in a natural way. For each task $i \in \mathcal{T}_P$, the UFP-path instance has a corresponding task whose path is $p_i \cap P$. Notice that each task $i \in \mathcal{T}_P$ uses the top edge of P, so we can assume w.l.o.g. that when traversing the edges of P from top to bottom, their capacities are non-increasing. We call such an instance a *one-sided staircase* instance.

We claim that for such an instance of UFP-path on a path P, each feasible subset of the tasks can be represented as a collection of non-overlapping rectangles drawn underneath the capacity profile, such that each task i has a corresponding rectangle of height d_i whose projection on P is the path of i. We interpret these rectangles as open sets. We call such a drawing a *representing drawing*.

Lemma 3. *Consider an instance of* UFP-path *on a path P in which all of the tasks use the first edge of P. Any feasible subset of the tasks admits a representing drawing.*

LP Relaxation. Using this geometric viewpoint, we write an LP relaxation for intersecting instances of UFP-tree as follows. Recall that we have an $O(k)$-nice splitting $\mathcal{P}$ of the tree T. We add constraints to the relaxation to enforce that there is a representing drawing for the selected tasks on each path $P \in \mathcal{P}$; we remark that these constraints will automatically enforce the capacity constraints.

Variables. The IP has the following variables. For each task i, we have a variable $x_i \in \{0, 1\}$ with the interpretation that $x_i = 1$ if task i is in the solution. For each path $P \in \mathcal{P}$, each task $i \in \mathcal{T}_P$, and each height h, we have a variable $y(i, h, P) \in \{0, 1\}$ with the interpretation that $y(i, h, P) = 1$ if the rectangle for task i is drawn at height h in the representing drawing for P. The allowed heights h are the ones satisfying $h + d_i \leq u_e$ for each edge $e \in p_i \cap P$, i.e., such that the rectangle fits under the capacity profile. We introduce variables $y(i, h, P)$ only for such heights.

Constraints. For each path $P \in \mathcal{P}$ and each task $i \in \mathcal{T}_P$, we have a constraint

$$\sum_{hs.t. \forall e \in p_i \cap P:\ h+d_i \leq u_e} y(i, h, P) = x_i \ . \tag{1}$$

For each path $P \in \mathcal{P}$, we add constraints enforcing that in the representing drawing for P the rectangles do not overlap. This is achieved by imposing constraints modeling that any point q underneath the capacity profile is covered by at most one rectangle. Since all tasks use the first edge of P, it suffices to consider only points q on a vertical line going through the first edge of P, i.e., points $q = (x_0, h)$ where x_0 is an arbitrary x-coordinate strictly between the first and the second vertex of P and h is an integral height that is at most the capacity of the first edge of P. We use $R(i, h, P)$ to denote a rectangle representing task i on P drawn at height h, i.e., $R(i, h, P)$ is a rectangle of height d_i, with a bottom y-coordinate h, and whose projection on the x-axis equal $p_i \cap P$.

For each path $P \in \mathcal{P}$ and each point $q = (x_0, h)$ as described above we have a constraint

$$\sum_{i \in \mathcal{T}_P} \sum_{(h':\ q \in R(i,h',P))} y(i, h', P) \leq 1. \tag{2}$$

We refer to the resulting LP relaxation as Rectangle-LP($\mathcal{P}$). It clearly has pseudo-polynomial complexity. In the following, we show an $O(k)$-approximation based on LP rounding.

Rounding. Let (x, y) be a feasible solution to Rectangle-LP($\mathcal{P}$). We use a randomized rounding with alteration strategy (as introduced in [7] to select a subset of the tasks and a representing drawing for them on each path $P \in \mathcal{P}$. We proceed in two phases. In the selection phase, we pick a subset of the tasks and determine a drawing for them. The drawing in this phase may contain overlapping rectangles. In the alteration phase, we pick a subset of the selected tasks whose corresponding rectangles do not overlap.

Selection phase. We select a (not necessarily feasible) set S of tasks. For each task i, we add i to S independently at random with probability $x_i/(c_1 \cdot k)$, where $c_1 > 1$ is a sufficiently large constant that will be determined later. We refer to the tasks in the random sample S as the *selected* tasks. Additionally, for each task $i \in S$ and each path $P \in \mathcal{P}$ such that $i \in \mathcal{T}_P$, we choose a rectangle representing the drawing of i on P, as follows. We choose a height h for the rectangle independently at random according to the probability distribution $\{y(i, h, P)/x_i\}_h$. Note that the constraints (1) ensure that the values $y(i, h, P)/x_i$ form a probability distribution over the allowed heights h.

Let $h(i, P)$ be the height chosen for task i on the path P; we use the rectangle $R(i, h(i, P), P)$ to represent task i on the path P. Let $\mathcal{R}$ denote the resulting drawing, i.e., $\mathcal{R}$ is the collection of rectangles selected for the tasks in S. Note that each rectangle $R(i, h, P)$ is in $\mathcal{R}$ with probability $x_i \cdot \frac{y(i,h,P)}{x_i} = y(i, h, P)$.

Alteration phase. In the alteration phase, we select a subset $S' \subseteq S$ of the tasks such that the rectangles $\mathcal{R}' \subseteq \mathcal{R}$ representing them on the paths are non-overlapping. Recall that we view the rectangles as open sets and thus two rectangles overlap iff they contain a common point in their interiors. We consider the paths of $\mathcal{P}$ in an arbitrary order. For each $P \in \mathcal{P}$, let $S(P) = \{i \in S : i \in \mathcal{T}_P\}$. Our goal is to choose a subset $S'(P) \subseteq S(P)$ such that the rectangles $\{R(i, h(i, P), P) : i \in S'(P)\}$ are non-overlapping. We choose the set of *accepted tasks* $S'(P)$ as follows.

We order the tasks in $S(P)$ in *non-increasing* order according to their demands, breaking ties arbitrarily. We consider the tasks in this order. Let i be the current task. We add i to $S'(P)$ if the rectangle $R(i, h(i, P), P)$ does not overlap with any of the rectangles $\{R(i', h(i', P), P) : i' \in S'(P)\}$ for the tasks we have accepted so far.

We refer to the tasks in $S'(P)$ as the tasks *accepted* on P, and we refer to the tasks in $S(P) - S'(P)$ as the tasks *rejected* on P. The following key lemma shows that each selected task $i \in S(P)$ is accepted with a constant probability. The main observation behind the lemma is that, for each task j that appears before i in the ordering, if the rectangles $R(i, h(i, P), P)$ and $R(j, h(j, P), P)$ overlap, then $R(j, h(j, P), P)$ contains the top left or the bottom left corner of $R(i, h(i, P), P)$ since $d_j \geq d_i$; this allows us to check the constraints only at two points.

Lemma 4. *For any path P and task $i \in \mathcal{T}_P$, $\mathbf{Pr}[i \notin S'(P) \mid i \in S(P)] \leq 2/(c_1 \cdot k)$.*

Finally, we use the sets $\{S'(P) : P \in \mathcal{P}\}$ to select a subset $S' \subseteq S$ such that the rectangles $\mathcal{R}' \subseteq \mathcal{R}$ representing S' on each path of $\mathcal{P}$ are non-overlapping. We set $S' = \{i \in S : \forall_{P \in \mathcal{P} : i \in \mathcal{T}_P} \, i \in S'(P)\}$, i.e., a task is accepted if it was accepted for all paths. It follows from Lemma 4 and the union bound that each selected task is rejected with probability at most $|\{P \in \mathcal{P} : i \in \mathcal{T}_P\}| \cdot \frac{2}{c_1 k} \leq 1/2$ if c_1 is sufficiently large[5].

We summarize the rounding step in the following lemma.

Lemma 5. *Consider an instance of UFP-tree. Suppose that the instance has a K-nice splitting $\mathcal{P}$ and let (x, y) be a feasible solution to Rectangle-LP$(\mathcal{P})$. Let S be a random sample of the tasks such that each task i is in S independently at random with probability $x_i/(4K)$. There is a polynomial-time algorithm that constructs a feasible solution $S' \subseteq S$ such that, for each task i, $\mathbf{Pr}[i \in S' \mid i \in S] \geq 1/2$.*

For linear objective functions, this yields a pseudo-polynomial LP-based $O(k)$-approximation for intersecting instances of UFP-tree and, with Lemma 1, a $O(k \log n)$-approximation for arbitrary instances of UFP-tree.

---

[5] More precisely, if $\mathcal{P}$ is ck-nice, then this happens when $c_1 \geq 4c$.

2.2   A Polynomial-Sized Relaxation

In this section, we show how to turn a pseudo-polynomial sized LP in the previous section to a polynomial sized one. Notice that the pseudo-polynomial running time is caused by the fact that the rectangles for the tasks in T can be drawn at pseudo-polynomially many heights. We show that restricting to a polynomial sized set of heights incurs only an $O(1)$ factor loss in the approximation ratio.

Task Classification. For a path $P \in \mathcal{P}$ and a task $i \in T_P$, let $b_P(i) := \min_{e \in p_i \cap P} u_e$ be the *bottleneck capacity of i on P*. We say that a task $i \in T_P$ is *big on P* if $d_i > \frac{1}{16} \cdot b_P(i)$. Otherwise we say that i is *small on P*.

Allowed Hights. For each path $P \in \mathcal{P}$ and task $i \in T_P$, we will now construct a set $\mathcal{H}(i, P)$ of *allowed heights* for drawing the rectangle corresponding to i on P. If i is big on P, we set $\mathcal{H}(i, P) = \{b_P(i) - d_i\}$, i.e., the only allowed height is obtained by drawing the rectangle for i as high as possible underneath the capacity profile. If i is small on P, for the integer j such that $b_P(i) \in [2^j, 2^{j+1})$, we set $\mathcal{H}(i, P) = \bigcup_{r \in \mathbb{N}_0 : \, r \lceil 2^{j-3}/n \rceil \le 2^{j-1}} \{2^{j-1} + r \lceil 2^{j-3}/n \rceil\}$. We have $|\mathcal{H}(i, P)| \le 8n$. Let $\mathcal{H}$ be the union of all sets $\mathcal{H}(i, P)$. By construction, $\mathcal{H}$ has polynomial size.

Restricted LP. Denote by Restricted-Rectangle-LP$(\mathcal{P}, \mathcal{H})$ the LP relaxation where we introduce variables $y(i, h, P)$ and the constraints (1) and (2) only for the heights $h \in \mathcal{H}$. As $\mathcal{H}$ has polynomial size, the size of Restricted-Rectangle-LP$(\mathcal{P}, \mathcal{H})$ is also polynomial. The following lemma argues that the LP restricted to these heights still admit a good fractional solution. Combining it with Lemma 5 yields the desired polynomial time approximation algorithm for linear UFP-tree.

Lemma 6. *For each feasible integral solution $T' \subseteq T$, there is a feasible fractional solution (x, y) for* Restricted-Rectangle-LP *$(\mathcal{P}, \mathcal{H})$ s.t. $(\forall i \in T') x_i = \frac{1}{64}$.*

2.3   Submodular Objective via the CR Scheme Framework

In this section, we extend our results to submodular objectives by combining the results from the previous section with the framework from [13].

Let N be a finite ground set. Let $\mathcal{I} \subseteq 2^N$ be a family of subsets of N, and $\mathbf{P}_\mathcal{I}$ a convex relaxation for the constraints imposed by $\mathcal{I}$, such that $\mathbf{P}_\mathcal{I}$ is down-monotone and solvable.[6] Let $x \in \mathbf{P}_\mathcal{I}$ and let support$(x) = \{i \in N : x_i > 0\}$. For any $b \in [0, 1]$, let $b \cdot \mathbf{P}_\mathcal{I} = \{bx : x \in \mathbf{P}_\mathcal{I}\}$. Let $\mathbf{R}(x)$ be a random sample of N such that each element $i \in N$ is in $\mathbf{R}(x)$ independently at random with probability x_i. For a set function $f : 2^N \to \mathbb{R}_+$ let $F : [0, 1]^N \to \mathbb{R}_+$ denote the *multilinear extension* of f, which is defined as $F(x) := \mathbb{E}[f(\mathbf{R}(x))]$.

Definition 2 ([13]). *For $b, c \in [0, 1]$, a (b, c)-balanced CR scheme π for a polytope $\mathbf{P}_\mathcal{I}$ is a procedure that for every $x \in b \cdot \mathbf{P}_\mathcal{I}$ and $A \subseteq N$ returns a random set $\pi_x(A)$ satisfying*

---

[6] We call a polytope $\mathbf{P} \subseteq [0, 1]^N$ *down-monotone* if for all $\mathbf{z}, \mathbf{z}' \in [0, 1]^N$ we have that $\mathbf{z} \le \mathbf{z}'$ and $\mathbf{z}' \in \mathbf{P}$ implies that $\mathbf{z} \in \mathbf{P}$. The polytope is *solvable* if one can optimize any linear function over $\mathbf{P}$ in polynomial time.

(i) $\pi_x(A) \subseteq \text{support}(x) \cap A$ *and* $\pi_x(A) \in \mathcal{I}$ *with probability* 1, *and*
(ii) for all $i \in \text{support}(x)$, $\mathbf{Pr}[i \in \pi_x(\mathbf{R}(x)) \mid i \in \mathbf{R}(x)] \geq c$.

We use the CR schemes as in [13]: first, we compute a vector x^* with $F(x^*) \geq \Omega(\max\{F(x'): x' \in \mathbf{P}_\mathcal{I}\})$. Then, we compute a random sample $\mathbf{R}(x)$ with $x := b \cdot x^*$. We apply the CR scheme π and obtain the set $\pi_x(\mathbf{R}(x))$. We know that for each element i we have that $\mathbf{Pr}[i \in \mathbf{R}(x)] = b \cdot x_i^*$ and $\mathbf{Pr}[i \in \pi_x(\mathbf{R}(x)) \mid i \in \mathbf{R}(x)] \geq c$. Thus, $\mathbf{Pr}[i \in \pi_x(\mathbf{R}(x))] \geq bc \cdot x_i^*$ which can be used to show that $\mathbb{E}[f(\pi_x(\mathbf{R}(x)))] \geq \Theta(bc) \cdot \max\{F(x'): x' \in \mathbf{P}_\mathcal{I}\}$.

Theorem 4 ([13]). *Let* $f : 2^N \to \mathbb{R}_+$ *be a submodular function. Let* $\mathcal{I} \subseteq 2^N$ *be a family of feasible solutions and let* $\mathbf{P}_\mathcal{I} \subseteq [0,1]^N$ *be a convex relaxation for* $\mathcal{I}$ *that is down-monotone and solvable. Suppose that there is a* (b,c)-*balanced CR scheme for* $\mathbf{P}_\mathcal{I}$. *Then there is a polynomial time randomized algorithm that constructs a solution* $I \in \mathcal{I}$ *such that*

$$\mathbb{E}[f(I)] \geq \Theta(bc) \cdot \max\{F(x): x \in \mathbf{P}_\mathcal{I}\}.$$

To apply the above framework, let $\mathbf{P}$ denote the set of points x for which there exists a vector y such that (x, y) is contained in the polytope defined by Restricted-Rectangle-LP($\mathcal{P}, \mathcal{H}$). Clearly, $\mathbf{P}$ is down-monotone and solvable. Similarly as in the case of linear objective functions, $\mathbf{P}$ contains a fractional point with large profit according to F: Let $\mathcal{T}^*$ be an optimal integral solution. By Lemma 6, $\frac{1}{64} \cdot \mathbf{1}_{\mathcal{T}^*} \in \mathbf{P}$. Moreover, it is straightforward to verify that $F\left(\frac{1}{64} \cdot \mathbf{1}_{\mathcal{T}^*}\right) \geq \frac{1}{64} f(\mathcal{T}^*)$. So $\max\{F(x) : x \in \mathbf{P}_\mathcal{I}\} = \Omega(\text{OPT})$.

By Lemma 5, there is a $(1/\Theta(k), 1/2)$-balanced CR scheme for $\mathbf{P}$. Therefore we can apply Theorem 4 to obtain our main result for Submodular UFP-tree.

Theorem 5. *There is a polynomial time* $O(k)$ *approximation algorithm for* Submodular UFP-tree *on intersecting instances and, therefore, an* $O(k \log n)$ *approximation for arbitrary instances, where* k *is the pathwidth of the tree.*

References

1. Adamaszek, A., Wiese, A.: Approximation schemes for maximum weight independent set of rectangles. In: 54th Annual IEEE Symposium on Foundations of Computer Science, FOCS 2013, pp. 400–409. IEEE Computer Society (2013)
2. Anagnostopoulos, A., Grandoni, F., Leonardi, S., Wiese, A.: A mazing 2+ε approximation for unsplittable flow on a path. In: Chekuri, C. (ed.) Proceedings of the Twenty-Fifth Annual ACM-SIAM Symposium on Discrete Algorithms, SODA 2014, pp. 26–41. SIAM (2014)
3. Bansal, N., Chakrabarti, A., Epstein, A., Schieber, B.: A quasi-PTAS for unsplittable flow on line graphs. In: Kleinberg, J.M. (ed.) Proceedings of the 38th Annual ACM Symposium on Theory of Computing, pp. 721–729. ACM (2006)
4. Batra, J., Garg, N., Kumar, A., Mömke, T., Wiese, A.: New approximation schemes for unsplittable flow on a path. In: Indyk, P. (ed.) Proceedings of the Twenty-Sixth Annual ACM-SIAM Symposium on Discrete Algorithms, SODA 2015, pp. 47–58. SIAM (2015)

5. Bonsma, P.S., Schulz, J., Wiese, A.: A constant-factor approximation algorithm for unsplittable flow on paths. SIAM J. Comput. **43**(2), 767–799 (2014)
6. Boykov, Y., Jolly, M.: Interactive graph cuts for optimal boundary and region segmentation of objects in N-D images. In: ICCV, pp. 105–112 (2001)
7. Călinescu, G., Chakrabarti, A., Karloff, H.J., Rabani, Y.: An improved approximation algorithm for resource allocation. ACM Trans. Algorithms **7**(4), 48 (2011)
8. Chakaravarthy, V.T., Choudhury, A.R., Gupta, S., Roy, S., Sabharwal, Y.: Improved algorithms for resource allocation under varying capacity. In: Schulz, A.S., Wagner, D. (eds.) ESA 2014. LNCS, vol. 8737, pp. 222–234. Springer, Heidelberg (2014)
9. Chakrabarti, A., Chekuri, C., Gupta, A., Kumar, A.: Approximation algorithms for the unsplittable flow problem. Algorithmica **47**(1), 53–78 (2007)
10. Chalermsook, P., Chuzhoy, J.: Maximum independent set of rectangles. In: Mathieu, C. (ed.) Proceedings of the Twentieth Annual ACM-SIAM Symposium on Discrete Algorithms, SODA 2009, pp. 892–901. SIAM (2009)
11. Chan, T.M., Har-Peled, S.: Approximation algorithms for maximum independent set of pseudo-disks. Discrete Comput. Geom. **48**(2), 373–392 (2012)
12. Chekuri, C., Ene, A., Korula, N.: Unsplittable flow in paths and trees and column-restricted packing integer programs. In: Dinur, I., Jansen, K., Naor, J., Rolim, J. (eds.) Approximation, Randomization, and Combinatorial Optimization. LNCS, vol. 5687, pp. 42–55. Springer, Heidelberg (2009)
13. Chekuri, C., Vondrák, J., Zenklusen, R.: Submodular function maximization via the multilinear relaxation and contention resolution schemes. SIAM J. Comput. **43**(6), 1831–1879 (2014)
14. Dueck, D., Frey, B.J.: Non-metric affinity propagation for unsupervised image categorization. In: IEEE 11th International Conference on Computer Vision, ICCV 2007, pp. 1–8. IEEE (2007)
15. Friggstad, Z., Gao, Z.: On linear programming relaxations for unsplittable flow in trees. In: Garg, N., Jansen, K., Rao, A., Rolim, J.D.P. (eds.) Approximation, Randomization, and Combinatorial Optimization. Algorithms and Techniques, APPROX/RANDOM 2015. LIPIcs, vol. 40, pp. 265–283. Schloss Dagstuhl - Leibniz-Zentrum fuer Informatik (2015)
16. Jegelka, S., Bilmes, J.A.: Submodularity beyond submodular energies: coupling edges in graph cuts. In: The 24th IEEE Conference on Computer Vision and Pattern Recognition, CVPR 2011, pp. 1897–1904. IEEE Computer Society (2011)
17. Kempe, D., Kleinberg, J.M., Tardos, É.: Maximizing the spread of influence through a social network. Theor. Comput. **11**, 105–147 (2015)
18. Korach, E., Solel, N.: Tree-width, path-width, and cutwidth. Discrete Appl. Math. **43**(1), 97–101 (1993)
19. Krause, A., Guestrin, C.: Near-optimal observation selection using submodular functions. In: Proceedings of the Twenty-Second AAAI Conference on Artificial Intelligence, pp. 1650–1654. AAAI Press (2007)
20. Krause, A., Guestrin, C.: Submodularity and its applications in optimized information gathering. ACM TIST **2**(4), 32 (2011)
21. Krause, A., Singh, A.P., Guestrin, C.: Near-optimal sensor placements in gaussian processes: theory, efficient algorithms and empirical studies. J. Mach. Learn. Res. **9**, 235–284 (2008)

22. Lin, H., Bilmes, J.A.: A class of submodular functions for document summarization. In: Lin, D., Matsumoto, Y., Mihalcea, R. (eds.) Proceedings of the Conference of 49th Annual Meeting of the Association for Computational Linguistics: Human Language Technologies, pp. 510–520. The Association for Computer Linguistics (2011)
23. Nemhauser, G.L., Wolsey, L.A., Fisher, M.L.: An analysis of approximations for maximizing submodular set functions-I. Math. Program. **14**(1), 265–294 (1978)

Strong Reductions for Extended Formulations

Gábor Braun[1]([⊠]), Sebastian Pokutta[1], and Aurko Roy[2]

[1] ISyE, Georgia Institute of Technology, Atlanta, GA, USA
{gabor.braun,sebastian.pokutta}@isye.gatech.edu
[2] College of Computing, Georgia Institute of Technology, Atlanta, GA, USA
aurko@gatech.edu

Abstract. We generalize the reduction mechanism between linear programming problems from [1] in two ways (1) relaxing the requirement of affineness, and (2) extending to fractional optimization problems.

As applications we provide several new LP-hardness and SDP-hardness results, e.g., for the SparsestCut problem, the BalancedSeparator problem, the MaxCut problem and the Matching problem on 3-regular graphs. We also provide a new, very strong Lasserre integrality gap for the IndependentSet problem, which is strictly greater than the best known LP approximation, showing that the Lasserre hierarchy does not always provide the tightest SDP relaxation.

1 Introduction

Linear and semidefinite programs are the main components in the design of many practical (approximation) algorithms and therefore understanding their expressive power is a fundamental problem. The complexity of these programs is measured by the number of constraints, ignoring all other aspects affecting the running time of an actual algorithm, in particular, these measures are independent of the P vs. NP question. We call a problem *LP-hard* if it does not admit an LP formulation with a polynomial number of constraints, and we define *SDP-hardness* similarly.

Recently, motivated by Yannakakis's influential work [2], a plethora of strong lower bounds have been established for many important optimization problems, such as e.g., the Matching problem [3] or the TravelingSalesman problem [4,5]. In [1], the authors introduced a reduction mechanism providing inapproximability results for large classes of problems. However, the reductions were required to be affine, and hence failed for e.g., the VertexCover problem, where intermediate Sherali–Adams reductions were employed in [6] due to this shortcoming.

In this work we extend the reduction mechanism of [1] in two ways, establishing several new hardness results both in the LP and SDP setting; both are special cases arising from reinterpreting LPs and SDPs as proof systems (which we explore in detail in the full-length version). First, by including additional 'computation' in the reduction, we allow non-affine relations between problems, eliminating the need of Sherali–Adams reduction in [6]. Second, we extend the framework to fractional optimization problems (such as e.g., SparsestCut) where

© Springer International Publishing Switzerland 2016
Q. Louveaux and M. Skutella (Eds.): IPCO 2016, LNCS 9682, pp. 350–361, 2016.
DOI: 10.1007/978-3-319-33461-5_29

ratios of linear functions have to be optimized. Here typically one optimizes the numerator and denominator at the same time.

Related Work

The immediate precursor to this work is [1] (generalizing [7,8]), introducing a reduction mechanism. Base hard problems are the Matching problem [3], as well as constraint satisfaction problems [5,9] based on hierarchy hardness results, such as e.g., [10,11].

Contribution

Generalized LP/SDP Reductions. We generalize the reduction mechanism in [1] by modeling additional computation, i.e., using extra LP or SDP constraints. Put differently, we allow for more complicated reduction maps as long as they have a small LP/SDP formulation. As a consequence, we can relax the affineness requirement and enable a weak form of *gap-amplification* and *boosting*. This overcomes a major limitation of the approach in [1], yielding significantly stronger reductions at a small cost.

Fractional LP/SDP Optimization. Second, we present a *fractional* LP/SDP framework and reduction mechanism, where the objective functions are ratios of functions from a low dimensional space, such as for the SparsestCut problem. For these problems the standard LP/SDP framework is meaningless as the ratios span a high dimensional affine space. The fractional framework models the usual way of solving fractional optimization problems, with strong statements about LP or SDP complexity.

Direct Non-linear Hardness Reductions. We demonstrate the power of our generalized reduction by establishing new LP-hardness and SDP-hardness for several problems of interest, i.e., these problems cannot be solved by LPs/SDPs of polynomial size; see Table 1. We establish various hardness results for the SparsestCut and BalancedSeparator problems even when one of the underlying graph has bounded treewidth. We also show the first explicit SDP-hardness for the Max-Cut problem, inapproximability within a factor of $15/16 + \varepsilon$, which is better than the algorithmic hardness of $16/17 + \varepsilon$. Finally, we prove a new, strong Lasserre integrality gap of $n^{1-\gamma}$ after $O(n^\gamma)$ rounds for the IndependentSet problem for any sufficiently small $\gamma > 0$. It not only significantly strengthens and complements the best-known integrality gap results so far [12–15], but also shows the suboptimality of Lasserre relaxations for the IndependentSet problem together with [6]. Our reduction mechanism also allows for direct reductions to intermediate CSP problems as used for optimal inapproximability results for the VertexCover problem over simple graphs and Q-regular hypergraphs in [6], eliminating Sherali–Adams reductions and we obtain a natural reduction to establish LP-hardness of matching over 3-regular graphs; we provide details in the full-length version.

Table 1. Inapproximability of optimization problems. tw denotes treewidth.

Problem	Factor	Source	Paradigm	Note
MaxCut	$\frac{15}{16} + \varepsilon$	Max-3-XOR/0	SDP	
SparsestCut(n), tw(supply) $= O(1)$	$2 - \varepsilon$	MaxCut	LP	opt. [9]
SparsestCut(n), tw(supply) $= O(1)$	$\frac{16}{15} - \varepsilon$	MaxCut	SDP	
BalancedSeparator(n, d), tw(demand) $= O(1)$	$\omega(1)$	UniqueGames	LP	
IndependentSet	$\omega(n^{1-\varepsilon})$	Max-k-CSP	Lasserre $O(n^\varepsilon)$ rounds	
Matching, 3-regular	$1 + \varepsilon/n^2$	Matching	LP	
1F-CSP $\bigg\}$ Q-$\neq$-CSP	$\omega(1)$	UniqueGames	LP	[6] w/o SA

Finally, inspired by our reductions we also obtain a new technique to derive small *uniform* linear programs for problems over graphs of bounded treewidth, where *the same* linear program is used for all instances of a problem (independent of the actual tree decomposition), complementing the nonuniform formulation in [16]; details are to be found in the full-length version.

Outline

We start by recalling and refining the linear programming framework in Sect. 2 to include fractional optimization problems. We provide the reduction mechanism in Sect. 3, with applications in Sects. 4, 5 and 6 to various problems. For missing proofs and additional details we refer the reader to the full-length version of the paper.

2   Preliminaries

2.1   Optimization Problems

Definition 1 (Optimization problem). *An* optimization problem *is a tuple* $\mathcal{P} = (\mathcal{S}, \mathfrak{I}, \mathrm{val})$ *consisting of a set* $\mathcal{S}$ *of feasible solutions, a set* $\mathfrak{I}$ *of instances, and a real-valued objective called measure* $\mathrm{val}\colon \mathfrak{I} \times \mathcal{S} \to \mathbb{R}$.

We shall write $\mathrm{val}_{\mathcal{I}}(s)$ *for the objective value of a feasible solution* $s \in \mathcal{S}$ *for an instance* $\mathcal{I} \in \mathfrak{I}$.

The SparsestCut problem is defined over a graph with two kinds of edges: *supply* and *demand* edges. The objective is to find a cut that minimizes the ratio of the capacity of cut supply edges to the total demand separated. For a weight function $f\colon E(K_n) \to \mathbb{R}_{\geq 0}$, we define the graph $[n]_f := ([n], E_f)$ where $E_f := \{(i,j) \mid i,j \in [n], f(i,j) > 0\}$. We study the SparsestCut problem with bounded-treewidth supply graph. Let $\mathrm{tw}(G)$ denote the treewidth of the graph G.

Definition 2 (SparsestCut(n,k)). *Let n be a positive integer. The minimization problem* SparsestCut(n,k) *consists of*

instances *nonnegative* demand $d\colon E(K_n) \to \mathbb{R}_{\geq 0}$ *and* capacity $c\colon E(K_n) \to \mathbb{R}_{\geq 0}$ *such that* $\operatorname{tw}([n]_c) \leq k$;
feasible solutions *all subsets s of $[n]$;*
measure *ratio of separated capacity and separated demand:*

$$\operatorname{val}_{\mathcal{I}}(s) = \frac{\operatorname{val}_{\mathcal{I}}^n(s)}{\operatorname{val}_{\mathcal{I}}^d(s)}$$

where $\operatorname{val}_{\mathcal{I}}^n(s) := \sum_{i \in s, j \notin s} c(i,j)$ *and* $\operatorname{val}_{\mathcal{I}}^d(s) := \sum_{i \in s, j \notin s} d(i,j)$ *for capacity c, demand d and set s.*

Recall that an *independent set* I of a graph G is a subset of pairwise non-adjacent vertices $I \subseteq V(G)$. The IndependentSet problem on a graph G asks for an independent set of G of maximum size. We formally define it as an optimization problem below.

Definition 3 (IndependentSet(G)). *Given a graph G, the maximization problem* IndependentSet(G) *consists of*

instances *all induced subgraphs H of G;*
feasible solutions *all independent subsets I of G;*
measure $\operatorname{val}_H(I) = |I \cap V(H)|$.

The MaxCut problem asks for a vertex set in a graph cutting a maximum number of edges. Here we formulate two versions of the problem differing in the largeness of the class of instances. Given a vertex set $X \subseteq V(G)$, let $\delta(X) := \{\{u,v\} \in E(G) \mid u \in X, v \notin X\}$ denote the set of edges of G with one end point in X and the other end point outside X.

Definition 4 (MaxCut(G)). *Given a graph G, the (non-uniform) maximization problem* MaxCut(G) *consists of*

instances *all induced subgraph H of G;*
feasible solutions *all vertex subsets $X \subseteq V(G)$;*
measure $\operatorname{val}_H(X) = |E(H) \cap \delta(X)|$.

For a positive integer n, we denote by MaxCut(n) *the (uniform) version of the problem where the instances are all subgraphs of the complete graph K_n (and the feasible solutions are all vertex subsets of $V(K_n)$).*

2.2   LP/SDP Complexity and Fractional Optimization

A *fractional optimization problem* is an optimization problem where the objectives have the form of a fraction $\operatorname{val}_{\mathcal{I}} = \operatorname{val}_{\mathcal{I}}^n / \operatorname{val}_{\mathcal{I}}^d$, such as for SparsestCut. In this case the affine space of the objective functions $\operatorname{val}_{\mathcal{I}}$ of instances is typically not low dimensional, immediately ruling out small linear and semidefinite formulations. Nevertheless, there are examples of efficient linear programming based algorithms for such problems, however here the linear programs are used to find an optimal value of a linear combination of $\text{val}_{\mathcal{I}}^n$ and $\text{val}_{\mathcal{I}}^d$ (see e.g., [17]). To be able to analyze the size of LPs or SDPs for such problems we refine the notion of formulation complexity from [1] to incorporate these types of linear programs, which reduces to the original definition with the choice of $\text{val}_{\mathcal{I}}^n = \text{val}_{\mathcal{I}}$ and $\text{val}_{\mathcal{I}}^d = 1$.

We first remind the reader of the notions of linear programming and semidefinite programming complexity of non-fractional optimization problems from [1]; precise definitions are to be found in the full-length version. Given a maximization problem $\mathcal{P} = (\mathcal{S}, \mathfrak{I}, \text{val})$, let C, S be real-valued functions on $\mathfrak{I}$, called *completeness guarantee* and *soundness guarantee*. The (C, S)-approximate linear programming formulation complexity $\text{fc}_{\text{LP}}(\mathcal{P}, C, S)$ of $\mathcal{P}$ is intuitively the number of linear inequalities needed to derive $\max_{\mathcal{I}} \leq C(\mathcal{I})$ for all instances $\mathcal{I}$ of $\mathcal{P}$ satisfying $\max_{\mathcal{I}} \leq S(\mathcal{I})$. The notion of (C, S)-approximate semidefinite programming formulation complexity $\text{fc}_{\text{SDP}}(\mathcal{P}, C, S)$ is defined similarly. Similar definitions hold for minimization problems.

The essential difference in the fractional case is that instead of deriving $\text{val}_{\mathcal{I}}^n / \text{val}_{\mathcal{I}}^d = \text{val}_{\mathcal{I}} \leq C(\mathcal{I})$, we require deriving the equivalent $\text{val}_{\mathcal{I}}^n \leq C(\mathcal{I}) \text{val}_{\mathcal{I}}^d$ and $\text{val}_{\mathcal{I}}^d \geq 0$, which is significantly easier in terms of linear programming proofs. This leads to definitions of LP formulation complexity $\text{fc}_{\text{LP}}(\mathcal{P}, C, S)$ and SDP formulation complexity $\text{fc}_{\text{SDP}}(\mathcal{P}, C, S)$ similar to the non-fractional case.

For both types of problems, the guarantees C and S will often be of the form $C = \alpha g$ and $S = \beta g$ for some constants α and β and an easy-to-compute function g. Then we shall write $\text{fc}_{\text{LP}}(\mathcal{P}, \alpha, \beta)$ instead of the more precise $\text{fc}_{\text{LP}}(\mathcal{P}, \alpha g, \beta g)$.

3   Reductions with Distortion

We now introduce a generalization of the affine reduction mechanism for LPs and SDPs as introduced in [1], answering an open question posed both in [1,6] and leading to many new reductions that were impossible in the affine framework.

Definition 5 (Reduction). *Let $\mathcal{P}_1 = (\mathcal{S}_1, \mathfrak{I}_1, \text{val})$ and $\mathcal{P}_2 = (\mathcal{S}_2, \mathfrak{I}_2, \text{val})$ be optimization problems with guarantees C_1, S_1 and C_2, S_2, respectively. Let $\tau_1 = +1$ if $\mathcal{P}_1$ is a maximization problem, and $\tau_1 = -1$ if $\mathcal{P}_1$ is a minimization problem. Similarly, let $\tau_2 = \pm 1$ depending on whether $\mathcal{P}_2$ is a maximization problem or a minimization problem.*

A reduction from $\mathcal{P}_1$ to $\mathcal{P}_2$ respecting the guarantees consists of

1. *two mappings: $*: \mathfrak{I}_1 \to \mathfrak{I}_2$ and $*: \mathcal{S}_1 \to \mathcal{S}_2$ translating instances and feasible solutions independently;*
2. *two nonnegative $\mathfrak{I}_1 \times \mathcal{S}_1$ matrices M_1, M_2*

subject to the conditions

$$\tau_1 \left[C_1(\mathcal{I}_1) - \text{val}_{\mathcal{I}_1}(s_1) \right] = \tau_2 \left[C_2(\mathcal{I}_1^*) - \text{val}_{\mathcal{I}_1^*}(s_1^*) \right] M_1(\mathcal{I}_1, s_1) + M_2(\mathcal{I}_1, s_1)$$

$$\text{(1-complete)}$$

$$\tau_2 \text{OPT}\,(\mathcal{I}_1^*) \leq \tau_2 S_2(\mathcal{I}_1^*) \qquad \textit{if } \tau_1 \text{OPT}\,(\mathcal{I}_1) \leq \tau_1 S_1(\mathcal{I}_1). \qquad \text{(1-sound)}$$

The matrices M_1 and M_2 provide extra freedom to add additional (valid) inequalities during the reduction. In fact, we might think of them as modeling more complex reductions. These matrices should have low computational overhead, which in our framework means LP or SDP rank.

Theorem 1. *Let $\mathcal{P}_1$ and $\mathcal{P}_2$ be optimization problems with a reduction from $\mathcal{P}_1$ to $\mathcal{P}_2$ respecting the completeness guarantees C_1, C_2 and soundness guarantees S_1, S_2 of $\mathcal{P}_1$ and $\mathcal{P}_2$, respectively. Then*

$$\text{fc}_{\text{LP}}(\mathcal{P}_1, C_1, S_1) \leq \text{rk}_{\text{LP}} M_2 + \text{rk}_{\text{LP}} M_1 + \text{rk}_+ M_1 \cdot \text{fc}_{\text{LP}}(\mathcal{P}_2, C_2, S_2), \qquad (2)$$

$$\text{fc}_{\text{SDP}}(\mathcal{P}_1, C_1, S_1) \leq \text{rk}_{\text{SDP}} M_2 + \text{rk}_{\text{SDP}} M_1 + \text{rk}_{\text{psd}} M_1 \cdot \text{fc}_{\text{SDP}}(\mathcal{P}_2, C_2, S_2), \qquad (3)$$

where M_1 and M_2 are the matrices in the reduction as in Definition 5.

The corresponding multiplicative inapproximability factors can be obtained as usual, by taking the ratio of soundness and completeness.

3.1   Reduction Between Fractional Problems

Reductions for fractional optimization problems are completely analogous to the non-fractional case:

Definition 6 (Reduction). *Let $\mathcal{P}_1 = (\mathcal{S}_1, \mathfrak{I}_1, \text{val})$ and $\mathcal{P}_2 = (\mathcal{S}_2, \mathfrak{I}_2, \text{val})$ be fractional optimization problems with guarantees C_1, S_1 and C_2, S_2, respectively. Let $\tau_1 = +1$ if $\mathcal{P}_1$ is a maximization problem, and $\tau_1 = -1$ if $\mathcal{P}_1$ is a minimization problem. Similarly, let $\tau_2 = \pm 1$ depending on whether $\mathcal{P}_2$ is a maximization problem or a minimization problem.*

A reduction from $\mathcal{P}_1$ to $\mathcal{P}_2$ respecting the guarantees consists of

1. *two mappings: $*: \mathfrak{I}_1 \to \mathfrak{I}_2$ and $*: \mathcal{S}_1 \to \mathcal{S}_2$ translating instances and feasible solutions independently;*
2. *four nonnegative $\mathfrak{I}_1 \times \mathcal{S}_1$ matrices $M_1^{(n)}$, $M_1^{(d)}$, $M_2^{(n)}$, $M_2^{(d)}$*

subject to the conditions

$$\tau_1 \left[C_1(\mathcal{I}_1)\text{val}_{\mathcal{I}_1}^d(s_1) - \text{val}_{\mathcal{I}_1}^n(s_1) \right]$$
$$= \tau_2 \left[C_2(\mathcal{I}_1^*)\text{val}_{\mathcal{I}_1^*}^d(s_1^*) - \text{val}_{\mathcal{I}_1^*}^n(s_1^*) \right] M_1^{(n)}(\mathcal{I}_1, s_1) + M_2^{(n)}(\mathcal{I}_1, s_1)$$
$$\text{(4-complete)}$$

$$\text{val}_{\mathcal{I}_1}^d(s_1) = \text{val}_{\mathcal{I}_1^*}^d(s_1^*) \cdot M_1^{(d)}(\mathcal{I}_1, s_1) + M_2^{(d)}(\mathcal{I}_1, s_1) \qquad \text{(4-denominator)}$$

$$\tau_2 \text{OPT}\,(\mathcal{I}_1^*) \leq \tau_2 S_2(\mathcal{I}_1^*) \qquad \textit{if } \tau_1 \text{OPT}\,(\mathcal{I}_1) \leq \tau_1 S_1(\mathcal{I}_1). \qquad \text{(4-sound)}$$

As the val^d are supposed to span a small dimensional subspace, the matrices $M_1^{(d)}$ and $M_2^{(d)}$ are not supposed to significantly influence the strength of the reduction even with the trivial choice $M_1^{(d)} = 0$ and $M_2^{(d)}(\mathcal{I}_1, s_1) = val_{\mathcal{I}_1}^d(s_1)$. However, as in the non-fractional case, the complexity of $M_1^{(n)}$ and $M_2^{(n)}$ could have a major influence on the strength of the reduction. The reduction theorem is a special case of the reinterpretation as proof system; see full-length version:

Theorem 2. *Let $\mathcal{P}_1$ and $\mathcal{P}_2$ be optimization problems with a reduction from $\mathcal{P}_1$ to $\mathcal{P}_2$ Then*

$$fc_{LP}(\mathcal{P}_1, C_1, S_1) \le rk_{LP}\begin{bmatrix} M_2^{(n)} \\ M_2^{(d)} \end{bmatrix} + rk_{LP}\begin{bmatrix} M_1^{(n)} \\ M_1^{(d)} \end{bmatrix} \tag{5}$$

$$+ rk_+\begin{bmatrix} M_1^{(n)} \\ M_1^{(d)} \end{bmatrix} \cdot fc_{LP}(\mathcal{P}_2, C_2, S_2),$$

$$fc_{SDP}(\mathcal{P}_1, C_1, S_1) \le rk_{SDP}\begin{bmatrix} M_2^{(n)} \\ M_2^{(d)} \end{bmatrix} + rk_{SDP}\begin{bmatrix} M_1^{(n)} \\ M_1^{(d)} \end{bmatrix} \tag{6}$$

$$+ rk_{psd}\begin{bmatrix} M_1^{(n)} \\ M_1^{(d)} \end{bmatrix} \cdot fc_{SDP}(\mathcal{P}_2, C_2, S_2),$$

where $M_1^{(n)}$, $M_1^{(d)}$, $M_2^{(n)}$, and $M_2^{(d)}$ are the matrices in the reduction as in Definition 6.

4   BalancedSeparator and SparsestCut

The SparsestCut problem is a high-profile problem that received considerable attention in the past. It is known that SparsestCut with general demands can be approximated within a factor of $O(\sqrt{\log n} \log \log n)$ [18] and the standard SDP has an integrality gap of $(\log n)^{\Omega(1)}$ [19]. In this section we will show that the SparsestCut problem cannot be approximated well by small LPs and SDPs by using the new reduction mechanism from Sect. 3.1, even if the supply graph has bounded treewidth, with the lower bound matching the upper bound in [17] in the LP case. The results are *unconditional* LP/SDP analog to [20], however for a different regime.

Theorem 3 (LP/SDP hardness for SparsestCut, tw(supply) = $O(1)$). *For any $\varepsilon \in (0, 1)$ there are $\eta_{LP} > 0$ and $\eta_{SDP} > 0$ such that for every large enough n the following hold*

$$fc_{LP}\left(SparsestCut(n, 2), \eta_{LP}(1 + \varepsilon), \eta_{LP}(2 - \varepsilon)\right) \ge n^{\Omega(\log n / \log \log n)},$$

$$fc_{SDP}\left(SparsestCut(n, 2), \eta_{SDP}\left(1 + \frac{4\varepsilon}{5}\right), \eta_{SDP}\left(\frac{16}{15} - \varepsilon\right)\right) \ge n^{\Omega(\log n / \log \log n)}.$$

In other words SparsestCut(n, 2) is LP-hard with an inapproximability factor of $2 - \epsilon$, and SDP-hard with an inapproximability factor of $\frac{16}{15} - O(\epsilon)$.

A complementary reduction proves the hardness of approximating Balanced Separator where the demand graph has constant treewidth.

Theorem 4 (LP-hardness for BalancedSeparator). *For any constant $c_1 \geq 1$ there is another constant $c_2 \geq 1$ such that for all n there is a demand function $d \colon E(K_n) \to \mathbb{R}_{\geq 0}$ satisfying $\mathrm{tw}([n]_d) \leq c_2$ so that BalancedSeparator(n, d) is LP-hard with inapproximability factor of c_1.*

Proof (Proof sketch of Theorem 3). We use the reduction from [17], reducing MaxCut to SparsestCut. Given an instance $\mathcal{I}$ of MaxCut(n) we first construct the instance $\mathcal{I}^*$ on vertex set $V = \{u, v\} \cup [n]$ where u and v are two special vertices. Let us denote the degree of a vertex i in $\mathcal{I}$ by $\deg(i)$ and let $m := \frac{1}{2} \sum_{i=1}^{n} \deg(i)$ be the total number of edges in $\mathcal{I}$. We define the capacity function $c \colon V \times V \to \mathbb{R}_{\geq 0}$ as

$$c(i, j) := \begin{cases} \frac{\deg(i)}{m} & \text{if } j = u, i \neq v \text{ or } j = v, i \neq u \\ 0 & \text{otherwise.} \end{cases}$$

Note that the supply graph has treewidth at most 2 being a copy of $K_{2,n}$. The demand function $d \colon V \times V \to \mathbb{R}_{\geq 0}$ is defined as

$$d(i, j) := \begin{cases} \frac{2}{m} & \text{if } \{i, j\} \in E(\mathcal{I}) \\ 0 & \text{otherwise.} \end{cases}$$

We map a solution s to MaxCut(n) to the cut $s^* := s \cup \{u\}$ of SparsestCut$(n+2, 2)$.

We remind the reader of the powering operation from [17] to handle the case of unbalanced and non u-v cuts. It successively adds for every edge of $\mathcal{I}^*$ a copy of itself, scaling both the capacities and demands by the capacity of the edge. After l rounds, we obtain an instance $\mathcal{I}_l^*$ on a fixed set of $O(N^{2l})$ vertices, and similarly the cuts s^* extend naturally to cuts s_l^* on these vertices, independent of the instance $\mathcal{I}$. Recall that the supply graph of $\mathcal{I}_l^*$ has the same treewidth as that of $\mathcal{I}^*$ [17, Observation 4.4], i.e., at most 2. Completeness follows by construction, see [17, Claim 4.2],

$$\mathrm{val}_{\mathcal{I}_l^*}^n(s_l^*) = 1, \qquad \mathrm{val}_{\mathcal{I}_l^*}^d(s_l^*) = l\,\mathrm{val}_{\mathcal{I}}(s).$$

Soundness follows from [17, Lemmas 4.3 and 4.7]:

$$\mathrm{OPT}\,(\mathcal{I}_l) \geq \frac{1}{1 + (l-1)\mathrm{OPT}\,(\mathcal{I}) / |E(\mathcal{I})|}.$$

Now the hardness result follow from Theorem 2 with matrices $M_1^{(n)}(\mathcal{I}_l, s_1) := C_1$ $(\mathcal{I}_1)$, $M_2^{(n)}(\mathcal{I}_l, s_1) := 0$, $M_1^{(d)}(\mathcal{I}_l, s_1) := 0$, $M_2^{(d)}(\mathcal{I}_l, s_1) := 1$. Hardness of the base problem MaxCut is provided by [9] (for the LP case) and Theorem 5 (for the SDP case), and leads to $\eta_{LP} = \frac{5\varepsilon}{3-\varepsilon}$ and $\eta_{SDP} = \frac{3\varepsilon}{1-4\varepsilon}$.

5   SDP Hardness of MaxCut

We now show that MaxCut cannot be approximated via small SDPs within a factor of $15/16 + \varepsilon$. As approximation guarantees for an instance graph H, we shall use $C(H) = \alpha\,|E(H)|$ and $S(H) = \beta\,|E(H)|$ for some constants α and β, and for brevity write only α and β.

Theorem 5. *For any $\delta, \varepsilon > 0$ there are infinitely many n such that there is a graph G with n vertices and*

$$\mathrm{fc}_{\mathrm{SDP}}(\mathsf{MaxCut}(G), 4/5 - \varepsilon, 3/4 + \delta) = n^{\Omega(\log n/\log\log n)}. \tag{7}$$

Proof (Proof sketch). By [10, Theorem 4.5] and [5, Theorem 6.4], for any $\delta, \varepsilon > 0$ we have $\mathrm{fc}_{\mathrm{SDP}}(\mathsf{Max\text{-}3\text{-}XOR}/0,1 -\varepsilon, 1/2 + \delta) = m^{\Omega(\log m/\log\log m)}$ for infinitely many m. We reuse the gadget based reduction from Max-3-XOR to MaxCut in [21, Lemma 4.2]. Let $x_1, \ldots, x_m$ be the variables for Max-3-XOR. For every possible clause $C = (x_i + x_j + x_k = 0)$, we shall use the gadget graph H_C from [21, Fig. 4.1], reproduced here in Fig. 1. We shall use the graph G, which is the union of all the gadgets $H(C)$ for all possible clauses. The vertices 0 and $x_1, \ldots, x_m$ are shared by the gadgets, the other vertices are unique to each gadget.

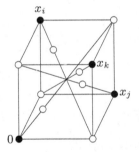

Fig. 1. The gadget H_C for the clause $C = (x_i + x_j + x_k = 0)$ in the reduction from Max-3-XOR to MaxCut. Solid vertices are shared by gadgets, the empty ones are local to the gadget.

A Max-3-XOR instance $\mathcal{I} = \{C_i\}_i$ is mapped to the union $G_{\mathcal{I}} = \bigcup_i H(C_i)$ of the gadgets of the clauses C_i in $\mathcal{I}$, which is an induced subgraph of G. A feasible solution, i.e., an assignment $s\colon \{x_1, \ldots, x_m\} \to \{0, 1\}$ is mapped to a vertex set s^* satisfying the following conditions: (1) $x_i \in s^*$ if and only if $s(x_i) = 1$, (2) $0 \notin s^*$ (3) on every gadget $H(C)$ the set s^* cuts the maximal number of edges subject to the previous two conditions. It is easy to see that s^* cuts 16 out of the 20 edges of every $H(C)$ if s satisfies C, and it cuts 14 edges if s does not satisfy C. Therefore $\mathrm{val}_{G_{\mathcal{I}}}(s^*) = (14 + 2\mathrm{val}_{\mathcal{I}}(s))/20$. It also follows from the construction that $\mathrm{val}_{G_{\mathcal{I}}}$ achieves its maximum on a vertex set of the form s^*, showing soundness of the reduction, and proving the claim, noting that G has $n = O(m^3)$ vertices.

6   Lasserre Relaxation is Suboptimal for IndependentSet(G)

As a curiosity, we will now derive a new lower bound on the Lasserre integrality gap for the IndependentSet problem, establishing that the Lasserre hierarchy is suboptimal: there exists a linear-sized LP formulation for the IndependentSet problem with approximation guarantee $2\sqrt{n}$, whereas there exists a family of graphs with Lasserre integrality gap $n^{1-\gamma}$ after $\Omega(n^\gamma)$ rounds for arbitrary small γ. While this is expected assuming P vs. NP, our result is unconditional. It also complements previous integrality gaps, like $n/2^{O(\sqrt{\log n \log \log n})}$ for $2^{\Theta(\sqrt{\log n \log \log n})}$ rounds in [12], and others in [22], e.g., $\Theta(\sqrt{n})$ rounds of Lasserre are required for deriving the exact optimum.

Theorem 6. *For any small enough $\gamma > 0$ there are infinitely many n, such that there is a graph G with n vertices with the largest independent set of G having size $\alpha(G) = O(n^\gamma)$ but there is a $\Omega(n^\gamma)$-round Lasserre solution of size $\Theta(n)$, i.e., the integrality gap is $n^{1-\gamma}$. However* $\mathrm{fc_{LP}}(\mathsf{IndependentSet}(G), 2\sqrt{n}) \leq 3n+1$.

Proof (Proof sketch). The statement $\mathrm{fc_{LP}}(\mathsf{IndependentSet}(G), 2\sqrt{n}) \leq 3n + 1$ is [6, Lemma 5.2]. For the integrality gap construction, we apply [23, Theorem 4.2], providing an instance $\mathcal{I}$ of Max-$\Theta(\log N)$-CSP on N variables with $m = N^{\Theta(1)}$ clauses, such that at most an $N^{-\omega(1)}$ fraction of the clauses are satisfiable but there is a $N^{1-o(1)}$-round Lasserre solution satisfying all the clauses. In particular, every clause has only a constant number a of satisfying partial assignments.

Let G be the conflict graph of $\mathcal{I}$, i.e., the vertices of G are pairs (i, s) with $i \in [m]$ and s a satisfying partial assignments s of clause C_i with domain the set of free variables of C_i. Two pairs (i, s) and (j, t) assignments are adjacent as vertices of G if and only if the partial assignments s and t conflict, i.e., $s(x_j) \neq t(x_j)$ for some variable x_j on which both s and t are defined. Thus G has $n = N^{\Theta(1)}$ vertices.

Given an assignment $t\colon \{x_1, \dots, x_N\} \to [q]$ we define the independent set t^* of G as the set of partial assignments s compatible with t. (Obviously, t^* is an independent set.) This provides a mapping $*$ from the set of assignments of the $x_1, \dots, x_N$ to the set of independent sets of G. Clearly, $\mathrm{val}_G(t^*) = m\mathrm{val}_{\mathcal{I}}(t)$, as t^* contains one vertex per clause satisfied by t. It is easy to see that every independent set I of G is a subset of some t^*, and hence OPT $(G) = $ mOPT $(\mathcal{I}) = O(N)$. The map $*$ has degree $O(\log N)$, and hence extends to Lasserre solutions the usual way, providing $\mathrm{SoS}_{N^{1-o(1)}/\log N}(G) \geq m$. It follows that with the right choice of parameters, G has the Lasserre integrality gap claimed in the theorem. $\square$

Acknowledgements. Research reported in this paper was partially supported by NSF CAREER award CMMI-1452463. Parts of this research was conducted at the CMO-BIRS 2015 workshop *Modern Techniques in Discrete Optimization: Mathematics, Algorithms and Applications* and we would like to thank the organizers for providing a stimulating research environment, as well as Levent Tunçel for helpful discussions on Lasserre relaxations of the IndependentSet problem.

References

1. Braun, G., Pokutta, S., Zink, D.: Inapproximability of combinatorial problems via small LPs and SDPs (2015)
2. Yannakakis, M.: Expressing combinatorial optimization problems by linear programs. J. Comput. Syst. Sci. **43**(3), 441–466 (1991)
3. Rothvoß, T.: The matching polytope has exponential extension complexity. In: Proceedings of STOC, pp. 263–272 (2014)
4. Fiorini, S., Massar, S., Pokutta, S., Tiwary, H.R., de Wolf, R.: Linear vs. semidefinite extended formulations: exponential separation and strong lower bounds. J. ACM (2015, to appear)
5. Lee, J.R., Raghavendra, P., Steurer, D.: Lower bounds on the size of semidefinite programming relaxations. arXiv preprint arXiv:1411.6317 (2014)
6. Bazzi, A., Fiorini, S., Pokutta, S., Svensson, O.: No small linear program approximates Vertex Cover within a factor $2 - \varepsilon$. arXiv preprint arXiv:1503.00753 (2015)
7. Pashkovich, K.: Extended Formulations for Combinatorial Polytopes. Ph.D. thesis. Magdeburg Universität (2012)
8. Braun, G., Fiorini, S., Pokutta, S., Steurer, D.: Approximation limits of linear programs (beyond hierarchies). In: 53rd IEEE Symposium on Foundations of Computer Science (FOCS 2012), pp. 480–489 (2012)
9. Chan, S.O., Lee, J.R., Raghavendra, P., Steurer, D.: Approximate constraint satisfaction requires large LP relaxations. In: IEEE 54th Annual Symposium on Foundations of Computer Science (FOCS 2013), pp. 350–359. IEEE (2013)
10. Schoenebeck, G.: Linear level Lasserre lower bounds for certain k-CSPs. In: IEEE 49th Annual IEEE Symposium on Foundations of Computer Science, FOCS 2008, pp. 593–602. IEEE (2008)
11. Charikar, M., Makarychev, K., Makarychev, Y.: Integrality gaps for Sherali-Adams relaxations. In: Proceedings of the Forty-First Annual ACM Symposium on Theory of Computing, pp. 283–292. ACM (2009)
12. Tulsiani, M.: CSP gaps and reductions in the Lasserre hierarchy. In: Proceedings of the Forty-First Annual ACM Symposium on Theory of Computing, pp. 303–312. ACM (2009)
13. Au, Y.H., Tunçel, L.: A comprehensive analysis of polyhedral lift-and-project methods. arXiv preprint (2013). arXiv:1312.5972
14. Lipták, L., Tunçel, L.: The stable set problem and the lift-and-project ranks of graphs. Math. Program. **98**(1–3), 319–353 (2003)
15. Stephen, T., Tuncel, L.: On a representation of the matching polytope via semidefinite liftings. Math. Oper. Res. **24**(1), 1–7 (1999)
16. Kolman, P., Koutecký, M., Tiwary, H.R.: Extension complexity, MSO logic, and treewidth. CoRR abs/1507.04907 (2015)
17. Gupta, A., Talwar, K., Witmer, D.: Sparsest cut on bounded treewidth graphs: algorithms and hardness results. In: Proceedings of the Forty-Fifth Annual ACM Symposium on Theory of Computing, pp. 281–290. ACM (2013)
18. Arora, S., Lee, J., Naor, A.: Euclidean distortion and the sparsest cut. J. Am. Math. Soc. **21**(1), 1–21 (2008)
19. Cheeger, J., Kleiner, B., Naor, A.: A $(\log n)^{\Omega(1)}$ integrality gap for the sparsest cut SDP. In: 50th Annual IEEE Symposium on Foundations of Computer Science, FOCS 2009, pp. 555–564. IEEE (2009)
20. Chawla, S., Krauthgamer, R., Kumar, R., Rabani, Y., Sivakumar, D.: On the hardness of approximating multicut and sparsest-cut. Comput. Complex. **15**(2), 94–114 (2006)

21. Trevisan, L., Sorkin, G.B., Sudan, M., Williamson, D.P.: Gadgets, approximation, and linear programming. SIAM J. Comput. **29**(6), 2074–2097 (2000)
22. Au, Y.H., Tunçel, L.: Complexity analyses of Bienstock–Zuckerberg and Lasserre relaxations on the matching and stable set polytopes. In: Günlük, O., Woeginger, G.J. (eds.) IPCO 2011. LNCS, vol. 6655, pp. 14–26. Springer, Heidelberg (2011)
23. Bhaskara, A., Charikar, M., Vijayaraghavan, A., Guruswami, V., Zhou, Y.: Polynomial integrality gaps for strong SDP relaxations of densest k-subgraph. In: Proceedings of the Twenty-Third Annual ACM-SIAM Symposium on Discrete Algorithms, SIAM, pp. 388–405 (2012)

Sum-of-Squares Hierarchy Lower Bounds
for Symmetric Formulations

Adam Kurpisz, Samuli Leppänen$^{(\boxtimes)}$, and Monaldo Mastrolilli

IDSIA, 6928 Manno, Switzerland
{adam,samuli,monaldo}@idsia.ch

Abstract. We introduce a method for proving Sum-of-Squares (SoS)/ Lasserre hierarchy lower bounds when the initial problem formulation exhibits a high degree of symmetry. Our main technical theorem allows us to reduce the study of the positive semidefiniteness to the analysis of "well-behaved" univariate polynomial inequalities.

We illustrate the technique on two problems, one unconstrained and the other with constraints. More precisely, we give a short elementary proof of Grigoriev/Laurent lower bound for finding the integer cut polytope of the complete graph. We also show that the SoS hierarchy requires a non-constant number of rounds to improve the initial integrality gap of 2 for the MIN-KNAPSACK linear program strengthened with cover inequalities.

1 Introduction

Proving lower bounds for the Sum-of-Squares (SoS)/Lasserre hierarchy [21,27] has attracted notable attention in the theoretical computer science community during the last decade, see e.g. [6,11,14,15,23–25,29,31]. This is partly because the hierarchy captures many of the best known approximation algorithms based on semidefinite programming (SDP) for several natural 0/1 optimization problems (see [24] for a recent result). Indeed, it can be argued that the SoS hierarchy is the strongest candidate to be the "optimal" meta-algorithm predicted by the Unique Games Conjecture (UGC) [18,28]. On the other hand, the hierarchy is also one of the best known candidates for refuting the conjecture since it is still conceivable that one could show that the SoS hierarchy achieves better approximation guarantees than the UGC predicts (see [5] for discussion). Despite the interest in the algorithm and due to the many technical challenges presented by semidefinite programming, only relatively few techniques are known for proving lower bounds for the hierarchy. In particular, several integrality gap results follow from applying gadget reductions to the few known original lower bound constructions.

Indeed, many of the known lower bounds for the SoS hierarchy originated in the works of Grigoriev [14,15] on the positivstellensatz proof system. We defer

Supported by the Swiss National Science Foundation project 200020-144491/1 "Approximation Algorithms for Machine Scheduling Through Theory and Experiments" and by Sciex Project 12.311.

Q. Louveaux and M. Skutella (Eds.): IPCO 2016, LNCS 9682, pp. 362–374, 2016.
DOI: 10.1007/978-3-319-33461-5_30

the formal definition of the hierarchy for later and only point out that solving the hierarchy after t rounds takes $n^{O(t)}$ time. In [15] Grigoriev showed that random 3XOR or 3SAT instances cannot be solved even by $\Omega(n)$ rounds of the SoS hierarchy (some of these results were later independently rediscovered by Schoenebeck [29]). Lower bounds, such as those of [6,31] rely on [15,29] combined with gadget reductions. Another important lower bound was also given by Grigoriev [14] for the KNAPSACK problem (a simplified proof can be found in [16]), showing that the SoS hierarchy cannot prove within $\lfloor n/2 \rfloor$ rounds that the polytope $\{x \in [0,1]^n : \sum_{i=1}^{n} x_i = n/2\}$ contains no integer point when n is odd. Using essentially the same construction as in [16], Laurent [23] independently showed that $\lfloor \frac{n}{2} \rfloor$ rounds are not enough for finding the integer cut polytope of the complete graph with n nodes, where n is odd.[1] By using several new ideas and techniques, but a similar starting point as in [16,23], Meka, Potechin and Wigderson [25] were able to show a lower bound of $\Omega(\log^{1/2} n)$ for the PLANTED-CLIQUE problem. Common to the works [16,23,25] is that the matrix involved in the analysis has a large kernel, and they prove that a principal submatrix is positive definite by applying the theory of association schemes [12]. For different techniques to obtain lower bounds, we refer for example to the recent papers [4,19,20] (see also Sect. 5.2) and the survey [11] for an overview of previous results.

In this paper we introduce a method for proving SoS hierarchy lower bounds when the initial problem formulation exhibits a high degree of symmetry. Our main technical theorem (Theorem 1) allows us to reduce the study of the positive semidefiniteness to the analysis of "well-behaved" univariate polynomial inequalities. The theorem applies whenever the solution and constraints are *symmetric*, informally meaning that all subsets of the variables of equal cardinality play the same role in the formulation (see Sect. 3 for the formal definition). For example, the solution in [14,16,23] for MAX-CUT is symmetric in this sense.

We note that exploiting symmetry reduces the number of variables involved in the analysis, and different ways of utilizing symmetry have been widely used in the past for proving integrality gaps for different hierarchies, see for example [7,13,15,17,20,30]. An interesting difference of our approach from others is that we establish several lower bounds without fully identifying the formula of eigenvectors. More specifically, the common task in this context is to identify the spectral structure to get a simple diagonalized form. In the previous papers the moment matrices belong to the Bose-Mesner algebra of a well-studied association scheme, and hence one can use the existing theory. In this paper, instead of identifying the spectral structure completely, we identify only possible forms and propose to test all the possible candidates. This is in fact an important point, since the approach may be extended even if the underlying symmetry is imperfect or its spectral property is not well understood.

---

[1] The two problems, KNAPSACK and MAX-CUT in complete graphs, considered respectively in [14,16] and in [23], are essentially the same and we will use MAX-CUT to refer to both.

The proof of Theorem 1 is obtained by a sequence of elementary operations, as opposed to notions such as big kernel in the matrix form, the use of interlacing eigenvalues, the machinery of association schemes and various results about hyper-geometric series as in [14,16,23]. Thus Theorem 1 applies to the whole class of symmetric solutions, even when several conditions and machinery exploited in [14,16,23] cannot be directly applied. For example the kernel dimension, which was one of the important key property used to prove the results in [14,16,23], depends on the particular solution that is used and it is not a general property of the class of symmetric solutions. The solutions for two problems considered in this paper have completely different kernel sizes of the analyzed matrices, one large and the other zero.

We demonstrate the technique with two illustrative and complementary applications. First, we show that the analysis of the lower bound for MAX-CUT in [14,16,23] simplifies to few elementary calculations once the main theorem is in place. This result is partially motivated by the open question posed by O'Donnell [26] of finding a simpler proof for Grigoriev's lower bound for the KNAPSACK problem.

As a second application we consider a constrained problem. We show that after $\Omega(\log^{1-\epsilon} n)$ levels the SoS hierarchy does not improve the integrality gap of 2 for the MIN-KNAPSACK linear program formulation strengthened with *cover inequalities* [9] introduced by Wolsey [32]. Adding cover inequalities is currently the most successful approach for capacitated covering problems of this type [1–3,8,10]. Our result is the first SoS lower bound for formulations with cover inequalities. In this application we demonstrate that our technique can also be used for suggesting the solution and for analyzing its feasibility.

2   The SoS Hierarchy

Consider a 0/1 optimization problem with $m \geq 0$ linear constraints $g_\ell(x) \geq 0$, for $\ell \in [m]$ and $x \in \mathbb{R}^n$. We are interested in approximating the convex hull of the integral points of the set $K = \{x \in \mathbb{R}^n \mid g_\ell(x) \geq 0, \forall \ell \in [m]\}$ with the SoS hierarchy defined in the following.

The form of the SoS hierarchy we use in this paper (Definition 1) is equivalent to the one used in literature (see e.g. [4,21,22]). It follows from applying a change of basis to the dual certificate of the refutation of the proof system [22] (see also [25] for discussion on the connection to the proof system). We use this change of basis in order to obtain a useful decomposition of the moment matrices as a sum of rank one matrices of special kind. This will play an important role in our analysis.

For any $I \subseteq N = \{1, \ldots, n\}$, let x_I denote the 0/1 solution obtained by setting $x_i = 1$ for $i \in I$, and $x_i = 0$ for $i \in N \setminus I$. We denote by $g_\ell(x_I)$ the value of the constraint evaluated at x_I. For each integral solution x_I, where $I \subseteq N$, in the SoS hierarchy defined below there is a variable y_I^N that can be interpreted as the "relaxed" indicator variable for the solution x_I. We point out that in this formulation of the hierarchy the number of variables $\{y_I^N : I \subseteq N\}$ is exponential in n, but this is not a problem in our context since we are interested in proving lower bounds rather than solving an optimization problem.

Let $\mathcal{P}_t(N)$ be the collection of subsets of N of size at most $t \in \mathbb{N}$. For every $I \subseteq N$, the q-zeta vector $Z_I \in \mathbb{R}^{\mathcal{P}_q(N)}$ is a 0/1 vector with J-th entry ($|J| \leq q$) equal to 1 if and only if $J \subseteq I$.[2] Note that $Z_I Z_I^\top$ is a rank one matrix and the matrices considered in Definition 1 are linear combinations of these rank one matrices.

Definition 1. *The t-th round SoS hierarchy relaxation for the set K, denoted by* $\mathrm{SoS}_t(K)$*, is the set of values* $\{y_I^N \in \mathbb{R} : \forall I \subseteq N\}$ *that satisfy*

$$\sum_{I \subseteq N} y_I^N = 1, \tag{1}$$

$$\sum_{I \subseteq N} y_I^N Z_I Z_I^\top \succeq 0, \text{ where } Z_I \in \mathbb{R}^{\mathcal{P}_{t+d}(N)} \tag{2}$$

$$\sum_{I \subseteq N} g_\ell(x_I) y_I^N Z_I Z_I^\top \succeq 0, \ \forall \ell \in [m], \text{ where } Z_I \in \mathbb{R}^{\mathcal{P}_t(N)} \tag{3}$$

where $d = 0$ if $m = 0$ (no linear constraints), otherwise $d = 1$.

It is straightforward to see that the SoS hierarchy formulation given in Definition 1 is a relaxation of the integral polytope. Indeed consider any feasible integral solution $x_I \in K$ and set $y_I^N = 1$ and the other variables to zero. This solution clearly satisfies Condition (1), Condition (2) because the rank one matrix $Z_I Z_I^\top$ is positive semidefinite (PSD), and Condition (3) since $x_I \in K$.

3 The Main Technical Theorem

The main result of this paper (see Theorem 1 below) allows us to reduce the study of the positive semidefiniteness for matrices (2) and (3) to the analysis of "well-behaved" univariate polynomial inequalities. It can be applied whenever the solutions and constraints are *symmetric*, namely they are invariant under all permutations π of the set N: $z_I^N = z_{\pi(I)}^N$ for all $I \subseteq N$ (equivalently when $z_I^N = z_J^N$ whenever $|I| = |J|$),[3] where z_I^N is understood to denote either y_I^N or $g_\ell(x_I) y_I^N$. For example, the solution for MAX-CUT considered by Grigoriev [14] and Laurent [23] belongs to this class.

Theorem 1. *For any $t \in \{1, \ldots, n\}$, let $\mathcal{S}_t$ be the set of univariate polynomials* $G_h(k) \in \mathbb{R}[k]$*, for $h \in \{0, \ldots, t\}$, that satisfy the following conditions:*

$$G_h(k) \in \mathbb{R}[k]_{2t} \tag{4}$$

$$G_h(k) = 0 \qquad \text{for } k \in \{0, \ldots, h-1\} \cup \{n-h+1, \ldots, n\}, \text{ when } h \geq 1 \tag{5}$$

$$G_h(k) \geq 0 \qquad \text{for } k \in [h-1, n-h+1] \tag{6}$$

---

[2] In order to keep the notation simple, we do not emphasize the parameter q as the dimension of the vectors should be clear from the context.

[3] We define the set-valued permutation by $\pi(I) = \{\pi(i) \mid i \in I\}$.

For any fixed set of values $\{z_k^N \in \mathbb{R} : k = 0, \ldots, n\}$, if the following holds

$$\sum_{k=h}^{n-h} \binom{n}{k} z_k^N G_h(k) \geq 0 \qquad \forall G_h(k) \in \mathcal{S}_t \tag{7}$$

then the matrix

$$\sum_{k=0}^{n} z_k^N \sum_{\substack{I \subseteq N \\ |I|=k}} Z_I Z_I^\top \qquad (\text{where } Z_I \in \mathbb{R}^{\mathcal{P}_t(N)}) \tag{8}$$

is positive-semidefinite.

Note that polynomial $G_h(k)$ in (6) is nonnegative in a *real interval*, and in (5) it is zero for a *finite set of integers*. Moreover, constraints (7) are trivially satisfied for $h > \lfloor n/2 \rfloor$.

Theorem 1 is a actually a corollary of a technical theorem that is not strictly necessary for the applications of this paper, and therefore deferred to a later section (see Theorem 3 in Sect. 6). The proof (given in Sect. 6) is obtained by exploiting the high symmetry of the eigenvectors of the matrix appearing in (8). Condition (7) corresponds to the requirement that the Rayleigh quotient being non-negative restricted to some highly symmetric vectors (which we show are the only ones we need to consider).

4   Max-Cut for the Complete Graph

In the MAX-CUT problem, we are given an undirected graph and we wish to find a partition of the vertices (a cut) which maximizes the number of edges whose endpoints are on different sides of the partition (cut value). For the complete graph with n vertices, consider any solution with ω vertices on one side and the remaining $n - \omega$ on the other side of the partition. This gives a cut of value $\omega(n - \omega)$. When n is odd and for *any* $\omega \leq n/2$, Grigoriev [14] and Laurent [23] considered the following solution (reformulated in the basis considered in Definition 1):

$$y_I^N = (n+1) \binom{\omega}{n+1} \frac{(-1)^{n-|I|}}{\omega - |I|} \qquad \forall I \subseteq N \tag{9}$$

It is shown [14,23] that (9) is a feasible solution for the SoS hierarchy of value $\omega(n-\omega)$, for *any* $\omega \leq n/2$ up to round $t \leq \lfloor \omega \rfloor$. In particular for $\omega = n/2$ the cut value of the SoS relaxation is strictly larger than the value of the optimal integral cut (i.e. $\lfloor \frac{n}{2} \rfloor (\lfloor \frac{n}{2} \rfloor + 1)$), showing therefore an integrality gap at round $\lfloor n/2 \rfloor$.

We note that the formula for the solution (9) is essentially implied by the requirement of having exactly ω vertices on one side of the partition (see [14,23,25] for more details) and the core of the analysis in [14,23] is in showing that (9) is a feasible solution for the SoS hierarchy. By taking advantage of Theorem 1, the proof that (9) is a feasible solution for the SoS relaxation follows by observing the fact below.

Lemma 1. *For any polynomial $P(x) \in \mathbb{R}[x]$ of degree $\leq n$ and $y_i^N = y_I^N$ defined in (9) we have*

$$\sum_{k=0}^{n} \binom{n}{k} y_k^N P(k) = P(\omega)$$

Proof. By the polynomial remainder theorem $P(k) = (\omega - k)Q(k) + P(\omega)$, where $Q(k)$ is a unique polynomial of degree at most $n - 1$. It follows that

$$\sum_{k=0}^{n} \binom{n}{k} y_k^N P(k) = \underbrace{\sum_{k=0}^{n} \binom{n}{k} y_k^N (\omega - k)Q(k)}_{=0} + P(\omega) \underbrace{\sum_{k=0}^{n} \binom{n}{k} y_k^N}_{=1} = P(\omega)$$

since $\sum_{k=0}^{n} (-1)^k \binom{n}{k} Q(k) = 0$ for any polynomial of degree at most $n - 1$. $\square$

Now by Lemma 1 we have $\sum_{k=0}^{n} y_k^N \binom{n}{k} G_h(k) = G_h(\omega)$ and the feasibility of (9) follows by Theorem 1, since we have that $G_h(\omega) \geq 0$ whenever $t \leq \omega$ for $\omega \leq n/2$.

5   Min-Knapsack with Cover Inequalities

The MIN-KNAPSACK problem is defined as follows: we have n items with costs c_i and profits p_i, and we want to choose a subset of items such that the sum of the costs of the selected items is minimized and the sum of the profits is at least a given demand P. Formally, this can be formulated as an integer program (IP) $\min \left\{ \sum_{j=1}^{n} c_j x_j : \sum_{j=1}^{n} p_j x_j \geq P, x \in \{0,1\}^n \right\}$. It is easy to see that the natural linear program (LP), obtained by relaxing $x \in \{0,1\}^n$ to $x \in [0,1]^n$ in (IP), has an unbounded integrality gap.

By adding the *Knapsack Cover* (KC) inequalities introduced by Wolsey [32] (see also [9]), the arbitrarily large integrality gap of the natural LP can be reduced to 2 (and it is tight [9]). The KC constraints are as follows: $\sum_{j \notin A} p_j^A x_j \geq P - p(A)$ for all $A \subseteq N$, where $p(A) = \sum_{i \in A} p_i$ and $p_j^A = \min \{p_j, P - p(A)\}$. Note that these constraints are valid constraints for integral solutions. Indeed, in the "integral world" if a set A of items is picked we still need to cover $P - p(A)$; the remaining profits are "trimmed" to be at most $P - p(A)$ and this again does not remove any feasible integral solution.

The following instance [9] shows that the integrality gap implied by KC inequalities is 2: we have n items of unit costs and profits. We are asked to select a set of items in order to obtain a profit of at least $1 + 1/(n - 1)$. The resulting linear program formulation with KC inequalities is as follows (for $x_i \in [0,1]$, $i = 1, \ldots, n$)

$$(LP^+) \min \sum_{j=1}^{n} x_j \quad \text{s.t.} \sum_{j=1}^{n} x_j \geq 1 + 1/(n - 1) \tag{10}$$

$$\sum_{j \in N'} x_j \geq 1 \quad \forall N' \subseteq N : |N'| = n - 1 \tag{11}$$

Note that the solution $x_i = 1/(n-1)$ is a valid fractional solution of value $1 + 1/(n-1)$ whereas the optimal integral solution has value 2. In the following we show that $\mathrm{SoS}_t(LP^+)$, with t arbitrarily close to a logarithmic function of n, admits the same integrality gap as the initial linear program (LP^+) relaxation.

Theorem 2. *For any $\delta > 0$ and sufficiently large n', let $t = \lfloor \log^{1-\delta} n' \rfloor$, $n = \lfloor \frac{n'}{t} \rfloor t$ and $\epsilon = o(t^{-1})$. Then the following solution is feasible for $\mathrm{SoS}_t(LP^+)$ with integrality gap of $2 - o(1)$*

$$y_I^N = \binom{n}{|I|}^{-1} \cdot \begin{cases} \frac{(1+\epsilon)n}{(n-1)\lfloor \log n \rfloor} & \text{for } |I| = \lfloor \log n \rfloor \\ \frac{\epsilon t}{jn} & \text{for } |I| = j\frac{n}{t} \text{ and } j \in [t] \\ 1 - \sum_{\emptyset \neq I \subseteq N} y_I^N & \text{for } I = \emptyset \\ 0 & \text{otherwise} \end{cases} \tag{12}$$

5.1 Overview of the Proof

An integrality gap proof for the SoS hierarchy can in general be thought of having two steps: first, choosing a solution to the hierarchy that attains a superoptimal value, and second showing that this solution is feasible for the hierarchy. We take advantage of Theorem 1 in both steps. Here we describe the overview of our integrality gap construction while keeping the discussion informal and technical details minimal in this extended abstract.

Choosing the solution. We make the following simplifying assumptions about the structure of the variables y_I^N: due to symmetry in the problem we set $y_I^N = y_J^N = y_k^N$ for each I, J such that $|I| = |J| = k$, and for every $I \subseteq N$ we set $y_I^N \geq 0$ in order to satisfy (2) for free. Furthermore, in order to have an integrality gap (i.e. a small objective function value), we guess that $y_0^N \approx 1$ forcing the other variables to be small due to (1).

We then show that satisfying (3) for every constraint follows from showing that

$$\sum_{k=0}^{n} \binom{n}{k} y_k^N (k-2) \prod_{i=1}^{t} (k - r_i)^2 \geq 0 \tag{13}$$

for every choice of t real variables r_i. We get this condition by observing similarities in the structure of the constraints and applying Theorem 1, then expressing the polynomial in root form.[4] If we set $y_1^N = 0$, the only negative term in the sum corresponds to y_0^N. Then, it is clear that we need at least $t+1$ non-zero variables y_k^N, otherwise the roots r_i can be set such that the positive terms in (13) vanish and the inequality is not satisfied. Therefore, we choose exactly $t + 1$ of the y_k^N to be strictly positive (and the rest 0 excluding y_0^N), and we distribute them "far away" from each other, so that no root can be placed such that the coefficient of two positive terms become small. To take this idea further, for one "very small" k' (logarithmic in n), we set $y_{k'}^N$ positive and space out the rest evenly.

---

[4] It can be shown that the roots r_i can be assumed to be real numbers.

Proving that the solution is feasible. We show that (13) holds for all possible r_i with our chosen solution by analysing two cases. In the first case we assume that all of the roots r_i are larger than $\log^3 n$. Then, we show that the "small" point k' we chose is enough to satisfy the condition. In the complement case, we assume that there is at least one root r_i that is smaller than $\log^3 n$. It follows that one of the evenly spaced points is "far" from any remaining root, and can be used to show that the condition is satisfied.

5.2 Further Results

In a recent paper [19] the authors characterize the class of the initial 0/1 relaxations that are *maximally hard* for the SoS hierarchy. Here, maximally hard means those relaxations that still have an integrality gap even after $n-1$ rounds of the SoS hierarchy.[5] An illustrative natural member of this class is given by the simple LP relaxation for the MIN KNAPSACK problem, i.e. the minimization of $\sum_{i=1}^{n} x_i$ such that $\sum_{i=1}^{n} x_i \geq P$. In [19] it is shown that at level $n-1$ the integrality gap is k, for any $k \geq 2$ if and only if $P = \Theta(k) \cdot 2^{2n}$. A natural question is to understand if the SoS hierarchy is able to reduce the gap when P is "small". The previous analysis can be used in a very similar vein to show that the integrality gap of the LP does not reduce at round $t = \Omega(\log^{1-\varepsilon} n)$ for any $P < 1$. We omit the details in this extended abstract.

6 Overview of the Proof of Theorem 1

Theorem 1 is actually a corollary of a stronger statement (see Theorem 3 below) that provides *necessary and sufficient* conditions for the matrix (8) to be positive-semidefinite. Theorem 3 uses a special family of polynomials $G_h(k) \in \mathbb{R}[k]$ whose definition is deferred to a later section (see Definition 3 in Sect. 6.1). We postpone the definition because it will become natural in the flow of the proof of Theorem 3. Here we remark that the polynomials $G_h(k)$ of Definition 3 satisfy the conditions (4), (5) and (6) of Theorem 1 (as shown in Lemma 7 to follow).

Theorem 3. *Let $z_k^N \in \mathbb{R}$ for $k \in \{0, \dots, n\}$. Then for any $t \in N$ the following matrix is positive-semidefinite*

$$\sum_{k=0}^{n} z_k^N \sum_{\substack{I \subseteq N \\ |I|=k}} Z_I Z_I^\top \qquad (\text{where } Z_I \in \mathbb{R}^{\mathcal{P}_t(N)}) \tag{14}$$

if and only if

$$\sum_{k=0}^{n} z_k^N \binom{n}{k} G_h(k) \geq 0 \qquad \text{for } h \in \{0, \dots, t\} \tag{15}$$

for every univariate polynomial $G_h(x) \in \mathbb{R}[x]$ of degree at most $2t$ as defined in Definition 3.

---

[5] Recall that at level n the integrality gap vanishes.

By Lemma 7, Theorem 1 is a straightforward corollary of Theorem 3. In the following we review the essential steps of the proof of Theorem 3. We suppress the proofs in this extended abstract due to page limitations.[6]

6.1   Overview of the Proof of Theorem 3

We study when the matrix $M = \sum_{k=0}^{n} z_k \sum_{I \subseteq N, |I|=k} Z_I Z_I^\top$, where $Z_I \in \mathbb{R}^{\mathcal{P}_t(N)}$ is positive-semidefinite. Theorem 3 allows us to reduce the condition $M \succeq 0$ to inequalities of the form $\sum_{k=0}^{n} \binom{n}{k} z_k p(k) \geq 0$, where $p(k)$ is a univariate polynomial of degree $2t$ with some additional remarkable properties.

A basic key idea that is used to obtain such a characterization is that the eigenvectors of M are "very well" structured. This structure is used to get $p(k)$ with the claimed properties.

The structure of the eigenvectors. Let Π denote the group of all permutations of the set N, i.e. the *symmetric* group. Let P_π be the permutation matrix of size $\mathcal{P}_t(N) \times \mathcal{P}_t(N)$ corresponding to any permutation π of set N, i.e. for any vector v we have $[P_\pi v]_I = v_{\pi(I)}$ for any $I \in \mathcal{P}_t(N)$ (see Footnote 3). Note that $P_\pi^{-1} = P_\pi^\top$.

Lemma 2. *For every $\pi \in \Pi$ we have $P_\pi^\top M P_\pi = M$ or, equivalently, M and P_π commute $M P_\pi = P_\pi M$.*

Corollary 1. *If $w \in \mathbb{R}^{\mathcal{P}_t(N)}$ is an eigenvector of M then $v = P_\pi w$ is also an eigenvector of M for any $\pi \in \Pi$.*

By using Corollary 1 we can show that the set of interesting eigenvectors have some "strong" symmetry properties that will be used in our analysis. In the simplest case, for any eigenvector w we could take the vector $u = \sum_{\pi \in \Pi} P_\pi w$ and observe that the elements of u have the form $u_I = u_J$ for each I, J such that $|I| = |J|$. If $\|u\| \neq 0$ then $u/\|u\|$ and $w/\|w\|$ are two eigenvectors corresponding to the same eigenvalue. The latter implies that by considering *only* eigenvectors having the form $u_I = u_J$ for each $|I| = |J|$ we would consider the eigenvalue corresponding to the "unstructured" eigenvector w as well. This is not the case in general, however, since it is possible that $\sum_{\pi \in \Pi} P_\pi w = 0$.

We overcome this obstacle by restricting the permutations in a way which guarantees u to be non-zero. Before going into the details, we introduce some notation.

Definition 2. *For any $H \subseteq N$, we denote by Π_H the permutation group that fixes the set H in the following sense: $\pi \in \Pi_H \Leftrightarrow \pi(H) = H$.*

Note that the definition is equivalent to saying that $\pi \in \Pi_H$ if and only if $\pi(i) \in H$ for every $i \in H$ and $\pi(i) \notin H$ for every $i \notin H$.

Now, we choose a subset $H \subseteq N$ such that $\sum_{\pi \in \Pi_I} P_\pi w = 0$ for each I such that $|I| < |H|$ and $u = \sum_{\pi \in \Pi_H} P_\pi w \neq 0$. Such a set H always exists, since

---

[6] The full version of the paper is available at http://arxiv.org/abs/1407.1746.

otherwise w is a zero vector, since if there is one non-zero entry w_J in w, we can take $H = J$ and the resulting u is non-zero. The choice of H is not unique, but we can always assume that it is the subset of the first $h = |H|$ elements from N, i.e. $H = \{1, \ldots, h\}$. Indeed, if it is not the case, there exists a permutation $\pi \in \Pi$ that maps H to the subset of the first $|H|$ elements from N and such that $P_\pi w$ is an eigenvector of M by Lemma 1. Now it holds that $u \neq 0$ and the vector $u/\|u\|$ is a unit eigenvector corresponding to the same eigenvalue as w and has many elements that are equal to each other.

Lemma 3. *Let $w \in \mathbb{R}^{\mathcal{P}_t(N)}$ be a unit eigenvector of M corresponding to eigenvalue λ, and H be the smallest subset of N such that $u = \sum_{\pi \in \Pi_H} P_\pi w \neq 0$. Then $u/\|u\|$ is also a unit eigenvector of M corresponding to eigenvalue λ.*

The following lemma shows the structure of eigenvectors obtained from summing the permutations of any "unstructured" eigenvector.

Lemma 4. *Let $u = \sum_{\pi \in \Pi_H} P_\pi w$. Then the vector u is invariant under the permutations of Π_H, namely $u_I = u_{\pi(I)}$ for $\pi \in \Pi_H$. Equivalently, $u_I = u_J$ for all $|I| = |J|$ such that $|I \cap H| = |J \cap H|$.*

Lemmas 3, 4 and the arguments above imply Lemma 5.

Lemma 5. *For any eigenvalue λ of M there exists an $h = 0, 1, \ldots, t$ such that the following is an eigenvector corresponding to λ:*

$$u_h = \sum_{i=0}^{t} \sum_{j=0}^{\min\{h,i\}} \alpha_{i,j} b_{i,j} \tag{16}$$

where $H = \{1, \ldots, h\}$, $\alpha_{i,j} \in \mathbb{R}$ and $b_{i,j} \in \mathbb{R}^{\mathcal{P}_t(N)}$ such that $[b_{i,j}]_Q = 1$ if $|Q| = i$ and $|Q \cap H| = j$, $[b_{i,j}]_Q = 0$ otherwise.

By Lemma 5, we have that the positive semidefiniteness condition of M follows by ensuring that for any $h = 0, 1, \ldots, t$ we have $u_h^\top M u_h \geq 0$, i.e.

$$u_h^\top M u_h = \sum_{k=0}^{n} z_k \underbrace{\sum_{\substack{I \subseteq N \\ |I|=k}} \left(u_h^\top Z_I\right)^2}_{A_k} = \sum_{k=0}^{n} z_k \sum_{\substack{I \subseteq N \\ |I|=k}} \left(\sum_{i=0}^{t} \sum_{j=0}^{\min\{h,i\}} \alpha_{i,j} b_{i,j}^\top Z_I\right)^2 \geq 0$$

In the following (Lemma 6) we show that the above values A_k are interpolated by the univariate polynomial $G_h(x)$ defined in Definition 3. In Lemma 7 we prove some remarkable properties of $G_h(x)$ as claimed in Theorem 1.

Definition 3. *For any $h \in \{0, \ldots, t\}$, let $G_h(k) \in \mathbb{R}[k]$ be a univariate polynomial defined as follows*

$$G_h(k) = \sum_{r=0}^{h} \binom{h}{r} h_r(k) \left(\sum_{j=0}^{h} \binom{r}{j} p_j(k-r)\right)^2 \tag{17}$$

where $h_r(k) = k^{\underline{r}} \cdot (n-k)^{\underline{h-r}}$ *and* $p_j(k-r) = \sum_{i=0}^{t-j} \alpha_{i+j,j}\binom{k-r}{i}$ *(for* $\alpha_{i,j} \in \mathbb{R}$*).*[7]

Lemma 6. *For every* $k = 0,\ldots,n$ *the following identity holds* $A_k = \binom{n}{k}\frac{1}{n^{\underline{h}}}G_h(k)$.

It follows that for any unit eigenvector u of the form (16) the corresponding eigenvalue is equal to $u^\top M u = \frac{1}{n^{\underline{h}}}\sum_{k=0}^{n} z_k \binom{n}{k}G_h(k)$. Theorem 3 requires that $\sum_{k=0}^{n} z_k \binom{n}{k}G_h(k) \geq 0$ which implies that the eigenvalue $u^\top M u$ is nonnegative.

In the following section we complete the proof by showing that the polynomials $G_h(k)$ of Definition 3 satisfy the conditions (4), (5) and (6) of Theorem 1.

Lemma 7. *For any* $h \in \{0,\ldots,t\}$, *the polynomials* $G_h(k)$ *as defined in Definition 3 have the following properties:*

(a) $G_h(k)$ *is a univariate polynomial of degree at most $2t$,*
(b) $G_h(k) \geq 0$ *for* $k \in [h-1, n-h+1]$
(c) $G_h(k) = 0$ *for every* $k \in \{0,...,h-1\} \cup \{n-h+1,...,n\}$.

Acknowledgements. The authors would like to express their gratitude to Ola Svensson for helpful discussions and ideas regarding this paper.

References

1. Bansal, N., Buchbinder, N., Naor, J.: Randomized competitive algorithms for generalized caching. In: STOC, pp. 235–244 (2008)
2. Bansal, N., Gupta, A., Krishnaswamy, R.: A constant factor approximation algorithm for generalized min-sum set cover. In: SODA, pp. 1539–1545 (2010)
3. Bansal, N., Pruhs, K.: The geometry of scheduling. In: FOCS, pp. 407–414 (2010)
4. Barak, B., Chan, S.O., Kothari, P.: Sum of squares lower bounds from pairwise independence. In: STOC (2015)
5. Barak, B., Steurer, D.: Sum-of-squares proofs, the quest toward optimal algorithms. In: Electronic Colloquium on Computational Complexity (ECCC), vol. 21, p. 59 (2014)
6. Bhaskara, A., Charikar, M., Vijayaraghavan, A., Guruswami, V., Zhou, Y.: Polynomial integrality gaps for strong SDP relaxations of densest k-subgraph. In: SODA, pp. 388–405 (2012)
7. Blekherman, G., Gouveia, J., Pfeiffer, J.: Sums of squares on the hypercube. In: CoRR, abs/1402.4199 (2014)
8. Carnes, T., Shmoys, D.B.: Primal-Dual schema for capacitated covering problems. In: Lodi, A., Panconesi, A., Rinaldi, G. (eds.) IPCO 2008. LNCS, vol. 5035, pp. 288–302. Springer, Heidelberg (2008)
9. Carr, R.D., Fleischer, L., Leung, V.J., Phillips, C.A.: Strengthening integrality gaps for capacitated network design and covering problems. In: SODA, pp. 106–115 (2000)

---

[7] Denote by $x^{\underline{m}} = x(x-1)\cdots(x-m+1)$ the falling factorial (with the convention that $x^{\underline{0}} = 1$).

10. Chakrabarty, D., Grant, E., Könemann, J.: On column-restricted and priority covering integer programs. In: Eisenbrand, F., Shepherd, F.B. (eds.) IPCO 2010. LNCS, vol. 6080, pp. 355–368. Springer, Heidelberg (2010)
11. Chlamtac, E., Tulsiani, M.: Convex relaxations and integrality gaps. In: Anjos, M.F., Lasserre, J.B. (eds.) Handbook on Semidefinite, Conic and Polynomial Optimization. International Series in Operations Research & Management Science, vol. 166, pp. 139–169. Springer, US (2011)
12. Godsil, C.: Association schemes (2010). Lecture Notes available at http://quoll.uwaterloo.ca/mine/Notes/assoc2.pdf
13. Goemans, M.X., Tunçel, L.: When does the positive semidefiniteness constraint help in lifting procedures? Math. Oper. Res. **26**(4), 796–815 (2001)
14. Grigoriev, D.: Complexity of positivstellensatz proofs for the knapsack. Comput. Complex. **10**(2), 139–154 (2001)
15. Grigoriev, D.: Linear lower bound on degrees of positivstellensatz calculus proofs for the parity. Theor. Comput. Sci. **259**(1–2), 613–622 (2001)
16. Grigoriev, D., Hirsch, E.A., Pasechnik, D.V.: Complexity of semi-algebraic proofs. In: Alt, H., Ferreira, A. (eds.) STACS 2002. LNCS, vol. 2285, p. 419. Springer, Heidelberg (2002)
17. Hong, S., Tunçel, L.: Unification of lower-bound analyses of the lift-and-project rank of combinatorial optimization polyhedra. Discrete Appl. Math. **156**(1), 25–41 (2008)
18. Khot, S.: On the power of unique 2-prover 1-round games. In: STOC, pp. 767–775 (2002)
19. Kurpisz, A., Leppänen, S., Mastrolilli, M.: A Lasserre lower bound for the min-sum single machine scheduling problem. In: Bansal, N., Finocchi, I. (eds.) ESA 2015. LNCS, vol. 9294, pp. 853–864. Springer, Heidelberg (2015). doi:10.1007/978-3-662-48350-3_71
20. Kurpisz, A., Leppänen, S., Mastrolilli, M.: On the hardest problem formulations for the 0/1 Lasserre hierarchy. In: Halldórsson, M.M., Iwama, K., Kobayashi, N., Speckmann, B. (eds.) ICALP 2015. LNCS, vol. 9134, pp. 872–885. Springer, Heidelberg (2015)
21. Lasserre, J.B.: Global optimization with polynomials and the problem of moments. SIAM J. Optim. **11**(3), 796–817 (2001)
22. Laurent, M.: A comparison of the Sherali-Adams, Lovász-Schrijver, and Lasserre relaxations for 0–1 programming. Math. Oper. Res. **28**(3), 470–496 (2003)
23. Laurent, M.: Lower bound for the number of iterations in semidefinite hierarchies for the cut polytope. Math. Oper. Res. **28**(4), 871–883 (2003)
24. Lee, J.R., Raghavendra, P., Steurer, D.: Lower bounds on the size of semidefinite programming relaxations. In: STOC, pp. 567–576 (2015)
25. Meka, R., Potechin, A., Wigderson, A.: Sum-of-squares lower bounds for planted clique. In: STOC, pp. 87–96 (2015)
26. O'Donnell, R.: Approximability proof complex. Talk at ELC Tokyo (2013). Slides available at http://www.cs.cmu.edu/~odonnell/slides/approx-proof-cxty.pps
27. Parrilo, P.: Structured Semidefinite Programs and Semialgebraic Geometry Methods in Robustness and Optimization. Ph.D. thesis. California Institute of Technology (2000)
28. Raghavendra, P.: Optimal algorithms and inapproximability results for every CSP? In: STOC, pp. 245–254 (2008)
29. Schoenebeck, G.: Linear level Lasserre lower bounds for certain k-CSPs. In: FOCS, pp. 593–602 (2008)

30. Stephen, T., Tunçel, L.: On a representation of the matching polytope via semi-definite liftings. Math. Oper. Res. **24**(1), 1–7 (1999)
31. Tulsiani, M., CSP gaps and reductions in the Lasserre hierarchy. In: STOC, pp. 303–312 (2009)
32. Wolsey, L.A.: Facets for a linear inequality in 0–1 variables. Math. Program. **8**, 168–175 (1975)

Approximation-Friendly Discrepancy Rounding

Nikhil Bansal[1] and Viswanath Nagarajan[2]([⊠])

[1] Department of Mathematics and Computer Science,
Eindhoven University of Technology, Eindhoven, The Netherlands
[2] Department of Industrial and Operations Engineering,
University of Michigan, Ann Arbor, USA
viswa@umich.edu

Abstract. Rounding linear programs using techniques from discrepancy is a recent approach that has been very successful in certain settings. However this method also has some limitations when compared to approaches such as randomized and iterative rounding. We provide an extension of the discrepancy-based rounding algorithm due to Lovett-Meka that (i) combines the advantages of both randomized and iterated rounding, (ii) makes it applicable to settings with more general combinatorial structure such as matroids. As applications of this approach, we obtain new results for various classical problems such as linear system rounding, degree-bounded matroid basis and low congestion routing.

1 Introduction

A very common approach for solving discrete optimization problems is to solve some linear programming relaxation, and then round the fractional solution into an integral one, without (hopefully) incurring much loss in quality. Over the years several ingenious rounding techniques have been developed (see e.g. [23,24]) based on ideas from optimization, probability, geometry, algebra and various other areas. Randomized rounding and iterative rounding are two of the most commonly used methods.

Recently, discrepancy-based rounding approaches have also been very successful; a particularly notable result is due to Rothvoss for bin packing [18]. Discrepancy is a well-studied area in combinatorics with several surprising results (see e.g. [16]), and as observed by Lovász et al. [14], has a natural connection to rounding. However, until the recent algorithmic developments [1,9,15,17,19], most of the results in discrepancy were non-constructive and hence not directly useful for rounding. These algorithmic approaches combine probabilistic approaches like randomized rounding with linear algebraic approaches such as iterated rounding [12], which makes them quite powerful.

Interestingly, given the connection between discrepancy and rounding, these discrepancy algorithms can in fact be viewed as meta-algorithms for rounding.

N. Bansal—Supported by a NWO Vidi grant 639.022.211 and an ERC consolidator grant 617951.

V. Nagarajan—Supported in part by a faculty award from Bloomberg Labs.

Q. Louveaux and M. Skutella (Eds.): IPCO 2016, LNCS 9682, pp. 375–386, 2016.
DOI: 10.1007/978-3-319-33461-5_31

We discuss this in Sect. 1.1 in the context of the Lovett-Meka (LM) algorithm [15]. This suggests the possibility of one single approach that generalizes both randomized and iterated rounding. This is our motivating goal in this paper.

While the LM algorithm is already an important step in this direction, it still has some important limitations. For example, it is designed for obtaining additive error bounds and it does not give good multiplicative error bounds (like those given by randomized rounding). This is not an issue for discrepancy applications, but crucial for many approximation algorithms. Similarly, iterated rounding can work well with exponentially sized LPs by exploiting their underlying combinatorial structure (e.g., degree-bounded spanning tree [20]), but the current discrepancy results [15,19] give extremely weak bounds in such settings.

Our Results: We extend the LM algorithm to overcome the limitations stated above. In particular, we give a new variant that also gives Chernoff type multiplicative error bounds (sometimes with an additional logarithmic factor loss). We also show how to adapt the above algorithm to handle exponentially large LPs involving matroid constraints, like in iterated rounding.

This new discrepancy-based algorithm gives new results for problems such as linear system rounding with violations [5,13], degree-bounded matroid basis [7,11], low congestion routing [10,13] and multi-budgeted matroid basis [8], These results simultaneously combine non-trivial guarantees from discrepancy, randomized rounding and iterated rounding and previously such bounds were not even known existentially.

Our results are described formally in Sect. 1.2. To place them in the proper context, we first need to describe some existing rounding approaches (Sect. 1.1). The reader familiar with the LM algorithm can directly go to Sect. 1.2.

1.1   Preliminaries

We begin by describing LM rounding [15], randomized rounding and iterated rounding in a similar form, and then discuss their strengths and weaknesses.

LM Rounding: Let A be a $m \times n$ matrix with 0–1 entries[1], $x \in [0,1]^n$ a fractional vector and let $b = Ax$. Lovett and Meka showed the following rounding result.

Theorem 1 (LM Rounding [15]). *Given A and x as above, For $j = 1, \ldots, m$, pick any λ_j satisfying*

$$\sum_j \exp(-\lambda_j^2/4) \leq n/16. \tag{1}$$

There is an efficient randomized algorithm to find a solution x' such that: (i) at most $n/2$ variables of x' are fractional (strictly between 0 and 1) and, (ii) $|a_j \cdot (x' - x)| \leq \lambda_j \|a_j\|_2$ for each $j = 1, \ldots, m$, where a_j denotes the j-th row of A.

---

[1] The results below generalize to arbitrary real matrices A and vectors x in natural ways, but we consider 0–1 case for simplicity.

Remark: The right hand side of (1) can be set to $(1 - \epsilon)n$ for any $\epsilon > 0$, at the expense of $O(1)$ factor loss in other parameters of the theorem; see e.g. [2].

Randomized Rounding: Chernoff bounds state that if $X_1, \ldots, X_n$ are independent Bernoulli random variables, and $X = \sum_i X_i$ and $\mu = \mathbb{E}[X]$, then

$$\Pr[|X - \mu| \geq \epsilon\mu] \leq 2\exp(-\epsilon^2\mu/4) \qquad \text{for } \epsilon \leq 1.$$

Then independent randomized rounding can be viewed as the following (by using Chernoff bounds and union bound, and denoting $\lambda_j = \epsilon_j\sqrt{b_j}$):

Theorem 2 (Randomized Rounding). *For $j = 1, \ldots, m$, pick any λ_j satisfying $\lambda_j \leq \sqrt{b_j}$, and*

$$\sum_j \exp(-\lambda_j^2/4) < 0.5 \qquad (2)$$

Then independent randomized rounding gives a solution x' such that: (i) All variables are 0–1, and (ii) $|a_j(x' - x)| \leq \lambda_j\sqrt{b_j}$ for each $j = 1, \ldots, m$.

Iterated Rounding [12]: This is based on the following linear-algebraic fact.

Theorem 3. *If $m < n$, then there is a solution $x' \in [0, 1]^n$ such that (i) x' has at least $n - m$ variables set to 0 or 1 and, (ii) $A(x' - x) = 0$ (i.e., $b = Ax'$).*

If $m > n$, some cleverly chosen constraints are dropped until $m < n$ and then (by Theorem 3) some integral variables are obtained. This is done repeatedly.

Strengths of LM Rounding: Note that if we set $\lambda_j \in \{0, \infty\}$ in LM rounding, then it gives a very similar statement to Theorem 3. E.g., if we only care about some $m = n/2$ constraints then Theorem 3 gives an x' with at least $n/2$ integral variables and $a_jx = a_jx'$ for all these m constraints. Theorem 1 (and the remark below it) give the same guarantee if we set $\lambda_j = 0$ for all constraints. In general, LM rounding can be much more flexible as it allows arbitrary λ_j.

Second, LM rounding is also related to randomized rounding. Note that (1) and (2) have the same left-hand-side. However, the right-hand-side of (1) is $\Omega(n)$, while that of (2) is $O(1)$. This actually makes a huge difference. In particular, in (2) one cannot set $\lambda_j = 1$ for more than a couple of constraints (to get an $o(\sqrt{b_j})$ error bound on constraints), while in (1), one can even set $\lambda_j = 0$ for $O(n)$ constraints. In fact, almost all non-trivial results in discrepancy [16, 21, 22] are based on this ability.

Weaknesses of LM Rounding: First, Theorem 1 only gives a partially integral solution instead of a fully integral one as in Theorem 2.

Second, and more importantly, it only gives additive error bounds instead of multiplicative ones. In particular, note the $\lambda_j\|a_j\|_2$ vs $\lambda_j\sqrt{b_j}$ error in Theorems 1 and 2. E.g., for a constraint $\sum_i x_i = \log n$, Theorem 2 gives $\lambda\sqrt{\log n}$ error but Theorem 1 gives a much higher $\lambda\sqrt{n}$ error. So, while randomized rounding can give a good multiplicative error like $a_j x' \leq (1 \pm \epsilon_j) b_j$, LM rounding is completely insensitive to b_j.

Finally, iterated rounding works extremely well in many settings where Theorem 1 does not give anything useful. E.g., in problems involving exponentially many constraints such as the degree bounded spanning tree problem. The problem is that if m is exponentially large, then the λ_j's in Theorem 1 need to be very large to satisfy (2).

1.2   Our Results and Techniques

Our first result is the following improvement over Theorem 1.

Theorem 4. *There is a constant $K_0 > 0$ and randomized polynomial time algorithm that given $x \in [0,1]^n$, m linear constraints $a_1, \ldots, a_m \in \mathbb{R}^n$, and $\lambda_1, \cdots, \lambda_m \geq 0$ with $\max_{j=1}^m \lambda_j \leq poly(n)$ and $\sum_{j=1}^m e^{-\lambda_j^2 / K_0} < \frac{n}{16}$, finds a solution $x' \in [0,1]^n$ such that:*

$$|\langle x' - x, a_j \rangle| \leq \lambda_j \cdot \sqrt{W_j(x)} + \frac{1+\lambda_j}{n^2} \cdot \|a_j\|, \quad \forall j \in [m] \tag{3}$$

$$x_i' \in \{0,1\}, \quad for \; \Omega(n) \; indices \; i \in [n] \tag{4}$$

Here $W_j(x) := \sum_{i=1}^n a_{ji}^2 \cdot \min\{x_i, 1-x_i\}^2$ for each $j \in [m]$.

Remarks: (1) The error $\lambda_j \sqrt{W_j(x)}$ is always smaller than $\lambda_j \|a_j\|$ in LM-rounding and $\lambda_j (\sum_{i=1}^n a_{ji}^2 \cdot x_i(1-x_i))^{1/2}$ in randomized rounding. In fact it could even be much less if the x_i are very close to 0 or 1.
(2) The term $n/16$ above can be made $(1-\epsilon)n$ for any constant $\epsilon > 0$, at the expense of worsening other constants (just as in LM rounding).
(3) The additional error term $\frac{1+\lambda_j}{n^2} \cdot \|a_j\|$ above is negligible and can be reduced to $\frac{1+\lambda_j}{n^c} \cdot \|a_j\|$ for any constant c, at the expense of running time $n^{O(c)}$.

We note that Theorem 4 can also be obtained in a "black box" manner from LM-rounding (Theorem 1) by rescaling the polytope given by (3) and (4) and using its symmetry.[2] However, such an approach does not work in the setting of matroid polytopes (Theorem 5 below), and in order to achieve that we modify LM-rounding as outlined below.

Applications: We focus on *linear system rounding* as the prime example. Here, given matrix $A \in \{0,1\}^{m \times n}$ and vector $b \in \mathbb{Z}_+^m$, the goal is to find a vector $y \in \{0,1\}^n$ satisfying $Ay = b$. As this is NP-hard, the focus has been on finding a $y \in \{0,1\}^n$ where $Ay \approx b$.

Given any fractional solution $x \in [0,1]^n$ satisfying $Ax = b$, using Theorem 4 iteratively we can obtain an integral vector $y \in \{0,1\}^n$ with

$$|a_j y - b_j| \leq \min\left\{O(\sqrt{n \log(2 + m/n)}), \; \sqrt{L \cdot b_j} + L\right\}, \quad \forall j \in [m], \tag{5}$$

---

[2] We thank an anonymous reviewer for pointing this out.

where $L = O(\log n \log m)$. Previously known algorithms could provide a bound of either $O(\sqrt{n}\log(m/n))$ for all constraints [15] (e.g., for $m = O(n)$ this gives nontrivial Spencer type bounds of $O(\sqrt{n})$ [21]) or $O(\sqrt{\log m} \cdot \sqrt{b_j} + \log m)$ for all constraints (Theorem 2). Note that this does not imply a $\min\{\sqrt{n\log(m/n)}, \sqrt{\log m} \cdot \sqrt{b_j} + \log m\}$ violation per constraint, as in general it is not possible to combine two integral solutions and achieve the better of their violation bounds on all constraints. To the best of our knowledge, even the existence of an integral solution satisfying the bounds in (5) was not known prior to our work.

In the setting where the matrix A is "column sparse", i.e., each variable appears in at most Δ constraints, we obtain a more refined error of

$$|a_j y - b_j| \leq \min\left\{O(\sqrt{\Delta}\log n),\ \sqrt{L \cdot b_j} + L\right\}, \quad \forall j \in [m], \qquad (6)$$

where $L = O(\log n \cdot \log m)$. Previous algorithms could separately achieve bounds of $\Delta - 1$ [5], $O(\sqrt{\Delta}\log n)$ [15] or $O(\sqrt{\log \Delta} \cdot \sqrt{b_j} + \log \Delta)$ [13]. For clarity, Fig. 1 plots the violation bounds achieved by these different algorithms as a function of the right-hand-side b when $m = n$ (we assume $b, \Delta \geq \log^2 n$). Note again that since there are multiple constraints we can not simply combine algorithms to achieve the smaller of their violation bounds.

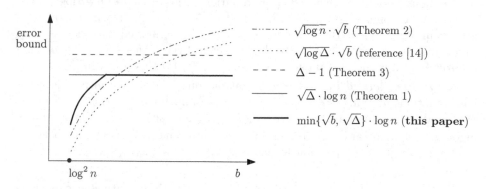

Fig. 1. Additive violation bounds for linear system rounding when $\Delta \geq \log^2 n$ and $b \geq \log^2 n$.

One can also combine the bounds in (5) and (6), and use some additional ideas from discrepancy to obtain $\forall j \in [m]$:

$$|a_j y - b_j| \leq O(1) \cdot \min\left\{\sqrt{j}, \sqrt{n\log(2 + m/n)}, \sqrt{L \cdot b_j} + L, \sqrt{\Delta}\log n\right\}. \quad (7)$$

Matroid Polytopes: Our main result is an extension of Theorem 4 where the fractional solution lies in a matroid polytope in addition to satisfying the linear constraints $\{a_j\}_{j=1}^m$. Recall that a matroid $\mathcal{M}$ is a tuple $(V, \mathcal{I})$ where V is the groundset of elements and $\mathcal{I} \subseteq 2^V$ is a collection of independent sets satisfying the hereditary and exchange properties. The rank function $r : 2^V \rightarrow \mathbb{Z}$ of a matroid is defined as $r(S) = \max_{I \in \mathcal{I}, I \subseteq S} |I|$. The matroid polytope (i.e. convex hull of all independent sets) is given by the following linear inequalities:

$$P(\mathcal{M}) \quad := \quad \{x \in \mathbb{R}^n \ : \ x(S) \leq r(S) \ \forall S \subseteq V, \ x \geq 0\}.$$

Theorem 5. *There is a randomized polynomial time algorithm that given matroid $\mathcal{M}$, $y \in P(\mathcal{M})$, linear constraints $\{a_j \in \mathbb{R}^n\}_{j=1}^m$ and values $\{\lambda_j\}_{j=1}^m$ satisfying the conditions in Theorem 4, finds a solution $y' \in P(\mathcal{M})$ satisfying (3) and (4).*

The fact that we can *exactly* preserve the matroid constraints leads to a number of additional improvements:

Degree-Bounded Matroid Basis (DEGMAT). Given a matroid on elements $[n]$ with costs $d : [n] \to \mathbb{Z}_+$ and m "degree constraints" $\{S_j, b_j\}_{j=1}^m$ where each $S_j \subseteq [n]$ and $b_j \in \mathbb{Z}_+$, the goal is to find a minimum-cost basis I in the matroid that satisfies $|I \cap S_j| \leq b_j$ for all $j \in [m]$. Since even the feasibility problem is NP-hard, we consider *bicriteria* approximation algorithms that violate the degree bounds. We obtain an algorithm where the solution costs at most the optimal and the degree bound violation is as in (7); here Δ denotes the maximum number of sets $\{S_j\}_{j=1}^m$ containing any element.

Previous algorithms achieved approximation ratios of $(1, b + O(\sqrt{b \log n}))$ [7], based on randomized swap rounding, and $(1, b + \Delta - 1)$ [11] based on iterated rounding. Again, these bounds could not be combined together as they used different algorithms. We note that in general the $(1, b + O(\sqrt{n \log(m/n)}))$ approximation is the best possible (unless $P = NP$) for this problem [3,6].

Multi-criteria Matroid Basis. Given a matroid on elements $[n]$ with k different cost functions $d_i : [n] \to \mathbb{Z}_+$ (for $i = 1, \cdots, k$) and budgets $\{B_i\}_{i=1}^k$, the goal is to find (if possible) a basis I with $d_i(I) \leq B_i$ for each $i \in [k]$. We obtain an algorithm that for any $\epsilon > 0$ finds in $n^{O(k^{1.5}/\epsilon)}$ time, a basis I with $d_i(I) \leq (1 + \epsilon)B_i$ for all $i \in [k]$. Previously, [8] obtained such an algorithm with $n^{O(k^2/\epsilon)}$ running time.

Low Congestion Routing. Given a directed graph $G = (V, E)$ with edge capacities $b : E \to \mathbb{Z}_+$, k source-sink pairs $\{(s_i, t_i)\}_{i=1}^k$ and a length bound Δ, the goal is to find an $s_i - t_i$ path P_i of length at most Δ for each pair $i \in [k]$ such that the number N_e of paths using any edge e is at most b_e. Using an LP-based reduction [7] this can be cast as an instance of DEGMAT. So we obtain violation bounds as in (7) which implies:

$$N_e \ \leq \ b_e + \min\left\{O(\sqrt{\Delta}\log n), O(\sqrt{b_e}\log n + \log^2 n)\right\}, \quad \forall e \in E.$$

Here $n = |V|$ is the number of vertices. Previous algorithms achieved bounds of $\Delta - 1$ [10] or $O(\sqrt{\log \Delta} \cdot \sqrt{b_j} + \log \Delta)$ [13] separately. We can also handle a richer set of routing requirements: given a laminar family $\mathcal{L}$ on the k pairs, with a requirement r_T on each set $T \in \mathcal{L}$, we want to find a multiset of paths so that there are at least r_T paths between the pairs in each $T \in \mathcal{L}$. Although this is not an instance of DEGMAT, the same approach works.

Overview of Techniques: Our algorithm in Theorem 4 is similar to the Lovett-Meka algorithm, and is also based on performing a Gaussian random walk at each step in a suitably chosen subspace. However, there some crucial differences. First, instead of updating each variable by the standard Gaussian $N(0,1)$, the variance for variable i is chosen proportional to $\min(y_i, 1 - y_i)$, i.e. proportional to how close it is to the boundary 0 or 1. This is crucial for getting the multiplicative error instead of the additive error in the constraints. However, this slows down the "progress" of variables toward reaching 0 or 1. To get around this, we add $O(\log n)$ additional constraints to define the subspace where the walk is performed: these restrict the total fractional value of variables in a particular "scale" to remain fixed. Using these we can ensure that enough variables eventually reach 0 or 1.

In order to handle the matroid constraints (Theorem 5) we need to incorporate them (although they are exponentially many) in defining the subspace where the random walk is performed. One difficulty that arises here is that we can no longer implement the random walk using "near tight" constraints as in [15] since we are unable to bound the dimension of near-tight matroid constraints. However, as is well known, the dimension of *exactly* tight matroid constraints is at most $n/2$ at any (strictly) fractional solution, and so we implement the random walk using exactly tight constraints. This requires us to truncate certain steps in the random walk (when we move out of the polytope), but we show that the effect of such truncations is negligible.

2   Matroid Partial Rounding

In this section we will prove Theorem 5 which also implies Theorem 4.

Let $y \in \mathbb{R}^n$ denote the initial solution. The algorithm will start with $X_0 = y$ and update this vector over time. Let X_t denote the vector at time t for $t = 1, \ldots, T$. The value of T will be defined later. Let $\ell = 2 \log_2 n$. We classify the n elements in to 2ℓ classes based on their initial values $y(i)$ as follows.

$$U_k := \begin{cases} \{i \in [n] : 2^{-k-1} < y(i) \le 2^{-k}\} & \text{if } 1 \le k \le \ell - 1 \\ \{i \in [n] : y(i) \le 2^{-\ell}\} & \text{if } k = \ell. \end{cases}$$

$$V_k := \begin{cases} \{i \in [n] : 2^{-k-1} < 1 - y(i) \le 2^{-k}\} & \text{if } 1 \le k \le \ell - 1 \\ \{i \in [n] : 1 - y(i) \le 2^{-\ell}\} & \text{if } k = \ell. \end{cases}$$

Note that the U_k's partition elements of value (in y) between 0 and $\frac{1}{2}$ and the V_k's form a symmetric partition of elements valued between $\frac{1}{2}$ and 1. This partition does not change over time, even though the value of variables might change. We define the "scale" of each element as:

$$s_i := 2^{-k}, \quad \forall i \in U_k \cup V_k, \quad \forall k \in [\ell].$$

Define $W_j(s) = \sum_{i=1}^n a_{ji}^2 \cdot s_i^2$ for each $j \in [m]$. Note that $W_j(s) \ge W_j(y)$ and

$$W_j(s) - 4 \cdot W_j(y) \le \sum_{i=1}^n a_{ji}^2 \cdot \frac{1}{n^4} = \frac{\|a_j\|^2}{n^4}.$$

So $\sqrt{W_j(y)} \leq \sqrt{W_j(s)} \leq 2\sqrt{W_j(y)} + \frac{\|a_j\|}{n^2}$. Our algorithm will find a solution y' such that

$$|\langle y' - y, a_j \rangle| \leq \lambda_j \cdot \sqrt{W_j(s)} + \frac{1}{n^2} \cdot \|a_j\|_2 \leq 2\lambda_j \cdot \sqrt{W_j(y)} + \frac{1 + \lambda_j}{n^2} \cdot \|a_j\|_2, \quad \forall j \in [m];$$

and y' has $\Omega(1)$ integral variables. This would suffice to prove Theorem 5.

Consider the polytope $\mathcal{Q}$ of points $x \in \mathbb{R}^n$ satisfying the following constraints.

$$x \in P(\mathcal{M}), \tag{8}$$

$$|\langle x - y, a_j \rangle| \leq \lambda_j \cdot \sqrt{W_j(s)} + \frac{1}{n^2} \cdot \|a_j\| \qquad \forall j \in [m], \tag{9}$$

$$x(U_k) = y(U_k) \qquad \forall k \in [\ell], \tag{10}$$

$$x(V_k) = y(V_k) \qquad \forall k \in [\ell], \tag{11}$$

$$0 \leq x_i \leq \alpha \cdot 2^{-k} \qquad \forall i \in U_k, \forall k \in [\ell], \tag{12}$$

$$0 \leq 1 - x_i \leq \alpha \cdot 2^{-k} \qquad \forall i \in V_k, \forall k \in [\ell]. \tag{13}$$

Here $\alpha > 1$ is some constant that will be fixed later. The algorithm will maintain the invariant that at any time $t \in [T]$, the solution X_t lies in $\mathcal{Q}$. In particular the constraint (8) requires that X_t stays in the matroid polytope. Constraint 9 controls the violation of the linear (degree) constraints over all time steps. The last two constraints (12) enforce that variables in U_k (and symmetrically V_k) do not deviate far beyond their original scale of 2^{-k}. The constraints (10) and (11) ensure that the total value of elements in U_k (and V_k) stay equal to the initial sum throughout the algorithm. These constraints will play a crucial role in arguing that the algorithm finds a partial coloring. Note that there are only 2ℓ such constraints.

The Algorithm: Let $\gamma = (n^5 \cdot \max_j \lambda_j)^{-1}$ and $T = K/\gamma^2$ where $K := 10\alpha^2$. The algorithm starts with solution $X_0 = y \in \mathcal{Q}$, and does the following at each time step $t = 0, 1, \cdots, T$:

1. Consider the set of constraints of $\mathcal{Q}$ that are tight at the point $x = X_t$, and define the following sets based on this.
   (a) Let $\mathcal{C}_t^{var}$ be the set of tight variable constraints among (12) and (13). This consists of:
      i. $i \in U_k$ (for any k) with $X_t(i) = 0$ or $X_t(i) = \min\{\alpha \cdot 2^{-k}, 1\}$; and
      ii. $i \in V_k$ (for any k) with $X_t(i) = 1$ or $X_t(i) = \max\{1 - \alpha \cdot 2^{-k}, 0\}$.
   (b) Let $\mathcal{C}_t^{deg}$ be the set of tight degree constraints from (9), i.e. those $j \in [m]$ with
      $$|\langle X^t - y, a_j \rangle| = \lambda_j \cdot \sqrt{W_j(s)} + \frac{1}{n^2} \|a_j\|.$$
   (c) Let $\mathcal{C}_t^{part}$ denote the set of the 2ℓ equality constraints (10) and (11).
   (d) Let $\mathcal{C}_t^{rank}$ be some linearly independent set of rank constraints that span the tight constraints among (8).

2. Let $\mathcal{V}_t$ denote the subspace orthogonal to all the constraints in $\mathcal{C}_t^{var}$, $\mathcal{C}_t^{deg}$, $\mathcal{C}_t^{part}$ and $\mathcal{C}_t^{rank}$. Let D be a $n \times n$ diagonal matrix with entries $d_{ii} = 1/s_i$, and let $\mathcal{V}_t'$ be the subspace $\mathcal{V}_t' = \{Dv : v \in \mathcal{V}_t\}$.
3. Let G_t be a random Gaussian vector in $\mathcal{V}_t'$. That is, $G_t := \sum_{h=1}^k g_h b_h$ where the g_h are iid $N(0,1)$, and $\{b_1, \ldots, b_k\}$ is some orthonormal basis of $\mathcal{V}_t'$.
4. Define $\overline{G}_t := D^{-1} G_t$. As $G_t \in \mathcal{V}_t'$, it must be that $G_t = Dv$ for some $v \in \mathcal{V}_t$ and thus $\overline{G}_t = D^{-1} G_t \in \mathcal{V}_t$.
5. Set $Y_t = X_t + \gamma \cdot \overline{G}_t$.
   (a) If $Y_t \in \mathcal{Q}$ then $X_{t+1} \leftarrow Y_t$ and continue to the next iteration.
   (b) Else $X_{t+1} \leftarrow$ the point in $\mathcal{Q}$ that lies on the line segment (X_t, Y_t) and is closest to Y_t. This can be found by binary search using a membership oracle for the matroid.

This completes the description of the algorithm. The analysis involves proving the following main lemma, which is our goal in the rest of this section.

Lemma 1. *With constant probability, the final solution X_T has $|\mathcal{C}_T^{var}| \geq \frac{n}{20}$.*

In the full version, we show how this lemma implies Theorem 5. All other missing proofs in this section can also be found in the full version [4].

Claim 1. *Given any $x \in P(\mathcal{M})$ with $0 < x < 1$, the subspace spanned by all the tight rank constraints has dimension at most $n/2$. Moreover, such a basis can be found in polynomial time.*

Claim 2. *The truncation Step 5b occurs at most n times.*

The statements of the following two lemmas are similar to those in [15], but the proofs require additional work since our random walk is different. The first lemma shows that the expected number of tight degree constraints at the end of the algorithm is not too high, and the second lemma shows that the expected number of tight variable constraints is large.

Lemma 2. $\mathbb{E}[|\mathcal{C}_T^{deg}|] < \frac{n}{4}$.

Proof: Note that $X_T - y = \gamma \sum_{t=0}^T \overline{G}_t + \sum_{q=1}^n \Delta_{t(q)}$ where Δs correspond to the truncation incurred during the iterations $t = t(1), \cdots, t(n)$ for which Step 5b applies (by Claim 2 there are at most n such iterations). Moreover for each q, $\Delta_{t(q)} = \delta \cdot \overline{G}_{t(q)}$ for some δ with $0 < |\delta| < \gamma$.

If $j \in \mathcal{C}_T^{deg}$, then $|\langle X_T - y, a_j \rangle| = \lambda_j \sqrt{W_j(s)} + \frac{1}{n^2} \cdot \|a_j\|$. As

$$|\langle X_T - y, a_j \rangle| \leq |\gamma \sum_{t=0}^T \langle \overline{G}_t, a_j \rangle| + \sum_{q=1}^n \gamma |\langle \overline{G}_{a(q)}, a_j \rangle| \leq |\gamma \sum_{t=0}^T \langle \overline{G}_t, a_j \rangle| + n\gamma \cdot \max_{t=0}^T |\langle \overline{G}_t, a_j \rangle|,$$

it follows that if $j \in \mathcal{C}_T^{deg}$, then one of the following events must occur:

$$|\gamma \sum_{t=0}^T \langle \overline{G}_t, a_j \rangle| \geq \lambda_j \sqrt{W_j(s)} \quad \text{or} \quad \max_{t=0}^T |\langle \overline{G}_t, a_j \rangle| \geq \frac{1}{\gamma n^3} \cdot \|a_j\|.$$

We bound the probabilities of these two events separately.

Event 1. In order to bound the probability of the first event, we consider the sequence $\{Z_t\}$ where $Z_t = \langle \overline{G}_t, a_j \rangle$, and note the following useful facts.

Observation 1. *The sequence $\{Z_t\}$ forms a martingale satisfying:*

1. $\mathbb{E}\left[Z_t \mid Z_{t-1}, \ldots, Z_0\right] = 0$ *for all t.*
2. $|Z_t| \leq n^2 \|a_j\|$ *whp for all t.*
3. $\mathbb{E}\left[Z_t^2 \mid Z_{t-1}, \ldots, Z_0\right] \leq \sum_{i=1}^{n} s_i^2 \cdot a_{ji}^2 = W_j(s)$ *for all t.*

Using a martingale concentration inequality, we obtain:

Claim 3. $\Pr\left[|\gamma \sum_{t=0}^{T} \langle \overline{G}_t, a_j \rangle| \geq \lambda_j \sqrt{W_j(s)}\right] \leq 2 \cdot \exp(-\lambda_j^2/3K).$

Event 2. Here we will just need simple conditional probabilities.

$$\Pr\left[\max_{t=0}^{T} |\langle \overline{G}_t, a_j \rangle| \geq \frac{1}{\gamma n^3} \cdot \|a_j\|\right] \leq \sum_{t=0}^{T} \Pr\left[|\langle \overline{G}_t, a_j \rangle| \geq n\|a_j\| \, \Big| \, \overline{G}_0, \cdots, \overline{G}_{t-1}\right],$$

which is at most $T \cdot \exp(-n)$. The first inequality uses $\gamma < n^{-4}$. The last inequality uses the fact that conditioned on previous $\overline{G}$s, $|\langle \overline{G}_t, a_j \rangle|$ is Gaussian with mean zero and variance at most $\|a_j\|^2$.

Combining the probabilities of the two events, $\Pr[j \in \mathcal{C}_T^{deg}] \leq 2\exp(-\lambda_j^2/3K) + T\exp(-n)$, we get

$$\mathbb{E}[|\mathcal{C}_T^{deg}|] < 2 \sum_{j=1}^{m} \exp\left(-\lambda_j^2/(30\alpha^2)\right) + \frac{Km}{\gamma^2 e^n} < 0.25n$$

To bound the first term we use the condition on the λ_j's in Theorem 5, with $K_0 = 30\alpha^2$. The latter term is negligible assuming $m < \gamma^2 2^n = 2^n/n^8$, and n is large enough.   ∎

We now prove that in expectation, at least $0.1n$ variables become tight at the end of the algorithm. This immediately implies Lemma 1.

Lemma 3. $\mathbb{E}[|\mathcal{C}_T^{var}|] \geq 0.1n.$

Proof: Define the following potential function, which will measure the progress of the algorithm toward the variables becoming tight.

$$\Phi(x) \quad := \quad \sum_{k=1}^{\ell} 2^{2k} \cdot \left(\sum_{i \in U_k} x(i)^2 + \sum_{i \in V_k} (1 - x(i))^2\right), \qquad \forall x \in \mathcal{Q}.$$

Note that since $X_T \in \mathcal{Q}$, we have $X_T(i) \leq \alpha \cdot 2^{-k}$ for $i \in U_k$ and $1 - X_T(i) \leq \alpha \cdot 2^{-k}$ for $i \in V_k$. So $\Phi(X_T) \leq \alpha^2 \cdot n$. We also define the "incremental function" for any $x \in \mathcal{Q}$ and $g \in \mathbb{R}^n$, $f(x, g) := \Phi(x + \gamma D^{-1}g) - \Phi(x)$

$$= \gamma^2 \sum_{i=1}^{n} g(i)^2 + 2 \sum_{k=1}^{\ell} 2^{2k} \cdot \left(\sum_{i \in U_k} x(i)\gamma s_i \cdot g(i) - \sum_{i \in V_k} (1 - x(i))\gamma s_i \cdot g(i)\right), \quad (14)$$

where we used $s_i = 2^{-k}$ for $i \in U_k \cup V_k$.

Recall that D^{-1} is the $n \times n$ diagonal matrix with entries $(s_1, \cdots, s_n)$. Suppose the algorithm was modified to never have the truncation step 5b, then in any iteration t, the increase $\Phi(Y_t) - \Phi(X_t) = f(X_t, G_t)$ where G_t is a random Gaussian in $\mathcal{V}'_t$. To deal with the effect of truncation, we consider the worst possible contribution truncation could have. We define the following quantity:

$$M \quad := \quad \max_{t=0}^{T} \left(\gamma^2 \|G_t\|_2^2 + 2\gamma\alpha\|G_t\|_1 \right).$$

Recall that $\Phi(X_{t+1}) - \Phi(X_t) = f(X_t, \delta_t G_t)$ for some $\delta_t \in (0,1]$, and $\delta_t < 1$ if and only if the truncation step 5b occurs in iteration t. The following is by simple calculation.

$$f(X_t, \delta G_t) \geq f(X_t, G_t) - M, \quad \forall \delta \in (0,1). \tag{15}$$

This implies that $\Phi(X_T) - \Phi(X_0)$ equals

$$\sum_{t=0}^{T} f(X_t, \delta_t G_t) \geq \sum_{t=0}^{T} f(X_t, G_t) - M \sum_{t=0}^{T} \mathbb{1}[\text{step 5b occurs in iteration } t]$$

$$\geq \sum_{t=0}^{T} f(X_t, G_t) - nM \quad \text{(by Claim 2)} \tag{16}$$

Claim 4. $\mathbb{E}[\Phi(X_T)] - \Phi(y) \geq \gamma^2 T \cdot \mathbb{E}[\dim(\mathcal{V}_T)] - 1$.

By Claim 1 and the fact that $|\mathcal{C}_T^{part}| = 2\ell$, we have

$$\dim(\mathcal{V}_T) \geq n - \dim(\mathcal{C}_T^{var}) - \dim(\mathcal{C}_T^{deg}) - \dim(\mathcal{C}_T^{rank}) - \dim(\mathcal{C}_T^{part})$$

$$\geq \frac{n}{2} - 2\ell - \dim(\mathcal{C}_T^{var}) - \dim(\mathcal{C}_T^{deg})$$

Taking expectations and by Claim 2, $\mathbb{E}[\dim(\mathcal{V}_T)] \geq \frac{n}{4} - 2\ell - \dim(\mathcal{C}_T^{var})$. Using $\Phi(X_T) \leq \alpha^2 n$ and Claim 4, we obtain:

$$\alpha^2 n \geq \mathbb{E}[\Phi_T] \geq \gamma^2 T \cdot \left(\frac{n}{4} - 2\ell - \mathbb{E}[\dim(\mathcal{C}_T^{var})] \right) - 1.$$

Rearranging and using $T = K/\gamma^2$, $K = 10\alpha^2$ and $\ell = \log n$ gives $\mathbb{E}[\dim(\mathcal{C}_T^{var})] \geq \frac{n}{4} - \frac{\alpha^2 n}{K} - 2\ell - \frac{1}{K}$, which proves the Lemma 3. ∎

References

1. Bansal, N.: Constructive algorithms for discrepancy minimization. In: Foundations of Computer Science (FOCS), pp. 3–10 (2010)
2. Bansal, N., Charikar, M., Krishnaswamy, R., Li, S.: Better algorithms and hardness for broadcast scheduling via a discrepancy approach. In: SODA, pp. 55–71 (2014)
3. Bansal, N., Khandekar, R., Könemann, J., Nagarajan, V., Peis, B.: On generalizations of network design problems with degree bounds. Math. Program. **141**(1–2), 479–506 (2013)

4. Bansal, N., Nagarajan, V.: Approximation-friendly discrepancy rounding. CoRR abs/1512.02254 (2015)
5. Beck, J., Fiala, T.: Integer-making theorems. Discrete Appl. Math. **3**, 1–8 (1981)
6. Charikar, M., Newman, A., Nikolov, A.: Tight hardness results for minimizing discrepancy. In: SODA, pp. 1607–1614 (2011)
7. Chekuri, C., Vondrak, J., Zenklusen, R.: Dependent randomized rounding via exchange properties of combinatorial structures. In: FOCS, pp. 575–584 (2010)
8. Grandoni, F., Ravi, R., Singh, M., Zenklusen, R.: New approaches to multi-objective optimization. Math. Program. **146**(1–2), 525–554 (2014)
9. Harvey, N.J.A., Schwartz, R., Singh, M.: Discrepancy without partial colorings. In: APPROX/RANDOM 2014, pp. 258–273 (2014)
10. Karp, R.M., Leighton, F.T., Rivest, R.L., Thompson, C.D., Vazirani, U.V., Vazirani, V.V.: Global wire routing in two-dimensional arrays. Algorithmica **2**, 113–129 (1987)
11. Király, T., Lau, L.C., Singh, M.: Degree bounded matroids and submodular flows. In: Lodi, A., Panconesi, A., Rinaldi, G. (eds.) IPCO 2008. LNCS, vol. 5035, pp. 259–272. Springer, Heidelberg (2008)
12. Lau, L.C., Ravi, R., Singh, M.: Iterative Methods in Combinatorial Optimization. Cambridge University Press, Cambridge (2011)
13. Leighton, F.T., Lu, C., Rao, S., Srinivasan, A.: New algorithmic aspects of the local lemma with applications to routing and partitioning. SIAM J. Comput. **31**(2), 626–641 (2001)
14. Lovasz, L., Spencer, J., Vesztergombi, K.: Discrepancy of set-systems and matrices. Eur. J. Combin. **7**, 151–160 (1986)
15. Lovett, S., Meka, R.: Constructive discrepancy minimization by walking on the edges. In: FOCS, pp. 61–67 (2012)
16. Matoušek, J.: Geometric Discrepancy: An Illustrated Guide. Springer, Heidelberg (2010)
17. Nikolov, A., Talwar, K.: Approximating hereditary discrepancy via small width ellipsoids. In: Symposium on Discrete Algorithms, SODA, pp. 324–336 (2015)
18. Rothvoss, T.: Approximating bin packing within o(log OPT * log log OPT) bins. In: FOCS, pp. 20–29 (2013)
19. Rothvoss, T.: Constructive discrepancy minimization for convex sets. In: IEEE Symposium on Foundations of Computer Science, FOCS, pp. 140–145 (2014)
20. Singh, M., Lau, L.C.: Approximating minimum bounded degree spanning trees to within one of optimal. In: STOC, pp. 661–670 (2007)
21. Spencer, J.: Six standard deviations suffice. Trans. Am. Math. Soc. **289**(2), 679–706 (1985)
22. Srinivasan, A.: Improving the discrepancy bound for sparse matrices: better approximations for sparse lattice approximation problems. In: Symposium on Discrete Algorithms (SODA), pp. 692–701 (1997)
23. Vazirani, V.V.: Approximation Algorithms. Springer-Verlag, Heidelberg (2001)
24. Williamson, D., Shmoys, D.: The Design of Approximation Algorithms. Cambridge University Press, Cambridge (2011)

Deciding Emptiness of the Gomory-Chvátal Closure is NP-Complete, Even for a Rational Polyhedron Containing No Integer Point

Gérard Cornuéjols[1] and Yanjun Li[2]($\boxtimes$)

[1] Tepper School of Business, Carnegie Mellon University,
Pittsburgh, PA 15213, USA
gc0v@andrew.cmu.edu
[2] Krannert School of Management, Purdue University,
West Lafayette, IN 47906, USA
li14@purdue.edu

Abstract. Gomory-Chvátal cuts are prominent in integer programming. The Gomory-Chvátal closure of a polyhedron is the intersection of all half spaces defined by its Gomory-Chvátal cuts. In this paper, we show that it is $\mathcal{NP}$-complete to decide whether the Gomory-Chvátal closure of a rational polyhedron is empty, even when this polyhedron contains no integer point. This implies that the problem of deciding whether the Gomory-Chvátal closure of a rational polyhedron P is identical to the integer hull of P is $\mathcal{NP}$-hard. Similar results are also proved for the $\{-1,0,1\}$-cuts and $\{0,1\}$-cuts, two special types of Gomory-Chvátal cuts with coefficients restricted in $\{-1,0,1\}$ and $\{0,1\}$, respectively.

Keywords: Integer programming · Gomory-Chvátal cuts · Gomory-Chvátal closure · Integer hull · Computational complexity

1 Introduction

Throughout this paper we will assume a knowledge of elementary integer programming definitions and results. One may use the book by Conforti et al. [4] as a reference.

We consider the integer program: min wx s.t. $Ax \leq b$, $x \in \mathbb{Z}^n$, where $b \in \mathbb{Z}^m, w \in \mathbb{Z}^n$ and $A \in \mathbb{Z}^{m \times n}$. The polyhedron associated with the linear programming relaxation of the integer program is denoted by $P \equiv \{x \in \mathbb{R}^n : Ax \leq b\}$. Polyhedra of this form, where $b \in \mathbb{Z}^m$ and $A \in \mathbb{Z}^{m \times n}$, are called *rational polyhedra*. The convex hull of all feasible solutions of the integer program is a polyhedron called the *integer hull* and denoted by P_I. An inequality of the form $cx \leq \lfloor d \rfloor$ is called a *Gomory-Chvátal cut* of P if $cx \leq d$ is valid for every $x \in P$, where $c \in \mathbb{Z}^n$. Gomory-Chvátal cuts were originally proposed by Gomory [9] as a method for solving integer programming and combinatorial optimization problems. Chvátal [3] introduced a notion of closure associated with these cuts. The *Gomory-Chvátal closure* of P is $P' \equiv \{x \in P : cx \leq \lfloor d \rfloor \ \forall c \in \mathbb{Z}^n$ and

© Springer International Publishing Switzerland 2016
Q. Louveaux and M. Skutella (Eds.): IPCO 2016, LNCS 9682, pp. 387–397, 2016.
DOI: 10.1007/978-3-319-33461-5_32

$d \in \mathbb{R}$ such that $cx \leq d$ is valid for P}. Clearly, $P_I \subseteq P' \subseteq P$, and, from the theory of Gomory-Chvátal cuts, the second inclusion is strict when $P \neq P_I$.

The *separation problem* for the Gomory-Chvátal closure of a rational polyhedron (GC-Sep) is the following: Given a rational polyhedron $P \subseteq \mathbb{R}^n$ and a point $x^* \in \mathbb{R}^n$, either give a Gomory-Chvátal cut of P such that x^* violates the cut, or conclude that $x^* \in P'$. The *optimization problem* over the Gomory-Chvátal closure of a rational polyhedron (GC-Opt) is: Given a rational polyhedron $P \subseteq \mathbb{R}^n$ and a vector $c \in \mathbb{Z}^n$, either find a point $x^* \in P'$ that optimizes the function cx, or conclude that the optimal value of cx over P' is unbounded, or conclude that $P' = \emptyset$. It follows from a general result of Grötschel, Lovász and Schrijver [10] that solving GC-Sep in polynomial time is equivalent to solving GC-Opt in polynomial time. Eisenbrand [6] proved that GC-Sep is $\mathcal{NP}$-hard, which implies the $\mathcal{NP}$-hardness of GC-Opt.

In this paper, we show a stronger result: Given a rational polyhedron P such that $P_I = \emptyset$, it is $\mathcal{NP}$-complete to decide whether $P' = \emptyset$.

A Gomory-Chvátal cut is called a $\{-1, 0, 1\}$-*cut* (or $\{0, 1\}$-*cut*) if the vector of variable coefficients $c \in \{-1, 0, 1\}^n$ (or $\{0, 1\}^n$). The $\{-1, 0, 1\}$-*closure* (or $\{0, 1\}$-*closure*) of P is $P'_{\{-1,0,1\}} \equiv \{x \in P : cx \leq \lfloor d \rfloor \; \forall c \in \{-1, 0, 1\}^n$ and $d \in \mathbb{R}$ such that $cx \leq d$ is valid for $P\}$ (or $P'_{\{0,1\}} \equiv \{x \in P : cx \leq \lfloor d \rfloor \; \forall c \in \{0, 1\}^n$ and $d \in \mathbb{R}$ such that $cx \leq d$ is valid for $P\}$). We show that, given a polyhedron P such that $P_I = \emptyset$, it is $\mathcal{NP}$-complete to decide whether $P'_{\{-1,0,1\}} = \emptyset$ (or $P'_{\{0,1\}} = \emptyset$).

We borrowed some ideas from Mahajan and Ralphs [13] to construct polyhedra P in the proof of our $\mathcal{NP}$-completeness results. In their paper the following disjunctive infeasibility problem is proved to be $\mathcal{NP}$-complete: Given a polyhedron $P \subset \mathbb{R}^n$, does there exist $\pi \in \mathbb{Z}^n$ and $\pi_0 \in \mathbb{Z}$ such that $\{x \in P : \pi x \leq \pi_0$ or $\pi x \geq \pi_0 + 1\} = \emptyset$? The polyhedra P used in [13] are simplices, whereas our constructed polyhedra are convex hulls of $n + 2$ or $n + 3$ vectors in $\mathbb{R}^n$. The well-known partition problem is reduced to the disjunctive infeasibility problem in Mahajan and Ralphs' proof, whereas, in our proofs, the single constraint integer programming feasibility problem is reduced to the emptiness problem of Gomory-Chvátal closure, and the partition problem is reduced to the emptiness problems of $\{-1, 0, 1\}$-closure and $\{0, 1\}$-closure.

The rest of the paper is organized as follows. In Sect. 2, we prove the $\mathcal{NP}$-completeness of deciding whether the Gomory-Chvátal closure is empty. In Sect. 3, we prove the $\mathcal{NP}$-completeness of deciding whether the $\{-1, 0, 1\}$-closure (or $\{0, 1\}$-closure) is empty. Lastly, in Sect. 4, we present conclusions and open questions.

2   Deciding Emptiness of the Gomory-Chvátal Closure

In this section, we prove two results. First we show that it is $\mathcal{NP}$-complete to decide whether the Gomory-Chvátal closure of a rational polyhedron P is empty.

We then show that this problem is $\mathcal{NP}$-complete, even when the polyhedron P is known to contain no integer point. We first observe that these problems are in the complexity class $\mathcal{NP}$.

Lemma 1. *Deciding whether the Gomory-Chvátal closure of a rational polyhedron is empty belongs to the complexity class $\mathcal{NP}$.*

Proof. Let $P \equiv \{x \in \mathbb{R}^n : Ax \leq b\}$ be a rational polyhedron, where $b \in \mathbb{Z}^m$, and $A \in \mathbb{Z}^{m \times n}$. Chvátal [3] showed that there is only a finite number of inequalities needed to describe the Gomory-Chvátal closure, namely inequalities $uAx \leq \lfloor ub \rfloor$ where $u \in \mathbb{R}^m$ is a vector satisfying $uA \in \mathbb{Z}^n$ and $0 \leq u < 1$. Note that the integer vectors $d \equiv uA$ in these inequalities have components satisfying $-\sum_{i=1}^m |a_{ij}| \leq d_j \leq \sum_{i=1}^m |a_{ij}|$. Similarly, $d_0 \equiv \lfloor ub \rfloor$ satisfies $-\sum_{i=1}^m |b_i| \leq d_0 \leq \sum_{i=1}^m |b_i|$. Therefore the above inequalities are described by coefficients whose encoding size is polynomial in the size of the input. To certify that the Gomory-Chvátal closure is empty, we appeal to Helly's theorem: If the Gomory-Chvátal closure is empty, there exist n+1 of these inequalities whose intersection is empty. A list of $n + 1$ such inequalities is a polynomial certificate that the Gomory-Chvátal closure is empty. $\qquad\square$

Theorem 1. *It is $\mathcal{NP}$-complete to decide whether the Gomory-Chvátal closure of a rational polyhedron is empty.*

Proof. The theorem will be proved by polynomially reducing the following single constraint integer programming feasibility problem, which is known to be $\mathcal{NP}$-complete [12], to the problem of deciding whether $P' = \emptyset$ for a rational polyhedron P.

Single Constraint Integer Programming Feasibility Problem: Given a finite set of non-negative integers $\{a_i\}_{i=1}^s$ and a non-negative integer b, is there a set of non-negative integers $\{x_i\}_{i=1}^s$ satisfying $\sum_{i=1}^s a_i x_i = b$?

We consider the Single Constraint Integer Programming Feasibility Problem with $s = n - 1$ and $n \geq 3$. We assume without loss of generality that the greatest common divisor of $a_1, a_2, \cdots, a_{n-1}$ is 1, and $2 < a_1 < a_2 < \cdots < a_{n-1} < b$. So $b \geq n + 2$. Let $r = n + 1 + \frac{1}{2b}$. So r is a rational number satisfying $r < b$ and $rb \notin \mathbb{Z}_+$. We will show:

Reduction:

The Single Constraint Integer Programming Feasibility Problem can be polynomially reduced to the problem of deciding whether $P' = \emptyset$ for the polyhedron $P \subseteq \mathbb{R}^n$ that is the convex hull of the following $n + 3$ vectors:
$v_1 = (\frac{1}{2b}, 0, \cdots, 0, \frac{1}{2b})$, $v_2 = (0, \frac{1}{2b}, 0, \cdots, 0, \frac{1}{2b})$, $\cdots, v_{n-1} = (0, \cdots, 0, \frac{1}{2b}, \frac{1}{2b})$, $v_n = (0, \cdots, 0, \frac{1}{2} + \frac{3}{2r})$, $v_{n+1} = (a_1, a_2, \cdots, a_{n-1}, -b + \frac{1}{2})$, $v_{n+2} = ((1-r)a_1, (1-r)a_2, \cdots, (1-r)a_{n-1}, (r-1)b + 1)$, and $v_{n+3} = (0, \cdots, 0, \frac{1}{2r})$.

To show that this reduction is correct, we prove the following two claims; we then observe that converting the vectors $v_1, v_2, \cdots, v_{n+3}$ into an inequality description of P can be done in polynomial time.

Claim 1. There is a set of non-negative integers $\{w_i\}_{i=1}^{n-1}$ satisfying $\sum_{i=1}^{n-1} a_i w_i = b$ only if $P' = \emptyset$.

Proof. Consider an inequality $cx \leq q$, where $c = (c_1, c_2, \cdots, c_n)$, $c_i = -w_i$ for $1 \leq i \leq n-1$, $c_n = -1$, and $q = \max\{-\frac{1}{2b}, -\frac{1}{2r}\} = -\frac{1}{2b}$. Because $1 < a_i < b$ for $1 \leq i \leq n-1$ and $\sum_{i=1}^{n-1} a_i w_i = b$, it is easy to verify that $-1 < cv_i \leq -\frac{1}{2b}$ for $1 \leq i \leq n-1$. In addition, $-1 < cv_n < -\frac{1}{2}$, $cv_{n+1} = -\frac{1}{2}$, $cv_{n+2} = -1$, and $cv_{n+3} = -\frac{1}{2r}$. So $cx \leq q$ is valid for P, and the associated Gomory-Chvátal inequality is $cx \leq \lfloor q \rfloor$ $(= -1)$. We can see that v_{n+2} is the only vector in P that satisfies the inequality $cx \leq \lfloor q \rfloor$. Consider the inequality $fx \leq g$, where $fx = x_n$ and $g = (r-1)b + 1$. It can be easily checked that every v_i satisfies $fx \leq g$, so $fx \leq g$ is valid for P, and $fx \leq \lfloor g \rfloor$ $(= \lfloor rb \rfloor - b + 1)$ is a Gomory-Chvátal inequality of P. Since $rb \notin \mathbb{Z}_+$, $fx \leq \lfloor g \rfloor$ is violated by v_{n+2}. Now we can conclude that $P' = \emptyset$ and Claim 1 is proved.

Claim 2. There is a set of non-negative integers $\{w_i\}_{i=1}^{n-1}$ satisfying $\sum_{i=1}^{n-1} a_i w_i = b$ if $P' = \emptyset$.

Proof. Let $v_0 \equiv (0, 0, \cdots, 0, \frac{1}{2r} + \frac{1}{2})$. So $v_0 \in P$ because $v_0 = \alpha v_n + (1-\alpha) v_{n+3}$ for some $0 < \alpha < 1$. Let $cx \leq \lfloor q \rfloor$ be a Gomory-Chvátal inequality of P that is violated by v_0, where $c = (c_1, c_2, \cdots, c_n) \in \mathbb{Z}^n$ and $cx \leq q$ is valid for P. Then $c_n \neq 0$, otherwise we would have $0 = cv_0 \leq q$, contradicting that $cv_0 > \lfloor q \rfloor$.

Let $\Delta \equiv \sum_{i=1}^{n-1} c_i a_i - c_n b$. First, we show that $c_n \leq -1$ by deriving contradiction in the following two cases:

Case 1. $c_n \geq 1$ and $\Delta \geq 1$. If $\frac{c_n}{2r} \leq \Delta - 1$, then $cv_0 = \frac{c_n}{2r} + \frac{c_n}{2} \leq \Delta - 1 + \frac{c_n}{2} < \lfloor \Delta + \frac{c_n}{2} \rfloor = \lfloor cv_{n+1} \rfloor \leq \lfloor q \rfloor$, which contradicts that $cv_0 > \lfloor q \rfloor$. If $\frac{c_n}{2r} > \Delta - 1$, then $cv_n - cv_0 = \frac{c_n}{r} > 2(\Delta - 1)$. If $\Delta \geq 2$, then $cv_n - cv_0 > 2$, so $cv_0 < cv_n - 2 < \lfloor cv_n \rfloor \leq \lfloor q \rfloor$, a contradiction to that $cv_0 > \lfloor q \rfloor$. If $\Delta = 1$, then $\frac{c_n}{r} \geq 1$. Otherwise, if $\frac{c_n}{r} < 1$, then $cv_0 = \frac{c_n}{2r} + \frac{c_n}{2} < \frac{1}{2} + \frac{c_n}{2} \leq \lfloor 1 + \frac{c_n}{2} \rfloor = \lfloor cv_{n+1} \rfloor \leq \lfloor q \rfloor$, a contradiction. Since $\frac{c_n}{r} \geq 1$, $cv_n - cv_0 \geq 1$, therefore $cv_0 \leq cv_n - 1 < \lfloor cv_n \rfloor \leq \lfloor q \rfloor$, a contradiction again.

Case 2. $c_n \geq 1$ and $\Delta \leq 0$. Because $r > n + 1 > 3$, $cv_0 = \frac{c_n}{2r} + \frac{c_n}{2} < c_n \leq (1-r)\Delta + c_n = cv_{n+2}$. Hence, $cv_0 < \lfloor cv_{n+2} \rfloor \leq \lfloor q \rfloor$, a contradiction.

It is easy to see that $c_n = -1$. Otherwise, if $c_n \leq -2$, then $cv_{n+3} - cv_0 = -\frac{c_n}{2} \geq 1$, which implies $cv_0 \leq cv_{n+3} - 1 < \lfloor cv_{n+3} \rfloor \leq \lfloor q \rfloor$, a contradiction.

Now we show that $\Delta = 0$. If $\Delta \geq 1$, then $cv_0 = -\frac{1}{2r} - \frac{1}{2} < 0 < \Delta - \frac{1}{2} = cv_{n+1}$, implying $cv_0 < \lfloor cv_{n+1} \rfloor \leq \lfloor q \rfloor$, a contradiction. If $\Delta \leq -1$, then, because $r > 3$, $cv_0 < 0 < (1-r)\Delta - 1 = cv_{n+2}$, a contradiction again.

We claim that $c_i \leq 0$ for $i = 1, 2, \cdots, n-1$. Otherwise, if $c_i \geq 1$ for some $1 \leq i \leq n-1$, then $cv_0 < 0 \leq \frac{c_i}{2b} - \frac{1}{2b} = cv_i$, a contradiction.

Now let $w_i = -c_i$ for $1 \leq i \leq n-1$. Then Claim 2 is proved.

To complete the proof, it suffices to show that a description of P in the form of $\tilde{A}x \leq \tilde{b}$, where $\tilde{A} \in \mathbb{Z}^{m \times n}$ and $\tilde{b} \in \mathbb{Z}^m$, can be obtained in polynomial time from the vectors $v_1, v_2, \cdots, v_{n+3}$. We can see from the coordinates of $v_1, v_2, \cdots, v_n$ and v_{n+3} that P is a n-dimensional polyhedron. Let i be a counter looping through $n, n+1$ and $n+2$. For every i vectors of $v_1, v_2, \cdots, v_{n+3}$, check if they are on a unique hyperplane by solving linear equations. If yes, then further check if the $n + 3 - i$ other vectors are all on one side of the hyperplane. If yes again, then the equation $\tilde{c}x = \tilde{d}$ of the hyperplane with integral $\tilde{c}$ and $\tilde{d}$ whose greatest common divisor is 1 yields a linear inequality of $\tilde{A}x \leq \tilde{b}$. One can easily see that this process takes polynomial time and the size of $\tilde{A}$ and $\tilde{b}$ is polynomial in the size of $v_1, v_2, \cdots, v_{n+3}$.   □

Theorem 2. *Given a rational polyhedron containing no integer point, it is $\mathcal{NP}$-complete to decide whether its Gomory-Chvátal closure is empty.*

Proof. We build on Theorem 1 and show that the polytope P that was used in the reduction contains no integer point.

This is equivalent to showing that $P^d \equiv P \cap \{x \in \mathbb{R}^n : x_n = d\}$ contains no integer points for every integer $d \in [-b+1, nb+1]$. Let P_1 be the convex hull of $v_1, v_2, \cdots, v_n, v_{n+1}$ and v_{n+3}, and let P_2 be the convex hull of $v_1, v_2, \cdots, v_n, v_{n+2}$ and v_{n+3}. Since $v_0 = \frac{r-1}{r} v_{n+1} + \frac{1}{r} v_{n+2}$ and $v_0 = \alpha v_n + (1 - \alpha) v_{n+3}$ for some $0 < \alpha < 1$, it is sufficient to show: (a) $P_1^d \equiv P_1 \cap \{x \in \mathbb{R}^n : x_n = d\}$ contains no integer points for every integer $d \in [-b+1, 0]$; (b) $P_2^d \equiv P_2 \cap \{x \in \mathbb{R}^n : x_n = d\}$ contains no integer points for every integer $d \in [1, nb+1]$.

We first prove (a). Because $v_1, v_2, \cdots, v_n, v_{n+3} \in \{x \in \mathbb{R}^n : x_n > 0\}$ and $v_{n+1} \in \{x \in \mathbb{R}^n : x_n < 0\}$, it is easy to verify by calculation that P_1^d, where d is an integer in $[-b+1, 0]$, is the convex hull of the $n + 1$ vectors:

$$\left(\frac{1-2bd}{2b^2-b+1} a_1 + \frac{b+d-\frac{1}{2}}{2b^2-b+1}, \frac{1-2bd}{2b^2-b+1} a_2, \frac{1-2bd}{2b^2-b+1} a_3, \cdots, \frac{1-2bd}{2b^2-b+1} a_{n-1}, d\right),$$

$$\left(\frac{1-2bd}{2b^2-b+1} a_1, \frac{1-2bd}{2b^2-b+1} a_2 + \frac{b+d-\frac{1}{2}}{2b^2-b+1}, \frac{1-2bd}{2b^2-b+1} a_3, \cdots, \frac{1-2bd}{2b^2-b+1} a_{n-1}, d\right),$$

$$\cdots\cdots\cdots\cdots$$

$$\left(\frac{1-2bd}{2b^2-b+1} a_1, \frac{1-2bd}{2b^2-b+1} a_2, \cdots, \frac{1-2bd}{2b^2-b+1} a_{n-2}, \frac{1-2bd}{2b^2-b+1} a_{n-1} + \frac{b+d-\frac{1}{2}}{2b^2-b+1}, d\right),$$

$$\left(\frac{r+3-2rd}{2rb+3} a_1, \frac{r+3-2rd}{2rb+3} a_2, \frac{r+3-2rd}{2rb+3} a_3, \cdots, \frac{r+3-2rd}{2rb+3} a_{n-1}, d\right),$$

$$\left(\frac{1-2rd}{2rb+1-r} a_1, \frac{1-2rd}{2rb+1-r} a_2, \frac{1-2rd}{2rb+1-r} a_3, \cdots, \frac{1-2rd}{2rb+1-r} a_{n-1}, d\right).$$

Indeed, the first n vectors above are obtained by intersecting the hyperplane $x_n = d$ with the line segment $v_i v_{n+1}$, for $i = 1, 2, \cdots, n$, and the last vector is obtained by intersecting the hyperplane $x_n = d$ with the line segment $v_{n+3} v_{n+1}$.

Since $P_1^d \subset \{x \in \mathbb{R}^n : x_n = d\}$, we only need to consider the convex hull of the following $n + 1$ vectors in $\mathbb{R}^{n-1}$: $\frac{1-2bd}{2b^2-b+1} a + \frac{b+d-\frac{1}{2}}{2b^2-b+1} e_1, \frac{1-2bd}{2b^2-b+1} a + \frac{b+d-\frac{1}{2}}{2b^2-b+1} e_2, \cdots, \frac{1-2bd}{2b^2-b+1} a + \frac{b+d-\frac{1}{2}}{2b^2-b+1} e_{n-1}, \frac{r+3-2rd}{2rb+3} a$ and $\frac{1-2rd}{2rb+1-r} a$, where $a \equiv (a_1, a_2, \cdots, a_{n-1})$ and e_i is the i-th unit vector. Let $\tilde{P}_1^d$ denote the convex hull. Since $r < b$ and $d \geq -b+1$, it is easy to verify that $0 < \frac{1-2bd}{2b^2-b+1} < \frac{1-2rd}{2rb+1-r} < \frac{r+3-2rd}{2rb+3} < 1$. So $\tilde{P}_1^d \subseteq Q_1^d$, where Q_1^d is the convex hull of the $n + 1$ vectors:

$$z_0 \equiv \frac{1-2bd}{2b^2-b+1} a,$$

$$z_1 \equiv \frac{1-2bd}{2b^2-b+1} a + \frac{b+d-\frac{1}{2}}{2b^2-b+1} e_1,$$

$$z_2 \equiv \frac{1-2bd}{2b^2-b+1} a + \frac{b+d-\frac{1}{2}}{2b^2-b+1} e_2,$$

$$\cdots\cdots\cdots\cdots$$

$$z_{n-1} \equiv \frac{1-2bd}{2b^2-b+1}a + \frac{b+d-\frac{1}{2}}{2b^2-b+1}e_{n-1},$$
$$z_n \equiv \frac{r+3-2rd}{2rb+3}a.$$

To prove (a), it suffices to show the following claim.

Claim 3. There is no integer point in Q_1^d.

Proof. By contradiction, suppose $\tilde{v} \in Q_1^d \cap \mathbb{Z}^{n-1}$. Then there must exist a vector $\tilde{v}' \equiv \beta_0 a$, where $0 < \beta_0 < 1$, such that $\|\tilde{v} - \tilde{v}'\|_\infty \equiv \max_{1 \le i \le n-1} |\tilde{v}_i - \tilde{v}_i'| \le \frac{b+d-\frac{1}{2}}{2b^2-b+1} \le \frac{b-\frac{1}{2}}{2b^2-b+1} < \frac{1}{2b}$. From the construction of P, it is easy to see that $0 < \tilde{v}_i < a_i$ for $i = 1, 2, \cdots, n-1$. Because the greatest common divisor of $a_1, a_2, \cdots, a_{n-1}$ is 1, there exists no integer point on the line segment connecting 0 and a except for the two end points. Therefore, there exists $1 \le i_0 \le n-2$ such that $(\tilde{v}_{i_0}, \tilde{v}_{n-1})$ is not on the line segment connecting $(0,0)$ and (a_{i_0}, a_{n-1}) in $\mathbb{R}^2$. To derive contradiction, we show below that $\|(\tilde{v}_{i_0}, \tilde{v}_{n-1}) - (\tilde{v}_{i_0}', \tilde{v}_{n-1}')\|_\infty = \max(|\tilde{v}_{i_0} - \tilde{v}_{i_0}'|, |\tilde{v}_{n-1} - \tilde{v}_{n-1}'|) \ge \frac{1}{2(b-1)}$.

Let L denote the line segment connecting $(0,0)$ and (a_{i_0}, a_{n-1}) in $\mathbb{R}^2$. We know that $(\tilde{v}_{i_0}', \tilde{v}_{n-1}')$ is on L. Because the integer points between 0 and (a_{i_0}, a_{n-1}) that are not on L are symmetric across $(\frac{a_{i_0}}{2}, \frac{a_{n-1}}{2})$, we may assume without loss of generality that $\frac{\tilde{v}_{n-1}}{\tilde{v}_{i_0}} < \frac{a_{n-1}}{a_{i_0}}$. It is not hard to see that the shortest distance under $\|\cdot\|_\infty$ between a point on L and $(\tilde{v}_{i_0}, \tilde{v}_{n-1})$ is attained at a point on the segment L connecting $(\frac{\tilde{v}_{n-1}a_{i_0}}{a_{n-1}}, \tilde{v}_{n-1})$ and $(\tilde{v}_{i_0}, \frac{\tilde{v}_{i_0}a_{n-1}}{a_{i_0}})$. Since $\tilde{v}_{i_0} > \frac{\tilde{v}_{n-1}a_{i_0}}{a_{n-1}}$, $\|(\tilde{v}_{i_0}, \tilde{v}_{n-1}) - (\frac{\tilde{v}_{n-1}a_{i_0}}{a_{n-1}}, \tilde{v}_{n-1})\|_\infty \ge \frac{1}{a_{n-1}} \ge \frac{1}{b-1}$. Because $a_{n-1} \ge a_{i_0}$, $\|(\tilde{v}_{i_0}, \tilde{v}_{n-1}) - (\tilde{v}_{i_0}, \frac{\tilde{v}_{i_0}a_{n-1}}{a_{i_0}})\|_\infty \ge \frac{1}{b-1}$. So it follows that the shortest distance under $\|\cdot\|_\infty$ between a point on L and $(\tilde{v}_{i_0}, \tilde{v}_{n-1})$ is no less than $\frac{1}{2(b-1)}$. Therefore, $\|(\tilde{v}_{i_0}, \tilde{v}_{n-1}) - (\tilde{v}_{i_0}', \tilde{v}_{n-1}')\|_\infty \ge \frac{1}{2(b-1)}$. Claim 3 is proved.

Next we prove (b). Because $v_1, v_2, \cdots, v_n, v_{n+3} \in \{x \in \mathbb{R}^n : x_n < 1\}$ and $v_{n+2} \in \{x \in \mathbb{R}^n : x_n > 1\}$, we know by calculation that P_2^d, where d is an integer in $[1, nb+1]$, is the convex hull of the $n+1$ vectors:

$$\left(\frac{(r-1)(1-2bd)}{2(r-1)b^2+2b-1}a_1 + \frac{(r-1)b+1-d}{2(r-1)b^2+2b-1}, \frac{(r-1)(1-2bd)}{2(r-1)b^2+2b-1}a_2, \frac{(r-1)(1-2bd)}{2(r-1)b^2+2b-1}a_3, \cdots, \right.$$
$$\left. \frac{(r-1)(1-2bd)}{2(r-1)b^2+2b-1}a_{n-1}, d\right),$$
$$\left(\frac{(r-1)(1-2bd)}{2(r-1)b^2+2b-1}a_1, \frac{(r-1)(1-2bd)}{2(r-1)b^2+2b-1}a_2 + \frac{(r-1)b+1-d}{2(r-1)b^2+2b-1}, \frac{(r-1)(1-2bd)}{2(r-1)b^2+2b-1}a_3, \cdots, \right.$$
$$\left. \frac{(r-1)(1-2bd)}{2(r-1)b^2+2b-1}a_{n-1}, d\right),$$
$$\cdots \cdots \cdots \cdots$$
$$\left(\frac{(r-1)(1-2bd)}{2(r-1)b^2+2b-1}a_1, \frac{(r-1)(1-2bd)}{2(r-1)b^2+2b-1}a_2, \cdots, \frac{(r-1)(1-2bd)}{2(r-1)b^2+2b-1}a_{n-2}, \frac{(r-1)(1-2bd)}{2(r-1)b^2+2b-1}a_{n-1} + \right.$$
$$\left. \frac{(r-1)b+1-d}{2(r-1)b^2+2b-1}, d\right),$$
$$\left(\frac{(r-1)(\frac{1}{2}+\frac{3}{2r}-d)}{(r-1)b+\frac{1}{2}-\frac{3}{2r}}a_1, \frac{(r-1)(\frac{1}{2}+\frac{3}{2r}-d)}{(r-1)b+\frac{1}{2}-\frac{3}{2r}}a_2, \frac{(r-1)(\frac{1}{2}+\frac{3}{2r}-d)}{(r-1)b+\frac{1}{2}-\frac{3}{2r}}a_3, \cdots, \frac{(r-1)(\frac{1}{2}+\frac{3}{2r}-d)}{(r-1)b+\frac{1}{2}-\frac{3}{2r}}a_{n-1}, d\right),$$
$$\left(\frac{(r-1)(\frac{1}{2r}-d)}{(r-1)b+1-\frac{1}{2r}}a_1, \frac{(r-1)(\frac{1}{2r}-d)}{(r-1)b+1-\frac{1}{2r}}a_2, \frac{(r-1)(\frac{1}{2r}-d)}{(r-1)b+1-\frac{1}{2r}}a_3, \cdots, \frac{(r-1)(\frac{1}{2r}-d)}{(r-1)b+1-\frac{1}{2r}}a_{n-1}, d\right).$$

So we just need to prove that the convex hull of the following $n+1$ vectors in $\mathbb{R}^{n-1}$, denoted by $\tilde{P}_2^d$, contains no integer points:

$$\frac{(r-1)(1-2bd)}{2(r-1)b^2+2b-1}a + \frac{(r-1)b+1-d}{2(r-1)b^2+2b-1}e_1, \ \frac{(r-1)(1-2bd)}{2(r-1)b^2+2b-1}a + \frac{(r-1)b+1-d}{2(r-1)b^2+2b-1}e_2, \cdots,$$

$$\frac{(r-1)(1-2bd)}{2(r-1)b^2+2b-1}a + \frac{(r-1)b+1-d}{2(r-1)b^2+2b-1}e_{n-1}, \ \frac{(r-1)(\frac{1}{2}+\frac{3}{2r}-d)}{(r-1)b+\frac{1}{2}-\frac{3}{2r}}a, \ \frac{(r-1)(\frac{1}{2r}-d)}{(r-1)b+1-\frac{1}{2r}}a.$$

The following properties can be verified by calculation, using $d \le nb+1$ and $b > r$:

1. $\frac{(r-1)(1-2bd)}{2(r-1)b^2+2b-1} < \frac{(r-1)(\frac{1}{2r}-d)}{(r-1)b+1-\frac{1}{2r}} < \frac{(r-1)(\frac{1}{2}+\frac{3}{2r}-d)}{(r-1)b+\frac{1}{2}-\frac{3}{2r}} < 0$, and each of the three terms strictly decreases as d increases.

2. For $d = kb + h$, where integers k and h satisfy $0 \le k \le \lfloor r-1 \rfloor = n$ and $0 < h < b$, $\frac{(r-1)(\frac{1}{2}+\frac{3}{2r}-d)}{(r-1)b+\frac{1}{2}-\frac{3}{2r}}a_i < -ka_i$ and $\frac{(r-1)(1-2bd)}{2(r-1)b^2+2b-1}a_i + \frac{(r-1)b+1-d}{2(r-1)b^2+2b-1} < -ka_i$ for $i = 1, 2, \cdots, n-1$.

3. For $d = kb$, where integer k satisfies $1 \le k \le \lfloor r-1 \rfloor = n$, $\frac{(r-1)(\frac{1}{2}+\frac{3}{2r}-d)}{(r-1)b+\frac{1}{2}-\frac{3}{2r}}a_i < -(k-1)a_i$, $-ka_i < \frac{(r-1)(1-2bd)}{2(r-1)b^2+2b-1}a_i$, and $\frac{(r-1)(1-2bd)}{2(r-1)b^2+2b-1}a_i + \frac{(r-1)b+1-d}{2(r-1)b^2+2b-1} < -(k-1)a_i$ for $i = 1, 2, \cdots, n-1$.

By the above property 1, $\tilde{P}_2^d \subseteq Q_2^d$, where Q_2^d is the convex hull of the $n+1$ vectors:

$$
\begin{aligned}
y_0 &\equiv \frac{(r-1)(1-2bd)}{2(r-1)b^2+2b-1}a, \\
y_1 &\equiv \frac{(r-1)(1-2bd)}{2(r-1)b^2+2b-1}a + \frac{(r-1)b+1-d}{2(r-1)b^2+2b-1}e_1, \\
y_2 &\equiv \frac{(r-1)(1-2bd)}{2(r-1)b^2+2b-1}a + \frac{(r-1)b+1-d}{2(r-1)b^2+2b-1}e_2, \\
&\qquad\qquad \cdots\cdots\cdots \\
y_{n-1} &\equiv \frac{(r-1)(1-2bd)}{2(r-1)b^2+2b-1}a + \frac{(r-1)b+1-d}{2(r-1)b^2+2b-1}e_{n-1}, \\
y_n &\equiv \frac{(r-1)(\frac{1}{2}+\frac{3}{2r}-d)}{(r-1)b+\frac{1}{2}-\frac{3}{2r}}a.
\end{aligned}
$$

To prove (b), it suffices to show that Q_2^d contains no integer points. Given the properties 2 and 3 and the fact that $\frac{(r-1)b+1-d}{2(r-1)b^2+2b-1} < \frac{1}{2b}$, the proof is very similar to that of Claim 3. The theorem is proved. $\qquad\square$

3 Deciding Emptiness of the $\{-1, 0, 1\}$-Closure (or $\{0, 1\}$-Closure) of a Rational Polyhedron with No Integer Point

In this section, we first prove $\mathcal{NP}$-completeness of deciding emptiness of the $\{-1, 0, 1\}$-closure of a rational polyhedron containing no integer point, and then prove the same for the $\{0, 1\}$-closure as a corollary.

Theorem 3. *Given a rational polyhedron containing no integer point, it is $\mathcal{NP}$-complete to decide whether its $\{-1, 0, 1\}$-closure is empty.*

Proof. We will prove the theorem by polynomially reducing the following partition problem, which is known to be $\mathcal{NP}$-complete [7], to the problem of deciding whether $P'_{\{-1,0,1\}} = \emptyset$ for a polyhedron P with no integer points.

Partition Problem: Given a finite set of positive integers $S = \{a_i\}_{i=1}^{s}$, is there a subset $K \subseteq S$ such that $\sum_{i \in K} a_i = \sum_{i \in S \setminus K} a_i$?

Let $b \equiv \frac{1}{2} \sum_{1 \le i \le s} a_i$. We assume without loss of generality that $a_i < b$ for $i = 1, 2, \cdots, s$ and that the greatest common divisor of $a_1, a_2, \cdots, a_s$ is 1. We also assume without loss of generality that $s \ge 8$ (because the Partition Problem with fixed s can be formulated as an integer program with only s binary variables, which can be solved in polynomial time [11]).

First, we prove that b can be assumed to be greater than or equal to $s + 3$. Note that a Partition Problem with $b < s + 3$ can be polynomially converted to another Partition Problem with $S' = S \cup \{\sum_{1 \le i \le s} a_i + 1, \sum_{1 \le i \le s} a_i + 1\}$. Let $s' = |S'|$. So $s' = s + 2$. It is easy to see that the Partition Problem with S has a feasible partition if and only if the Partition Problem with S' has a feasible partition. Let $b' \equiv b + (\sum_{1 \le i \le s} a_i + 1)$. Then $b' = \frac{3}{2} \sum_{1 \le i \le s} a_i + 1 \ge \frac{3s}{2} + 1 = s + 5 + (\frac{s}{2} - 4) \ge (s + 2) + 3 = s' + 3$.

Now we consider the Partition Problem with $s = n - 1$, $n \ge 9$ and $b \ge n + 2$. Let $r = n + 1 + \frac{1}{2b}$. So r is a rational number satisfying $r < b$ and $rb \notin \mathbb{Z}_+$. We only need to show that the Partition Problem can be polynomially reduced to the problem of deciding whether $P'_{\{-1,0,1\}} = \emptyset$ for the same polyhedron P as constructed in the proof of Theorem 2, *i.e.*, the convex hull of the $n + 3$ vectors:
$v_1 = (\frac{1}{2b}, 0, \cdots, 0, \frac{1}{2b})$, $v_2 = (0, \frac{1}{2b}, 0, \cdots, 0, \frac{1}{2b})$, $\cdots$, $v_{n-1} = (0, \cdots, 0, \frac{1}{2b}, \frac{1}{2b})$,
$v_n = (0, \cdots, 0, \frac{1}{2} + \frac{3}{2r})$, $v_{n+1} = (a_1, a_2, \cdots, a_{n-1}, -b + \frac{1}{2})$, $v_{n+2} = ((1-r)a_1, (1-r)a_2, \cdots, (1-r)a_{n-1}, (r-1)b + 1)$, and $v_{n+3} = (0, \cdots, 0, \frac{1}{2r})$. It suffices to prove the following two claims.

Claim 1. There is a subset $K \subseteq S$ such that $\sum_{i \in K} a_i = \sum_{i \in S \setminus K} a_i$ only if $P'_{\{-1,0,1\}} = \emptyset$.

Proof. Consider an inequality $cx \le q$, where $c = (c_1, c_2, \cdots, c_n)$, $c_i = -1$ for $i \in K$, $c_i = 0$ for $i \in S \setminus K$, $c_n = -1$, and $q = -\frac{1}{2b}$. One can easily verify: $-1 < cv_i < -\frac{1}{2b}$ for $1 \le i \le n - 1$, $cv_n = -\frac{1}{2} - \frac{3}{2r}$, and because $\sum_{i \in K} a_i = b$, $cv_{n+1} = -\frac{1}{2}$ and $cv_{n+2} = -1$. Hence, $cx \le q$ is valid for P. Since $c_i = -1$ or 0, the inequality $cx \le \lfloor q \rfloor$ ($= -1$) is a $\{-1,0,1\}$-cut of P. From the value of cv_i for $1 \le i \le n + 2$, we see that v_{n+2} is the only vector in P that satisfies the inequality $cx \le \lfloor q \rfloor$. Since $rb \notin \mathbb{Z}_+$ and $b = \sum_{i \in K} a_i$, there exists some $j \in K$ satisfying $ra_j \notin \mathbb{Z}_+$. Consider the inequality $fx \le g$, where $fx = -x_j$ and $g = (r - 1)a_j$. Apparently, $g > 0$ and $g \notin \mathbb{Z}_+$. In addition, $fv_i < 0$ for $1 \le i \le n + 1$ and $fv_{n+2} = g > 0$. It is obvious that the inequality $fx \le \lfloor g \rfloor$ ($= \lfloor (r - 1)a_j \rfloor$) is a $\{-1,0,1\}$-cut of P and that v_{n+2} violates this inequality. Therefore, $P'_{\{-1,0,1\}} = \emptyset$ and Claim 1 is proved.

Claim 2. There is a subset $K \subseteq S$ such that $\sum_{i \in K} a_i = \sum_{i \in S \setminus K} a_i$ if $P'_{\{-1,0,1\}} = \emptyset$.

Proof. Let $v_0 \equiv (0, 0, \cdots, 0, \frac{1}{2r} + \frac{1}{2}) \in P$. Since $P'_{\{-1,0,1\}} = \emptyset$, v_0 violates an inequality $cx \le \lfloor q \rfloor$, where $c = (c_1, c_2, \cdots, c_n) \in \{-1, 0, 1\}^n$ and $cx \le q$ is valid for P. It is easy to see that $c_n \ne 0$. Otherwise, $0 = cv_0 \le q$, which contradicts that v_0 violates $cx \le \lfloor q \rfloor$.

Now we show that $\sum_{1 \le i \le n-1} c_i a_i = c_n b$. If $c_n = 1$, then $cv_0 = \frac{1}{2r} + \frac{1}{2}$. In this case, if $\sum_{1 \le i \le n-1} c_i a_i \ge c_n b + 1$, then $\frac{3}{2} \le cv_{n+1} \le q$; if $\sum_{1 \le i \le n-1} c_i a_i \le c_n b - 1$, then $n + 1 < r \le cv_{n+2} \le q$. Hence, $cv_0 < 1 < q$, contradicting $cv_0 > \lfloor q \rfloor$. If $c_n = -1$, then $cv_0 = -\frac{1}{2r} - \frac{1}{2}$. In this case, if $\sum_{1 \le i \le n-1} c_i a_i \ge c_n b + 1$, then $\frac{1}{2} \le cv_{n+1} \le q$; if $\sum_{1 \le i \le n-1} c_i a_i \le c_n b - 1$, then $n - 1 < r - 2 \le cv_{n+2} \le q$. So, $cv_0 < 0 < q$, a contradiction to $cv_0 > \lfloor q \rfloor$.

It is true that $c_n = -1$. Otherwise, $cv_0 = \frac{1}{2r} + \frac{1}{2}$ and $cv_{n+2} = 1 \le q$, contradicting that v_0 violates $cx \le \lfloor q \rfloor$.

We now claim that $c_i \le 0$ for $1 \le i \le n - 1$. Otherwise, suppose $c_j = 1$ for some $1 \le j \le n - 1$. Then $cv_j = 0$. Since $cv_0 = -\frac{1}{2r} - \frac{1}{2}$ and $0 = cv_j \le q$, a contradiction similar to the early ones can be derived. Therefore, Claim 2 is proved.

Using the same approach as shown in the end of the proof of Theorem 2, it is straight to polynomially obtain a description of P in the form of $\tilde{A}x \le \tilde{b}$, where $\tilde{A} \in \mathbb{Z}^{m \times n}$ and $\tilde{b} \in \mathbb{Z}^m$, from the vectors $v_1, v_2, \cdots, v_{n+3}$. The theorem is proved. □

Corollary 1. *Given a rational polyhedron containing no integer point, it is $\mathcal{NP}$-complete to decide whether its $\{0, 1\}$-closure is empty.*

Proof. The proof is similar to that of Theorem 3, hence we omit the details and only point out the differences.

The Partition Problem is polynomially reduced to the problem of deciding whether $P'_{\{0,1\}} = \emptyset$ for the polyhedron $P \subseteq \mathbb{R}^n$ that is the convex hull of the $n + 3$ vectors: $v_1 = (-\frac{1}{2b}, 0, \cdots, 0, -\frac{1}{2b})$, $v_2 = (0, -\frac{1}{2b}, 0, \cdots, 0, -\frac{1}{2b})$, $\cdots$, $v_{n-1} = (0, \cdots, 0, -\frac{1}{2b}, -\frac{1}{2b})$, $v_n = (0, \cdots, 0, -\frac{1}{2} - \frac{3}{2r})$, $v_{n+1} = (-a_1, -a_2, \cdots, -a_{n-1}, b - \frac{1}{2})$, $v_{n+2} = ((r-1)a_1, (r-1)a_2, \cdots, (r-1)a_{n-1}, (1-r)b - 1)$, and $v_{n+3} = (0, \cdots, 0, -\frac{1}{2r})$.

Claim 1. There is a subset $K \subseteq S$ such that $\sum_{i \in K} a_i = \sum_{i \in S \setminus K} a_i$ only if $P'_{\{0,1\}} = \emptyset$.

The proof of Claim 1 is similar to that of Claim 1 in the proof of Theorem 3 except that $c_i = 1$ for $i \in K$, $c_i = 0$ for $i \in S \setminus K$, $c_n = 1$, and $fx = x_j$.

Claim 2. There is a subset $K \subseteq S$ such that $\sum_{i \in K} a_i = \sum_{i \in S \setminus K} a_i$ if $P'_{\{0,1\}} = \emptyset$.

The proof of Claim 2 is similar to but simpler than that of Claim 2 in the proof of Theorem 3. Here are two differences: First, we let $v_0 \equiv (0, 0, \cdots, 0, -\frac{1}{2r} - \frac{1}{2})$. Second, to show by contradiction that $\sum_{1 \le i \le n-1} c_i a_i = c_n b$, we only consider the case that $c_n = 1$, and a contradiction can be derived due to $cv_0 < 0 < q$. □

4 Conclusions

In this paper, we proved that the problem of deciding whether the Gomory-Chvátal closure of a rational polyhedron P is empty is $\mathcal{NP}$-complete, even when

P is known to contain no integer point. Similar results are also proved for the $\{-1, 0, 1\}$-closure and $\{0, 1\}$-closure of polyhedron.

There are several questions to which we have not found an answer yet. First, what if our attention is restricted to the polyhedra in the unit cube (denoted by $[0, 1]^n$)? Namely,

(i) Is it $\mathcal{NP}$-complete to decide whether $P' = \emptyset$ for $P \subseteq [0, 1]^n$ that contains no integer point?
(ii) Is it $\mathcal{NP}$-complete to decide whether $P'_{\{-1, 0, 1\}} = \emptyset$ for $P \subseteq [0, 1]^n$ that contains no integer point?
(iii) Is it $\mathcal{NP}$-complete to decide whether $P'_{\{0, 1\}} = \emptyset$ for $P \subseteq [0, 1]^n$ that contains no integer point?

An interesting class of rational polyhedra is those for which $P' = P_I$. A well-known example in this family is due to Edmonds [5] for 1-matchings of undirected graphs $G = (V, E)$: $P = \{x \in \mathbb{R}^{|E|} : x(\delta(v)) \leq 1 \; \forall v \in V, \; x_e \geq 0 \; \forall e \in E\}$, where $\delta(v)$ is the set of edges incident on node v. Edmonds proposed a polynomial-time algorithm for solving GC-Opt for 1-matchings, and Padberg and Rao [14] devised a polynomial-time separation algorithm for b-matching polytopes, implying polynomial-time solvability of GC-Sep for 1-matchings. The question of deciding whether $P' = P_I$ for a rational polytope P is not known to be in $\mathcal{NP}$ and Theorem 2 implies that it is $\mathcal{NP}$-hard. But it is an open question whether the separation problem for the Gomory-Chvátal closure of polyhedra P that satisfy $P' = P_I$ is polynomially solvable, and similarly for the associated optimization problem.

(iv) Is there a polynomial algorithm to find a point in P_I or show that $P_I = \emptyset$ when we know that $P' = P_I$?
(v) Is there a polynomial algorithm to optimize over P_I when we know that $P' = P_I$?

We believe that the answers to the last two questions are positive. As evidence, we observe that the problem of deciding whether $P_I = \emptyset$ when we know that $P' = P_I$ is in the complexity class $\mathcal{NP} \cap$ co-$\mathcal{NP}$. We already observed (Lemma 1) that the problem is in $\mathcal{NP}$. To prove that it is in co-$\mathcal{NP}$, it suffices to exhibit a point $x \in \mathbb{Z}^n$ that satisfies $Ax \leq b$. It is well known that, if such a point exists, there is one whose encoding is polynomial in the size of the input [1]. Therefore a polynomial co-$\mathcal{NP}$ certificate exists for $P' = \emptyset$ when $P' = P_I$. On the other hand no obvious co-$\mathcal{NP}$ certificate is known for $P' = \emptyset$ in general.

As an example, consider the maximum weight stable set problem in a graph $G = (V, E)$, $\max\{wx : x \in P_I\}$ where $P = \{x \in \mathbb{R}_+^V : x_i + x_j \leq 1 \text{ for } ij \in E\}$. We note that this problem is $\mathcal{NP}$-hard in general, but that it can be solved in polynomial time when $P' = P_I$. Indeed, Campelo and Cornuéjols [2] showed that P' is entirely described by the inequalities defining P together with the odd circuit inequalities $\sum_{i \in C} x_i \leq \frac{|C| - 1}{2}$ for vertex sets C of odd cardinality that induce a circuit of G. The graphs for which these inequalities completely describe the stable set polytope P_I are called t-perfect graphs. These graphs are discussed in Chap. 68 of Schrijver's book [15]. Theorem 68.1 states that a maximum-weight stable set in a t-perfect graph can be found in polynomial time. This follows from the equivalence of optimization and separation [10] and the fact that the separation of odd circuit inequalities can be done in polynomial time by reduction to shortest path problems [8].

Acknowledgement. We thank Michele Conforti for pointing out to us that the problem of deciding whether $P_I = \emptyset$ is in $\mathcal{NP} \cap$ co-$\mathcal{NP}$ when we know that $P' = P_I$.

References

1. Borosh, I., Treybig, L.B.: Bounds on positive integral solutions to linear Diophantine equations. Proc. Am. Math. Soc. **55**, 299–304 (1976)
2. Campelo, M., Cornuéjols, G.: Stable sets, corner polyhedra and the Chvátal closure. Oper. Res. Lett. **37**, 375–378 (2009)
3. Chvátal, V.: Edmonds polytope and a hierarchy of combinatorial problems. Discrete Math. **4**, 305–337 (1973)
4. Conforti, M., Cornuéjols, G., Zambelli, G.: Integer Programming. Springer, Switzerland (2014)
5. Edmonds, J.: Maximum matching and a polyhedron with 0,1-vertices. J. Res. Natl. Bur. Stan. B **69**, 125–130 (1965)
6. Eisenbrand, F.: On the membership problem for the elementary closure of a polyhedron. Combinatorica **19**, 297–300 (1999)
7. Garey, M.R., Johnson, D.S.: Computers and Intractability: A Guide to the Theory of NP-Completeness. W.H. Freeman, San Francisco (1979)
8. Gerards, A.M.H., Schrijver, A.: Matrices with the Edmonds-Johnson property. Combinatorica **6**, 365–379 (1986)
9. Gomory, R.E.: Outline of an algorithm for integer solutions to linear programs. Bull. Am. Math. Soc. **64**, 275–278 (1958)
10. Grötschel, M., Lovász, L., Schrijver, A.: The ellipsoid method and its consequences in combinatorial optimization. Combinatorica **1**, 169–197 (1981)
11. Lenstra Jr., H.W.: Integer programming with a fixed number of variables. Math. Oper. Res. **8**, 538–548 (1983)
12. Lueker, G.S.: Two NP-complete Problems in Non-negative Integer Programming. Report No. 178, Department of Computer Science, Princeton University, Princeton, N.J. (1975)
13. Mahajan, A., Ralphs, T.: On the complexity of selecting disjunctions in integer programming. SIAM J. Optim. **20**, 2181–2198 (2010)
14. Padberg, M.W., Rao, M.R.: Odd minimum cut-sets and b-matchings. Math. Oper. Res. **7**, 67–80 (1982)
15. Schrijver, A.: Combinatorial Optimization: Polyhedra and Efficiency. Springer, Berlin (2003)

On the Quantile Cut Closure
of Chance-Constrained Problems

Weijun Xie and Shabbir Ahmed[(⊠)]

School of Industrial and Systems Engineering,
Georgia Institute of Technology, Atlanta, GA 30332, USA
sahmed@isye.gatech.edu

Abstract. A chance constrained problem involves a set of scenario constraints from which a small subset can be violated. Existing works typically consider a mixed integer programming (MIP) formulation of this problem by introducing binary variables to indicate which constraint systems are to be satisfied or violated. A variety of cutting plane approaches for this MIP formulation have been developed. In this paper we consider a family of cuts for chance constrained problems in the original space rather than those in the extended space of the MIP reformulation. These cuts, known as quantile cuts, can be viewed as a projection of the well known family of mixing inequalities for the MIP reformulation, onto the original problem space. We show the following results regarding quantile cuts: (i) the closure of all quantile cuts is a polyhedral set; (ii) separation of quantile cuts is in general NP-hard; (iii) successive application of quantile cut closures achieves the convex hull of the chance constrained problem in the limit; and (iv) in the pure integer setting this convergence is finite.

1 Introduction

A chance constrained problem (CCP) involves optimization over constraints (specified by stochastic data) which are required to be satisfied with a prescribed probability level. A generic formulation of CCP is

$$\min_{x} \left\{ c^{\top} x : \ x \in S, \ \mathbb{P}[\boldsymbol{\xi} : x \in \mathcal{X}(\boldsymbol{\xi})] \geq 1 - \epsilon \right\}. \tag{1}$$

In the above formulation, S denotes a set of deterministic constraints, $\boldsymbol{\xi}$ denotes a random data vector, and $\mathcal{X}(\boldsymbol{\xi})$ denotes a system of stochastic constraints whose data is specified by the random vector $\boldsymbol{\xi}$. The CCP (1) seeks a solution $x \in S$ that minimizes the cost $c^{\top} x$ and satisfies the stochastic constraints $\mathcal{X}(\boldsymbol{\xi})$ with probability at least $(1 - \epsilon)$ where $\epsilon \in (0, 1)$ is a prespecified risk level.

We consider a mixed integer CCP under finite distribution, where we assume that

- $S = \{x \in \mathbb{R}^{n-\tau} \times \mathbb{Z}^{\tau} : Dx \geq d\}$ is a nonempty, compact deterministic mixed integer set;

© Springer International Publishing Switzerland 2016
Q. Louveaux and M. Skutella (Eds.): IPCO 2016, LNCS 9682, pp. 398–409, 2016.
DOI: 10.1007/978-3-319-33461-5_33

– $\boldsymbol{\xi}$ is a random vector with a finite distribution supported on $\Xi = \{\xi^1, \ldots, \xi^N\}$, where each ξ^i for $i \in [N] := \{1, \ldots, N\}$ corresponds to a scenario with a probability mass p_i; and

– for a given scenario i, the vector ξ^i defines a nonempty, compact mixed integer constraint system $\mathcal{X}^i := \mathcal{X}(\xi^i) = \{x \in \mathbb{R}^{n-\tau} \times \mathbb{Z}^\tau : A^i x \geq b^i\}$.

In this setting, the chance constraint in (1) corresponds to satisfying a subset $\mathcal{C} \subseteq [N]$ of the scenario constraints, i.e. $x \in \cap_{i \in \mathcal{C}} \mathcal{X}^i$, such that $\sum_{i \in \mathcal{C}} p_i \geq 1 - \epsilon$. Let

$$\mathcal{Z} := \left\{ \mathcal{C} \subseteq [N] : \sum_{i \in \mathcal{C}} p_i \geq 1 - \epsilon \right\}, \qquad (2)$$

be the collection of all feasible subsets of scenarios. Then the feasible region of (1), denoted by X, can be written in the disjunctive normal form:

$$X = \bigcup_{\mathcal{C} \in \mathcal{Z}} \left[S \bigcap_{i \in \mathcal{C}} \mathcal{X}^i \right]. \qquad (3)$$

We assume throughout that CCP is feasible, and hence X is nonempty. From the above disjunctive normal form it is clear that, even in the absence of integrality restrictions, i.e. $\tau = 0$, the set X is nonconvex, and not surprisingly CCP is strongly NP-hard [9,10].

Since $\mathcal{X}^i$ is compact for all $i \in [N]$ we can introduce binary variables z_i for $i \in [N]$ and reformulate (1) as the mixed integer program (MIP):

$$\min_{x,z} \left\{ c^\top x : x \in S, \ A^i x \geq b^i - M_i(1 - z_i), \ z \in Z \right\}, \qquad (4)$$

where

$$Z := \left\{ z \in \{0,1\}^N : \sum_{i \in [N]} p_i z_i \geq 1 - \epsilon \right\},$$

and M_i for all $i \in [N]$ are suitable big-M coefficients. Since the linear programming (LP) relaxation of (4) is typically very weak, there has been a great deal of work in deriving strong valid inequalities for this MIP. One popular approach is to derive a relaxation of (4) in the form of the well-studied mixing set [5] and add the corresponding mixing inequalities [1,6–9,15].

We consider a family of valid inequalities for the nonconvex feasible region X of the CCP (1) in the original x-space, rather than those for the MIP formulation (4) in the (x, z)-space. These valid inequalities known as *quantile cuts* are obtained as follows. We first optimize a linear function $\alpha^\top x$ over each scenario constraint, and record the optimal values $\beta_i^\alpha = \min\{\alpha^\top x : x \in S \cap \mathcal{X}^i\}$ for $i \in [N]$. This approach and resulting β_i^α values were used in [8] to derive a mixing set relaxation for (4). Notice that each β_i^α has the associated probability p_i. Next we compute the $(1 - \epsilon)$-quantile of $\{\beta_i^\alpha\}_{i \in [N]}$ based on these probabilities – denote this by β_q^α. The quantile cut is then given by $\alpha^\top x \geq \beta_q^\alpha$. Such inequalities were studied in [10] where it is shown that a single quantile cut represents the projection of the convex hull of a mixing set relaxation of (4) in the (x, z)-space onto the x space. Quantile cuts have been used in computational studies of chance constrained problems with good results [2,10,14].

In this paper we undertake a theoretical study of quantile cuts. In particular we study properties of the quantile closure, i.e. the intersection of all quantile cuts. Quantile cuts represent an infinite family of inequalities - one for each α vector - and so polyhedrality of and separation over the quantile closure are interesting questions. By replacing the deterministic constraint system S in (1) with the (first) quantile closure we obtain a stronger formulation for which we can apply another round of quantile cuts and derive the second quantile closure and so on. We investigate how the sequence of sets produced by such successive quantile closure operations relates to the convex hull of the feasible region of (1). Our main results are summarized below.

1. The set obtained after a finite number of quantile closure operations is a polyhedral set.
2. Separation over the first quantile closure is NP-hard.
3. The sequence of sets obtained by successive quantile closure operations converges to the convex hull of X with respect to the Hausdorff metric.
4. In the pure integer setting, i.e. $X \subseteq \mathbb{Z}^n$, there exists a finite number of quantile closure operations after which we can recover the convex hull of X.

The remainder of this paper is organized as follows. In Sect. 2 we discuss the connection of quantile cuts for conv(X) to the mixing set inequalities for the MIP (4). In Sect. 3 we prove the polyhedrality of the quantile closure and establish complexity of separation over it. In Sect. 4 we discuss the convergence properties of successive quantile closures. We provide some concluding remarks in Sect. 5.

2   Quantile Cuts and Mixing Inequalities

We first formally define the quantile cut for CCP (1). Recall that S is the set of deterministic constraints, $\mathcal{X}^i$ are constraints associated with scenario $i \in [N]$, $\mathcal{Z}$ defined in (2) is the collection of all feasible scenario sets, and X given by (3) is the set of feasible solutions of (1).

Definition 1. *Given $\alpha \in \mathbb{R}^n$ let $\{\beta_i^\alpha(S)\}_{i \in [N]}$ be the optimal values of*

$$\beta_i^\alpha(S) = \min \left\{ \alpha^\top x : x \in S \cap \mathcal{X}^i \right\} \ \forall i \in [N]. \tag{5}$$

The quantile $\beta_q^\alpha(S)$ is given by

$$\beta_q^\alpha(S) := \min_{\mathcal{C} \in \mathcal{Z}} \max_{i \in \mathcal{C}} \beta_i^\alpha(S) \tag{6}$$

and the associated "quantile cut" is

$$\alpha^\top x \geq \beta_q^\alpha(S). \tag{7}$$

Note that the above definition depends on S so as to allow for successive applications with changing S. Since S and $\mathcal{X}^i$ are assumed to be compact we have that $\beta_i^\alpha(S) \in (-\infty, +\infty]$ where the value of $+\infty$ is taken when the problem (5) is infeasible. When $p_i = \frac{1}{N}$ for all $i \in [N]$, $\beta_q^\alpha(S)$ is the $(\lfloor \epsilon N \rfloor + 1)$st largest value among $\{\beta_i^\alpha(S)\}_{i \in [N]}$.

From the definition above and the disjunctive normal form (3) of X it should be clear that the quantile cut (7) is valid for $\text{conv}(X)$. We next reveal the connection between quantile cuts and mixing inequalities for CCP, which also establishes the validity of these cuts.

A mixing set [5] is a mixed-integer set of the form

$$P = \{(v, z) \in \mathbb{R}_+ \times \{0,1\}^s : v + h_i z_i \geq h_i \ i = 1, \ldots, s\} \tag{8}$$

with $h_1 \geq \ldots \geq h_s$. The following exponential family of mixing inequalities are valid for P

$$v + \sum_{j=1}^l (h_{t_j} - h_{t_{j+1}}) z_{t_j} \geq h_{t_1} \ \forall \ T = \{t_1, \ldots, t_l\} \subseteq \{1, \ldots, s\}, \tag{9}$$

where $h_{t_1} \geq \ldots \geq h_{t_l}$, $h_{t_{l+1}} = 0$. These inequalities are facet defining for P when $t_1 = 1$ and are sufficient to describe the convex hull of P (see [3,5]).

Using the β-values as defined in (5), Luedtke [8] constructed the following mixing set relaxation of the MIP formulation (4) of CCP

$$Y^\alpha = \{(x, z) \in \mathbb{R}^n \times \{0,1\}^N : \alpha^\top x + (\beta_i^\alpha(S) - \beta_q^\alpha(S))(1 - z_i) \geq \beta_i^\alpha(S),$$
$$i \in \mathcal{B}_q^\alpha, z \in Z\}. \tag{10}$$

where $\mathcal{B}_q^\alpha := \{i \in [N] : \beta_i^\alpha(S) \geq \beta_q^\alpha(S)\}$ is a subset of scenarios each of whose β-value is at least as large as the quantile $\beta_q^\alpha(S)$.

Proposition 1 *(Theorem 1, [8]). For any α, the system Y^α is a relaxation of the feasible region of the MIP (4), and hence $X \subseteq \text{Proj}_x(Y^\alpha)$, where $\text{Proj}_x(\cdot)$ denotes the projection of a set onto x space.*

Note that Y^α is a mixing system with a knapsack side constraint defined by Z, thus the mixing inequalities of the form (9) are valid. These inequalities were used within a branch and cut scheme for solving the MIP (4) in [8]. Note that the mixing inequalities are in the (x, z)-space while the quantile cuts are in the original x-space. The next result shows that a single quantile cut in the x-space captures the effect of the entire exponential family of mixing inequalities.

Proposition 2 *(Proposition 5, [10]). For any α,*

$$\text{Proj}_x(\text{conv}(Y^\alpha)) = \{x \in \mathbb{R}^n : \alpha^\top x \geq \beta_q^\alpha(S)\}.$$

Inspired by the above result we investigate, in the remainder of the paper, the strength of the quantile closure, i.e. the intersection of all quantile cuts.

3   Quantile Closure

In this section we define the quantile closure, prove that it is polyhedral, and establish its hardness of separation.

Definition 2. *The first quantile closure of S is defined as*

$$S^1 := \bigcap_{\substack{\alpha \in \mathbb{R}^n \\ \beta_q^\alpha(S) < \infty}} \{x \in \mathbb{R}^n : \alpha^\top x \geq \beta_q^\alpha(S)\}.$$

Inductively, we can define rth round quantile closure S^r as

$$S^r := \bigcap_{\substack{\alpha \in \mathbb{R}^n \\ \beta_q^\alpha(S^{r-1}) < \infty}} \{x \in \mathbb{R}^n : \alpha^\top x \geq \beta_q^\alpha(S^{r-1})\} \quad r \geq 2.$$

Next we characterize $\text{conv}(X)$ and S^1 in conjunctive normal form. Let us begin with the following definition.

Definition 3. *A set $G \subseteq [N]$ is a "partial covering subset" if it intersects with all of feasible scenario subsets in $\mathcal{Z}$, i.e., for any $\widehat{\mathcal{C}} \in \mathcal{Z}$, we have $G \cap \widehat{\mathcal{C}} \neq \emptyset$. Also, a set G is a "minimal" partial covering subset if there does not exist another partial covering subset $G' \subseteq [N]$ such that $G' \subsetneq G$. We let $\mathcal{G}$ denote the collection of all of the minimal partial covering subsets.*

Note that when $p_i = \frac{1}{N}$ for all $i \in [N]$, then the collection of minimal partial covering subsets is $\mathcal{G} = \{G \subseteq [N] : |G| = \lfloor \epsilon N \rfloor + 1\}$.

Proposition 3.

$$X = \bigcap_{G \in \mathcal{G}} \left[\bigcup_{i \in G} \left(S \cap \mathcal{X}^i \right) \right]. \tag{11}$$

Proof. Define $X' = \bigcap_{G \in \mathcal{G}} \left[\bigcup_{i \in G} (S \cap \mathcal{X}^i) \right]$. We need to show that $X = X'$.

Let $x \in X$. Then, there exists a feasible subset $\mathcal{C} \in \mathcal{Z}$ such that $x \in \bigcap_{i \in \mathcal{C}} (S \cap \mathcal{X}^i)$. For an arbitrary minimal partial covering $G \in \mathcal{G}$, we must have $x \in \bigcup_{i \in G} (S \cap \mathcal{X}^i)$ since from Definition 3, G intersects with all of feasible subsets (i.e., $G \cap \mathcal{C} \neq \emptyset$). Thus, $X \subseteq X'$.

Suppose that there exists an $x' \in X'$ such that $x' \notin X$. Define a subset $\mathcal{C}' := \{i \in [N] : x' \in S \cap \mathcal{X}^i\} \notin \mathcal{Z}$. Let G' be the complement of $\mathcal{C}'$, i.e., $G' = [N] \backslash \mathcal{C}'$. We claim that for all $\mathcal{C} \in \mathcal{Z}$, we have $G' \cap \mathcal{C} \neq \emptyset$. Suppose not, then there must exist a $\widehat{\mathcal{C}} \in \mathcal{Z}$ such that $G' \cap \widehat{\mathcal{C}} = \emptyset$. This implies that $\widehat{\mathcal{C}} \subseteq \mathcal{C}'$, and thus

$$x' \in \bigcap_{i \in \mathcal{C}'} \left(S \cap \mathcal{X}^i \right) \subseteq \bigcap_{i \in \widehat{\mathcal{C}}} \left(S \cap \mathcal{X}^i \right) \subseteq X,$$

which contradicts $x' \notin X$. Hence, G' is a partial covering subset of $[N]$, and thus $x' \in X' \subseteq \bigcup_{i \in G'} (S \cap \mathcal{X}^i)$; i.e., this contradicts $G' = [N] \backslash \mathcal{C}'$.   □

Next we provide a conjunctive normal form for S^1. We will need the following preliminary observations.

Lemma 1. *The set $\mathcal{B}_q^\alpha = \{i \in [N] : \beta_i^\alpha(S) \geq \beta_q^\alpha(S)\}$ is a partial covering subset.*

Proof. From the definition of $\beta_q^\alpha(S)$, for any subset $\widehat{\mathcal{C}} \in \mathcal{Z}$, there must exist an $i_0 \in \widehat{\mathcal{C}}$ such that $\beta_{i_0}^\alpha(S) \geq \beta_q^\alpha(S)$. Thus $\mathcal{B}_q^\alpha$ is a partial covering subset. $\qquad\square$

Lemma 2. *There exist a $G \in \mathcal{G}$ such that $\beta_q^\alpha(S) = \min_{i \in G} \beta_i^\alpha(S)$.*

Proof. By the definition of $\beta_q^\alpha(S)$, there exists a $\widehat{\mathcal{C}} \in \mathcal{Z}$ with $\beta_q^\alpha(S) = \max_{\widehat{\mathcal{C}}} \beta_i^\alpha(S) \geq \beta_j^\alpha(S)$ for all $j \in \widehat{\mathcal{C}}$. From Definition 3, for each $\widehat{G} \in \mathcal{G}$, we have $\widehat{G} \cap \widehat{\mathcal{C}} \neq \emptyset$. Hence, $\beta_q^\alpha(S)$ must be no smaller than the smallest value in set $\{\beta_i^\alpha(S)\}_{i \in \widehat{G}}$; i.e.,

$$\beta_q^\alpha(S) \geq \min_{i \in \widehat{G} \cap \widehat{\mathcal{C}}} \beta_i^\alpha(S) \geq \min_{i \in \widehat{G}} \beta_i^\alpha(S).$$

From Lemma 1, $\mathcal{B}_q^\alpha$ is a partial covering subset. Now let G be a minimal partial covering subset such that $G \subseteq \mathcal{B}_q^\alpha$. Thus,

$$\min_{i \in G} \beta_i^\alpha(S) \geq \min_{i \in \mathcal{B}_q^\alpha} \beta_i^\alpha(S) \geq \beta_q^\alpha(S). \qquad\square$$

Proposition 4.

$$S^1 = \bigcap_{G \in \mathcal{G}} \text{conv} \left[\bigcup_{i \in G} \left(S \bigcap \mathcal{X}^i \right) \right]. \tag{12}$$

Proof. Let $W^G := \text{conv} \left[\bigcup_{i \in G} \left(S \bigcap \mathcal{X}^i \right) \right]$, $W := \bigcap_{G \in \mathcal{G}} W^G$. We need to show that $S^1 = W$.

$[S^1 \subseteq W]$. Consider $G \in \mathcal{G}$, and take any valid inequality $\alpha^\top x \geq \beta$ for W^G. Let $\widehat{\mathcal{C}} \in \mathcal{Z}$ such that $\beta_q^\alpha(S) = \max_{i \in \widehat{\mathcal{C}}} \beta_i^\alpha(S)$. Since $G \cap \widehat{\mathcal{C}} \neq \emptyset$ by Definition 3, hence

$$\beta_q^\alpha(S) \geq \min_{i \in G \cap \widehat{\mathcal{C}}} \beta_i^\alpha(S) \geq \min_{i \in G} \beta_i^\alpha(S) \geq \beta.$$

Thus, $\alpha^\top x \geq \beta$ is a valid inequality of S^1. This holds for any valid inequality of W^G, we have that $S^1 \subseteq W^G$. Since G was arbitrary, it follows that $S^1 \subseteq W^G$ for all $G \in \mathcal{G}$; i.e., $S^1 \subseteq W$.

$[S^1 \supseteq W]$. For any given α, from Lemma 2, there exist a $G \in \mathcal{G}$ such that $\beta_q^\alpha(S) = \min_{i \in G} \beta_i^\alpha(S)$. Clearly, $\alpha^\top x \geq \beta_q^\alpha(S)$ is a valid inequality for W^G; and so it is valid for W. Thus, $S^1 \supseteq W$. $\qquad\square$

Next we show that the above conjunctive normal form of S^1 which is independent of α implies the polyhedrality of the quantile closures. We will need the following elementary facts on the convex hull of a finite union of nonempty compact sets.

Lemma 3. *Let $R_1, \ldots, R_k$ be nonempty compact sets in $\mathbb{R}^n$.*

(i) $\operatorname{conv}(\cup_{i=1}^k R_i) = \operatorname{conv}(\cup_{i=1}^k \operatorname{conv}(R_i))$.
(ii) Suppose, for all i, $\operatorname{conv}(R_i)$ is a polytope, then $\operatorname{conv}(\cup_{i=1}^k R_i)$ is a polytope.
(iii) If x is an extreme point of $\operatorname{conv}(\cup_{i=1}^k R_i)$ then x is an extreme point of $\operatorname{conv}(R_{i'})$ for some $i' \in \{1, \ldots, k\}$, and hence $x \in R_{i'}$.

Theorem 1. *For each $r \in \mathbb{Z}_{++}$, S^r is a polytope and*

$$S^r = \bigcap_{G \in \mathcal{G}} \operatorname{conv} \left[\bigcup_{i \in G} \left(S^{r-1} \bigcap \mathcal{X}^i \right) \right].$$

Proof. From Lemma 3 it follows that, for any $G \in \mathcal{G}$, $\operatorname{conv} \left[\bigcup_{i \in G} \left(S \bigcap \mathcal{X}^i \right) \right] = \operatorname{conv} \left[\bigcup_{i \in G} \operatorname{conv} \left(S \bigcap \mathcal{X}^i \right) \right]$ and is a polytope. Since $\mathcal{G}$ is a finite set, it follows from Proposition 4 that S^1 is a polytope.

By induction, suppose S^r is a polytope for $r \leq t$. Now let $r = t + 1$, by Proposition 4, we have

$$S^r = \bigcap_{G \in \mathcal{G}} \operatorname{conv} \left[\bigcup_{i \in G} \left(S^{r-1} \bigcap \mathcal{X}^i \right) \right]$$

and S^{r-1} is a polytope, hence S^r is a polytope. □

We close this section by showing that separating over the first quantile closure even in the absence of integrality restrictions is NP-hard. Our proof is based on the constructions in [9, 10].

Theorem 2. *The separation over S^1 is, in general, NP-hard*

Proof. We consider a covering CCP where $\mathcal{X}^i = \{x \in \mathbb{R}^n : (a^i)^\top x \geq 1\}$ and $p_i = 1/N$ for all $i \in [N]$, and $S = [0, M]^n$ with $M \geq \max_{i \in [N], j \in [n]: a_j^i \neq 0} \frac{1}{a_j^i}$. From [10] it can be shown that

$$S^1 = \bigcap_{G \in \mathcal{G}} \left\{ x \in S : a_G^\top x \geq 1 \right\}, \tag{13}$$

where $(a_G)_j = \max_{i \in G} a_j^i, \forall j \in [n]$, $\mathcal{G} = \{G \subseteq [N] : |G| = k + 1\}$ and $k = \lfloor \epsilon N \rfloor$ (see Definition 3).

For a given solution $\widehat{x} \in S$, to separate it from S^1 is equivalent to solving the following problem

$$\delta^* = \min_{G \in \mathcal{G}} \max_{i \in G} \sum_{j \in [n]} a_j^i \widehat{x}_j - 1, \tag{14}$$

i.e., find a violated constraint of the form $a_G^\top x \geq 1$ in the description (13). If $\delta^* < 0$, then $\widehat{x} \notin S^1$; otherwise, $\widehat{x} \in S^1$. Consider the decision version of this separation problem:

(SepCCP). Given nonnegative integers $\{a_j^i\}_{i\in[N],j\in[n]}$ and a rational vector $\widehat{x} \in S$, does there exist a $G \subseteq [N]$ with $|G| = k + 1(k < N)$ such that $\sum_{j\in[n]} \max_{i\in G} a_j^i \widehat{x}_j < 1$?

Following [9] we can show that SepCCP is NP-complete via reduction from the NP-complete problem CLIQUE which asks

(CLIQUE). Given a graph with nodes V and edges E, does it contain a clique of size C?

Given an instance of CLIQUE we can construct an instance of SepCCP as $[n] = V, [N] = E, \widehat{x}_j = \frac{1}{C+1}$ for all $j \in [n]$, $k + 1 = \frac{1}{2}C(C - 1)$ and $a_j^i = 1$ if edge i contains nodes j and $a_j^i = 0$ otherwise. It is easy to verify that if CLIQUE has an answer Yes, then there exists a subgraph with edges $G \subseteq [N]$ and $|G| = \frac{1}{2}C(C - 1)$ such that

$$\sum_{j\in[n]} \max_{i\in G} a_j^i \widehat{x}_j = \frac{C}{C+1} < 1.$$

Hence, SepCCP has an answer Yes. Conversely, if SepCCP has an answer Yes, this implies that

$$\sum_{j\in[n]} \max_{i\in G} a_j^i < C + 1;$$

i.e., there exists a subgraph with edges $G \subseteq [N]$ and $|G| = \frac{1}{2}C(C - 1)$, which contains at most C nodes. Clearly, thus CLIQUE has an answer Yes.   □

4   Convergence of Quantile Closures

In this section, we investigate convergence of successive rounds of quantile closure operations. Our convergence notions are with respect to the Hausdorff distance [13]. For two closed convex sets $K_1, K_2 \in \mathbb{R}^n$, the Hausdorff distance $d_H(K_1, K_2)$ is defined as

$$d_H(K_1, K_2) := \min\{\delta : K_1 \subseteq K_2 + B(0, \delta), K_2 \subseteq K_1 + B(0, \delta)\},$$

where $B(0, \delta)$ denotes the ball centered at origin with radius δ. We will need the following fact on the limit of a set sequence.

Lemma 4 (Proposition 2, [12]). *Let $\{R^r\}$ be a sequence of nonempty compact convex sets such that $R^{r+1} \subseteq R^r$ for all r. Then R^r converges to $\bar{R} := \lim_{r\to\infty} R^r = \bigcap_{r=1}^\infty R^r$ with respect to the Hausdorff distance, and $\bar{R}$ is also a compact convex set.*

The following lemma reveals the convergence properties of a sequence of sets produced by successive quantile closure operations.

Lemma 5. *Let $\{S^r\}$ be a sequence of quantile closures. Then*

(i) there exists a $\bar{S} := \lim_{r \to \infty} S^r$;
(ii) for each $G \in \mathcal{G}$, we have

$$\mathrm{conv}\left[\bigcup_{i \in G} \mathrm{conv}\left(\bar{S} \bigcap \mathcal{X}^i\right)\right] = \bar{S}. \tag{15}$$

Proof.

(i) This directly follows from Lemma 4 since $\{S^r\}$ is an inclusion-wise monotone sequence of compact convex sets.
(ii) Let $\bar{S}^1$ be the quantile closure operation applied to set $\bar{S}$. Since $\bar{S} = \bar{S}^1$ by the limiting operation, we have that

$$\bar{S} = \bar{S}^1 = \bigcap_{G \in \mathcal{G}} \mathrm{conv}\left[\bigcup_{i \in G}\left(\bar{S} \bigcap \mathcal{X}^i\right)\right] = \bigcap_{G \in \mathcal{G}} \mathrm{conv}\left[\bigcup_{i \in G} \mathrm{conv}\left(\bar{S} \bigcap \mathcal{X}^i\right)\right],$$

where the second equality is due to Theorem 1 and the third equality follows from Lemma 3. Since $\mathrm{conv}\left[\bigcup_{i \in G}\left(\bar{S} \bigcap \mathcal{X}^i\right)\right] \subseteq \bar{S}$ for all $G \in \mathcal{G}$, we have that (15) holds.

$\square$

Now, we are ready to prove the convergence of the quantile closure procedure to the convex hull of X.

Theorem 3. *The set sequence $\{S^r\}$ converges to $\mathrm{conv}(X)$ with respect to the Hausdorff distance; i.e., $\bar{S} = \lim_{r \to \infty} S^r = \mathrm{conv}(X)$.*

Proof. From Lemma 4, we know that there exists an $\bar{S} = \lim_{r \to \infty} S^r$. Since $\mathrm{conv}(X) \subseteq S^r$ for all r, if follows that $\mathrm{conv}(X) \subseteq \bar{S}$. Thus, we only need to show that $\mathrm{conv}(X) \supseteq \bar{S}$. We will show that any extreme point of the compact convex set $\bar{S}$ belongs to X which will complete the proof.

Consider an extreme point $\bar{x}$ of $\bar{S}$. By the identity (15) in Lemmas 3(iii) and 5 and the fact $\bar{S} \subseteq S$, it follows that there exists an $i_G \in G$ such that $\bar{x} \in \bar{S} \bigcap \mathcal{X}^{i_G} \subseteq S \bigcap \mathcal{X}^{i_G}$ for each $G \in \mathcal{G}$. Let $\bar{\mathcal{C}} := \{i \in [N] : \bar{x} \in S \bigcap \mathcal{X}^i\}$. We make the following claim.

Claim: $\bar{\mathcal{C}} \in \mathcal{Z}$.

Proof. Suppose not. Let $\bar{G}$ be the complement of $\bar{\mathcal{C}}$, i.e., $\bar{G} = [N] \backslash \bar{\mathcal{C}}$. First of all, note that we have $\bar{G} \bigcap \mathcal{C} \neq \emptyset$ for all $\mathcal{C} \in \mathcal{Z}$. Otherwise, there must exist a $\widehat{\mathcal{C}} \in \mathcal{Z}$ and $\bar{G} \bigcap \widehat{\mathcal{C}} = \emptyset$, which implies that $\widehat{\mathcal{C}} \subseteq \bar{\mathcal{C}}$, a contradiction that $\bar{\mathcal{C}} \notin \mathcal{Z}$. Hence, $\bar{G}$ is a partial covering subset of $[N]$. Let $\widehat{G}$ be a minimal partial covering subset such that $\widehat{G} \subseteq \bar{G}$. Since we know that $\bar{x} \in S \bigcap X^{i_{\widehat{G}}}$ for some $i_{\widehat{G}} \in \widehat{G}$ (i.e., $i_{\widehat{G}} \in \bar{\mathcal{C}}$), we have a contradiction that $\widehat{G} \cap \bar{\mathcal{C}} = \emptyset$. $\diamond$

It then follows that $\bar{x} \in \bigcap_{i \in \bar{\mathcal{C}}} S \bigcap \mathcal{X}^i \subseteq X$. This completes the proof. $\square$

Next we show that in the pure integer setting the convex hull of X can be obtained after a finite number of quantile closure operations.

Theorem 4. *Suppose that $S \cap \mathcal{X}^i \subseteq \mathbb{Z}^n$ for all $i \in [N]$ (i.e., $\tau = n$), then there exists a finite $\bar{r}$ such that*

$$\bar{S} = S^{\bar{r}} = \text{conv}(X).$$

Proof. From Theorem 3, we know that $\bar{S} = \text{conv}(X)$. Now we only need to show the finite convergence.

Claim 1: If $\text{conv}(S^r \cap \mathbb{Z}^n) \neq \text{conv}(X)$, then there exists a $\delta > 0$ (irrespective of r) such that $d_H(S^r, \text{conv}(X)) \geq \delta$.

Proof. Note that $\text{conv}(X)$ is an integral polytope, thus we let $\text{conv}(X) = \{x : Hx \leq h\}$ with integral matrix H and vector h defining T inequalities. Let $\delta = \min_{t=1,\ldots,T} \frac{1}{\|H_t\|_2}$, where H_t is tth row of H. Since $\text{conv}(S^r \cap \mathbb{Z}^n) \neq \text{conv}(X)$, thus there exists an integer point $\hat{x} \in \text{conv}(S^r \cap \mathbb{Z}^n) \backslash \text{conv}(X)$. Since $\hat{x} \notin \text{conv}(X)$, there exists a half space $L = \{x : H_t^\top x \leq h_t\}$ such that $H_t \hat{x} > h_t$. Thus,

$$d_H(S^r, \text{conv}(X)) \geq d_H(\hat{x}, \text{conv}(X)) \geq d_H(\hat{x}, L) = \frac{H_t \hat{x} - h_t}{\|H_t\|_2} \geq \frac{1}{\|H_t\|_2} \geq \delta,$$

where the first inequality is due to $\hat{x} \in S^r \backslash \text{conv}(X)$, the second inequality is because of $\text{conv}(X) \subseteq L$, the third equality is because $H_t \hat{x} > h_t$, the fourth inequality is because $H_t, \hat{x}, h_t$ are all integral, and the last inequality is due to the choice of δ. ◇

It then follows that there must exist a $\bar{r} \in \mathbb{Z}_{++}$ such that $\text{conv}(S^{\bar{r}-1} \cap \mathbb{Z}^n) = \text{conv}(X)$; otherwise, by Claim 1, $d_H(S^r, \text{conv}(X)) \geq \delta$ for all r, contradicting the fact that $\lim_{r \to \infty} S^r = \text{conv}(X)$. Since $\text{conv}(S^{\bar{r}-1} \cap \mathbb{Z}^n) = \text{conv}(X)$, then by Theorem 1, we have $S^{\bar{r}} = \text{conv}(X) := \bar{S}$. ☐

We close this section with two examples. The first shows the necessity of the compactness assumption for the convergence of the quantile closure to the convex hull, and second shows the necessity of the pure integer setting for finite convergence.

Example 1. Let $S = \mathbb{R}^2, \mathcal{X}^1 = \{x \in \mathbb{R}^2 : 0 \leq x_1 \leq 2, x_2 = 0\}, \mathcal{X}^2 = \{x \in \mathbb{R}^2 : x_1 = 0, x_2 \geq 0\}, \mathcal{X}^3 = \{x \in \mathbb{R}^2 : 1 \leq x_1 \leq 2, x_2 \geq 0\}, \epsilon = \frac{1}{3}, p_i = \frac{1}{3}, i = 1, 2, 3$. Since each feasible set contains at least two scenarios, by (3), we have $\text{conv}(X) = \{x \in \mathbb{R}^2 : 0 \leq x_1 \leq 2, x_2 = 0\}$. As there are exactly two scenarios in each minimal partial covering subset, according to Theorem 1, we have $S^1 = \ldots = S^r = \ldots = \bar{S} = \{x \in \mathbb{R}^2 : 0 \leq x_1 \leq 2, x_2 \geq 0\}$. Hence, in this example, the scenario constraints do not define bounded feasible regions, and the quantile closures do not converge to the convex hull of the feasible region X; i.e., $\bar{S} \neq \text{conv}(X)$. ◇

Example 2. Suppose $S = [0, 2]^2, \mathcal{X}^1 = \{x \in \mathbb{R}_+^2 : 2x_1 + 0.5x_2 \geq 1\}, \mathcal{X}^2 = \{x \in \mathbb{R}_+^2 : 0.5x_1 + 2x_2 \geq 1\}, \mathcal{X}^3 = \{x \in \mathbb{R}_+^2 : x_1 + x_2 \geq 1\}, \epsilon = \frac{1}{3}, p_i = \frac{1}{3}, i = 1, 2, 3.$
Since each feasible set contains at least two scenarios, by (3), we have

$$\text{conv}(X) = \text{conv}\{(1, 0), (0.4, 0.4), (0, 1), (0, 2), (2, 0), (2, 2)\},$$

which contains the set $\mathcal{X}^3$. By induction, we can show that

$$S^r = \text{conv}\{(1, 0), (w_r, w_r), (0, 1), (0, 2), (2, 0), (2, 2)\},$$

where $0 < w_r < 0.4$ for all $r \in \mathbb{Z}_{++}$; i.e., $S^r \neq \text{conv}(X)$ whenever $r < \infty$.

Indeed, when $r = 1$, as there are exactly two scenarios in each minimal partial covering subset, according to (12), we have

$$S^1 = \text{conv}\{(1, 0), (1/3, 1/3), (0, 1), (0, 2), (2, 0), (2, 2)\},$$

where $w_1 = 1/3 \in (0, 0.4)$. Suppose for $\gamma = r \geq 1$, the hypothesis holds; i.e.,

$$S^r = \text{conv}\{(1, 0), (w_r, w_r), (0, 1), (0, 2), (2, 0), (2, 2)\},$$

where $0 < w_r < 0.4$. Now let $\gamma = r + 1$, then by Theorem 1, we have $S^{r+1} = \text{conv}\{(1, 0), (w_{r+1}, w_{r+1}), (0, 1), (0, 2), (2, 0), (2, 2)\}$, where $w_{r+1} = 0.3 + 1/(30 - 50w_r) \in (0, 0.4)$. ◇

5   Conclusion

In this paper, we studied a family of cuts known as quantile cuts for chance constrained mixed integer linear program with bounded feasible region and finite support. We showed the following results (i) the closure of all quantile cuts can be described in a conjunctive normal form, and hence is a polyhedral set; (ii) separation of quantile cuts is in general NP-hard; (iii) successive application of quantile closure operation achieves the convex hull of the chance constrained problem in the limit; and (iv) in the pure integer setting this convergence is finite. The boundedness assumption on the feasible region can be relaxed with some restrictions on the recession directions of the scenario feasible sets. We are currently working on extending some of the results to the setting of chance constrained mixed-integer convex programs. Future research questions include analyzing the strength of the quantile closure and establishing the quantile rank, i.e. the minimum number of closure operations required to achieve the convex hull, for some structured chance constrained problems. Finally, note that some of the results can also be applied to generalized disjunctive programming in [4, 11] by choosing each $\mathcal{X}^i$ as a disjunctive set and Z as logic constraints.

References

1. Abdi, A., Fukasawa, R.: On the mixing set with a knapsack constraint (2012). http://arxiv.org/pdf/1207.1077v1.pdf

2. Ahmed, S., Luedtke, J., Song, Y., Xie, W.: Nonanticipative duality and mixed-integer programming formulations for chance-constrained stochastic programs (2014). http://www.optimization-online.org/DB_FILE/2014/07/4447.pdf
3. Atamtürk, A., Nemhauser, G.L., Savelsbergh, M.W.: The mixed vertex packing problem. Math. Program. **89**(1), 35–53 (2000)
4. Grossmann, I.E., Ruiz, J.P.: Generalized disjunctive programming: a framework for formulation and alternative algorithms for minlp optimization. In: Lee, J., Leyffer, S. (eds.) Mixed Integer Nonlinear Programming, pp. 93–115. Springer, New York (2012)
5. Günlük, O., Pochet, Y.: Mixing mixed-integer inequalities. Math. Program. **90**(3), 429–457 (2001)
6. Küçükyavuz, S.: On mixing sets arising in chance-constrained programming. Math. Program. **132**(1–2), 31–56 (2012)
7. Luedtke, J.: An integer programming and decomposition approach to general chance-constrained mathematical programs. In: Eisenbrand, F., Shepherd, F.B. (eds.) IPCO 2010. LNCS, vol. 6080, pp. 271–284. Springer, Heidelberg (2010)
8. Luedtke, J.: A branch-and-cut decomposition algorithm for solving chance-constrained mathematical programs with finite support. Math. Program. **146**(1–2), 219–244 (2014)
9. Luedtke, J., Ahmed, S., Nemhauser, G.L.: An integer programming approach for linear programs with probabilistic constraints. Math. Program. **122**(2), 247–272 (2010)
10. Qiu, F., Ahmed, S., Dey, S.S., Wolsey, L.A.: Covering linear programming with violations. INFORMS J. Comput. **26**(3), 531–546 (2014)
11. Raman, R., Grossmann, I.E.: Modelling and computational techniques for logic based integer programming. Comput. Chem. Eng. **18**(7), 563–578 (1994)
12. Salinetti, G., Wets, R.J.B.: On the convergence of sequences of convex sets in finite dimensions. SIAM Rev. **21**(1), 18–33 (1979)
13. Schneider, R.: Convex Bodies: The Brunn-Minkowski Theory, vol. 151. Cambridge University Press, Cambridge (2013)
14. Song, Y., Luedtke, J.R., Küçükyavuz, S.: Chance-constrained binary packing problems. INFORMS J. Comput. **26**, 735–747 (2014)
15. Zhao, M., Huang, K., Zeng, B.: Strong inequalities for chance-constrained program (2014). http://www.optimization-online.org/DB_FILE/2014/11/4634.pdf

Author Index

Printed in the United States
By Bookmasters